Is the current state of our Galaxy primarily the result of its initial conditions or is it the product of a lifetime of complex interactions with its environment? Gathered in this volume are papers presented at an international meeting in Granada (Spain) dedicated to exploring this fundamental question.

This timely review examines all the key physical processes involved in the formation and evolution of the Milky Way. A dozen invited review articles by international experts summarise our understanding to date; whilst more than 50 topical research papers present the latest results. Together, these papers provide a state-of-the-art view on topical issues such as disk instabilities, large-scale star formation, large-scale structure formation in our Galaxy, chemical evolution, disk-halo feedback, the galactic globular cluster system, stellar populations, and the formation of galaxies. Also included are three panel sessions identifying key routes for critical future research.

For graduate students and researchers, this volume provides a valuable and pertinent review of our new vision of the formation and evolution of our Galaxy.

T0185486

The Formation of the
Milky Way

The Formation of the Milky Way

Proceedings of the IAA–IAC–University of Pisa Workshop,
held in Granada, Spain
September 4–9, 1994

Edited by
E. J. ALFARO AND A. J. DELGADO
Instituto de Astrofísica de Andalucía (CSIC), Granada, Spain

CAMBRIDGE
UNIVERSITY PRESS

CAMBRIDGE UNIVERSITY PRESS
Cambridge, New York, Melbourne, Madrid, Cape Town, Singapore,
São Paulo, Delhi, Dubai, Tokyo, Mexico City

Cambridge University Press
The Edinburgh Building, Cambridge CB2 8RU, UK

Published in the United States of America by Cambridge University Press, New York

www.cambridge.org
Information on this title: www.cambridge.org/9780521174916

First published 1995
First paperback edition 2011

A catalogue record for this publication is available from the British Library

ISBN 978-0-521-48177-9 Hardback
ISBN 978-0-521-17491-6 Paperback

Contents

Contents

LOCAL ORGANISING COMMITTEE

Antonio Alberdi
Emilio Javier Alfaro
Jesús Cabrera Caño
Antonio Jesús Delgado
Milagros Estepa
Susana Gómez
Josefina Molina
José María Torrelles

SCIENTIFIC ORGANISING COMMITTEE

Emilio Javier Alfaro (Spain, Chair)
Bruce Elmegreen (U.S.A.)
Yuri Efremov (Russia)
Federico Ferrini (Italy)
José Franco (Mexico)
Kenneth Janes (U.S.A)
Guillermo Tenorio Tagle (Spain)
Monica Tosi (Italy)

SPONSORS

Dirección General de Investigación Científica y Técnica DGICYT
Consejo Superior de Investigaciones Científicas CSIC
Junta de Andalucía
Ayuntamiento de Granada
Sol–Meliá Hotels

Participants

R.J. AGOSTINHO — Universidade de Lisboa, Portugal
A. ALBERDI — Instituto de Astrofísica de Andalucía, Spain
E.J. ALFARO — Instituto de Astrofísica de Andalucía, Spain
J. ANDERSEN — Niels Bohr Institute for Astronomy, Denmark
A. APARICIO — Instituto de Astrofísica de Canarias, Spain
M.A. AVILLEZ — Universidade de Evora, Portugal
T. BANIA — Boston University, U.S.A.
C. BARCELÓ — Instituto de Astrofísica de Andalucía, Spain
B. BATES — The Queen's University of Belfast, Northern Ireland
E. BATTANER — Universidad de Granada, Spain
M. BELLAZZINI — Università di Bologna, Italy
D.L. BERRY — Universidade de Evora, Portugal
R. BUONANNO — Osservatorio Astronomico di Roma, Italy
N.V. BYSTROVA — St. Petersburg Branch of the Special Astrophysical Observatory, Russia

J. CABRERA CAÑO — Universidad de Sevilla, Spain
C. CACCIARI — Osservatorio Astronomico di Bologna, Italy
V. CALOI — Istituto di Astrofisica Spaziale, Italy
F. CAPUTO — Istituto di Astrofisica Spaziale, Italy
B.W. CARNEY — University of North Carolina, U.S.A.
G. CARRARO — Università di Padova, Italy
J. CEPA — Instituto de Astrofísica de Canarias, Spain
B. CHABOYER — CITA University of Toronto, Canada
A.D. CHERNIN — Sternberg Astronomical Institute, Russia
C. CHIAPPINI — Universidade de São Paulo, Brasil
F. COMERÓN — Universitat de Barcelona, Spain
L. DANLY — Space Telescope Science Institute, U.S.A.
A.J. DELGADO — Instituto de Astrofísica de Andalucía, Spain
A. DEL OLMO — Instituto de Astrofísica de Andalucía, Spain
S. DEL RÍO — Instituto de Astrofísica de Canarias, Spain
YU.N. EFREMOV — Sternberg Astronomical Institute, Russia
B.G. ELMEGREEN — IBM Research Division, T.J. Watson Research Center, U.S.A.

A. FEINSTEIN — Observatorio Astronómico de La Plata, Argentina
J. FERNLEY — IUE, Spain
F.R. FERRARO — Osservatorio Astronomico di Bologna, Italy
F. FERRINI — Università di Pisa, Italy
L. FESTIN — Astronomiska Observatoriet i Uppsala, Sweden
T. FOGLIZZO — CE-Saclay, France
J. FRANCO — IAUNAM, Mexico
E.D. FRIEL — Maria Mitchell Observatory, U.S.A.
L. FULLTON — University of North Carolina, U.S.A.
R. FUX — Observatoire de Genève, Switzerland
C. GARCÍA GÓMEZ — Universitat Rovira i Virgili, Spain
R. GARRIDO — Instituto de Astrofísica de Andalucía, Spain
F. GARZÓN — Instituto de Astrofísica de Canarias, Spain
M.A. GÓMEZ FLECHOSO — Universidad Autónoma de Madrid, Spain
M.D. GUARNIERI — Università degli Studi di Torino, Italy

T. HANAWA	Nagoya University, Japan
G. HENSLER	Universität Kiel, Germany
J. ISERN	Centre d'Estudis Avançats de Blanes, Spain
H. IZUMIURA	SRON Groningen, The Netherlands
K.A. JANES	Boston University, U.S.A.
C. JONES	Harvard-Smithsonian Center for Astrophysics, U.S.A.
J. KALUZNY	Warsaw University Observatory, Poland
I.G. KOLESNIK	Ukrainian Academy of Sciences, Ukraine
C. KRAMER	Universität zu Köln, Germany
J. LAIRD	Bowling Green State University, U.S.A.
L. LARA	Instituto de Astrofísica de Andalucía, Spain
S. LEON	IRAM, Spain
H. LESCH	MPI für Radioastronomie, Germany
S. LEVINE	IAUNAM, Mexico
E. MAGNIER	University of Amsterdam, The Netherlands
S. MAJEWSKI	The Observatories of the Carnegie Institution of Washington, U.S.A.
M.L. MALAGNINI	Università degli Studi di Trieste, Italy
S. MALHOTRA	Princeton University, U.S.A.
I. MÁRQUEZ	Instituto de Astrofísica de Andalucía, Spain
E.L. MARTÍN	University of Amsterdam, The Netherlands
D. MARTÍNEZ	Instituto de Astrofísica de Andalucía, Spain
J. MASEGOSA	Instituto de Astrofísica de Andalucía, Spain
D. MINNITI	European Southern Observatory, Germany
C. MUÑOZ TUÑÓN	Instituto de Astrofísica de Canarias, Spain
T.W.B. MUXLOW	N.R.A.L. Jodrell Bank, England
B. NORDSTRÖM	Niels Bohr Institute for Astronomy, Denmark
D.K. OJHA	Observatoire de Besançon, France
L. ORIGLIA	Osservatorio Astronomico di Torino, Italy
J. PALOUŠ	Academy of Sciences of the Czech Republic
J.D. PEREA	Instituto de Astrofísica de Andalucía, Spain
E. PESCE	Università di Roma, Italy
R.L. PHELPS	Phillips Laboratory/GPOB, U.S.A.
C. PORCEL	Universidad de Granada, Spain
J.E. RHOADS	Princeton University, U.S.A.
J. RODRÍGUEZ QUINTERO	Universidad de Sevilla, Spain
J. SÁNCHEZ SALCEDO	Instituto de Astrofísica de Andalucía, Spain
W.J. SCHUSTER	IAUNAM, Mexico
A. SERRANO	INAOEP, Mexico
S. SHORE	Indiana University South Bend, U.S.A.
V. SURDIN	Sternberg Astronomical Institute, Russia
M. TAGA	University of Tokyo, Japan
G. TENORIO TAGLE	Instituto de Astrofísica de Canarias, Spain
V. TESTA	Università di Bologna, Italy
J.M. TORRELLES	Instituto de Astrofísica de Andalucía, Spain
M. TOSI	Osservatorio Astronomico di Bologna, Italy
H. UNGERECHTS	IRAM, Spain
M. VIETRI	Osservatorio Astronomico di Roma, Italy
J.M. VÍLCHEZ	Instituto de Astrofísica de Canarias, Spain
S. VON LINDEN	MPI für Radioastronomie, Germany

Preface

In recent years, our knowledge about the basic processes involved in the formation and evolution of our Galaxy has seen important advances, and, as often happens, new ideas on the status of the galactic system have given rise to a wealth of open questions, rather than settling old ones.

The goal of our workshop was twofold: to resume the reasonably well-established starting points, and to depict the framework in which the main lines of research should develop in the near future. A handful of basic questions were discussed, and conform to the general sequence of this book: the structure of the galactic disk and its relationship and evolutionary interplay with the other galactic components; the large-scale pattern of star-forming regions and its connection with the disk structure; the kinematical and metallicity structures of the various galactic components, with due attention paid to the old and intermediate age populations; the age and metallicity structures in the old thin disk, thick disk and halo; and theoretical and observational knowledge on the general topic of galaxy formation.

The contents of the book have not been arranged in separate chapters, a choice which reflects our primary idea of the topic "The Formation of the Milky Way" as an intricate puzzle, with significant connections between apparently unrelated pieces. The order of talks during the Workshop has been somewhat altered, but conceptually it follows the list of issues in the previous paragraph. The matters are presented in 14 review papers, and 57 contributed papers. Three panel sessions mark out the development of the book, reproducing the live discussions and exposition of ideas on general questions. There was a first panel session chaired by Guillermo Tenorio Tagle, which was regrettably not tape-recorded, making its inclusion in the proceedings unfeasible. However, many of the ideas and suggestions that arose there have wisely been taken into account by the other three panellists.

To finish this introduction, some special mentions deserve to be made. While working on a collaborative project some two years ago, Pepe Franco "suggested" to us that a meeting was necessary, where the basic questions on the formation of our Galaxy were approached from a multidisciplinary point of view, and he proposed Granada as an excellent venue for it to take place. His interest and persuasive powers led us to establish contacts with Guillermo Tenorio Tagle and Federico Ferrini, who joined us to make this project happen. They provided experience, friendship and the support of the Instituto de Astrofísica de Canarias and the University of Pisa, their respective host institutions. We are particularly indebted to them all.

Finally, as editors we wish to warmly thank every participant, most especially the members of the Scientific Organising Committee, whose interest and commitment contributed to making the IAA–IAC–UP workshop on the Formation of the Milky Way a most enjoyable and successful event.

Emilio Javier Alfaro, Antonio Jesús Delgado
Granada,
March, 1995

Acknowledgements

On behalf of the LOC, we wish to thank our host institution, Instituto de Astrofísica de Andalucía, who supported our work, not only prior to, and during, the Workshop, but further during the somewhat lengthy stage of editing the book. In particular, an acknowledgment must go to Carlos Barceló, Lucas Lara, David Martínez, and Javier Sánchez, for their assistance in the many technical details during the different sessions.

Ross Howard's collaboration in the editing tasks deserves a comment here. He took care not only of the linguistic accuracy, but also of the innumerable stylistic details. Our thanks are given to him.

We also wish to express our acknowledgement to the International Science Foundation, whose financial help made possible the attendance of several scientists from the former Soviet Union.

Last, but not least, particular thanks are due to the only contributor whose manuscript arrived in perfect accordance with the time, size and CUP-format requirements. In contrast to the biblical episode, the act of a righteous person kept alive our confidence in the whole community.

Gravitational and Magnetic Instabilities in Disks

By TOMOYUKI HANAWA

Department of Astrophysics, School of Science, Nagoya University, Chikusa-ku, Nagoya 464–01, Japan

We study the instabilities of the galactic disk induced by magnetic fields and self-gravity of the disk. Based on numerical computations, we discuss the linear and non-linear evolution of the instabilities. Gravity, rotation, gas pressure and magnetic force commonly have the timescale of 10^8 yr in the galactic disk. Accordingly, all these forces participate in the instabilities of the galactic gaseous disk. Our numerical computations take account of the rotation and magnetic shear, and also of self-gravity and non-uniform magnetic fields. It is shown that the stability of the present gas disk is very different from that of a gas-rich (presumably young) disk.

1. Introduction

The galactic disk is inhomogeneous to a high degree. Up to the present, 90% of the disk matter has been converted into stars. The rest of the disk consists of molecular clouds, HI clouds and coronal gas, the densities and temperatures of which range from 10^{-2} cm^{-3} to 10^5 cm^{-3} and from 10 to 10^6 K, respectively. The inhomogeneity of the gas disk will be attributed in part to instability of a homogeneous disk. Magnetic fields and self-gravity of the gas induce the Parker and the Jeans instabilities. In this paper we discuss these instabilities, taking account of the rotation and gravity of stars. In Section 2 we give an overview of the dynamical processes in the galactic disk and their timescales. In Section 3 we review the linear stability of the galactic disk according to Hanawa *et al.* (1992, hereafter HNN). In Section 4 the non-linear evolution of the instability is discussed based on numerical simulations. A brief summary is given in Section 5.

2. Timescale

First we review the dynamical processes forming dense clouds in the galactic disk. Figure 1 summarises main factors and processes involved. Gravity acts for the formation of dense clouds through the Jeans and Parker instabilities. Dense cloud-formation is suppressed by the pressure of thermal motion and centrifugal force due to rotation. Although the magnetic field suppresses the contraction of clouds by its pressure, it forms clouds through the Parker instability and extracts angular momentum from the cloud's contraction. The thick and thin arrows denote the processes initiating and suppressing dense cloud-formation, respectively. The dashed arrow denotes the different process.

These various processes have almost the same timescale of 10^8 yr in the galactic disk (see the comment by Shore in the Discussion). The galactic rotation has the frequency

$$\Omega = 8.40 \times 10^{-15} \left(\frac{v_\varphi}{220\,\mathrm{km\,s^{-1}}}\right) \left(\frac{r}{8.5\,\mathrm{kpc}}\right)^{-1} \mathrm{Hz} . \tag{2.1}$$

The timescale for gravitational collapse is evaluated to be

$$\nu_{\mathrm{J}} = \sqrt{4\pi G\rho} = 1.18 \times 10^{-15} \left(\frac{n}{1\,\mathrm{cm^{-3}}}\right)^{-1/2} \mathrm{Hz} , \tag{2.2}$$

1

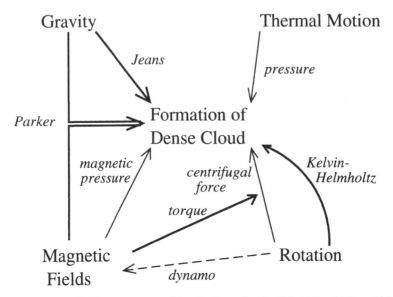

FIGURE 1. Dynamical processes working in the galactic disk. The thick and thin arrows denote those initiating and suppressing the formation of dense clouds, respectively. The dashed arrow denotes otherwise.

for the homogeneous medium having the mass density, ρ, and the equivalent nucleon number density, n. The magnetic force propagates with the frequency,

$$\nu_m = 1.34 \times 10^{-15} \left(\frac{B}{3\,\mu G}\right) \left(\frac{n}{1\,\text{cm}}\right)^{-1/2} \left(\frac{\lambda}{1\,\text{kpc}}\right)^{-1} \text{Hz} , \qquad (2.3)$$

where B and λ are the magnetic flux density and the wavelength, respectively. Similarly the sound wave has the frequency

$$\nu_s = 1.43 \times 10^{-15} \left(\frac{c_s}{7\,\text{km s}^{-1}}\right) \left(\frac{\lambda}{1\,\text{kpc}}\right)^{-1} \text{Hz} , \qquad (2.4)$$

where c_s denotes the sound speed. The timescale of 10^8 yr corresponds to the frequency of 1.77×10^{-15} Hz. Thus the gas dynamics is controlled by many equally dominant factors in the galactic disk.

3. Linear stability

Recently HNN analysed the stability of the galactic gaseous disk, taking account of the self-gravity of the gas and stars as well as rotation and magnetic fields. We approximated the equilibrium galactic disk by a uniformly rotating gaseous sheet in which the density and magnetic field are uniform in the horizontal (x- and y-) directions. The vertical ($z-$) component of the gravity in equilibrium was taken to be

$$g_{z0} = \left\{6.8\tanh\left[3.2\left(\frac{z}{1\,\text{kpc}}\right)\right] + 1.7\left(\frac{z}{1\,\text{kpc}}\right)\right\} \times 10^{-9}\,\text{cm s}^{-2} , \qquad (3.5)$$

in order to reproduce the observed value in the solar neighbourhood (Oort 1965). The ratio of the gas pressure to the magnetic pressure (β) was for simplicity assumed to be constant. The gas was assumed to be isothermal in the region of $z \leq z_d$ and the sound speed (c_s) was taken to be constant there. The HNN equilibrium model is specified by

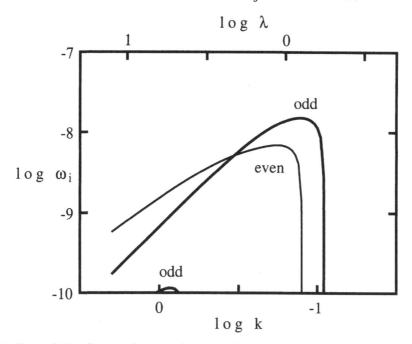

FIGURE 2. The stability diagram for perturbations with a wave number parallel to the magnetic field, k. The ordinate and abscissa are the wave number in units of kpc^{-1} and the growth rate (ω_i) in units of yr^{-1}. The thick and thin curves denote the dispersion relations for odd and even modes against the galactic plane, respectively. The equilibrium model parameters are set $n_c = 1$ cm^{-3}, $B_c = 3.23\,\mu$G, $v_\varphi = 220$ km s^{-1}, $c_s = 5$ km s^{-1}, and $z_d = 200$ pc. Reproduced from Figure 1a of HNN.

five model parameters, the density on the galactic plane (n_c), the magnetic flux density on the plane (B_c), the rotation velocity, (v_φ), the sound speed (c_s), and z_d.

Figure 2 shows the dispersion relation of HNN Model 1 of which parameters are set $n_c = 1$ cm^{-3}, $B_c = 3.23\,\mu$G, $c_s = 5$ km s^{-1}, and $z_d = 200$ pc. Since these values are typical in the solar neighbourhood, this model can be regarded as a standard one. In the equilibrium the magnetic field runs in the x-direction and the perturbation is set to be proportional to $\sin(kx)$. The thin and thick curves denote the growth rate (ω_i) as a function of the wave number (k) for even and odd modes, respectively. The even mode fragments the galactic disk, maintaining its symmetry with respect to the galactic plane, while the perturbation is anti-symmetric in the odd mode. The even mode is mainly driven by the Jeans instability while the odd mode is mainly driven by the Parker instability. In this standard model the growth of the odd mode is dominant (cf. Horiuchi *et al.* 1988; Giz & Shu 1993). The odd mode has the maximum, 1.52×10^{-8} yr^{-1}, when the wavelength is $2\pi/k = \lambda = 800$ pc.

Figure 3 shows the dispersion relation for a gas-rich disk (model 3 of HNN). The model parameters are set $n_c = 4$ cm^{-3}, $B_c = 6.46\,\mu$G, $v_\varphi = 220$ km s^{-1}, $c_s = 5$ km s^{-1}, and $z_d = 200$ pc. The even mode is dominant over the even mode in this model. The growth rate is maximum $(\omega_i = 2.80 \times 10^{-8}$ y$^{-1})$ at $\lambda = 890$ pc. We presume that our galactic disk was gas-rich in the early stage. The stability of the early galactic disk was presumably very different from that of the present one.

The dispersion relation depends also on the other parameters. When z_d is larger, the growth rate is also larger, mainly because the Parker instability is enhanced by a higher g_z. When magnetic fields are weaker (smaller B_c), both the odd mode and the

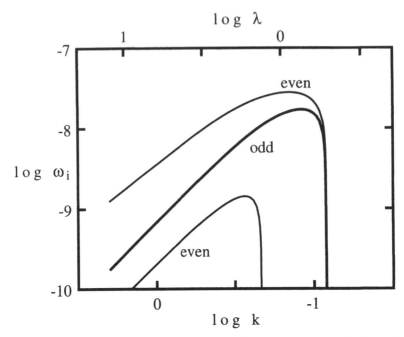

FIGURE 3. The same as Figure 2 but for $n_c = 4$ cm^{-3} and $B_c = 6.46$ μG. Reproduced from Figure 1b of HNN.

even mode have a lower growth rate. This is because the stabilising effect of rotation is cancelled by angular-momentum transfer through magnetic torque (Lynden-Bell 1966). These dependences are summarised in an approximate dispersion relation

$$\omega^2 = \frac{\nu_m{}^2(\nu_s{}^2 - \nu_P{}^2)}{(\nu_s^2 + \nu_P{}^2)(1 + \nu_m^2/\nu_s{}^2)} \, , \tag{3.6}$$

for the odd mode where the Parker frequency ν_P is defined as

$$\nu_P = \frac{g}{2c_s}(1 + \nu_m{}^2/2\nu_s{}^2)^{-1/2} \tag{3.7}$$

$$= 2.97 \times 10^{-15}\left(\frac{g_z}{4.2 \times 10^{-9}\,\mathrm{cm\,s^{-1}}}\right)\left(\frac{c_s}{5\,\mathrm{km\,s^{-1}}}\right)(1 + \nu_m{}^2/2\nu_s{}^2)^{-1/2}\,\mathrm{Hz}\,. \tag{3.8}$$

Similarly the even mode has an analytic approximate dispersion relation (see HNN).

4. Non-linear evolution

In this section we show the non-linear evolution of the instabilities for both odd and even modes. The non-linear evolution was followed with two-dimensional numerical simulations. By assuming the glide reflection symmetry for the odd mode and the mirror symmetry for the even mode, the computation domain is reduced to $0 \le x \le \lambda/2$ and $0 \le z \le z_d$. The corresponding boundary conditions are set at $x = 0$ & $\lambda/2$ and $z = 0$. Out of the boundary of $z = z_d$ buffer layers are set to mimic a hot dilute gas there. In the buffer layers the gravity is set to vanish. The spatial resolution is $\Delta x = \Delta z = 2$ pc in the simulations. The magnetohydrodynamical equations are integrated with an upwind scheme (see Hanawa *et al.* 1994) for the difference scheme. It is a virtue of an upwind scheme that the obtained numerical solutions are free from artificial oscillations.

In the numerical simulations, horizontal velocity perturbation is imposed on an equi-

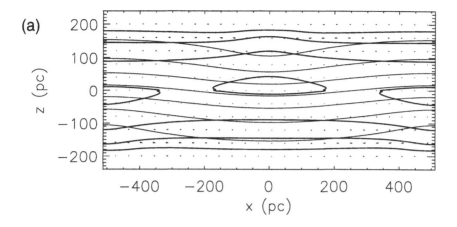

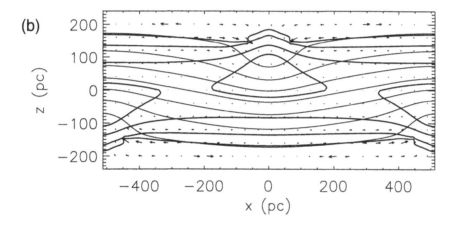

FIGURE 4. The evolution of the perturbation with the glide reflection symmetry. The upper and lower panels show the stages at which the time passed 9.2×10^7 and 1.4×10^8 y, respectively. The thick curves denote the density contours of $n = 0.1$, 0.2, 0.5, and 1.0 cm^{-3}. The thin curves denote the magnetic lines of force. The mass and magnetic field densities are $n = 1$ cm^{-3} and B = 3.23 μG, respectively, on the galactic plane at the initial model.

librium model. Figure 4 shows the evolution of a velocity perturbation with the glide reflection symmetry. Initially, the density has its maximum $(n = 1\,\mathrm{cm}^{-3})$ on the galactic plane. The sound speed is set as uniform and constant $(c_s = 5\,\mathrm{km\,s}^{-1})$. The rotation is not taken into account in this model. The thick curves are the contours of equal densities of $n = 0.1$, 0.2, 0.5, and 1.0 cm^{-3} while the thin curves denote lines of magnetic force. Figures 4a and b show the stages at which the time passed 9.2×10^7 yr and 1.4×10^8 y, respectively, from the initial. The maximum density is 1.05 and 1.52 cm^{-3} in figures 4a and b, respectively. The instability forms density enhancements off the galactic plane. They are located in the magnetic valleys and supported in part by the magnetic fields. Gas slides down to the valley along the field line. The density enhancements are bounded by shock waves in the regions of high $|z|$ (Matsumoto *et al.* 1988).

Figure 5 shows the evolution of a velocity perturbation with the mirror symmetry. The initial density and magnetic field distributions are the same as those for the glide

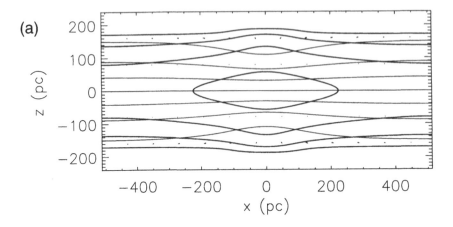

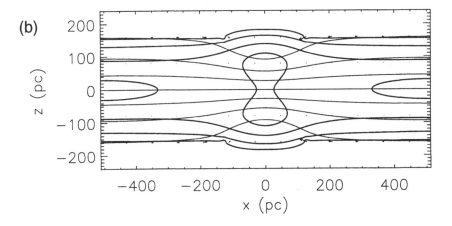

FIGURE 5. The evolution of the perturbation with the mirror symmetry. The upper and lower panels show the stages at which the time passed 1.13×10^8 and 1.67×10^8 y, respectively, from the initial model. The thick curves denote the density contours of $n = 0.1$, 0.2, 0.5, and 1.0 cm^{-3}. The thin curves denote the magnetic lines of force. The mass and magnetic field densities are $n = 1$ cm^{-3} and B $= 3.23$ μG, respectively, on the galactic plane at the initial model.

reflection symmetry. Figures 5a and b show the stages at which the time passed 1.13×10^8 and 1.67×10^8 y, respectively. The density has its maximum at $(x, z) = (0, 0)$ in early stages while it has the maximum at $(x, z) = (512\,\mathrm{pc}, 0)$ in late stages. The second density enhancement is formed by the reverse flow from the first one, which is not gravitationally bound. The second density enhancement may also be dissolved to form the third density enhancement. (See Matsumoto *et al.* 1990 for the nonlinear oscillation of the Parker instability).

Figure 6 shows the evolution of a velocity perturbation with the mirror symmetry in a gas-rich disk. At the initial state the mirror symmetric velocity perturbation was superimposed on an equilibrium model of $n_c = 4\,\mathrm{cm}^{-3}$, $B_c = 6.46\,\mu$G, $v_\varphi = 0$, $c_s = 5\,\mathrm{km\,s}^{-1}$, and $z_d = 200\,\mathrm{pc}$. Figures 6a and b show the stages at which the time passed 4.94×10^7 and 6.88×10^7 yr, respectively. The maximum density is 6.52 and 12.23 cm^{-3} in figures 6a and b, respectively. In this model the maximum density in-

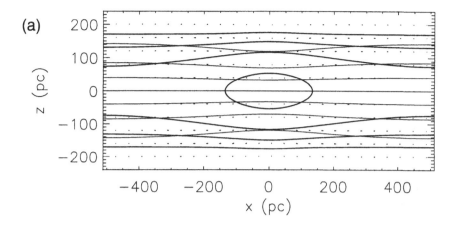

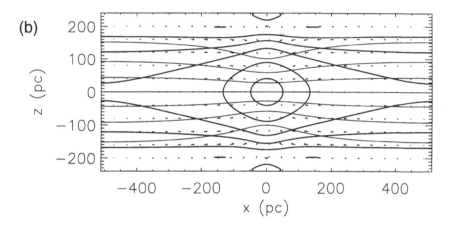

FIGURE 6. The evolution of the perturbation with the mirror symmetry in a gas-rich disk. The upper and lower panels show the stages at which the time passed 1.13×10^8 and 1.67×10^8 y, respectively, from the initial model. The thick curves denote the density contours of $n = 0.5$, 1.0, 2.0, 5.0, and 10.0 cm^{-3}. The thin curves denote the magnetic lines of force. The mass and magnetic field densities are $n = 4$ cm^{-3} and B $= 6.46$ μG, respectively, on the galactic plane at the initial model.

creases monotonically and unboundedly. The density enhancement formed by instability collapses unboundedly when its line density exceeds a critical value,

$$\ell_{\mathrm{cr}} = 2c_s^2/G$$
$$= 7.5 \times 10^{18} \left(\frac{c_s}{5\,\mathrm{km\,s^{-1}}}\right)^2 \mathrm{g\,cm^{-1}} . \qquad (4.9)$$

Otherwise the density enhancement remains gravitationally unbound. The even-mode instability produces the density enhancement having the line density,

$$\ell = 3.15 \times 10^{18} \left(\frac{n}{1\,\mathrm{cm^{-3}}}\right) \left(\frac{\lambda}{1\,\mathrm{kpc}}\right) \left(\frac{H}{100\,\mathrm{pc}}\right) \mathrm{g\,cm^{-1}} , \qquad (4.10)$$

where λ and H are the wavelength of the perturbation and the half thickness of the gas disk. Note that ℓ is close to ℓ_{cr}.

5. Summary

As shown above, the instability of the galactic gaseous disk is influenced by many factors, i.e. gravity of stars and gas clouds, magnetic fields, rotation, thermal pressure, and turbulence. It is shown that the stability of the galactic disk depends sensitively on the model parameters. This is because gravity, rotation and magnetic fields have coupled effects and the result is the delicate balance between the competing processes. In this sense, minor effects may also have significant influences on the instability. Although our model takes account of many factors, it is still simple and lacks other factors such as differential rotation and spiral arms (cf. Foglizzo & Tagger 1994; Foglizzo, this volume). Incorporating these effects is a problem for the future.

The author would like to acknowledge the collaborations of Drs. Fumitaka Nakamura and Yasushi Nakajima and thank them for their permission to include their unpublished simulations in this paper.

REFERENCES

FOGLIZZO, T. & TAGGER, M. 1994 *Astron. Astrophys.* **287**, 297.

GIZ, A. T. & SHU, F. H. 1993 *Astrophys. J.* **404**, 185.

HANAWA, T., NAKAJIMA, Y. & NOBUTA, K. 1994 *J. Comp. Phys.* submitted (DPNU-94-34).

HANAWA, T., NAKAMURA, F. & NAKANO, T. 1992 *Publ. Astron. Soc. Jpn.* **44**, 509.

HORIUCHI, T., MATSUMOTO, R., HANAWA, T. & SHIBATA, K. 1988 *Publ. Astron. Soc. Jpn.* **40**, 147.

LYNDEN-BELL, D. 1966 *Observatory* **86**, 57.

MATSUMOTO, R., HORIUCHI, T., HANAWA, T. & SHIBATA, K. 1990 *Astrophys. J.* **356**, 259.

MATSUMOTO, R., HORIUCHI, T., SHIBATA, K. & HANAWA, T. 1988 *Publ. Astron. Soc. Jpn.* **40**, 171.

Elmegreen: When does the even mode dominate?

Hanawa: The even mode dominates when the mass of the gas disk is more than half the total mass of the disk including disk stars.

Lesch: What is the role of turbulent diffusion in your model?

Hanawa: At the present stage, we restrict ourselves to the dynamical processes of short timescales and our model does not take account of turbulent diffusion.

Shore: First a comment. Your first viewgraph shows wonderfully that all of the growth times are wonderfully close—a little miracle of self-regulation. Maybe this is only because gravity is the source of everything. Now, to questions. First, where are the shocks in your model? And the second, you have large shear in some regions so the Kelvin–Helmholtz instability should lead to a turbulent cascade. Why is turbulence not inculded in the models? It will change the critical line density, ℓ_{crit}, as a function of density. It should be easy to include this in your model since you have a map of the velocity field.

Hanawa: Shocks are formed at the interfaces between the down flows and the density enhancements. The shocks are stronger in the regions of higher z. Figures 4b and 5b show the density and velocity discontinuities at the shock fronts. Now, to the second question. If the turbulent velocity increases, the critical line density increases. I do not, however, agree with your opinion that velocity shear induces the Kelvin–Helmholtz instability in our model simulations.

Galactic Disk Dynamics and Magnetic Field Evolution

By HARALD LESCH

Max-Planck-Institut für Radioastronomie, Auf dem Hügel 69, D–53121 Bonn, Germany

The Milky Way and the nearby disk galaxies contain large-scale magnetic fields. This is clearly shown by the high-frequency radio polarisation measurements, which also show that the large-scale magnetic field structure is connected to the overall velocity field of the interstellar gas. The magnetic field in the conductive plasma of the interstellar medium is subject to turbulent diffusion, i.e. it has to be amplified and sustained against diffusive losses via gas motions. The galactic dynamo describes the interplay of amplification via differential shear and turbulence on one side and turbulent diffusion on the other side. Such an axisymmetric dynamo takes a few gigayears to amplify the magnetic field. In the light of observations at high redshifts ($z = 2$) which already indicate a μGauss field the axisymmetric dynamos are too slow. Thus, we consider the influence of large-scale non-axisymmetric gravitational instabilities (bars, spirals, oval distortions, etc.) on the gas motion and the coupled magnetic field. The timescale for the field amplification due to non-axisymmetric velocity fields is related to the timescale of angular momentum transport. Due to its dissipation properties, the gas plays the major role for the excitation of non-axisymmetric features. Since the magnetic field amplification also takes place in the interstellar gas we consider the interplay of gas dynamical processes triggered by gravitational instabilities and magnetic fields. A comparison of the timescales shows that field amplification by non-axisymmetric velocity fields is faster than an axisymmetric dynamo.

1. Introduction

The Milky Way is a magnetised disk. It contains a large-scale magnetic field, which exhibits several reversals, as indicated by Faraday-rotation measures of pulsars and extragalactic objects. Although there is no general consensus about such field reversals, evidence is now accumulating that within the central 5 kpc the magnetic field indeed changes its direction several times, indicating a so-called bisymmetric field structure (Rand & Lyne 1994). The regular field resembles the spiral structure and is mainly confined in the galactic plane. The strength of the large-scale component is about 3 μGauss (Wielebinski 1993; Han & Qiao 1993). Radio surveys of the galactic plane offer even more details about the magnetic field structure. However, we cannot delineate much of the information because of our "insider" view of the Galaxy. Thus, radio observations of other galaxies obviously help to understand our own magnetised galaxy.

The magnetic field structure in other galaxies is most easily visible at high frequencies, where the Faraday rotation is negligible. At λ 2.8 cm, the Faraday rotation is less than 5 degrees. Therefore, a polarisation measurement at this wavelength gives a direct picture of the magnetic field configuration. (For a review about the status of high-frequency polarisation measurements, see Beck 1993.) I will briefly summarise the main results of the observations. The lines of force are well aligned with characteristic features of a disk galaxy such as spiral and bars, and in only a few of all edge-on galaxies prevail large-scale vertical fields perpendicular to a disk. Attempts to derive actual directions of magnetic vectors are rather delicate tasks with a lot of ambiguities, and are not decisive in revealing whether or not field directions change along the azimuth in a galactic disk. However, at least three examples have been found which show some evidence for a well-defined field structure: M31 and IC342 exhibit an axisymmetric field, and M81 contains a bisymmetric field structure (M. Krause 1990). High-frequency measurements (Neininger *et al.* 1991; Neininger 1992) suggest that the magnetic morphology is principally associated with non-axisymmetric galactic structures, such as bars or spiral arms, accompanying a large-scale gas flow.

It has been argued that the present magnetic field structure and typical strength of galactic magnetic fields are organised by induction processes called $\alpha - \Omega$-dynamo (Parker 1971; Vainshtein & Ruzmaikin 1972). We critically review this model in the next section. The problems of this axisymmetric description will lead us to investigate non-axisymmetric amplification mechanisms, which may not only explain the present status of galactic magnetic fields, but also describe the evolution of magnetic fields in high-redshift objects, which seem to contain magnetic field strengths already $\sim \mu$Gauss, comparable with the present field strengths in nearby spirals (Kronberg 1994).

2. Galactic plasmaphysics I

The spatial and time evolution of magnetic fields is described by the *induction equation*, which can be derived from the Maxwell's equation and Ohm's law:

$$\frac{\partial \mathbf{B}}{\partial t} = \nabla \times (\mathbf{v} \times \mathbf{B} - \eta(\nabla \times \mathbf{B})), \qquad (2.1)$$

where $\eta = \frac{c^2}{4\pi\sigma}$ is the magnetic diffusivity, with the electrical conductivity σ. If σ is determined by Coulomb collisions the magnetic field is strongly coupled to plasma flow with velocity $\mathbf{v}$ (often stated as "frozen-in"-field), since $\sigma \propto T^{3/2}$ is very large, which means η is very small and the characteristic diffusion timescale $t_d \propto \frac{L^2}{\eta}$ is very large. For galactic length scales (few kiloparsecs) the diffusion timescale due to classical diffusivity is of the order 10^{27} yr !

There is no source term in Equation (2.1), i.e. there is no outright creation of magnetic field in the hydromagnetic description. The contemporary magnetic fields can be sustained against resistive decay only by amplification of the existing magnetic field flux, rather than by the fresh creation of magnetic fields. Hence, if at any point in time the Universe was devoid of magnetic fields, then as far as hydromagnetic effects are concerned, there would be no magnetic field at any other time.

Biermann (1950) made the point that there are microscopic thermal and inertial "battery" effects in a moving ionised gas. These battery effects are overlooked by the hydromagnetic approximation of the gas as a classical resistive fluid. The importance of battery effects is that they guarantee that, if all else fails, the stars and galaxies are seeded with magnetic fields. The typical seed field values are of the order of 10^{-18} G (Lesch & Chiba 1995).

Equation (2.1) gives a qualitative description about the mechanisms involved: the velocity field of the plasma increases the field strength, whereas the resistance reduces the field strength. Only little is known about the resistance of the galactic plasma. Observationally we have much more information about the motions in the interstellar medium (ISM) of the Milky Way. The gas in the disk rotates differentially, with velocities of about 200 km s^{-1} and exhibits a velocity dispersion of more than 10 km s^{-1} at the solar neighbourhood, increasing towards the centre (Stark *et al.* 1991).

Nevertheless, the controversial discussion about the origin and fate of galactic magnetic fields concentrates on the question of what kind of diffusivity is acting in galaxies. If the diffusivity is very small ($\leq 10^{23} cm^2 s^{-1}$), the existence of primordial magnetic fields is enough to explain the contemporary magnetic fields, since they do not simply decay

(Cattaneo & Vainshtein 1991). On the other hand, we know from the Sun that turbulent motions in a plasma considerably increase the diffusivity via kinematic swirling and mixing (Parker 1979 and references therein). The rapid decay of the solar magnetic field is necessary to explain the solar cycle.

Since observations of the ISM of the Milky Way indicate the presence of turbulent motions, Parker (1971) and Vainshtein & Ruzmaikin (1971) applied the formalism of the solar dynamo to explain the existence of large-scale magnetic fields in disks of galaxies. The size of the turbulent eddies l_t in the ISM is about 100 pc and the typical turbulence velocity v_t is of about 10 km s^{-1}, which led to a *turbulent diffusivity* $\eta_t \sim 0.1 v_t l_t \sim 3 \cdot 10^{25}$ cm^2 s^{-1}. The diffusion time is then of the order 10^9 yr (i.e. to sustain the magnetic field strength, continuous amplification is necessary).

However, besides the decaying effect of turbulence, there is also an amplification of B via turbulence in a stratified, rotating medium: *the α-effect* (F. Krause & Rädler 1980). The central assumption is that turbulent magnetised gas flows perpendicular to the disk, (driven, for example, by stellar activity, SN explosions, etc.) will create an expanding "loop", which, due to its poloidal B_p, will be twisted by the Coriolis force as it moves out of the plane. The rotation of the loop induces a current parallel to the disk magnetic field, thereby increasing the poloidal magnetic field. The influence of the turbulence leads to the mean-field dynamo, in which the correlated action of velocity and magnetic field fluctuations is taken into account via averaging over correlation times and length scales. The induction equation of the mean-field electrodynamics is (F. Krause & Rädler 1980):

$$\frac{\partial \mathbf{B}}{\partial t} = \nabla \times (\mathbf{v} \times \mathbf{B} + \alpha \mathbf{B} - \eta(\nabla \times \mathbf{B})), \tag{2.2}$$

$\alpha \simeq l_t^2 \Omega / h$ denotes the helicity (F. Krause & Rädler 1980). The standard galactic dynamo model describes the amplification of the magnetic field in an *axisymmetric* disk through the following chain of α and Ω (differential rotation) actions: the poloidal field is generated from the toroidal component via the α effect; then the poloidal component is generated through shear motions by the overall galactic differential rotation. The strength of both amplification processes compared with turbulent diffusion (which decrease the field strength) is measured by the *dynamo numbers*

$$R_\alpha = \frac{\alpha h}{\eta_t}$$

and

$$R_\Omega = \frac{R \frac{Rd\Omega}{dR} h^2}{\eta_t}$$

where h is the half-thickness of the disk and Ω is the rotation frequency of the galaxy. Repeating the chain of α and Ω actions, magnetic fields in galaxies are amplified and ordered on large scales. Since $R_\Omega \gg R_\alpha$, the toroidal field in the disk is much stronger than the field in the halo. The formal solving procedure of Equation (2.2) is that of an eigenvalue problem. For an ansatz of the magnetic field varying in space and time, one obtains a spectrum of unstable eigenmodes, some of which are exponentially growing. The mode with the largest growth rate (smallest growth time) will dominate the evolution. The maximum growth rate of an $\alpha - \Omega$ dynamo is given by (Wielebinski & F. Krause 1993)

$$\gamma_{\text{max}}^{\alpha\Omega} \simeq \left[\frac{\alpha^2}{2\eta_{\text{t}}} \left(R \frac{d\Omega}{dR} \right)^2 \right]^{0.33}. \tag{2.3}$$

This growth rate corresponds to a growth time of about $5 \cdot 10^8$ yr.

The fastest growing mode in a thin galactic disk is of axisymmetric type. This is a very important result, since at least in M31 and IC342 an axisymmetric magnetic field is observed. In principle the application of the dynamo idea to galactic disks was useful. The dynamo presents a scenario in which large-scale magnetic field can be amplified on timescales comparable to a few galactic rotations and it explains the existence of axisymmetric magnetic field structures.

However, the axisymmetric dynamo cannot explain the generation and sustainment of dominant non-axisymmetric magnetic field structures, such as in M81. From the theoretical point of view, the standard dynamo model contains some critical ingredients: first, the concept of turbulent diffusivity is highly questionable (see Rosner & Deluca 1989; Kulsrud & Anderson 1992). It was discussed that turbulent magnetic field amplification on large scales already implies a sort of equipartition between turbulence and magnetic fields on small scales. This would lead to very high field strength (orders of magnitudes stronger than the large-scale field) on small scales ($\sim 10^{11-15}$ cm) which are accessible to observation. Such fields have never been detected (Lee & Jokipii 1976).

Second, the origin of the α effect in the ISM is still unclear. The theoretical description contains an α tensor instead of a simple scalar. However, the idea that the convective turbulence is driven by stellar disk activity failed, since the values for α are two orders of magnitude too small (Ferriere 1993).

A possible solution of these difficulties has been proposed by Hanasz & Lesch (1993). We considered the Parker instability as a source for turbulent motions ascending from the disk. If the magnetic field reaches a threshold value, the field lines become unstable and form loops whose length scales are comparable to the disk thickness. Due to drag by the surrounding medium, the ascending velocity of the flux tubes is rather small. The field lines can be twisted and rotated and produce the necessary α effect. However, this would lead to a new mechanism—the flux-tube dynamo well known from the solar dynamo (Schüssler 1993).

A third problem is the dynamo timescale for field amplification in the light of the recent detections of large-scale magnetic fields of the order of a μG in high-redshift galaxies (Kronberg *et al.* 1992). These detections lead to the requirement that, on timescales comparable to one rotation period, magnetic fields have to be amplified exponentially. This condition is incomparable with the axisymmetric galactic dynamo. Numerical simulations of the time and spatial evolution of axisymmetric dynamos in galaxies show that the timescale for amplification—including open, diffusive boundary conditions (diffusion allowed across the disk)—is of the order of 20 rotation periods $\sim 3 \cdot 10^9$ yr (Camenzind & Lesch 1994) instead of the growth time given by the inverse of Equation (2.3) $\sim 5 \cdot 10^8$ yr. Since a galaxy is connected with its immediate surrounding medium, the diffusive boundary condition is much more realistic than closed boundaries, where no magnetic flux is lost.

The axisymmetric dynamo is too slow to explain the strong magnetic fields at high redshift, and it neglects the existence of an essential component of the Milky Way and disk galaxies in general: the non-axisymmetric structures and gas flows, such as spiral arms, bars, tidal tails, etc.

3. Galactic plasma physics II

The large-scale galactic velocity field is the essential input parameter for the modelling of galactic magnetic fields. The motion of the conducting interstellar gas is accessible to observations, whereas the Ohmic properties can only be guessed theoretically. Thus a profound weakness of the mean field theory of galactic magnetic fields is the unknown magnetic diffusivity of the ISM. Certainly, this parameter deserves detailed discussion; however, I will proceed here by investigating the influence of the *real galactic velocity field* as indicated by innumerable observations of disk galaxies and the Milky Way.

3.1. *Non-axisymmetric gas flows in disk galaxies*

3.1.1. *Spiral structure*

Spiral patterns are most prominent in disk galaxies which contain gas and are forming stars. In many cases they have an underlying "grand design", but the symmetry is almost always broken: fragmentary patterns and branching arms are very common. Such features present essential deviations from the axisymmetric velocity distribution of differentially rotating galactic disks. Since the velocity field is an essential constituent of the induction equation (2.1), we consider now the influence of non-axisymmetric gas flows onto large-scale magnetic fields. First of all, we should describe the physical reason for the appearance of non-axisymmetric features in an axisymmetric disk:

The central twin problems of theories for spiral structure are the origin and persistence of the grand design. There is now little doubt that bars and companions are capable of driving grand-design spiral responses in gaseous and stellar disks (see Athanassoula 1984 for a review).

Attempts to establish a theory of long-lived spiral density waves began with Lin & Shu (1964), who showed that tightly wrapped spiral density waves are admissible free oscillations of a stellar disk. However, it was demonstrated that such waves require continuous regeneration, because they propagate radially with a substantial group velocity (Toomre 1977 and references therein). For recurrent transient spirals, substantial—but limited—growth was found as leading spirals swing to trailing in a shearing disk. The decline in the strength of the patterns is caused by a steady rise in random motion (heating) amongst the disk stars, brought about by the fluctuating potential perturbations of the spirals themselves (Binney & Tremaine 1987, Chapter 6).

The velocity dispersion in a stellar disk is conveniently expressed in terms of the parameter Q, which measures the ratio of radial component of velocity dispersion v_s to the minimum required to suppress axisymmetric gravitational instabilities (Toomre 1964); viz.

$$Q = \frac{v_s \kappa}{3.36 G \Sigma} \tag{3.1}$$

where κ is the local epicyclic frequency, and Σ is the local surface density. A Q of 1 is sufficient for axisymmetric stability, but non-axisymmetric instabilities continue until Q has risen to somewhere between 2 and 2.5. "The value of Q has to be maintained under 2.5 if recurrent spiral activity is to be prolonged" (Von Linden *et al.*, this volume). This may be achieved in two different ways:

(i) Dissipation of the gas layer. Collisions between gas clouds will prevent them from acquiring high random velocities.

(ii) Accretion. As material is added to the disk, the surface density will rise, which reduces the value of Q.

Both effects are intrinsic properties of non-axisymmetric instabilities:

Since the clouds crowd in the spiral arms they collide more often and since this kind of dissipation leads to a transfer of angular momentum, accretion will set in as a result of the excitation of spiral arms (see next section).

Sellwood & Carlberg (1984) (SC) presented computer simulations of a self-regulation mechanism for spiral instabilities in galactic disks. They allowed for cooling and accretion and argued that spirals are transient features in an evolving disk and that dissipation in the gas component is essential for the continued recurrence of spiral instabilities.

SC observed continuous spiral activity for as long as the models run: "Each individual spiral pattern lasts for less than a rotation period $\tau \simeq \frac{2\pi}{\Omega}$, but patterns succeed each other so rapidly that models always have spiral appearance."

Thus, the role of gas is evident: only this component can cool fast enough to sustain the onset condition for non-axisymmetric instabilities. The Q value of the stellar and gaseous component is only loosely connected. Thomasson *et al.* (1991) reported numerical simulations in which the stellar Q stays constant, whereas the gaseous Q varies considerably with radius and time.

3.1.2. *Bars*

Bars spontaneously form if, for self-gravitating masses, the criteria $T/W \geq 0.14$ and $Q < 3$ are fulfilled (Ostriker & Peebles 1973). T is the rotational kinetic energy and W is the potential energy. In other words, if much of the kinetic energy of the disk is in rotational rather than in random motion, the disk is strongly unstable to large-scale barlike modes.

The bar leads to a gas response which is faster than the rotation outside corotation CR (where the rotation frequency Ω is equal to the bar rotation frequency Ω_p) and lags it inside CR, thus giving rise to torques which act on the gas to drive inwards inside CR and outwards outside CR. The inflow timescale for any non-axisymmetric distortion, which is equal to the timescale for a redistribution of angular momentum in the disk, is given by (Lynden-Bell & Kalnajs 1972; Larson 1984):

$$\tau \simeq 2 \left(\frac{M}{M_{disk}} \right) \left(\frac{\Sigma}{\delta\Sigma} \right)^2 \frac{2\pi}{\Omega} \tag{3.2}$$

where $M \simeq v_{rot}^2 R/G$ is the total mass interior to radius R; $M_{disk} \simeq 2\pi R^2 \Sigma$ is the disk mass interior to R and $v_{rot} = R\Omega$ is the rotation velocity. The non-axisymmetric disturbance of the gravitational potential produces a disturbance in the surface density $\delta\Sigma$.

Equation (3.2) shows that if most of the mass is in the disk, $(M/M_{disk} \sim 1)$ and if a non-axisymmetric disturbance of large amplitude and wavelength is present $\Sigma/\delta\Sigma \sim 1$, the timescale for redistribution of angular momentum is of the same order as the orbital period, in accordance with the results of SC.

The condition $\delta\Sigma \sim \Sigma$ shows again the importance of the gaseous component. It has been shown by Lubow *et al.* (1986) that for a gas-to-star ratio of only 15%, the gas contributes seven times more to self-gravity than the stars. The main reason for the importance of gas gravity is that a stellar disturbance is a fractionally small perturbation on an otherwise axisymmetric stellar disc, while the gas is fully participating. In that case only a small fraction of the stellar density generates a bar or a spiral wave, while much of the gas density generates a spiral field or a bar; so the assumption $\delta\Sigma \sim \Sigma$ is easily fulfilled for the gas, for even weak disturbances in the stars.

3.2. *Magnetic field amplification and spiral arms*

Any density disturbance in a magnetised disk, which moves with high velocity, intensifies the component of magnetic field which is perpendicular to the disturbance. From the dynamical considerations it is clear that any non-axisymmetric perturbation leads to angular momentum transfer and thus to a radial inflow.

Thus, it seems reasonable to approximate the dynamics of a disk galaxy which contains spiral arms or bars or is interacting, by a two-dimensional velocity field $(v_r, v_{rot}, 0)$. We use the induction equation, with dissipation neglected (Chiba & Lesch 1994),

$$\frac{\partial \mathbf{B}}{\partial t} = \nabla \times (\mathbf{V} \times \mathbf{B}). \tag{3.3}$$

For the azimuthal field component B_φ one gets

$$\frac{\partial B_\varphi}{\partial t} = -\frac{\partial}{\partial r}(v_r B_\varphi) + B_r r \frac{d\Omega}{dr}. \tag{3.4}$$

with $B_r \sim const.$

The solution to this equation grows exponentially as

$$B_\varphi = \left[B_{\varphi 0}(r_0) + \tau B_r r \frac{d\Omega}{dr} \right] exp\left(\frac{t}{\tau} \right) - \tau B_r r \frac{d\Omega}{dr}, \tag{3.5}$$

where the amplification time is

$$\tau = -\left(\frac{\partial v_r}{\partial r} \right)^{-1}. \tag{3.6}$$

Equations (3.6) and (3.2) both describe the timescale for angular momentum transfer. Thus, we compare τ with the timescale an $\alpha - \Omega$-dynamo needs to amplify a magnetic field. Here we use the numerical simulations of Camenzind & Lesch (1994) which describe the time evolution of axisymmetric magnetic fields in galactic disks. As mentioned above, axisymmetric fields have the highest growth rate, i.e. the smallest timescale. Its typical growth time is (Camenzind & Lesch 1994)

$$t_{dynamo} \simeq \frac{R_{bulge}^2}{\eta_T(R_{bulge})} \simeq 10^9 \, yrs \left[\frac{R_{bulge}}{3 \, kpc} \right]^2 \left[\frac{\eta_T}{3 \cdot 10^{26} \, cm^2 s^{-1}} \right]^{-1}. \tag{3.7}$$

R_{bulge} is the bulge radius, where the differential rotation of the disk evolves into rigid rotation of the central bulge. There, the gradient of differential rotation is highest and thus, it acts as the source for the dynamo action. The amplified flux diffuses radially outwards.

In Figure 1 we show the timescales τ (Equations 3.2 and 3.6), t_{dynamo} (Equation 3.7) and the rotation period $P(R) = 2\pi/\Omega(R)$ for a rotation curve of the form (Ω_0 is the rotation frequency of the rigidly rotating central region)

$$\Omega(R) = \frac{\Omega_0}{\left[1 + \left(\frac{R}{R} \right)^{1/3} \right]^3}, \tag{3.8}$$

which presents a general rotation curve of a disk galaxy. Furthermore we assume an exponential for Σ. Obviously the amplification by non-axisymmetric gas flows is much faster than an axisymmetric dynamo.

Like conventional galactic dynamos, the mechanism for magnetic field amplification considers destructive processes such as turbulent diffusion. When the combination of velocity shears attained in highly non-axisymmetric galaxies overcomes the destructive effects of turbulent diffusion, the strength of magnetic fields increases exponentially with time. This defines the threshold value for the combination of velocity shears. Thus,

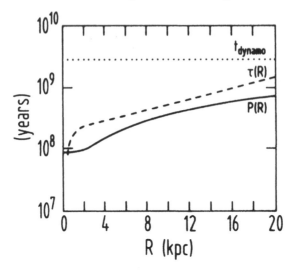

FIGURE 1. The dynamo timescale t_{dynamo} (Equation 3.7; dotted line); the timescale τ (Equations 3.2 and 3.6) for exponential growth of the toroidal field due to amplification by non-axisymmetric perturbations (broken line); and the rotation period $P(R)$ (solid line) in a "standard" galactic disc

in the sense that the field generation requires its overwhelming effect against turbulent diffusion, the mechanism is similar to the conventional approach.

However, as Hanasz & Lesch (1993) have shown, the Parker instability may play an essential role for the magnetic diffusivity. With the assumption that the magnetic field drives its own saturation via buoyant flux tubes ascending from the disk, if the magnetic pressure is comparable to turbulent pressure in the disk, we obtained an $\alpha \simeq 1$ km s^{-1}, sufficient for dynamo action and a diffusivity $\eta_t \simeq 2 \cdot 10^{24}$ cm^{2}s^{-1}, which is two orders of magnitude smaller than the generally adopted value. Thus, the diffusion time is also much longer than in conventional dynamos. The picture of a magnetic diffusivity produced by magnetic instabilities instead of external driven diffusivity (via SN explosions, stellar winds, etc.) has profound implications for the field amplification in galaxies. Since the field strengths during the formation of galaxies is far away from the critical value for the Parker-instability, the fields are amplified until the magnetic pressure is comparable to the turbulent pressure in the protodisks (Lesch & Chiba 1995). Then buoyant flux tubes start to transport magnetic flux into the halo and saturate the field amplification. This may explain the equipartition between magnetic and turbulent and cosmic ray pressure in galactic disks (Parker 1992).

This holds also for the creation of magnetic fields via battery processes and subsequent amplification by strong non-axisymmetric gas flows in protogalaxies, certainly triggered by intensive tidal interactions at high redshift. This scenario has been investigated by Lesch & Chiba (1995). We show that the relatively strong magnetic fields ($\geq 1\mu$G) in high redshift objects can be explained by the combined action of an evolving protogalactic fluctuation and electrodynamic processes providing the magnetic seed fields. Three different seed field mechanisms are incorporated into a spherical A "top-hat" model and tidal torque theory for the fate of a forming galaxy in an expanding universe (Padmanabhan 1993). Very weak fields $10^{-19} \sim 10^{-23}$G created in an expanding over-dense region are strongly enhanced due to the dissipative disk formation by a factor $\sim 10^4$, and subsequently amplified by strong non-axisymmetric flow by a factor $\sim 10^{6-10}$, depending on the cosmological parameters and the epoch of galaxy formation. The resulting field strength at $z = 0.395$ can be of the order of a few μG and be close to this value at $z \sim 2$.

4. Preliminary summary

The magnetic field in the Milky Way is not easy to investigate. Since we are situated inside the magnetised disk, it is not yet clear what kind of structure the large-scale magnetic field has. A view onto neighbours generally reveals more information: large-scale fields are similar to the overall galactic structure. The field lines resemble the spiral structure and/or continuously follow the bar through the central region, even mirroring galactic winds (Wielebinski & F. Krause 1993; Kronberg 1994 and references therein). Thus, galactic magnetic fields are connected with the galactic large-scale velocity field. As a first attempt, the dynamo takes into account turbulent motions and the axisymmetric differential rotation of the disk. This self-exciting mechanism increases the field strength of very weak seed fields via a chain of amplifying interactions. Unfortunately, this process proceeds on a timescale of the order of a few gigayears. This timescale is too long to account for the field strength in high-redshift objects of the order of more than μG. To understand these fields, a timescale for the exponential amplification of the field strength of the order of one rotation period is necessary. The consideration of non-axisymmetric gas flows lead to exactly this timescale, since the field amplification is now connected with radial inflow, which is triggered by angular momentum transfer via non-axisymmetric instabilities like spirals, bars, etc. Since the field amplification is very effective, the buoyancy of the field lines will saturate the field growth, until equipartition is reached.

Summarising, the field strength and structure of magnetic fields in galaxies can be explained by consistent consideration of plasma processes in the early Universe, galaxy formation and galactic dynamics.

I would like to thank M. Krause for the critical reading of the manuscript, and M. Chiba and P.P. Kronberg for many stimulating discussions and suggestions. I express my sincere thanks to the organisers for this enlightening conference.

REFERENCES

ATHANASSOULA, E. *Phys. Rept.* **114**, 321.

BECK, R. 1993 in *The Cosmic Dynamo, IAU 157* (ed. F. Krause, K.-H. Rädler & G. Rüdiger), p. 283. Kluwer.

BIERMANN, L. 1950 *Z. f. Naturf.* **5a**, 65.

BINNEY, J. J. & TREMAINE, S. 1987 *Galactic Dynamics*. Princeton University Press.

CAMENZIND, M. & LESCH, H. 1994 *Astron. Astrophys.* **284**, 411.

CATTANEO, F. & VAINSHTEIN, S. I. 1991 *Astrophys. J., Lett.* **376**, L21.

CHIBA, M. & LESCH, H. 1994 *Astron. Astrophys.* **284**, 731.

FERRIERE, K. M. 1993 in *The Cosmic Dynamo, IAU 157* (ed. F. Krause, K.-H. Rädler & G. Rüdiger), p. 383. Kluwer.

HANASZ, M. & LESCH, H. 1993 *Astron. Astrophys.* **278**, 561.

KRAUSE, F. & RÄDLER, K.-H. 1980 Mean Field Electrodynamics. Academic Press.

KRAUSE, M. 1990 in *Galactic and Intergalactic Magnetic Fields, IAU 140* (ed. R. Beck, P.P. Kronberg & R. Wielebinski), p. 187. Kluwer.

KRONBERG, P. P. 1994 *Rept. Prog. Phys.* **88**, 325.

KRONBERG, P. P., PERRY, J. & ZUKOWSKI, E. L. H. 1992 *Astrophys. J.* **387**, 528.

KULSRUD, R. M. & ANDERSON, S. W. 1992 *Astrophys. J.* **396**, 606.

LARSON, R. B. 1984 *Mon. Not. R. Astron. Soc.* **206**, 197.

LEE, L. C. & JOKIPII, J. R. 1976 *Astrophys. J.* **206**, 735.

LESCH, H. & CHIBA, M. 1995 *Astron. Astrophys.* in press.

LIN, C. C. & SHU, F. H. 1964 *Astrophys. J.* **140**, 646.

LUBOW, S. H., BALBUS, S. A. & COWIE, L. L. 1986 *Astrophys. J.* **309**, 496.

LYNDEN-BELL, D. & KALNAJS, A. J. 1972 *Mon. Not. R. Astron. Soc.* **157**, 1.

NEININGER, N. 1992 *Astron. Astrophys.* **263**, 30.

NEININGER, N., KLEIN, U., BECK, R. & WIELEBINSKI, R. 1991 *Nature* **352**, 781.

OSTRIKER, J. P. & PEEBLES, P. J. E. 1973 *Astrophys. J.* **186**, 467.

PADMANABHAN, T. 1993 *Structure Formation in the Universe.* Cambridge University Press.

PARKER, E. N. 1971 *Astrophys. J.* **163**, 252.

PARKER, E. N. 1979 *Cosmical Magnetic Fields.* Clarendon Press.

PARKER, E. N. 1992 *Astrophys. J.* **401**, 137.

RAND, R. J. & LYNE, A. G. 1994 *Mon. Not. R. Astron. Soc.* **268**, 497.

ROSNER, R. & DeLUCA, E. 1989 in *The Galactic Centre, IAU 136* (ed. M. Morris), p. 319. Kluwer.

SCHÜSSLER, M. 1993 in *The Cosmic Dynamo, IAU 157* (ed. F. Krause, K.-H. Rädler & G. Rüdiger), p. 27. Kluwer.

SELLWOOD, J. A. & CARLBERG, R. G. 1984 *Astrophys. J.* **282**, 61.

STARK, A. A., GERHARD, O. E., BINNEY, J. J. & BALLY, J. 1991 *Mon. Not. R. Astron. Soc.* **248**, 14P.

THOMASSON, M., DONNER, K. J. & ELMEGREEN, B. G. 1991 *Astron. Astrophys.* **250**, 316.

TOOMRE, A. 1964 *Astrophys. J.* **139**, 1217.

TOOMRE, A. 1977 *Annu. Rev. Astron. Astrophys.* **15**, 437.

VAINSHTEIN, S. I. & RUZMAIKIN, A. A. 1972 *Sov. Astron.* **15**, 714.

WIELEBINSKI, R. 1993 in *The Cosmic Dynamo, IAU 157* (ed. F. Krause, K.-H. Rädler & G. Rüdiger), p. 271. Kluwer.

WIELEBINSKI, R. & KRAUSE, F. 1993 *Astron. Astrophys. Rev.* **4**, 449.

Tenorio-Tagle: How does one get the Gauss fields observed in the nucleus of some Seyfert galaxies (e.g. Venturi *et al.* 1993, MNRAS)?

Lesch: The gas velocity field in the very centre of a galaxy considerably differs from the disk velocity. The rotation speeds within the central parsec of a Seyfert nucleus easily account for field strengths of the order of a few Gauss.

Franco: If I understand the second part of your talk, the strongest field amplification is achieved at the places where torques operate most effectively (such as spiral arms, bars, places of colliding streams, etc.). Is this seen in polarisation studies? Can we map these places where torques are acting in a galaxy, measuring the field strengths?

Lesch: Exactly, this is the idea behind our model—that the magnetic field tells us something about the effectiveness of the torques acting in a galactic disk. Of course, we already mapped regions with strong torques. (See the papers by Neininger *et al.* 1992, 1993.)

Palouš: We found in our N-body simulations that the viscosity forming rings in galaxies works very slowly in an axisymmetric disk. It substantially speeds up when we involve non-axisymmetric large-scale perturbations. May they also help to amplify the B field?

Lesch: Yes, I agree completely; non-axisymmetric velocity fields are much more effective for B-field amplification than axisymmetric ones.

The Parker–Shearing Instability in Azimuthally Magnetised Disks

By THIERRY FOGLIZZO† AND MICHEL TAGGER

CEA/DSM/DAPNIA, Service d'Astrophysique, CE-Saclay, 91191 Gif-sur-Yvette Cedex,
France

We study the linear stability of a disk embedded in a purely azimuthal magnetic field, subject to both Parker and shearing instabilities. We describe how differential rotation couples the Parker instability to the spiral density wave in a galaxy. The interplay of Parker and shearing instabilities is explained. Both occur on the slow MHD branch of the dispersion relation, but in different ranges of radial wavenumbers, i.e. at different times in the shearing evolution.

1. Introduction

The recent re-discovery by Balbus & Hawley (1991) of a powerful shearing instability (1991) has renewed the interest taken in differentially rotating magnetised disks: an axisymmetric instability was found to occur in any magnetic configuration which is not purely azimuthal. The configuration with a purely azimuthal magnetic field is stable to axisymmetric perturbations, but transiently unstable to non-axisymmetric ones. This different instability was very briefly mentioned by Acheson (1978), and described in more detail by Foglizzo (1991) and Balbus & Hawley (1992b) when applied to galaxies and accretion disks. Such a configuration is also unstable to the Parker instability (hereafter PI) because of the vertical gradient of magnetic pressure in the disk (Parker 1966). Both magnetic buoyancy and differential rotation were taken into account in two recent papers by Foglizzo & Tagger (1994, 1995, hereafter Papers I and II).

One of the most promising results of these shearing instabilities, combined with the Parker instability, could be a single-model description of both the turbulent viscosity and the dynamo action in accretion disks (Vishniac & Diamond 1992; Tout & Pringle 1992). Unfortunately, these attempts are still mainly qualitative, because of the analytical difficulties introduced by differential rotation.

We give here a general overview of the evolution of the Parker and shearing instabilities in a disk embedded in an azimuthal magnetic field, at the linear stage. We invite the reader to refer to Papers I and II for the technical aspects, and to Foglizzo (1994) for illustrations. We first recall the two effects of differential rotation in Section 2, and show in Section 3 how they lead to the coupling of waves. We describe the interplay between the Parker and shearing instabilities in Section 4, and conclude in Section 5.

2. The two effects of differential rotation

2.1. The shearing of waves

In a differentially rotating disk, the radial wavelength of a plane wave must evolve with time because of the shearing motion. More quantitatively, the effects of differential rotation can be described in the "shearing-sheet" model by a single parameter, the Oort constant: $A \equiv (R_0/2)(\partial\Omega/\partial R) < 0$. In this approximation, the radial wavenumber $k_x = 2\pi/\lambda_x$ is simply a linear function of time:

† Present address: MPIfA, Karl-Schwarzschild-Str. 1, Postfach 15 23, 85740 Garching bei
München, Germany

$$k_x(t) = k_{x0} - 2Ak_y t. \tag{2.1}$$

This time dependence is responsible for both the linear coupling between waves (Section 3), and the transience of the shearing instability (Section 4).

2.2. *The differential force*

The linearised inertial forces acting on the perturbed fluid are also modified by the differential rotation: written in Lagrangian displacements $\boldsymbol{\xi}$, they can be decomposed into a classical Coriolis force ($\mathbf{F}_C = -2\rho_0 \Omega \times d\boldsymbol{\xi}/dt$), and a *differential force* acting radially on the radially displaced elements of fluid:

$$\mathbf{F}_A = -4\rho_0 A\Omega\xi_x \mathbf{e}_x. \tag{2.2}$$

This tidal force is the result of the combined action of both the centrifugal and the gravitational radial forces. In Section 4 we show how the slow magnetosonic waves can be destabilised by this radial force.

3. The linear coupling of waves

3.1. *Qualitative approach*

In the classical normal mode analysis, any perturbation can be projected, in the $(\boldsymbol{\xi}, \mathbf{v})$ space, onto a basis of six eigenvectors which depend on the wavevector $\mathbf{k}$. When differential rotation is taken into account, the time-dependence of the wavevector makes each eigenvector rotate with time. If this rotation is slow enough, the solution can adiabatically "follow" the time-dependent eigenvectors, with each of the six modes keeping its identity during its sheared evolution from $k_x = -\infty$ to $k_x = +\infty$.

The linear coupling occurs when the temporal variation of the basis of eigenvectors is so fast that the solution cannot "follow" the time-dependent eigenvectors. The normal mode approach is then no longer valid, because it becomes impossible to identify six distinct modes of evolution.

According to Equation (2.1), the timescale of the evolution of the wave vector is:

$$\left(\frac{1}{|k|}\frac{dk}{dt}\right)^{-1} = |2A|^{-1}\left[1 + \left(\frac{k_z}{k_y}\right)^2 + \left(\frac{k_{x0}}{k_y} - 2At\right)^2\right]^{\frac{1}{2}} \geq |2A|^{-1}.. \tag{3.3}$$

This formula allows us to deduce the following properties:

(i) A weak differential rotation ($A/\Omega \ll 1$), a short vertical wavelength ($k_z \gg k_y$), or a short radial wavelength ($|k_x(t)| \gg k_y$) all favour a slow evolution of the basis of eigenvectors, and therefore an adiabatic evolution.

(ii) The fastest evolution is reached for perturbations such that $k_z = 0$.

(iii) Since the evolution is always adiabatic for large enough $|k_x(t)| \gg k_y$, the phase of linear coupling must be transient. This justifies the introduction in Paper I of coupling constants describing how one well-identified mode at $k_x(t) \ll -k_y$ projects itself onto each of the six modes at $k_x(t) \gg k_y$.

3.2. *The Parker instability triggered by spiral density waves*

Given the high value of the Oort constant in a galactic disk ($A/\Omega \sim -1/2$), all waves with $k_z = 0$ become coupled together when the sheared motion leads to $|k_x(t)| < k_y$. The spiral density wave propagating in the gas is a fast magnetosonic wave modified by self-gravity, and is naturally coupled to the unstable slow magnetosonic mode (the PI). The amplitude of the PI generated by this coupling is comparable to the amplitude of the initial spiral density wave. The density contrast in a spiral arm being of order 100%, the PI starts growing from vertical displacements comparable to the disk scale height! This efficient mechanism provides a basis for the understanding of how the PI is triggered in the spiral arms to form molecular clouds, in the spirit of Blitz & Shu (1980).

4. The Parker-shearing instability

4.1. *Polarisation properties of slow magnetosonic waves*

Since the differential force is proportional to the radial displacement, it will most influence waves whose displacements are essentially radial. It is therefore useful to bear in mind some polarisation properties of the slow magnetosonic waves in a disk. Using a normal mode analysis in a *uniformly* rotating disk, it is straightforward to show that the displacements involved in a slow magnetosonic wave are essentially (i) vertical, if $\lambda_x < \lambda_z$; (ii) horizontal, if $\lambda_z < \lambda_x$. The first case can lead to the PI if the magnetic buoyancy dominates the magnetic tension, i.e. if the azimuthal wavelength is long enough. The second case is always stable as far as uniform rotation is concerned.

4.2. *The shearing instability mechanism*

Let us now consider the two effects of differential rotation, as described in Section 2:

(i) The strongest effect of the differential force is to be expected at short λ_z, where displacements are horizontal. The shearing instability occurs if the radial differential force dominates the magnetic tension, i.e. again, if the azimuthal wavelength is long enough. The instability criterium is written:

$$4A\Omega + k_y^2 V_A^2 < 0. \tag{4.4}$$

The minimum azimuthal wavelength is consequently $\pi V_A |A\Omega|^{-1/2} \sim 1.5$ kpc in our Galaxy.

(ii) Because waves are sheared, the condition $\lambda_z < \lambda_x(t)$ can only be fulfilled *transiently*. According to the time dependence of $\lambda_x(t)$ expressed in Equation (2.1), the maximum available time for a horizontal polarisation is:

$$T_{\text{horiz.}} = \frac{\lambda_y}{\lambda_z} |A|^{-1}. \tag{4.5}$$

Ultimately, any perturbation will always be sufficiently sheared so that $\lambda_x < \lambda_z$. In this sense the PI always dominates in the end, as was shown by Shu (1974).

Hence the Parker and shearing instabilities correspond to the same slow magnetosonic mode, in different ranges of wavelengths. Both mechanisms suppose that the azimuthal wavelengths are large enough. Although transient, the shearing instability can last an arbitrarily long time, according to Equation (4.5), if λ_z is small enough.

If $\lambda_z \ll \lambda_y$, the evolution of the slow magnetosonic mode is *adiabatic* according to Equation (3.3), and can be sketched as follows:

(i) $k_x(t) < -k_z$: the PI occurs, producing vertical loops of magnetic field;

(ii) $|k_x(t)| < k_z$: the vertical loops rotate towards the horizontal plane, and keep on expanding in this plane because of the shearing instability;

(iii) $k_x(t) > k_z$: the horizontal loops rotate again towards the vertical plane, and follow their expansion due to the Parker mechanism again.

As is the case of the shearing instability in a vertical magnetic field (Balbus & Hawley 1992a), the optimum growth rate of the shearing instability in a purely azimuthal field is exactly bounded by the Oort constant $|A|$, corresponding to a minimum timescale of 70 million years in the galactic case. Hence the Parker and Shearing instabilities have comparable timescales in our Galaxy.

5. Conclusions

We have decomposed the effects of differential rotation into a differential force responsible for the shearing instability, and a shearing effect responsible for the linear coupling of waves and the transience of the shearing instability.

The coupling mechanism must favour the formation of molecular clouds along the spiral arms of disk galaxies.

At short λ_z, the shearing instability favours turbulence in the interstellar medium, in the same way as the PI does at short λ_x (Parker 1967).

REFERENCES

ACHESON, D. J. 1978 *Phil. Trans. R. Soc. London Ser. A* **289**, 459.

BALBUS, S. A. & HAWLEY, J. F. 1991 *Astrophys. J.* **376**, 214.

BALBUS, S. A. & HAWLEY, J. F. 1992a *Astrophys. J.* **392**, 662.

BALBUS, S. A. & HAWLEY, J. F. 1992b *Astrophys. J.* **400**, 610.

BLITZ, L. & SHU, F. H. 1980 *Astrophys. J.* **238**, 148.

FOGLIZZO, T. 1991 Champ magnétique azimutal dans les galaxies spirales. DEA thesis, Université Paris 7.

FOGLIZZO, T. 1994 L'instabilité de Parker et les phńomènes MHD associés dans un disque en rotation différentielle. PhD thesis, Université Paris 7.

FOGLIZZO, T. & TAGGER, M. 1994 *Astron. Astrophys.* **287**, 297.

FOGLIZZO, T. & TAGGER, M. 1995 *Astron. Astrophys.* in press.

LACHIÈZE-REY, M., ASSÉO, E., CESARSKY, C. J. & PELLAT, R. 1980 *Astrophys. J.* **238**, 175.

PARKER, E. N. 1966 *Astrophys. J.* **145**, 811.

PARKER, E. N. 1967 *Astrophys. J.* **149**, 535.

SHU, F. H. 1974 *Astron. Astrophys.* **33**, 55.

TOUT, C. A. & PRINGLE, J. E. 1992 *Mon. Not. R. Astron. Soc.* **259**, 604.

VISHNIAC, E. T. & DIAMOND, P. H. 1992 *Astrophys. J.* **398**, 561.

ZWEIBEL, E. G. & KULSRUD, R. M. 1975 *Astrophys. J.* **201**, 63.

Lesch: Does turbulence slow down the Parker instability significantly?

Foglizzo: Zweibel & Kulsrud (1975) and Lachièze-Rey *et al.* (1980) took into account a turbulent component of the magnetic field. They showed that the PI is strongly stabilised by turbulence for radial wavelengths shorter than the disk scale height.

Angular Momentum Transport in Galaxies

By SUSANNE VON LINDEN[1,2], H. LESCH[2]
AND F. COMBES[3]

[1]Landessternwarte Heidelberg, Königstuhl, D–69117 Heidelberg, Germany

[2]MPIfR, Auf dem Hügel 69, D–53121 Bonn, Germany

[3]DEMIRM, Observatoire de Paris, 61 Av. de l'Observatoire, F–75014 Paris, France

We present a model for the angular momentum transport in disk galaxies. Our theoretical investigation emphasises the role of the dissipational behaviour of molecular clouds: while stars in the disk do not collide during a galactic lifetime, molecular clouds collide inelastically several times within one rotation period. This dissipation cools down the gas component, whose Toomre Q parameter then falls below 1. This triggers dynamical instabilities. Since the latter are non-axisymmetric, they produce torques, i.e. angular momentum is transported outwards, thereby implying a gas inflow. This inflow is accompanied by variations in the gas surface density, mainly a central concentration, which in turn influences the evolution and stability of the disk. While instabilities increase the velocity dispersion v_s, heating up the gas component towards higher Q, the increase of gas surface density Σ has the reverse effect, since $Q \propto v_s/\Sigma$. When Q falls lower than ~ 2.5, the disk is again unstable with respect to nonaxisymmetric gravitational instabilities (bars, spiral arms, etc.). This self-regulating mechanism implicitly depends on the cloud–cloud collisions, and on the local physical conditions. To model galaxies in the framework of this nonlinear scenario we use self-consistent 2D N-body simulations.

1. Galactic disk structure

The Milky Way is a disk galaxy with non-axisymmetric (NAX) features such as spiral arms or a bar (Kuijken & Tremaine 1991). Disk galaxies are the result of a collapse of a rotating protogalactic cloud. In principle the actual disk structure is the result of the interplay of centrifugal force and gravity. Thus, the physics of angular momentum transport are an important ingredient of any dynamical model of a disk galaxy. In a stellar system this transfer can only be provided by NAX structures; such a system produces gravitational couples between the inner and outer parts of the disk (Lynden-Bell & Kalnajs 1972).

Possible mechanisms acting to transport angular momentum include turbulence, gravitational torques and magnetic torques (Larson 1988). Shakura & Sunyaev (1973) and Lynden-Bell & Pringle (1974) proposed that accretion is driven by an effective viscosity produced by some kind of turbulence which is present throughout the disk. It was shown that on galactic scales the turbulent viscosity cannot be the dominant mechanism responsible for angular momentum transport (Combes 1991). Likewise, magnetic fields are too weak to play an important role (Chiba & Lesch 1994). Gravitational torques, however, driven by tidal interaction or intrinsic instabilites, are always present and they effectively transfer angular momentum towards large radii whenever the rotating system contains a trailing spiral structure and/or a bar. Thus, we concentrate on gravitational torques and examine their significance for angular momentum transport on galactic scales.

The spatial and time evolution of the surface mass density Σ can be described as being driven by a viscosity. This transport parameter depends on the microphysics of the cloud–cloud collisions within the macroscopic NAX features. Such NAX perturbations (bars, spirals) are effectively capable of transferring angular momentum from the inner to the outer part of the disk.

A structure with a rotating bar or spiral pattern in a disk system with a nearly circular orbit will generate three strong resonances: the inner Lindblad resonance (ILR), the outer Lindblad resonance (OLR), and a resonance at the location of corotation (CR). Orbits between ILR and CR will be elongated in a direction parallel to the bar, while orbits outside OLR will be elongated perpendicular to the bar. Depending on the pattern speed and the shape of the rotation curve, there may be two, one or no ILR. Since the spiral waves are damped at the ILR, a NAX structure can be sustained over a galactic lifetime only if the disk is unstable with respect to NAX gravitational instabilities. The stability of a differentially rotating disk sensitively depends on the time-dependent parameter Q, which denotes the ratio of the velocity dispersion v to the surface density Σ. The Q value is different for gas and stars (Toomre 1964):

$$Q_{gas} = \frac{v_{gas}\kappa}{3.36G\Sigma_{gas}} \qquad Q_{star} = \frac{v_{star}\kappa}{\pi G\Sigma_{star}} \qquad (1.1)$$

with the epicycle frequency

$$\kappa = \sqrt{4\Omega^2 + r\frac{d\Omega^2}{dr}} \qquad (1.2)$$

where $\Omega = \frac{v}{r}$.

An enhancement of the velocity dispersion v_{gas} or v_{star} via cloud–cloud collisions stabilises a disk, whereas the increase of surface mass density—i.e. accretion during galactic evolution—destabilizes. Obviously, due to the different dissipation properties, the gaseous component determines whether a galactic disk is unstable with respect to NAX gravitational instabilities or not (Sellwood & Carlberg 1984).

Due to the angular momentum transfer, Σ changes, since mass is spiralling inwards, and the mass transport can therefore be translated into a viscosity. As mentioned above, an increase of Σ destabilises the disk, and NAX gravitational instabilities thereby build up a positive feedback mechanism, whose final result is the concentration of gas in the central region.

2. Numerical tools for simulating a disk galaxy

The galaxy model consists of a rigidly rotating bulge and halo potential chosen as a Plummer sphere ($\Phi_{bulge}(r) = GM_h/(r_d^2 + a^2)^{-1/2}$ and $\Phi_{halo}(r) = GM_h/(r_h^2 + b^2)^{-1/2}$) . The halo mass within the disk radius r_d is M_h, its scale length is a; the bulge mass is M_b (within the radius r_h) with a scale length b.

We use self-consistent 2D simulations with stars ($N_s = 38000$ particles) and gas ($N_g = 19000$ particles). The particles are initially distributed in a Toomre disk (Toomre 1962) of mass M_d within the disk radius r_d and a scale length $d \geq a$.

Collisions between molecular clouds are highly inelastic, and thus we use a particle model which is suitable for simulating energy dissipation due to cloud collisions. To model a collision in the cloud ensemble, we adopted the same scheme of local processes as described by Casoli & Combes (1982) and Combes & Gerin (1985).

Giant molecular clouds (GMCs) have a finite lifetime. After about 10^7 yr the clouds are destroyed by the tidal forces and/or star formation (stellar winds, supernova explosions, etc.). To describe this destruction numerically we explicitly focus on the GMC (mass of a GMC $M_{GMC} > 2 \cdot 10^5 M_\odot$). After $4 \cdot 10^7$ yr GMCs are reinjected as cloudlets into the interstellar space with a steep power-law spectrum and a two-dimensional random velocity distribution (Combes & Gerin 1985).

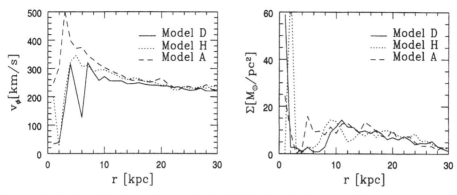

FIGURE 1. Surface density and rotational velocity for the three models at timestep 1000 (1 timestep equals $1.2 \, 10^7$ yr)

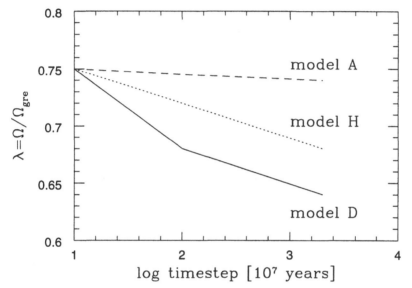

FIGURE 2. Angular momentum transport during the run of the simulation for the different models ($\lambda = \Omega/\Omega_{gre}$)

3. Numerical results

To calculate the viscosity we simulated various galaxy models with different mass distributions:

- model D: the disk dominated model; this galaxy develops a strong bar during the simulation. Because of the dominating bar, in this model the largest amount of mass is transported into the inner part of the galaxy.
- model H: the halo dominated model; the mean feature is a spiral structure. In this model less material/angular momentum than in model D is transported.
- to compare the NAX model with an axisymmetric (AX) one we use model A: this model develops only AX structures.

The rotational velocities and surface densities for all three models after $1.2 \, 10^{10}$ yr are shown in Figures 1a,b. Model D transports more angular momentum into the outer parts and more mass into the inner part of the disk than model H and A.

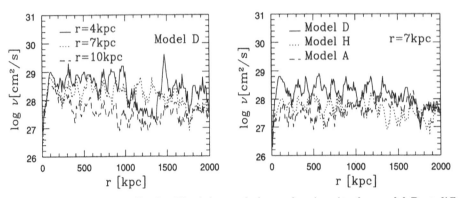

FIGURE 3. The viscosity amplitude. The left panel shows the viscosity for model D at different radii; the right panel shows the viscosity at a radius of 7 kpc for all models.

A comparison of the angular momentum transport in the three models is presented in Figure 2. The angular velocity of the system $\Omega \simeq (\frac{J}{Mr^2})$ (J = angular momentum, r = radius, M = mass of the system) divided by the angular velocity necessary to support the system by rotation $\Omega_{gre} \simeq (\frac{GM}{r^3})^{1/2}$, gives information about the angular momentum of the system. In the disk-dominated model the angular momentum transport is much more effective than in the halo-dominated model.

Figure 3a shows the amplitude of the viscosity of the models D at three different radii. At a radius of 10 kpc the viscosity has values of the order of $5 \cdot 10^{27}$ cm^2 s^{-1}, but increases with radius. The disk viscosity is also a function of time. For example, at radius 4 kpc, initially the viscosity has typical values of the order 10^{29} cm^2 s^{-1} and at the end of our simulations ($2.4 \cdot 10^{10}$ yr) of $2 \cdot 10^{28}$ cm^2 s^{-1}. Also the value for the viscosity is affected by the structural changes in the disk. To give an example, the viscosity decreases for model D at a radius of 4 kpc between $12 \cdot 10^9$ and $14 \cdot 10^9$ yr. This is easy to understand, because model D exhibits a ring structure at the ILR (5 kpc) during this interval. Thus, there is a low mass density and the values for the viscosity are therefore smaller.

If one compares the viscosity for the three models one realises that the model which is able to transport the largest amount of mass/angular momentum (model D) also shows the highest viscosity. Model A shows the smallest values of viscosity during the whole simulation.

The gravitational viscosity is an important element in the understanding of the transport of angular momentum. We have seen that the viscosity is really related to the spiral- and/or bar-driven gas flows. Furthermore, the viscosity in the inner part of the disk is higher than in the outer part.

REFERENCES

CASOLI, F. & COMBES, F. 1982 *Astron. Astrophys.* **110**, 287.

CHIBA, M. & LESCH, H. 1994 *Astron. Astrophys.* **284**, 731.

COMBES, F. 1991 in *Dynamics of Galaxies and Their Molecular Cloud Distribution* (ed. F. Combes & F. Casoli). IAU Symposium 146, p. 255. Kluwer.

COMBES, F. & GERIN, M., 1985 *Astron. Astrophys.* **150**, 327.

KUIJKEN, K. & TREMAINE, S. 1991 in *Dynamics of Disc Galaxies* (ed. B. Sundelius), p. 71. Göteborg University.

LARSON, R. B. 1988 in *Galactic and extragalactic star formation* (ed. R. E. Pudritz & M. Fich). NATO ASI Series C, vol. 232, p. 31. Kluwer.

LYNDEN-BELL, D. & KALNAJS, A. H. 1972 *Mon. Not. R. Astron. Soc.* **157**, 1.

LYNDEN-BELL, D. & PRINGLE, J. E. 1974 *Mon. Not. R. Astron. Soc.* **168**, 603.

SELLWOOD, J. A. & CARLBERG, R. G. 1984 *Astrophys. J.* **282**, 61.

SHAKURA, N. I. & SUNYAEV R. A. 1973, *Astron. Astrophys.* **24**, 337.

TOOMRE, A. 1964 *Astrophys. J.* **139**, 1217.

Elmegreen: How much do your results depend on the assumptions regarding cloud collisions?

Linden: If you change the cloud–cloud collision so much that you affect the structure formation in the disk—by collision without cooling, e.g. the spiral structure cannot be maintained for long—then it will affect the results for the viscosity, of course.

Elmegreen: How much of the total effective viscosity is from the spiral torques and how much from cloud collisions?

Linden: By forming, for example, a massive bar in the disk, we get similar values for the viscosity whatever different collision parameters are used. So the cloud–cloud collision viscosity appears negligible. But we must test more extreme models to conclude.

Density Waves and Star Formation: Is There Triggering?

By BRUCE G. ELMEGREEN

IBM Research Division, T.J. Watson Research Center, P.O. Box 218, Yorktown Heights, NY 10598, U.S.A.

Modern observations related to density-wave triggering of star formation are reviewed. These observations include age gradients through spiral arms, variations in the arm/interarm contrasts for star formation and gas, variations with spiral phase in the sizes of HII regions and star-forming clouds, and differences in the global star-formation rates for galaxies with and without spiral waves. There is some evidence now for triggering as measured by an enhancement in the star-formation rate per unit gas mass in the arms compared to the interarms, although most of the published evidence, particularly regarding age gradients, suffers from phase-dependent extinction or other problems. There is also evidence for arm-to-interarm variations in HII region and cloud sizes. There is, however, no apparent difference in the global star-formation rate per unit gas mass for galaxies with and without spiral arms, at least in the main disks, and there is evidence that the star-formation rate per unit molecular mass is about the same whether or not spiral arms are present. Thus it appears as if spiral waves do three things: they organise star formation into the arms by putting most of the gas there, they cause the gas to collect into giant clouds and star complexes in the arms, and they change, perhaps by a small factor, the rate of star formation per unit gas mass averaged over the arm compared to the interarm region. This change could be consistent with a star-formation rate per unit gas mass proportional to a power of density, ρ^α, with α in the range 0.5 to 1 (spanning the expected range for gravitational instabilities and cloud–cloud collisional triggering, respectively), but α could also be closer to 0 considering the uncertainties in the observations.

1. Introduction

Over 40 years ago, Baade & Mayall (1951) discovered that most of the HI regions in M31 are located in spiral arms. Morgan *et al.* (1952, 1953) found a similar clustering of star-formation in the Sagittarius spiral arm. These observations led to the idea that star-formation traces spiral structure.

A decade later, the wave theory for galactic spirals was advanced by Lin & Shu (1964, 1966). According to this theory, galactic spirals are moving regions of enhanced stellar density responding to their own gravitational perturbations. The perturbations were thought to be weak in the original linear theory, perhaps only 5% in the potential, and because of this weakness, the spiral response had to be amplified in order to be seen. This amplification was based on the idea that the waves trigger star formation (Roberts 1969). In this theory, the interstellar gas does not form many stars while it is in the interarm regions, but it is suddenly changed when it enters a spiral arm to a state in which stars rapidly form.

Numerous observations that claimed to support the triggering hypothesis were summarised in an earlier review (Elmegreen 1987), which pointed out that there was no unambiguous evidence for direct triggering and that instead the available evidence suggested that global star formation had a constant rate per unit gas mass, independent of the presence or lack of a spiral wave in the disk. The suspicion that triggering was not important was also based on the recent discovery that spiral-arm amplitudes are intrinsically strong, not just 5% perturbations in the disk, in which case triggering was not necessary for the arms to be seen. The review concluded that star formation could be just following the gas with a constant efficiency.

Since this earlier review, there has been a substantial amount of work on the triggering problem, centered on four main questions: (i) Are there age gradients in spiral arms? (ii) Does the star-formation rate per unit gas mass vary from the arms to the interarm regions? (iii) Are HII regions, star-forming regions or interstellar clouds any different (e.g. larger, more self-gravitating, etc.) in the arms compared to the interarms? (iv) Is the global efficiency of star formation in a galaxy influenced by the presence of a spiral density wave? We review the available answers to these questions here. A previous review on this topic, covering these and other points, was in D. Elmegreen (1993). Other reviews of star formation in spiral arms, including more details about the instability model and spiral-arm structure, are in Elmegreen (1994a, 1995).

1.1. *Age gradients*

The mean ages of stars often vary through a spiral arm. For example, young clusters are more concentrated in the arms of our Galaxy than old clusters (Sitnik 1991; Feinstein, this volume). For Cepheid variables, the periods decrease, and therefore the ages increase with increasing distance from the inner edges of spiral arms in M31 (Efremov 1988) and the Carina arm in the Milky Way (Avedisova 1988; Berdnikov 1986). There is also a slight (on the order of an arcsecond) colour shift in NGC 7479 with the blue peak concentrated more towards the steep part of the spiral arms than the near-infrared peak (Beckman & Cepa 1990). Other galaxies have a similar shift (Hodge *et al.* 1990; del Rio, this conference). The ultraviolet shows an even larger shift in spiral phase from the Hα emission (Courtes *et al.* 1993; Bersier *et al.* 1994).

Some of these shifts may be influenced by extinction, which should follow the non-linear spiral response of the gas. For example, if the dust is preferentially on one side of the arm, as is often the case for dust lanes, then the red light will be relatively stronger there and the blue or ultraviolet light stronger on the other side of the arm. Inside corotation, the dust lanes tend to be on the insides of the arms, so the blue and uv light should be on the outsides of the arms, which is downstream from the dustlane. This will appear as a colour gradient similar to what is expected from triggering, but it may be only an extinction gradient.

Even if there is a real gradient in the mean ages of stars across an arm, does that necessarily imply there is triggering? By triggering I mean an increase in the star-formation rate per unit gas mass, represented here by R_{SFR}. Triggering then implies that a gram of gas with a low star-formation rate in the interarm region, measured in stars per year, or solar masses of stars per year in that gram, changes inside an arm so that it has a higher star-formation rate per gram in the arm.

The answer to this question is tricky because the gas moves a lot relative to the stars in a spiral arm. Then some of these apparent age gradients could result from the relative offset between the gas and the old stars. Imagine corotating around with the stars in a galaxy in which there is no spiral-arm triggering. We begin in an interarm region where the star and gas densities are low and the star-formation rate is proportionally low too. This will give a certain colour of light and a certain Hα intensity. Suppose now that a wave starts to pass by. Before the crest arrives, the gas will leave our neighbourhood and move towards the approaching wave faster than we will following the stars, and this will lower the gas/star density ratio locally. It will also lower the star-formation rate per unit old star mass, but only because the gas mass per unit old star mass has decreased. The region becomes red as a result, just on the inside edge of a spiral arm. Next suppose a dense ridge of gas passes by our stellar neighbourhood; this is the shock front from the density wave. At the same time, our local star density increases a bit, but by a much smaller amount than the gas density. When the dense gas is with us, the star-formation rate per unit old star mass goes up a lot, in proportion to the gas mass per unit old star mass. Then our region has a lot of Hα, dust, CO emission, and blue light. The Hα and blue light are absorbed by the high local density of dust, but some Hα gets out, particularly in the giant, expanded HII regions that easily poke out of the dust layer. When the gas ridge passes, the gas density drops quickly, but the local star density drops more slowly (the stellar arm is wider than the gaseous arm). Now we are in a region where star formation has nearly stopped (only because of the low gas density), but there are still many young stars around from the star formation that recently occurred in our neighbourhood. This combination makes our region very blue, downstream from the shock and Hα ridge. Soon, we return to the initial gas and star density and to our initial colours in the interarm region. All of these colour variations occurred without triggering, as defined above. They result entirely from the relative motion between the gas and stars, and this relative motion results from the greater compression and hydrodynamic response of the gas compared to the stellar compression that is constrained by epicyclic motions.

Note that this non-triggering scenario has all of the essential characteristics of the triggering scenario, except for the triggering. In this sense, spiral arms are sharply delineated by star formation because they are sharply delineated by gas. This is the essential point of the Roberts (1969) calculations of the gas response to a spiral wave. It gives an age gradient and phase-dependent triggering morphology without any need for triggering. To check whether the observed morphologies require triggering too, we must measure the star- formation rate per unit gas mass, R_{SFR}, as a function of spiral phase.

There are other morphologies of gas and young stars in spiral arms too. Wiklind *et al.* (1990) found in a spiral-arm section of M83 that the CO and HI emission were about 300 pc downstream from the dust lane. Casoli *et al.* (1990) found in NGC 6946 that the HI, CO and Hα are all together. Tilanus & Allen (1991, 1993) found in M51 and M83 that the HI is downstream form the CO. These are unconventional morphologies, not predicted by the triggering hypothesis.

The most thorough study of gas morphology in a spiral galaxy is by Garcia-Burillo *et al.* (1993), who compared HI, radio continuum, Hα and a complete single-dish map of CO for M51. They found that the CO and radio continuum peaks are at the same location even though the Hα and CO peaks are shifted in spiral-arm phase. A very high resolution (3.5") study of CO in part of M51 by Rand (1993a) showed these CO to Hα shifts even better. The shift between the Hα and the radio continuum in M51 is apparently the result of extinction. Garcia-Burillo *et al.* also found that the CO and HI are at the same locations.

1.2. *Spiral variations in the star-formation rate per unit gas mass*

Observations of an increase in R_{SFR} in spiral arms compared to the interarms would prove the triggering hypothesis unambiguously. There are several publications that discuss such an increase. For example, in the Milky Way, Sievers *et al.* (1991) showed that the star-formation efficiency in W49, which is in a main spiral arm, is larger than in Orion, which is in an interarm region or an arm spur.

Similar studies have been made for other galaxies. Cepa & Beckman (1989, 1990a) found variations in the Hα/ HI ratios with spiral-arm phase in several galaxies. Tacconi & Young (1990) found that the ratio of the Hα to the CO intensity is larger in the northeast spiral arm of NGC 6946 than in the adjacent interarm region, but the southwest arm did not show much increase. The arm/interarm contrast of the Hα/CO ratio also increased with radius in the NE arm as the arm got stronger, and to a lesser extent in the SW arm, suggesting that star-formation triggering depends on spiral arm strength. Tacconi & Young corrected the Hα for extinction using the local HI+H$_2$ column density and a conversion from column density to extinction. Lord & Young (1991) did a similar study for M51, as did Knapen *et al.* (1992) using different data. The first of these papers had a relative large CO beam size, 45 arcseconds, and deconvolved the Hα to match this beam. The Knapen *et al.* study had a slightly smaller CO beam, based on a Nobeyama map in Nakai *et al.* (1991).

These variations in Hα/CO were confirmed by Garcia-Burillo *et al.* (1993), who had much better CO data for M51. These authors also found that the Hα/CO ratio was larger in the arms than in the interarms, although at midradius, (50"–100"), where the spiral arms are most sharply defined, the arm/interarm contrasts in CO and non-thermal radio continuum were equal. If the non-thermal radio continuum is unrelated to star formation, and if the CO is tracing the molecular mass, then the Hα/CO increase in the arms supports the triggering hypothesis. If the non-thermal radio continuum is related to the local star-formation rate, then triggering could be small. Arm-to-interarm variations in the ratio of the *thermal* radio continuum to CO would be important to obtain, because the thermal radio continuum is more directly connected to star formation then the non-thermal radio continuum, and it does not suffer from extinction problems like Hα.

Spiral-arm efficiency variations are not always found, however. The efficiency was found to the be same by Wilson & Scoville (1991) for the molecular clouds inside and outside the weak arms in M33. Perhaps, as suggested by Elmegreen (1987, 1988, 1994b) and Tacconi & Young (1990), the triggering efficiency depends on spiral-arm strength. M51 has unusually strong arms and evidence of spiral-arm triggering, but M33 has weak arms and apparently no triggering.

1.3. *Spiral variations in the sizes of HII regions and cloud complexes*

Since Mezger (1970) showed that the spiral arms in our Galaxy contain giant HII regions, which are much larger than those in the interarm regions, the size difference for star-forming regions inside and outside spiral arms has been widely discussed. Georgelin & Georgelin (1976) found this result again from optical data of our Galaxy, while Kennicutt & Hodge (1980) and Rumstay & Kaufman (1983) found it for HII regions in other galaxies.

What is more important for the triggering hypothesis than the size variation is a variation in the slope of the HII region distribution function. If the arm and interarm luminosity function slopes are the same, with the arm function merely shifted towards larger sizes, then there need not be selectively larger HII regions in the arms but only more HII regions in the arms. More HII regions allow us to sample further into the large end of the distribution function, and so larger HII regions are found in the same proportion to small HII regions as in the interarms. If the slope of the HII region luminosity function gets shallower in the arms, then proportionally larger star-forming regions are forming there.

Today it seems clear that HII regions are selectively larger in the arms than in the interarm regions. This was first shown for other galaxies by Kennicutt & Hodge (1980). Cepa & Beckman (1990b) also found a shallower HII luminosity function in the arms than in the interarms for NGC 4321, Rand (1992) found the same for M51, and Banfi *et al.* (1993) got this result for several galaxies. Garcia-Gomez & Athanassoula (1993) separated the arm and interarm HII regions by Fourier transform techniques and got essentially this same result, although they emphasised the average sizes rather than the slopes. Knapen *et al.* (1993) got a different result for NGC 6814, where they found the same slope of the HII region luminosity function in the arms and interarms. Perhaps the preponderance of giant HII regions in spiral arms varies from galaxy to galaxy.

Giant cloud complexes could also be more gravitationally bound in the arms than in the interarms. Rand (1993b) suggests this is the case for the giant molecular associations in M51. This result follows theoretically from the larger disruptive tidal forces in the interarm regions compared to the arms (Elmegreen 1992).

1.4. *Global efficiencies of star formation*

1.4.1. *Star-formation rates per unit gas mass*

The star-formation rate per unit gas mass, R_{SFR}, averaged over a whole galaxy disk, is relatively easy to measure from the ratio of the total Hα, radio continuum, uv, or far infrared luminosity to the total gas mass. This rate per unit mass is related to the inverse of the gas consumption time. Comparisons between galaxies with and without spiral waves might clarify the role the waves have in elevating the star-formation rate. As it turns out, R_{SFR} is about constant for all galaxies, except for those with obvious interactions (Young *et al.* 1986) or with very low surface brightness disks, and it is not so clear what this result implies about density-wave triggering.

To check the triggering hypothesis by looking for variations in R_{SFR}, the first step is to find galaxies without density waves. This is not possible from blue photographs alone because a spiral arm in a blue photograph could be either from a wave or pure star formation. The task is made easier with images in the near-infrared (I band) or J, H, and K bands. Then it is apparent, from B-I colour analyses and stellar evolution models, that most long spiral arms, whether or not they are symmetric, are density waves in the moderately old disk, while short and patchy arms are pure star formation (Elmegreen & Elmegreen 1984, 1985). Recent K-band photographs suggest the same thing without requiring such colour corrections for contamination by star formation (Cepa *et al.* 1988; Rix & Rieke 1993; Block *et al.* 1991, 1994). Thus galaxies with only short spiral arms in an irregular pattern, such as NGC 2841, NGC 5055, or NGC 7793 (a sequence of later and later Hubble types), presumably have no spiral waves at the present time. Such galaxies are called flocculent, in reference to their fleece-like appearance. Galaxies with waves can be either grand design or multiple arm, but in all cases they have at least one long arm, covering a large fraction of the visible disk.

Based on these morphological differences, two independent studies published in the same month found that galaxies with and without spiral waves have the same specific star-formation rates: McCall & Schmidt (1986) showed that the fraction of galaxies with grand-design spirals is the same whether they have Type I, Type II, or no recorded supernovae, which implies that this fraction is independent of the supernova rate; and Elmegreen & Elmegreen (1986) showed that the star-formation rate per unit area, from uv, FIR, Hα, integrated colour, and surface brightness data, is the same for grand-design and flocculent galaxies. A related study in the same year, using FIR from *IRAS* and CO luminosities, suggested that the star-formation rate per unit gas mass (H$_2$ here) is the same for all galaxies, even independent of Hubble type, although the spiral-arm type was not considered (Rengarajan & Verma 1986). This contrasts with the rate per unit area, which does depend on Hubble type (e.g. Elmegreen & Elmegreen 1986). Later studies showed a related lack of correlation between spiral-arm class and CO luminosity per unit area (Stark *et al.* 1987), and spiral-arm class and radio continuum emission or X-ray emission per unit light (Giuricin *et al.* 1989).

This surprising result was not without precedent. Talbot (1980) found that the star-formation rate per unit molecular mass in the Milky Way is independent of radius, and Young & Scoville (1982a,b) found the same for several other galaxies, comparing total blue light to CO emission. A detailed study of NGC 6946 by DeGioia-Eastwood *et al.* (1984) found the same thing, and they concluded that the rate of interaction between the gas and a spiral wave (this rate is larger at small radii) does not influence the star-formation rate, and therefore that the wave does not influence star formation. Further studies of the same galaxy by Tacconi & Young (1986) agreed, and demonstrated also that the star-formation rate follows the CO luminosity alone; neither the HI surface density nor the total HI+H_2 surface densities correlated with star formation as well as the H_2 surface density. In the same year, Rana & Wilkinson (1986) reanalysed the Talbot (1980) study with better data and obtained about the same result, i.e. that the star-formation rate per unit molecular mass is constant with radius in our Galaxy. Two years later, Wouterloot *et al.* (1988) found the same rate per unit molecular mass for molecular clouds in the far-outer disk.

Many more studies of CO emission in galaxies are now available and they all suggest that the star-formation rate per unit H_2 gas mass is about constant in the active parts of the disk. Wiklind & Henkel (1989), Thronson *et al.* (1989), Bajaja *et al.* (1991), and Li *et al.* (1993) showed this for early type galaxies, mostly with Hubble types S0 or S0/a, Devereux & Young (1991) got this result for all types Sa–Scd. Sage (1993) got it also for a distance- (rather than brightness-) limited sample. Blue compact galaxies have about the same R_{SFR} (Sage *et al.* 1992), as does a hot spot galaxy (Jackson *et al.* 1991) and the main disk of a starburst galaxy (Sage & Solomon 1991), although the nuclear region of the latter may have a higher efficiency (*ibid.*).

Buat *et al.* (1989) and Buat (1992) suggested that the star-formation rate per unit area scales better with the total gas mass per unit area, HI+H_2, than with the H_2 mass per unit area alone. These are global measures of star-formation rate or mass, all divided by the area inside the radius of the 25th mag arcsec^{-1} isophote, even though some of the emission may extend beyond that radius (e.g. HI) or end inside of it (CO). Kenney & Young (1988) also found a better correlation between star- formation rate and total HI+H_2 mass for dwarf galaxies, in which the ISM is dominated by HI. Kennicutt (1989) discussed the star- formation rate per unit area in terms of the total HI+H_2 surface density, which is a local measure, unlike the total HI+H_2 mass.

A nagging problem with all of these R_{SFR} determinations is that both the FIR and the CO luminosities could result from photon excitation from all stars. The FIR emission is generally from the warm dust, whose emission depends on the total starlight in the disk, and the CO, although usually assumed to be proportional to H_2, may in fact just be emitted from (or strongly influenced by) photon-heated cloud boundaries (Wolfire *et al.* 1993). In that case, FIR/CO will be about constant in all regions regardless of R_{SFR}. This observational selection has been illustrated well by the recent discovery of cold CO in the inner disk of M31, which has very little star formation (Allen & Lequeux 1993). The CO is presumably cold because the uv radiation field there is weak. The same might be said of the constant ratio of Hα to CO luminosities: that both come from excitation by the nearby massive stars, which dominate the interstellar radiation field. These selection effects might explain other observations as well. The constant ratio of H_2 mass to dynamical mass in the molecular region of a spiral galaxy (Sage 1993) might result from CO excitation by the blue starlight which is also proportional to the disk mass.

1.4.2. *Surface density thresholds*

If star formation follows from large-scale instabilities in galactic disks, then stars should not form with the same R_{SFR} in regions where the gas surface density is less than a threshold value that comes from the $Q > 2$ stability condition for non-axisymmetric, or spiral, instabilities (Goldreich & Lynden-Bell 1965; Quirk 1972; Larson 1983). Surface density thresholds for star formation were first suggested on the basis of observations by Skillman *et al.* (1986, 1988) and Guiderdoni (1987), who did not consider this Q threshold. Zasov & Simakov (1988) and Kennicutt (1989) first showed that Q is important. Kennicutt (1989) derived a critical column density for star formation of $\sigma_{crit} = 0.7\kappa c/(3.36G)$. This corresponds to $Q < 1.4$. Kennicutt (1989) also suggested that spiral arms might trigger star formation by raising the surface density above the local threshold. A recent theoretical analysis confirms this suggestion (Elmegreen 1994b).

Several other papers now point to a Q threshold as well: for star formation in the gas disks of elliptical galaxies (Vader & Vigroux 1991); for early type galaxies (Caldwell *et al.* 1992); for low surface brightness galaxies (van der Hulst *et al.* 1993); and in individual star-formation regions in M33 (Buat *et al.* 1994), where the average Q is too high for global instabilities (Wilson *et al.* 1991). The local velocity dispersion of the gas may play an additional role in regulating star formation (Arsenault *et al.* 1990). The idea that spirals trigger star formation in regions of a disk that would have too high a Q without the spiral, and therefore little star formation without the spiral, has not been confirmed by observations other than those initially suggested by Kennicutt (1989).

2. Putting it all together

An increase in the ratio of Hα to CO in spiral arms compared to the interarm regions suggests that the rate of star formation per unit gas mass is higher in the arms than in the interarms. Other observations of triggering, such as age or colour gradients, global efficiency variations from galaxy to galaxy, or variations in the size distribution of HII regions, are still somewhat ambiguous. The size variation of HII regions could be the result of cloud coagulation in the shock fronts of spiral arms, where the filling factor of diffuse clouds could be nearly unity. Size variations could also result from large-scale instabilities in the arms, which make the clouds there, followed by a gradual erosion of the clouds in the interarm regions. But coagulation or cloud growth in spiral arms does not alone imply that R_{SFR} increases.

The coagulation model has been proposed by Casoli & Combes (1982), Kwan & Valdes (1983) and others. The instability model follows from Mouschovias *et al.* (1974), Elmegreen (1979, 1994b), Balbus & Cowie (1985) and others. The coagulation model suggests that spiral-arm cloud complexes form from random collisions of smaller interarm clouds. Usually this model has been applied directly to the formation of molecular clouds, often by the collision of other molecular clouds, without first forming any $10\times$ larger clouds inside which the molecular clouds appear as dense cores. The instability model forms the larger clouds first, and then forms GMCs by dissipation and collapse inside of them (Kolesnik 1991), as a secondary process.

If the large spiral-arm clouds form by instabilities in the arms, driven presumably by self-gravity and magnetic forces in the gas, and if these clouds condense because of self-gravity and energy dissipation while they are in the arms, then the star-formation rate per unit gas mass should be proportional to the square root of the local density, ρ. This follows from the dependence of the gravitational instability rate on $\rho^{1/2}$. The star-formation rate per unit volume would then scale with $\rho^{3/2}$. With a magnetic field and some contribution from the Parker instability in the growth of self-gravitating perturbations, the power in this expression is slightly smaller (Elmegreen 1991). Such a simple density dependence is consistent with the observations of R_{SFR} variations in spiral arms. Another model in which star formation is triggered by cloud collisions (Scoville *et al.* 1986) is also consistent with the observations, although then the power of density should be higher. Presumably both of these triggering mechanisms operate because the arms are unstable to collapse into giant cloud complexes similar to those observed, and because interarm clouds should collide in the spiral shock front.

After the gas leaves a spiral arm, the mechanism of star formation should change from one involving coagulation and instabilities on a large scale to one involving secondary processes, such as propagating star formation in shells or rings, on a small scale. Eventually the giant clouds that form in the spiral arms disperse in the interarm regions and the process of assembly into large clouds begins again with the approach of the next arm.

Thanks to Drs. Alfaro, Delgado, and Alberdi for their hospitality in Granada during the week before this conference, when most of this material was compiled.

REFERENCES

ALLEN, R. J. & LEQUEUX, J. 1993 *Astrophys. J., Lett.* **410**, L15.

ARSENAULT, R., ROY, J. R. & BOULESTEIX, J. 1990 *Astron. Astrophys.* **234**, 23.

AVEDISOVA, V. C. 1988 in *Structure of Galaxies & Star Formation* (ed. J. Palouš), p. 219. Czechoslovakian Academy of Sciences, Prague.

BAADE, W. & MAYALL, N. U. 1951 in *Problems of Cosmical Aerodynamics*, p. 165. Control Air Document Office, Dayton Ohio.

BAJAJA, E., KRAUSE, M., DETTMAR, R. J. & WIELEBINSKI, R. 1991 *Astron. Astrophys.* **241**, 411.

BANFI, M., RAMPAZZO, R., CHINCARINI, G. & HENRY, R. B. C. 1993 *Astron. Astrophys.* **280**, 373.

BECKMAN, J. E. & CEPA, J. 1990 *Astron. Astrophys.* **229**, 37.

BERDNIKOV, L. N. 1986 *AJ Letters* **13**, 110.

BERSIER, D., BLECHA, A., GOLAY, M. & MARTINET, L. 1994 *Astron. Astrophys.* **286**, 37.

BLOCK, D. L., BERTIN, G., STOCKTON, A., GROSBOL, P., MOORWOOD, A. F. M. & PELETIER, R. F. 1994 *Astron. Astrophys.* **288**, 365.

BLOCK, D. L. & WAINSCOAT, R. J. 1991 *Nature* **353**, 48.

BUAT, V. 1992 *Astron. Astrophys.* **264**, 444.

BUAT, V., DEHARVENG, J. M. & DONAS, J. 1989 *Astron. Astrophys.* **223**, 42.

BUAT, V., VUILLEMIN, A., BURGARELLA, D., MILLIARD, B. & DONAS, J. 1994 *Astron. Astrophys.* **281**, 666.

CALDWELL, N., KENNICUTT, R., PHILLIPS, A. C. & SCHOMMER, R. A. 1992 *Astrophys. J.* **370**, 526.

CASOLI, F., CALUSSET, F., VIALLEFOND, F., COMBES, F. & BOULANGER, F. 1990 *Astron. Astrophys.* **233**, 357.

CASOLI, F. & COMBES, F. 1982 *Astron. Astrophys.* **110**, 287.

CEPA, J. & BECKMAN, J. E. 1989 *Astrophys. Space Sci.* **156**, 289.

CEPA, J. & BECKMAN, J. E. 1990a *Astrophys. J.* **349**, 497.

CEPA, J. & BECKMAN, J. E. 1990b *Astron. Astrophys., Suppl. Ser.* **83**, 211.

CEPA, J., PRIETO, M., BECKMAN, J. & MUÑOZ-TUÑÓN, C. 1988 *Astron. Astrophys.* **193**, 15.

COURTES, G., PETIT, H., HUA, C. T., MARTIN, P., BLECHA, A., HUGUENIN, D. & GOLAY, M. 1993 *Astron. Astrophys.* **268**, 419.

DEGIOIA-EASTWOOD, K., GRASDALEN, G. L., STROM, S. E. & STROM, K. M. 1984 *Astrophys. J.* **278**, 564.

DEVEREUX, N. A. & YOUNG, J. S. 1991 *Astrophys. J.* **371**, 515.

EFREMOV, YU. N. 1988 *Sov. Sci. Rev., Sect. E* **7**, 105.

ELMEGREEN, B. G. 1987 in *Star Forming Regions* (ed. M. Peimbert & J. Jugaku), p. 457. Reidel.

ELMEGREEN, B. G. 1988 *Astrophys. J.* **326**, 616.

ELMEGREEN, B. G. 1991 *Astrophys. J.* **378**, 139.

ELMEGREEN, B. G. 1992 in *Interstellar Gas Dynamics* (ed. D. Pfenniger & P. Bartholdi), p. 157. Springer.

ELMEGREEN, B. G. 1994a in *Physics of Gaseous and Stellar Disks of Galaxies* (ed. I. King). ASP Conference Series, vol 66. ASP.

ELMEGREEN, B. G. 1994b *Astrophys. J.* **433**, 39.

ELMEGREEN, B. G. 1995 in *Physics of the Interstellar Medium & Intergalactic Medium* (ed. C. McKee), ASP Conference Series, in press.

ELMEGREEN, B.G . & ELMEGREEN, D. M. 1985 *Astrophys. J.* **288**, 438.

ELMEGREEN, B. G. & ELMEGREEN, D. M. 1986 *Astrophys. J.* **311**, 554.

ELMEGREEN, D. M. 1993 in *Star Formation, Galaxies and the Interstellar Medium* (ed. J. Franco, F. Ferrini & G. Tenorio-Tagle). 4th EIPC Workshop, p. 108. Cambridge University Press.

ELMEGREEN, D. M. & ELMEGREEN, B. G. 1984 *Astrophys. J., Suppl. Ser.* **54**, 127.

GARCÍA-BURILLO, S., GUELIN, M. & CERNICHARO, J. 1993 *Astron. Astrophys.* **274**, 123.

GARCÍA-GÓMEZ, C. & ATHANASSOULA, E. 1993 *Astron. Astrophys., Suppl. Ser.* **100**, 431.

GEORGELIN, Y. M. & GEORGELIN, Y. P. 1976 *Astron. Astrophys.* **49**, 5.

GIURICIN, G., MARDIROSSIAN, F. & MEZZETTI, M. 1989 *Astron. Astrophys.* **208**, 27.

GOLDREICH, P. & LYNDEN-BELL, D. 1965 *Mon. Not. R. Astron. Soc.* **130**, 7.

GUIDERDONI, B. 1987 *Astron. Astrophys.* **172**, 27.

HODGE, P., JADERLUND, E. & MEAKES, M. 1990 *Publ. Astron. Soc. Pac.* **102**, 1263.

JACKSON, J. M., ECKART, A., CAMERON, M., WILD, W., HO, P. T. P., POGGE, R. W. & HARRIS, A. I. 1991 *Astrophys. J.* **375**, 105.

KENNICUTT, R. C. 1989 *Astrophys. J.* **344**, 685.

KENNICUTT, R. C. & HODGE, P. W. 1980 *Astrophys. J.* **241**, 573.

KENNEY, J. D. & YOUNG, J. S. 1988 *Astrophys. J.* **326**, 588.

KNAPEN, J. H., ARNTH JENSEN, N., CEPA, J. & BECKMAN, J. E. 1993 *Astron. J.* **106**, 56.

KNAPEN, J. H., BECKMAN, J. E., CEPA, J., VAN DER HULST, T. & RAND, R. J. 1992 *Astrophys. J., Lett.* **385**, L37.

KOLESNIK, I. G. 1991 *Astron. Astrophys.* **243**, 239.

KWAN, J. & VALDES, F. 1983 *Astrophys. J.* **271**, 604.

LARSON, R. B. 1983 *Highlights Astron.* **6**, 191.

LI, J. G., SEAQUIST, E. R., WROBEL, J. M., WANG, Z. & SAGE, L. J. 1993 *Astrophys. J.* **413**, 150.

LIN, C. C. & SHU, F. H. 1964 *Astrophys. J.* **140**, 646.

LIN, C. C. & SHU, F. H. 1967 in *Radio Astronomy & the Galactic System* (ed. H. van Woerden), p. 313. Academic Press.

LORD, S. D. & YOUNG, J. S. 1990 *Astrophys. J.* **356**, 135.

MCCALL, M. L. & SCHMIDT, F. H. 1986 *Astrophys. J.* **311**, 548.

MEZGER, P. G. 1970 in *The Spiral Structure of the Galaxy* (ed. W. Becker & G. Contopoulos), p. 107. Reidel.

MORGAN, W. W., SHARPLESS, S. & OSTERBROCK, D. 1952 *Astron. J.* **57**, 3.

MORGAN, W. W., WHITFORD, A. E. & CODE, A. D. 1953 *Astrophys. J.* **118**, 318.

NAKAI, N., KUNO, N., HANDA, T. & SOFUE, Y. 1991 in *Dynamics of Galaxies & their Molecular Cloud Distributions* (ed. F. Combes & F. Casoli), p. 63. Kluwer.

QUIRK, W. J. 1972 *Astrophys. J., Lett.* **176**, L9.

RANA, N. C. & WILKINSON, D. A. 1986 *Mon. Not. R. Astron. Soc.* **218**, 497.

RAND, R. J. 1992 *Astron. J.* **103**, 815.

RAND, R. J. 1993a *Astrophys. J.* **404**, 593.

RAND, R. J. 1993b *Astrophys. J.* **410**, 68.

RENGARAJAN, T. N. & VERMA, R. P. 1986 *Astron. Astrophys.* **165**, 300.

RIX, H. W. & RIEKE, M. J. 1993 *Astrophys. J.* **418**, 123.

ROBERTS, W. W. 1969 *Astrophys. J.* **158**, 123.

RUMSTAY, K. S. & KAUFMAN, M. 1983 *Astrophys. J.* **274**, 611.

SAGE, L. J. 1993 *Astron. Astrophys.* **272**, 123.

SAGE, L. J., SALZER, J. J., LOOSE, H. H. & HENKEL, C. 1992 *Astron. Astrophys.* **265**, 19.

SAGE, L. & SOLOMON, P. M. 1991 *Astrophys. J.* **379**, 392.

SCOVILLE, N. Z., SANDERS, D. B. & CLEMENS, D. P. 1986 *Astrophys. J., Lett.* **310**, L77.

SIEVERS, A. W., MEZGER, P. G., GORDON, M. A., KREYSA, E., HASLAM, C. G. T. & LEMKE, R. 1991 *Astron. Astrophys.* **251**, 231.

SITNIK, T. G. 1991 *Sov. Astron. Letters* **17**, 61.

SKILLMAN, E. D. & BOTHUN, G. D. 1986 *Astron. Astrophys.* **165**, 45.

SKILLMAN, E. D., TERLEVICH, R., TEUBEN, P. J. & VAN WOERDEN, H. 1988 *Astron. Astrophys.* **198**, 33.

STARK, A. A., ELMEGREEN, B. G. & CHANCE, D. 1987 *Astrophys. J.* **322**, 64.

TACCONI, L. J. & YOUNG, J. S. 1986 *Astrophys. J.* **308**, 600.

TACCONI, L. J. & YOUNG, J. S. 1990 *Astrophys. J.* **352**, 595.

TALBOT, R. J. 1980 *Astrophys. J.* **235**, 821.

THRONSON, H. A., TACCONI, L., KENNEY, J., GREENHOUSE, M. A., MARGULIS, M., TACCONI-GARMAN, L. & YOUNG, J. 1989 *Astrophys. J.* **344**, 747.

TILANUS, R. P. J. & ALLEN, R. J. 1991 *Astron. Astrophys.* **244**, 8.

TILANUS, R. P. J. & ALLEN, R. J. 1993 *Astron. Astrophys.* **274**, 707.

VADER, J. P. & VIGROUX, L. 1991 *Astron. Astrophys.* **246**, 32.

VAN DER HULST, J. M., SKILLMAN, E. D., SMITH, T. R., BOTHUN, G. D., McGAUGH, S. S. & DE BLOK, W. J. G. 1993 *Astron. J.* **106**, 548.

WIKLIND, T. & HENKEL, C. 1989 *Astron. Astrophys.* **225**, 1.

WIKLIND, T., RYDBECK, G., HJALMARSON, A. & BERGMAN, P. 1990 *Astron. Astrophys.* **232**, L11.

WILSON, C. D. & SCOVILLE, N. 1991 *Astron. Astrophys.* **101**, 1293.

WILSON, C. D., SCOVILLE, N. & RICE, W. 1991 *Astron. J.* **101**, 1293

WOLFIRE, M. G., HOLLENBACH, D. & TIELENS, A. G. G. M. 1993 *Astrophys. J.* **402**, 195.

WOUTERLOOT, J. G. A., BRAND, J. & HENKEL, C. 1988 *Astron. Astrophys.* **191**, 323.

YOUNG, J. S., KENNEY, J. D., TACCONI, L., CLAUSSEN, M. J., HUANG, Y. L., TACCONI-GARMAN, L., XIE, S. & SCHLOERB, F. P. 1986 *Astrophys. J., Lett.* **311**, L17.

YOUNG, J. S. & SCOVILLE, N. Z. 1982a *Astrophys. J.* **258**, 467.

YOUNG, J. S. & SCOVILLE, N. Z. 1982b *Astrophys. J., Lett.* **260**, L11.

ZASOV, A. V. & SIMAKOV, S. G. 1988 *Astrophys.* **29**, 190.

Chernin: Can we look a little more closely at the Milky Way. You've mentioned the arm in Carina, and what can you say about the thickness of the arm there, the corotation—-where is it—resonances and so on?

Elmegreen: The position of corotation in our Galaxy is not known. Studies of the Carina arm could help to determine corotation because we can observe part of that arm in a transverse direction, perpendicular to the spiral. Then we can look for age gradients or other indications of star-formation triggering, streaming motions, or resonances. But this has not been done yet.

Cepa: It is a problem to determine whether or not spiral arms trigger star formation. However, the location of corotation radius is also controversial, since some numerical simulations can reproduce spiral structure until corotation but not beyond. Could you please comment on that?

Elmegreen: Contopoulis *et al.* have suggested in a number of papers, including one this year (1994) with Patsis, that the stellar and gaseous density response to a spiral potential is most coherent and organised when the potential spiral ends at its inner 4:1 resonance. We believe the spirals end at the outer Lindblad resonance (OLR), which is a very different radius by a factor of two or three. Dynamical and kinematical studies of M51, M81, and M100 have agreed with our resonance locations so far, and clearly not the 4:1 resonance location. The recent paper by Patsis *et al.* suggests that although the density response is less organised for spirals ending at the OLR than at the inner 4:1 resonance, the response is still relatively well organised and strong inside corotation for the OLR case. The problem with spiral coherence in this model seems to lie between CR and the OLR. Now, Debra and I have studied the observational aspects of this problem lately and find that most spirals are indeed far more coherent in the inner half of their radius than the outer half. We think this is more evidence that the total spiral ends at the OLR case, because then CR is midway in the disk where the strong response ends. This is consistent with the Patsis *et al.* results, but not the conclusions they actually state.

Spiral Modes in Galaxies

By SOLEDAD DEL RIO AND JORDI CEPA

Instituto de Astrofísica de Canarias, E–38200 La Laguna, Tenerife, Spain

Via Fourier analysis, applied to broad-band images of a sample of spiral galaxies, the relative contribution of each spiral mode to the spiral arms, as a function of the galactocentric radius, is evaluated and compared with the position of the corotation radius, as determined from independent methods by the present authors (this volume). The relative contribution of each mode before and after corotation should lead to a better understanding of the physical role played by the corotation radius in propagating the spiral waves throughout it.

1. Introduction

The detection of resonances and corotation is an excellent indicator of the presence of a density-wave system responsible for driving and inducing star formation. Also, the corotation radius is in itself a dynamical peculiarity, frequently related to changes in the pitch angle of the arms and/or splitting of the arms at this radius (Cepa 1988; Cepa & Beckman 1990a; del Río & Cepa, this volume).

Several methods have been proposed to detect resonances in barred or unbarred spirals (e.g. Cepa 1988; Elmegreen et al. 1989; Elmegreen & Elmegreen 1990; Cepa & Beckman 1990a; Elmegreen et al. 1992; Canzian 1993; García-Burillo et al. 1994). Some arise as a result of the search for star formation induced by a density wave (e.g. Cepa 1988; Cepa & Beckman 1990a; del Río & Cepa, this volume), because the mechanism(s) causing the star formation in the arms, and the location of resonances, are linked and can be studied separately from a formal point of view only. However, it is difficult to develop an unambiguous method to locate resonances and corotation. So, ideally, by combining various independent methods, it should be possible to draw a coherent scenario of the mechanism causing the spiral structure and its properties. The contributions presented by the authors in this volume are an attempt to achieve this aim.

One of the methods utilised to analyse the spiral structure is based on bidimensional Fourier transform methods (see Puerari 1993 and references therein). This technique is used to show numerically the predominant spiral mode or number of arms. It has also been employed by Puerari & Dottori (1992) and Puerari (1993) to determine the corotation radius on the assumption that the resonances occur in places where the surface density of HII regions is lower (although the reverse is not necessarily the case). However, other authors (Vogel et al. 1993) have suggested that the largest star-forming complexes are located close to corotation, and hence a minimum should not be expected, neither in the surface density of HII regions, nor in the star-formation efficiency.

2. The Fourier transform

By means of the Fourier transform technique, in this work we analyse the galaxies NGC 157, NGC 4321, NGC 6814 and NGC 7479, for which the position of the resonances and corotation radius were estimated using a method independent of bidimensional Fourier transforms or star formation efficiencies, and which is based on the statistical moments of the azimuthal flux distribution (Cepa 1988; del Río & Cepa, this volume). From the sample studied it turns out that the position of the corotation coincides with the relative minima of the spatial distribution $S_m(r)$ (m is the spiral mode),

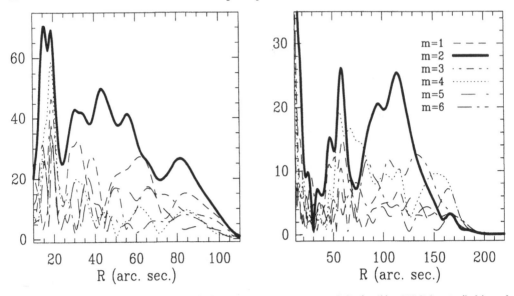

FIGURE 1. Spatial distribution $S_m(r)$ (the real, positive part of $S_m(r,\theta)$) of NGC 157 (left) and NGC 4321 (right). Modes are represented as shown in the key. In NGC 157, mode $m = 2$ dominates throughout the disk, with a relative minimum at $\sim40''$ which corresponds to corotation (del Río & Cepa, this volume). In NGC 4321, mode $m = 2$ dominates beyond corotation, which is located at $\sim80''$ (del Río & Cepa, this volume). Before corotation, modes $m = 2$ and $m = 4$ coexist.

obtained from the inverse Fourier transform of the Fourier spectrum. In NGC 157 the minimum is within a maximum. This indicates that indeed, although in some cases star formation could be enhanced near corotation, there is a relative minimum at corotation.

This method also shows the singularity exhibited by corotation: In two cases (NGC 4321 and NGC 7479) the predominant spiral mode before and after corotation is different, while in NGC 157 and NGC 6814 the mode before and after corotation is the same ($m = 2$ and $m = 4$, respectively). For example, in NGC 157, the mode $m = 2$ dominates (Figure 1) from the beginning until the end of the spiral structure at some $70''$. Corotation occurs at $40''$, where a relative minimum can be seen. In NGC 4321, corotation is located at some $80''$ as inferred from the positions of the inner Lindblad resonances in the circumnuclear arms (Cepa & Beckman 1990b), from the minimum in star-formation efficiency (Cepa & Beckman 1990c) and from the statistical analysis of the azimuthal flux profiles (del Río & Cepa, this volume).

Beyond this radius, mode $m = 2$ is predominant, while inside, $m = 4$ makes an important contribution to the spiral structure (Figure 1). The possible coexistence of both modes inside corotation was already pointed out by Cepa & Beckman (1990c), using the radial dependence of the relative arm/interarm star formation efficiency.

3. Conclusions

It has been shown that the corotation radius is located at a relative minimum of the spatial distribution $S_m(r)$ obtained from the inverse Fourier transform of the Fourier spectrum of the image of the galaxy, even though star formation could be enhanced near corotation. This would imply that, contrary to Vogel *et al.* (1993), a minimum in the star-formation efficiency in the arm with respect to the interarm disk should be expected at corotation.

Some galaxies are shown to change their predominant spiral mode beyond corotation, thus reinforcing their singularity, which in some cases is reflected in morphological terms (Cepa & Beckman 1990a; Cepa & Beckman 1990c).

REFERENCES

CANZIAN, B. 1993 *Astrophys. J.* **414**, 487.

CEPA, J. 1988 PhD thesis, Universidad de La Laguna.

CEPA, J. & BECKMAN, J. E. 1990a *Astrophys. J.* **349**, 497.

CEPA, J. & BECKMAN, J. E. 1990b *Astron. Astrophys., Suppl. Ser.* **83**, 211.

CEPA, J. & BECKMAN, J. E. 1990c *Astrophys. Space Sci.* **170**, 209.

ELMEGREEN, B. G., ELMEGREEN, D. M. & SEIDEN, P. E. 1989 *Astrophys. J.* **343**, 602.

ELMEGREEN, B. G. & ELMEGREEN, D. M. 1990 *Astrophys. J.* **355**, 52.

ELMEGREEN, B. G., ELMEGREEN, D. M. & MONTENEGRO, L. 1992 *Astrophys. J., Suppl. Ser.* **79**, 37.

GARCÍA-BURILLO, S., SEMPERE, M. J. & COMBES, F. 1994 *Astron. Astrophys.* **287**, 419.

PUERARI, I. & DOTTORI, H. 1992 *Astron. Astrophys., Suppl. Ser.* **93**, 469.

PUERARI, I. 1993 *Publ. Astron. Soc. Pac.* **105**, 693.

VOGEL, S. N., RAND, R. J., GRUENDL, R. A. & TEUBEN, P. J. 1993 *Publ. Astron. Soc. Pac.* **105**, 666.

Coherent Star Formation in Spiral Galaxies

By SOLEDAD DEL RIO AND JORDI CEPA

Instituto de Astrofísica de Canarias, E-38200 La Laguna, Tenerife, Spain

A sample of spiral galaxies is examined, applying moments analysis to their broad-band azimuthal profiles, in order to search for evidence of coherent large-scale star-forming processes in the form of star-formation fronts on one side of the arms. The methods employed enable us to determine the position of Lindblad resonances and corotation, showing that the corotation radius is frequently a physical discontinuity of the morphology and properties of the spiral arms. The authors present a further communication at this workshop, which is intended as complementary to the present paper.

1. Introduction

Our aim in this work is to detect star-formation fronts in spiral arms, and to determine their location and the position of resonances and corotation, if any. The existence of these fronts would imply that the density waves somehow trigger star formation in the arms, which is a controversial hypothesis (Elmegreen 1992).

2. The azimuthal profiles

The method of analysing azimuthal flux profiles is based on a set of parameters defined by Cepa (1988). An azimuthal profile taken at a radial distance r from the centre of a galaxy represents the flux distribution F_i^λ at a certain wavelength λ of the points $i = 1, \cdots, n$, located at radii, *in the plane of the galaxy*, ordered according to the angle θ_i^λ of each point with respect to a defined origin in azimuthal coordinates. The origin of the angles was taken at the major axis, with positive sense anticlockwise. The origin of the radius is the centre of each galaxy. For a particular feature (one arm in this case), characterised by a flux increase at a wavelength λ between the points $j = k, k+1, \cdots, l$, where $l \subset i, k \geq 1$ and $l \leq n$, we defined the following parameters :

$$\overline{\theta}^\lambda = \frac{\sum_{j=k}^{l} F_j^\lambda \theta_j^\lambda}{F^\lambda} \quad ; \quad \sigma^\lambda = \sqrt{\frac{\sum_{j=k}^{l} F_j^\lambda (\theta_j^\lambda - \overline{\theta}^\lambda)^2}{D^\lambda}} \quad ; \quad S^\lambda = \frac{\sum_{j=k}^{l} F_j^\lambda (\theta_j^\lambda - \overline{\theta}^\lambda)^3}{D^\lambda \sigma^\lambda 2^3}$$

where

$$F^\lambda \equiv \sum_{j=k}^{l} F_j^\lambda \quad \text{and} \quad D^\lambda \equiv F^\lambda - \frac{\sum_{j=k}^{l} (F_j^\lambda)^2}{F^\lambda}$$

Here $\overline{\theta}^\lambda$ is the mean azimuthal angle (first order momentum in the flux distribution); σ^λ is the full width at half maximum of the particular feature under study (second order momentum) and S^λ is it *skewness* (third order momentum), which can be 0 (if the profile is totally symmetric), negative (if the steeper side is before the maximum) or positive otherwise (Cepa 1988).

According to the spiral density-wave theory (SDW) the wave rotates rigidly with Ω_p— the angular speed—while the disk matter (gas and stars) is subject to differential rotation,

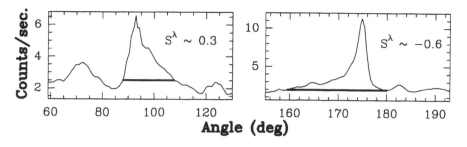

FIGURE 1. Different signs of the asymmetry

$\Omega(r)$. Gas is much more affected by the passage of the density waves than stars are. If density waves trigger star formation in the arms, directly or indirectly (via the formation of GMCs, for example), a star-formation front should be present on one of the sides of the arm and all along its length from the ILR till corotation, changing over to the other side from corotation till OLR. This would be related to the differences in velocity between the spiral pattern and the disk in its differential rotation. Also, the column density is directly related to luminosity, so that observable profiles reproduce this behaviour of stars. In a "trailing" spiral, before corotation, the front is located on the concave side of the arm, disk matter flowing through the convex one (the opposite for "leading" spirals). After corotation, the front is on the convex side. In this case, a sudden increase in luminosity occurs in the azimuthal profiles, followed by a smooth decrease as a region evolves (in a galaxy rotating anticlockwise). The main observable feature from these events is a steep slope in the azimuthal profiles, on the side of the arm where the front is located, and a bluer colour on that side (Schweizer 1976). The parameters defined above will allow us to study these features. In particular, S^λ will give us an idea about the location of the steeper side of the profile (the skewness degree), and $\overline{\theta}^\lambda$ indicates where the maximum would be if the flux distribution were symmetric. Although we want to know the position of the maxima, they can be masked by the presence of any (field) star. So the difference in positions between I and B maxima ($\theta^B_{arm} - \theta^I_{arm}$)—which indicates the direction of colour gradients across the arms—could yield erroneus information. Hence the use of media instead of maxima ($\overline{\theta}^B_{arm}$ and $\overline{\theta}^I_{arm}$) will provide more reliable information.

If the gas attains a critical density value, star formation is triggered and it is possible to observe a luminosity increase due to star formation (B band) and concentration of disk stars (I band). We expect to find asymmetric flux profiles with the steeper side where the front is located.

In corotation we do not expect a large amount of star formation to occur, because both $\Omega(r)$ and Ω_p velocities are equal, and the front should be weaker at that point. This implies that asymmetry should not have a sign, and its value should be very close to zero.

Also, when asymmetry is zero—a strong candidate for the corotation to be placed at this radius—note how the pitch angle sometimes changes, as is reflected in $\overline{\theta}(R)$. At this radius it is also possible to observe changes in the main spiral components when analysing the images with Fourier transforms.

The density wave is better represented by red bands (I–K); especially in K, since the contribution of newly formed stars is less important (Elmegreen & Elmegreen 1984). The passage of the wave leads to enhancement of the flux, which reaches a maximum

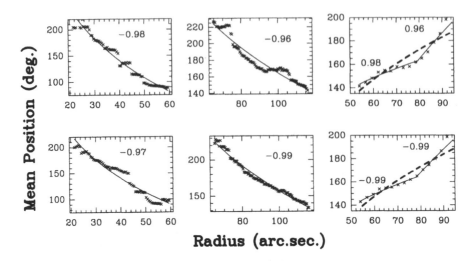

FIGURE 2. Fits of logarithmic spirals and their correlation coeficient to the mean position of one arm from NGC 157 (left, northern arm), NGC 4321 (middle, southern arm) and NGC 7479 (right, eastern arm). The best fit for NGC 7479 occurs when the pitch angle changes in corotation. The upper graphs correspond to the B band and the lower ones to the I band.

and finally falls, so media and maxima positions should be pretty close and skewedness should be small or null.

In the bluer bands the picture is very different. Gas density is enhanced up to a certain point, where the critical density is reached and stars are formed.

If the potential well is not strong enough, dust and gas probably need to attain the maximum of the wave to reach the density required to light up stars. So, for a trailing, anticlockwise-rotaing galaxy, we expect to find $S^B > 0$, $S^I \gtrsim 0$ and $\overline{\theta_B} - \overline{\theta_I} > 0$. But if the potential well is strong, stars can, in principle, be formed before the maximum of the wave, so that we measure $S^B \sim S^I$ and $\overline{\theta_B} - \overline{\theta_I} \lesssim 0$.

NGC 157 and NGC 4321 are "S"-shaped, trailing, anticlockwise-rotating galaxies, so— as expected—before corotation $S^B > 0$, but, whereas in NGC 157 $\overline{\theta_B} - \overline{\theta_I}$ is clearly greater than zero (i.e. the spiral wave is not very strong), in NGC 4321—also a 12 class—$S^B > 0$ but $\overline{\theta_B} - \overline{\theta_I} \sim 0$, i.e. the wave is stronger. NGC 7479 is a "Z"-shaped, trailing, clockwise-rotating galaxy. After a first change in sign, due to the end of the bar, $S^B \sim S^I < 0$ and $\overline{\theta_B} - \overline{\theta_I} < 0$ (it is bluer in the convex side, as expected).

3. Conclusions

By means of the azimuthal profiles technique, in this work we analyse the galaxies NGC 157, NGC 895, NGC 4321, NGC 6814, NGC 6951 and NGC 7479. All of them are grand design, and all but NGC 895 present at least one change in the asymmetry sign in the B band. The I band follows the same behaviour in all four cases. In every case in this sample $|S^I| \lesssim |S^B|$, as expected (see Schweizer 1976). Changes in the sign of asymmetry are probably due to the presence of corotation, because the front passes from one side to the other in the arms. When no change or random changes are detected—as is the case for NGC 895—there are two possibilities. One could be that the arms end at the corotation radius, so the asymmetry should have a plus or minus sign, because the front is only on one side, but this is not what we see. In fact S^B oscillates between -0.4

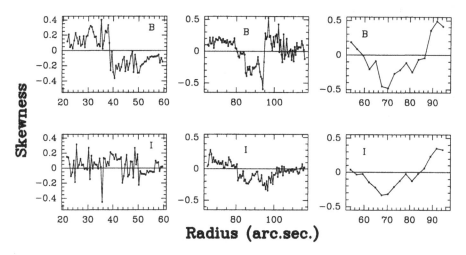

FIGURE 3. Asymmetry *vs.* radius of one of the arms of NGC 157 (left, northern arm), NGC 4321 (middle, southern arm) and NGC 7479 (right, eastern arm). The upper graphs correspond to the B band and the lower ones to the I band.

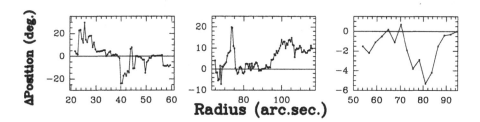

FIGURE 4. Difference of media positions *vs.* radius of one of the arms of NGC 157 (left, northern arm), NGC 4321 (middle, southern arm) and NGC 7479 (right, eastern arm).

and 0.4. Another more likely possibility is that spiral density waves do not trigger star formation in the arms in a significant manner.

The good fits in all cases ($|r| \gtrsim 0.96$ in every case) guarantee that the decomposition in a base of logarithmic spirals, for Fourier transforms, converges fast enough with very few terms.

Finally, all but one of the galaxies in our sample show that SDW is triggering star formation, while in NGC 895 we find no clear evidence.

REFERENCES

CEPA, J. 1988 PhD thesis, Universidad de La Laguna.

ELMEGREEN, B. G. 1992 in *Star Formation in Stellar Systems* (ed. G. Tenorio-Tagle, M. Prieto & F. Sánchez). III Canary Islands Winter School, p. 381. Cambridge University Press.

ELMEGREEN, D. M. & ELMEGREEN, B. G. 1984 *Astrophys. J., Suppl. Ser.* **54**, 127.

SCHWEIZER, F. 1976 *Astrophys. J., Suppl. Ser.* **31**, 313.

Stellar Complexes in the Galactic Spiral Arms

By YURI N. EFREMOV[1,2]

[1] Sternberg Astronomical Institute, Universitetskij Prospect, 13, 119899 Moscow, Russia

[2] Instituto de Astrofísica de Andalucía (CSIC), Apdo. 3004, E–18080, Granada, Spain

1. Introduction

We consider here the properties of giant star complexes of our Galaxy and peculiarities of their locations in the spiral arms. The term "star complexes" has been proposed to designate the giant groupings of star clusters, associations and individual high luminosity stars, measuring 300–1000 pc across and with ages up to 100 Myr (Efremov 1979). They were picked out by van den Bergh (1964) in M31 (under the usual name of OB associations) as groupings of blue stars and by Efremov (1978) in our Galaxy using the Cepheid variables as indicators. Groupings detected by van den Bergh are similar to the complexes of the Galaxy in every respect—size, age, and star content (Efremov 1979, 1982)—and only a dozen of smallest of them should be considered as genuine, classical OB associations (Efremov *et al.* 1987).

It was recognized that the Local System, whose brightest stars form the well-known Gould Belt, is nothing but a fairly typical (yet not populous) star complex (Efremov 1979), and that vast star-gas clouds distributed along spiral arms of the galaxies are just bright populous star complexes (Elmegreen & Elmegreen 1983; Efremov 1984).

The general description and history of the study of these groupings is given elsewhere, (Efremov 1989, 1995a) yet we add here some remarks, mainly to demonstrate once more the need to introduce the concept of complexes as entities different from the associations. There is some controversy on this point and sometimes complexes are considered as simply large old associations. Large sizes of blue star groupings in M31, as compared with five-fold smaller sizes of the Milky Way OB associations, were explained (van den Bergh 1964) as a consequence of the higher density of background stars in the Galaxy.

In our opinion, detection of complexes in the distribution of blue stars is not a result of higher signal-to-noise ratio in M31. Complexes appear in the distribution of older stars as well, such as Cepheids, and also as groupings of clusters with ages up to 100 Myr. OB associations are just brighter and younger subregions inside complexes. One could get a larger size for the Ori OB1 association by including up to the late B stars, but then one just gets the Local System complex, which includes other associations and a dozen open clusters as well.

Furthermore, no association is known that could give rise to a star complex by becoming older and larger. The stellar mass of a complex is by order of magnitude larger than the mass of an association, and while a typical complex includes a dozen Cepheids, a fossil association might contain a couple of them, and such twin Cepheids with similar periods (i.e. ages) and radial velocities, at mutual distances of some 100 pc, were indeed detected by Efremov (1971). Thus, the observations suggest that a star complex cannot be generated by the evolution of a single association. Single young-star groupings with masses similar to that of a stellar complex are surely known as well—they are bound young globular clusters, such as NGC 1866 in the LMC, with about two dozen Cepheids.

A sharp borderline between young small complexes and normal associations may not exist. This issue should be investigated further with more data on the sizes, ages and masses of these objects. Yet the existence of an embedded sequence of young stellar groupings is what is essential: more than 90% of associations are inside stellar complexes. They often form groupings of a few members, for which the term "aggregates" was suggested, and aggregates are always inside complexes. Complexes themselves sometimes form groups, usually of two members, and the term "supercomplexes" or "regions" was proposed to designate such structures (Efremov 1989).

2. Complexes and associations in the Andromeda galaxy

The Andromeda galaxy provides the best data to discriminate complexes *vs.* associations. Classical OB associations, as groupings of the bluest stars with a mean size of 80 pc (just as in the LMC), were isolated here by Efremov *et al.* (1987), and 95% of them are inside star complexes. They used plates with higher resolution and deeper magnitudes limit than those of van den Bergh (1964), who found 200 pc as the size of the brighter cores inside his large associations.

Objective detection of associations in M31 was achieved recently by Magnier *et al.* (1993) who used a statistically based version of cluster analysis, elaborated by Battinelli (1991) and selected some 7000 bluest stars from the CCD photometry . They found 174 asociations with an average size of 90 pc. It is worthwhile noting that many of their associations form groupings of a few members, located just inside the star complexes of Efremov *et al.* (1987).

Star complexes should also appear in the distribution of fainter stars, with a larger scale of clumpiness as compared to associations, which can be checked with the use of Battinelli's (1991) method. Recently, Battinelli *et al.* (in preparation) have found that this is the case, at least for the two strips along spiral arms of the M31 galaxy. These complexes do include associations as denser subclumps.

It is interesting to note that a minority of M31 associations located outside complexes are also located mainly outside a spiral arm. They are probably remnants of complexes that are dissolved outside arms, owing to the large shearing there, and also probably after star formation, that pulls out the gas of complexes (Elmegreen 1994, this volume). According to the observed age spread inside complexes of some 50 Myr, the late star formation in the GMC may start when the whole complex is already well outside an arm, and former members of associations are unnoted within the general galactic disk field. Surely there is some star formation outside spiral arms as well, but massive star formation needs a GMC as progenitor (Elmegreen 1983) and such rare interarm GMC probably formed upstream in the supercloud–star complexes that disappeared as a whole while getting out of an arm.

A striking regularity is seen along the NW arms of the galaxy (Figure 1)—many complexes are spaced at about equal distances along the arms. While being well seen in the van den Bergh (1964) figures, this regularity was noted only recently in M31 (Efremov 1989), after such regularity was discovered in a number of grand-design spiral galaxies and explained as result of formation of HI superclouds (that form star complexes) by the action of gravitational instability (Elmegreen & Elmegreen 1983).

Furthermore, the alternation of young and old complexes suggested by Avedisova (1989) for the Car–Sgr arm might also be guessed in the distribution of complexes in M31. The complexes labelled 21 and 22, on the S3-arm, (see Figure 1), are bracketing an HI supercloud, where star formation should be in the very beginning (Lada *et al.* 1988). In a couple of megayears we will see a young complex instead of this supercloud, and about

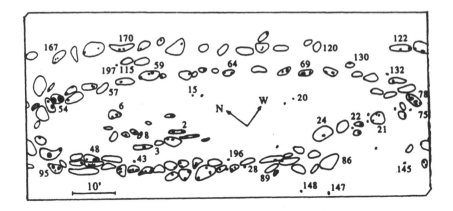

FIGURE 1. Distribution of star complexes and associations (shaded) detected in the
Andromeda galaxy by Efremov et al. (1987)

5 Myr later the OB21 complex will become too old to be seen with our contemporary tools.

The galaxy of Andromeda suggests the most important property of star complexes: they may well be omnipresent in the young disks of galaxies, yet often not easily observable. In M31, under lower resolution, Hodge (1986) found 42 groupings of 300 pc in size, instead of our 203 complexes of mean size 650 pc. In distant galaxies with a modest rate of star formation, like the one observed in M31, only the brightest parts of complexes, namely the aggregates, should be seen.

In many other galaxies with a lot of gas and strong spiral arms, such as M83 or NGC 6946, the bright complexes are filled with OB stars and are adjacent irregularly to each other in spiral arms, which may be connected with the high gas density there (Elmegreen 1994).

3. Star complexes in the solar neighbourhood

Some two dozen complexes are picked out within 3 kpc from the Sun in distribution of star clusters and associations (Efremov & Sitnik 1988; Avedisova 1989; Alfaro *et al.* 1991, 1992) and Cepheids (Berdnikov & Efremov 1989, 1993; see also Efremov 1989, 1995 and references therein). Isodensities of Cepheids outlining star complexes are shown in Figure 2. These stars, having reliable distance and age determinations, are even better than open clusters as indicators of star complexes in the Galaxy, because their density is larger than that of young clusters which delineate complexes.

Figures 2 and 3 show the strong complexes of Cepheids spaced at about 1 kpc intervals along the Car–Sgr arm. They go along the same pitch angle, about 20°, found by Alfaro *et al.* (1992) for this arm from the distribution of young clusters, yet well behind them, probably pointing to an age gradient across the arm. This angle is essentially larger than 10°, the value estimated by Grabelsky *et al.* (1988) for the whole 40 kpc of the Car–Sgr arm delineated by the GMCs. This probably reflects the local meandering of the arm, if not the peculiar position of just one young complex in Sgr.

It is interesting to note that outside this arm we have just two complexes in the Cas–Per arm and one in the Cyg–Ori arm. In this last arm, more complexes (including one of the Local System) are picked out with the young cluster data, though they are far from

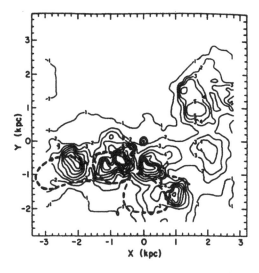

FIGURE 2. Distribution of star complexes of the Galaxy, outlined by Cepheid density isolines (Berdnikov & Efremov 1993) and by open clusters (broken lines, according to Avedisova 1989). Note that the complex of older clusters around (–1, –1) coincides with the complex of older Cepheids.

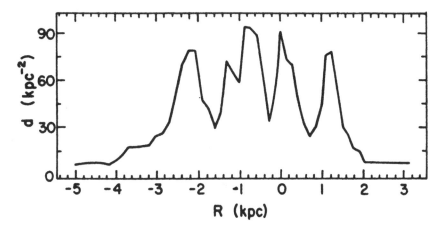

FIGURE 3. Density of Cepheids across the Car–Sgr arm (Berdnikov & Efremov 1993).

being as populous as those located in the Car–Sgr arm. This is not the case for the Per arm, which, as is well known, stops completely just at longitudes bordering the $h + \chi$ Per complex. The absence of Cepheid complexes in these locations is still more impressive if one takes into consideration the lower influence of observational selection outside the Car–Sgr arm. This distribution of Cepheid complexes suggests that observed part of the Car–Sgr arm is really a part of the galactic grand design, the density wave arm with a high density of stars of any age (Efremov 1994).

There is an increasing amount of observational evidence, showing that density wave arms are strong enough to produce density enhancement even for old disk stars (Elmegreen & Elmegreen 1984; Rix & Rieke 1993; Elmegreen 1994, and this volume). If so, old clusters, in disagreement with the common opinion (Becker 1963; Janes *et al.* 1988; Feinstein, this volume) should also concentrate in the grand-design arms, as the Car–Sgr arm is proposed to be. And this really is the case.

A concentration of older clusters, named complex B, was noted by Avedisova (1989) between younger complexes A and C in the arm, and it is also seen in the Efremov & Sitnik (1988) pictures. This older complex is certainly detected in the Cepheid distribution (Figures 2 and 3), and the average period of Cepheids there is smaller (i.e. their ages larger) than in the two adjacent complexes, that correspond more or less to the young cluster complexes (Berdnikov & Efremov 1993; Efremov 1994).

The lack of detection of older clusters concentrations in more distant parts of the Car–Sgr arm could be caused by observational selection. There is also the possibility of tidal destruction of old clusters located within young gas-star complexes, owing to the close passage of the LMCs, as Bruce Elmegreen noted in the discussion after Avedisova's talk at the Praga European Astronomical meeting (1987). However, this process demands much more time than is available for a complex to stay in a spiral arm. According to Danilov & Seleznev (1989), close-by passagges of a cluster near a GMC with a velocity of 10 km s^{-1} (the typical velocity dispersion in a complex) cause disruption of the cluster after some 260 Myr (see also Danilov 1993). The real alternation of young and old complexes along the arm therefore remains plausible. It is necessary to search for older (and therefore fainter) clusters farther away along this arm.

With regard to the Cyg–Ori and Per arms, the absence of a concentration of older clusters in these arms is plausibly real. It may be well explained if both arms are not part of the all-galactic spiral pattern but long spurs connected with star formation and shearing, or with the local instabilities. This is a fairly general view of the Cyg–Ori arm and may be also the case for the Per arm, which gives a natural explanation for its disappearance at longitudes larger than some 140°(Efremov 1995b).

4. The tilt of star complexe planes

Alfaro *et al.* (1991, 1992) discovered the intriguing peculiarity of the location of young cluster complexes in the Car–Sgr arm. They are both located below the mean galactic plane at some 50 pc. The same displacement, though a little smaller, is displayed by the corresponding Cepheid complexes, although no special pattern of z coordinates just along the arm was found (Berdnikov & Efremov 1993). They also found that the largest depression in z in the Cepheid distribution nearly coincides with the location of the largest depression in the open cluster distribution, discovered by Alfaro *et al.* (1991), proposed by the latter authors to be connected with the impact of a massive cloud with the galactic plane.

It is surely more probable, as suggested by Alfaro *et al.* (1992), that the correlation of the density of young clusters with z along the Car–Sgr arm is not connected with events of this kind, but with the general properties of large-scale star formation along the density wave spiral arm. This phenomenon surely has something to do with the inclination of "shingles" of open clusters to the galactic equator noted by Schmidt-Kaler & Schlosser (1973), and also with the similar inclination of large axes (in projection to the sky plane) of some star complexes, as noted by Efremov (1988). The same phenomenon for HI superclouds is also seen in Figure 15 (especially 15d) of Grabelsky *et al.* (1987), where the spatial maps of HI emission, integrated over respective velocity ranges, are presented.

The theories about the formation of such tilts, corrugations or depressions will throw more light on the origin of star complexes along a density wave spiral arm, as well as of the arms themselves. The well-known inclination (18°) of the plane of the Local System to the galactic plane is surely a manifestation of the same general phenomena, though a lot of *ad hoc* hypotheses were suggested to explain it. For instance, Comeron & Torra (1994) explained the origin of the Gould Belt and its tilt being the result of an HVC impact in the gas of the galactic disk.

The only difference between the Local System complex and most complexes is the lower density of clusters (at least populous ones) and Cepheids there (Efremov 1979, 1989, 1995a). Lindblad & Westin (1985) noted that stars of the Gould Belt are younger than 30 Myr and go to the galactic plane at distances coinciding with a discontinuity in the velocity gradient, though simple expansion does not explain the kinematics of the complex (Westin 1985).

The total mass of the Local System complex seems to be smaller than the average value, according to the smaller density of indicator objects there. Danilov & Seleznev (1995) obtained values of some 10 million solar units for four complexes of young clusters and associations; they also found the average tilt of 15° for ten complexes, a value in agreement with our preliminary results for a number of Cepheid complexes.

Strauss & Poeppel (1976) proposed that the Gould Belt system arose from a rotating supercloud that transformed into a disk perpendicular to the galactic plane by the action of the density wave when the supercloud was in the Car arm. Rotation or oscillation around an axis parallel to that of a spiral arm then brought the supercloud into its present position with a tilt of 18°. Expansion in the plane of the Gould Belt would then be a consequence of the compression of the initial supercloud in a disk in the spiral density wave (Strauss *et al.* 1979).

This model may be well applied to the bulk of star complexes, suggesting that all the node lines, intersections of the complex planes with the mean Galactic plane, should be aligned along the spiral arms. The existing data seem not to be in contradiction with such a proposition, yet they are still too unreliable to be considered as confirmation of such an hypothesis.

If the orientation of inclined complex planes really follows this pattern, and assuming the existence of an age gradient across the arm, the location of young clusters below the mean galactic plane could simply mean that complexes planes go to below the galactic plane in the direction of the galactic centre, because there are signs of a concentration of younger objects along the inner borderline of the Car–Sgr arm (see the references in Efremov 1989). Yet the sense of inclination of the Local System plane is just the opposite, and all these considerations surely depend on the time-scale of possible oscillations of the complex plane and the location of the corotation, which is probably near the Sun's distance (Efremov 1989). The quite possible rotation of complexes, at least of those located in the Car–Sgr arm (which may be one more indication of the strong density wave there, see Chernin, this volume; Efremov & Chernin 1995) adds more uncertainty to the picture.

A three-dimensional helical magnetic field along an arm together with a Parker-Jeans instability might be the best approach to the structure of the density-wave spiral arm (Alfaro, private communication; see also Shibata & Matsumoto 1991) and indeed the magnetic field structure with deviations from the arm with a scale length along the arm of about 2 kpc is observed in the NW arm of M31 (Beck *et al.* 1989). Unfortunately, these important observations concern only a small part of the arm. In any case, even with the consideration of the magnetic field, it is difficult to understand the location of young objects just to one side of the mean galactic plane.

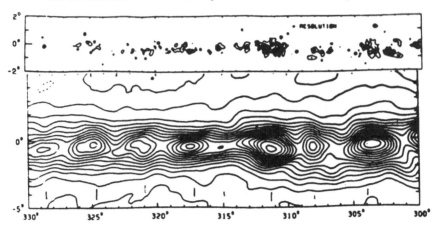

FIGURE 4. Spatial map of CO emission (top) and HI emission (bottom), integrated over the velocity range 7–50 km s^{-1} within a part of the Carina arm (composed with two figures of Grabelsky *et al.* 1987).

5. The grand design of the Milky Way galaxy

The stellar complexes and superclouds provide us with the best possibility of outlining the spiral pattern of the Galaxy. With the optical data the sample of star complexes may be complete only to 3–4 kpc from the Sun, although the IR observations should soon improve the situation considerably. With data on the GMCs, Grabelsky *et al.* (1988) were able to outline the whole Car–Sgr arm with a pitch angle of 10° and an extension of some 40 kpc. Data on HI superclouds are also available over the whole Galaxy and they should be even better indicators of the spiral arms. However, only recently they were used for this purpose (Efremov 1995b).

Distances and longitudes of HI superclouds in the first quadrant of galactic longitudes were taken from the article of Elmegreen & Elmegreen (1987). Such data do not exist for the fourth quadrant, although HI superclouds are beautifully seen in the figures of Grabelsky *et al.* (1987)—see Figure 4. The GMCs are concentrated within these superclouds though often not in their very centres (Figure 4) and the closer to the centre of the Galaxy, the larger the percent of the total supercloud-mass in the form of molecular hydrogen (Elmegreen & Elmegreen, 1987). The longitudes of superclouds were taken from Figure 15 of Grabelsky *et al.* (1987) and it was assumed that their distances are those of the GMCs located closer to the supercloud centres. Those distances were taken from Table 2 of Grabelsky *et al.* (1988), and are mainly obtained as the optical distances of HII regions connected with the respective GMCs.

The resulting picture of the complete Car–Sgr arm as outlined by the superclouds is shown as Figure 5. The arm is outlined even better than with the GMCs and the striking regularity in spacing of the superclouds along the arm, unconspicuous in the GMCs distribution, is seen here. The distribution in distances along the arm between adjacent superclouds is shown in Figure 6. It seems that intrinsic distances are about 1 kpc, as suggested by distances between Cepheid complexes closer to the Sun (Figures 2, 3). Note also that all four superclouds with masses in excess of 20 million solar units (data from Elmegreen & Elmegreen 1987) are aligned along the Sgr arm at mutual distances just of 1 kpc (Figure 5).

We then conclude that along this arm an alternation of younger and older supercloud-star complexes may exist, as it is observed in certain arms of the Andromeda galaxy and

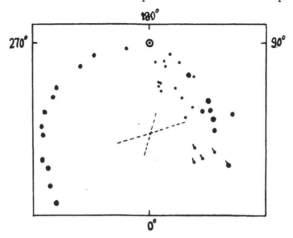

FIGURE 5. Distribution of HI superclouds along the Car–Sgr arm and possible extremal positions of the Galactic bar.

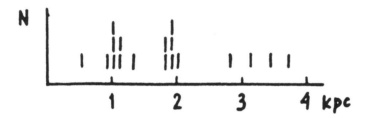

FIGURE 6. Distribution in spacing of HI superclouds along the Car–Sgr arm.

this regularity implies that the Jeans length scale in the gravitational instability model the of formation of supercloud along an arm (Elmegreen 1994) is here about 1 kpc. Older star complexes should be discovered in some positions between superclouds. In the Car arm this goal should soon be achievable with IR searches for Cepheids and older clusters, whereas in the Andromeda galaxy, to check the presence of such complexes between ones already detected with blue stars, it is possible even with the existing tools—yet not data...

The regular spacing between star complexes is a clear signature of the grand-design (or at least multiarmed) galaxies and this result for the Car–Sgr arm proves that the Galaxy belongs to this class. The location of the arm is in good agreement with two-armed grand-design modeles of the Galaxy suggested recently by Hammersley *et al.* (1994) on the basis of IR galactic survey data and by Han & Qiao (1994) with the configuration of the magnetic field in the Galaxy; it is also compatible with the existence of a (small) bar (Efremov 1995b).

The useful discussions with Bruce Elmegreen and Emilio Alfaro are appreciated. Thanks are also expressed to the members of the I.A.A. in Granada, for their hospitality during my stay there, and to the Spanish D.G.I.C.Y.T., whose grant made this stay possible.

REFERENCES

ALFARO, E. J., CABRERA-CAÑO, J. & DELGADO, A. J. 1991 *Astrophys. J.* **378**, 106.

ALFARO, E. J., CABRERA-CAÑO, J. & DELGADO, A. J. 1992 *Astrophys. J.* **399**, 576.

AVEDISOVA, V. S. 1989 *Astrofisica* **30**, 83.

BATTINELLI, P. 1991 *Astron. Astrophys.* **244**, 69.

BECK, R., LOISEAU, N., HUMMEL, E., BERKHUIJSEN, E. M., GRAVE, R. & WIELEBINSKI, R. 1989 *Astron. Astrophys.* **222**, 58.

BECKER, W. 1963 *Z. Astrophys.* **250**, 551.

BERDNIKOV, L. N. & EFREMOV, YU. N. 1989 *Sov. Astron.* **33**, 274.

BERDNIKOV, L. N. & EFREMOV, YU. N. 1993 *Sov. Astron. Lett.* **19**, 389.

COMERÓN, F. & TORRA, J. 1994 *Astron. Astrophys.* **281**, 35.

DANILOV, V. M. 1993 *Sov. Astron. Lett.* **19**, 175.

DANILOV, V. M. & SELEZNEV, A. C. 1989 *Astron. Circ.* No. 1534, 9

DANILOV, V. M. & SELEZNEV, A. F. 1995 *Astron. Astrophys. Trans.* in press.

EFREMOV, YU. N. 1971 *Sov. Astr. Circ.*, No. 639.

EFREMOV, YU. N. 1978 *Sov. Astron. Lett.* **4**, 66.

EFREMOV, YU. N. 1979 *Sov. Astron. Lett.* **5**, 12.

EFREMOV, YU. N. 1984 Vestnik Acad. Sci. USSR **12**, 56.

EFREMOV, YU. N. 1988 *Astrophys. Space Phys. Rev.* **7**, 2.

EFREMOV, YU. N. 1989 *Origins of Star Formation in Galaxies: Stellar Complexes and Spiral Arms* (in Russian). Nauka, Moscow.

EFREMOV, YU. N. 1994 in *Physics of Gaseous and Stellar Disks of the Milky Way* (ed. I. R. King). ASP Conference Series, vol. 66, p. 157. ASP.

EFREMOV, YU. N. 1995a *Astron. J.* in press.

EFREMOV, YU. N. 1995b *Mon. Not. R. Astron. Soc.* submitted.

EFREMOV, YU. N. & CHERNIN, A. D. 1995 *Astron. Astrophys.* **293**, 69.

EFREMOV, YU. N., IVANOV, G. R. & NICOLOV, N. S. 1987 *Astrophys. Space Sci.* **135**, 119.

EFREMOV, YU. N. & SITNIK, T. G. 1988 *Sov. Astron. Lett.* **14**, 347.

ELMEGREEN, B. G. 1993 *Mon. Not. R. Astron. Soc.* **203**, 1111.

ELMEGREEN, B. G. 1994 *Astrophys. J.* **433**, 39.

ELMEGREEN, B. G. & ELMEGREEN, D. M. 1983 *Mon. Not. R. Astron. Soc.* **203**, 31.

ELMEGREEN, B. G. & ELMEGREEN, D. M. 1987 *Astrophys. J.* **320**, 182.

ELMEGREEN, D. M. & ELMEGREEN, B. G. 1984 *Astrophys. J., Suppl. Ser.* **54**, 127.

GRABELSKY, D. A., COHEN, R. S., BRONFMAN, L. & THADDEUS, P. 1987 *Astrophys. J.* **315**, 122.

GRABELSKY, D. A., COHEN, R. S., BRONFMAN, L. & THADDEUS, P. 1988 *Astrophys. J.* **331**, 181.

HAMMERSLEY, P. L., GARZÓN, F., MAHONEY, T. & CALBET, X. 1994 *Mon. Not. R. Astron. Soc.* **269**, 75.

HAN, J. L. & QIAO, G. L. 1994 *Astron. Astrophys.* **288**, 759.

HODGE, P. 1986 in *Luminous Stars and associations* (ed. C. W. H. de Loore, A. J. Willis & L. Laskarides). IAU Symposium 116, p. 369. Reidel.

JANES, K. A., TILEY, T. & LYNGA, G. 1988 *Astron. J.* **95**, 771.

LADA, C. J., MARGULIS, M., SOFUE, Y., NAKAI, N. & HANDA, T. 1988 *Astrophys. J.* **285**, 141.

LYNDBLAD, P. O. & WESTIN, T. N. G. 1985 in *Birth and Evolution of Massive Stars and Stellar Groups* (ed. W. Boland & H. van Woerden), p. 33. Reidel.

MAGNIER, E. A., BATTINELLI, P., HAIMAN, Z., LEWIN, W. H. G., VAN PARADIJS, J.,

HASINGER, G., PIETSCH, W., SUPPER, R. & TRÜMPER, J. 1993 *Astron. Astrophys.* **278**, 36.

RIX, H.-W. & RIEKE, M. J. 1993 *Astrophys. J.* **418**, 123.

SCHMIDT-KALER, TH. & SCHLOSSER, W. 1973 *Astron. Astrophys.* **25**, 191.

SHIBATA, K. & MATSUMOTO, R. 1991 *Nature* **353**, 633.

STRAUSS, F. M. & POEPPEL, W. 1976 *Astrophys. J.* **204**, 94.

STRAUSS, F. M., POEPPEL, W. & VIELRA, E. R. 1979 *Astron. Astrophys.* **71**, 319.

WESTIN, T. N. 1985 *Astron. Astrophys., Suppl. Ser.* **60**, 99.

Elmegreen: You are observing stars with ages comparable to one-half the full oscillation time through the disk. How can you be sure that vertical displacements (shingles) or vertical age gradients reflect true z gradients in star-formation locations?

Alfaro: Most papers dealing with vertical displacements from the galactic plane make use of objects younger than 10^7 yr. Assuming a velocity dispersion for those objects of about 10 km s^{-1}, it can be expected that their present locations are not farther than 10 pc away from the original birthplaces.

Palouš: If the bar ends at 3–4 kpc, as argued from *COBE* experiment, to connect the Sgr–Car arm with the bar end you have to cross the molecular belt between 4–5 kpc. How do you do that?

Efremov: I suggest that the bar should end just at the distance of the inner border-line of the molecular ring.

Shore: It is possible that you have biased your search for young superassociations and complexes using Cepheids. Why not also use the IR recombination HI and HII lines for WR stars? This way, you could easily, and unambiguously, cover a lot of the Galaxy. Also, when you do this for the LMC, it shows that complexes and supercomplexes exist at the same time, as you require—they are not stages of a development of an OB association.

Efremov: A very good comment. Emilio Alfaro and coworkers found concentration of the WN stars spatially associated with complexes delineated by young clusters, and such an approach should be further developed. The large young complexes are indeed observed, and these clearly cannot be a product of the evolution of a single OB association.

The Spiral Arms of the Milky Way from the Distribution of Open Clusters

By ALEJANDRO FEINSTEIN[1,2]

[1]Observatorio Astronómico, La Plata 1900, Argentina

[2]PROFOEG, CONICET, Buenos Aires, Argentina

The open clusters define in the Milky Way a system of objects distributed in the so-called disk of our Galaxy. If we take a look at those very young objects, we expect to see where their spiral arms are located. There are many papers related to this problem. The first related to this subject was presented by Bok in 1959. Many others followed using different collections of observed data. The purpose of this paper is to determine with all those known open clusters belonging to the Milky Way what new information we can derive about their distribution. All the information about their main characteristics—especially their distances, colour excesses $E(B-V)$ and ages—was compiled from the data on open clusters available in the literature. The total number of clusters with some data known is 468. Those clusters with ages and distances classified according to their ages in several groups is presented in Table 1. The total number of clusters listed in the above groups amounts to 298. It is clear that in many other clusters there are not enough available references to enable us to include them in these groups.

1. The analysis of the data

It is well known that some kinds of open clusters, especially the realatively young ones, appear to be located in the spiral arms of the Milky Way (Lyngå, 1987; Janes *et al.* 1988). Our aim here is to use these newly compiled data to check what type of clusters fulfill this condition.

The distribution was plotted of each group of open clusters in the plane of the Milky Way. Diagrams of galactic longitude *vs.* distance are presented in Figures 1 and 2.

The plot of those clusters with ages of about 10^6 yr clearly shows the location of several spiral arms, i.e. the Sagittarius–Carina arm in the inner side of the Sun, and the Orion and Perseus arms in the outer side. We may also assume the presence of two more arms:, one in the inner side, i.e the Scutum arm, and another beyond the Perseus arm in the outer side.

This particular distribution of clusters is not seen at all in the diagrams of slightly older groups. But, in the diagram which corresponds to open clusters with approximate ages of $1.10^7 - 4.10^7$ yr (Figure 2), a few dots which denote clusters some 1.10^7 yr old appear still to be located in the corresponding spiral arms.

According to this diagram, it seems evident that within a period of about 10^7 yr the clusters run away from the spiral arms, which makes sense if we assume a space velocity of about 20 km s -1. With this velocity and this time interval (10^7 yr) the clusters are moving some 100 pc away from their birth places. In this condition they are no longer located in the present spiral arms. Most of them lie in the interarm zone.

The separation between the mean of the spiral arms is of about 2 kpc. The Sun is found to be located at a distance of 1.6 kpc from the Sagittarius arm and about 0.8 kpc from the Orion arm. The Perseus arm appears to be slightly more than 1.6 kpc from the Orion arm. A more external arm at 1.5 kpc from the Perseus arm appears to be present.

It should be mentioned here that the distribution in the Milky Way of 42 open clusters with O-type stars has already been analysed by Feinstein (1994). From their position

range of ages	number of open clusters
10^6	53
$1.10^7 < 4.10^7$	58
$4.10^7 < 1.10^8$	46
10^8	93
10^9	48
	total 298

TABLE 1. Number of open clusters according to their ages

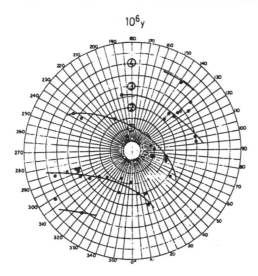

FIGURE 1. Open clusters with ages in the range of 10^6 yr. The lines suggest the position of the spiral arms.

in the galactic plane those open clusters also very clearly show the location of the spiral arms.

2. The colour excess E(B–V).

The plots of the colour excesses show that there are regions where these $E(B-V)$ values are very small, less than 0.2 magnitudes, but in other regions they become larger. This means that the dust distribution is quite non-uniform between the nearest spiral arms. Closer to the position of the spiral arms we find larger values, but between the two arms where the Sun is located the colour excesses are quite small in a large region: about 0.1 magnitudes.

Figure 3 shows the colour excesses for open clusters within 2.0 kpc of the Sun. It shows that in the direction between galactic longitude 230° and 320° a large number of open clusters are found, with quite low colour-excess values. In a few directions in this zone and up to 1 kpc the colour excesses give very low values, about 0.1 or 0.2 magnitudes. This is not the case when we look in the other direction of the interarm region, i.e. between 0° and 70°.

In this region the number of open clusters is smaller, but the colour excesses becomes

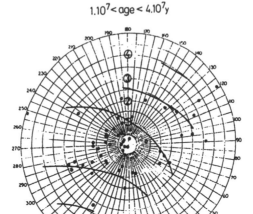

FIGURE 2. Open clusters with ages in the range of $1x10^7$ to $4x10^7$ yr. The dots are open clusters with ages of 10^7 yr.

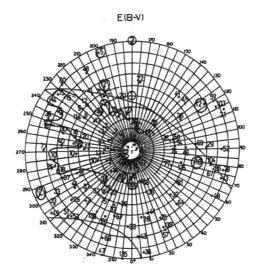

FIGURE 3. Colour excesses $E(B-V)$ for open clusters of 10^6 and 10^7 yr in units of 0.01 mag. in the solar neighbourhood

greater starting at a distance of about 0.5 kpc from the Sun. A similar conclusion was drawn by FitzGerald(1968). Is the non-uniform distribution of open clusters real or merely a selection effect? A large selection effect appears to be present in the open clusters distribution. The astronomers tend to observe those objects which are more or less easy to see, i.e. those which lie in regions with relatively low absorption, and this complicates the determination of the spiral arms of the Milky Way. The solution to this problem may be found by looking for faint open clusters in some particular regions of the Milky Way. Although their number may be low, they would give a clearer picture of the position of the spiral arms.

3. Conclusion

The spiral arms appear to be well defined by the location of open clusters with ages of around 10^6 yr. Some of the clusters with scarcely more than 1.10^7 yr still remain in the spiral arms, although those which are somewhat older have already moved out. Thus, an age of about 1.10^7 yr looks like being the limit for a cluster to be located in a spiral arm.

The distribution of dust is non-uniform; between galactic longitude 230° and 290° (the interarm region), the colour excess is between 0.1 and 0.2 mag. up to 1 kpc, or more in certain regions. In the other direction of the interarm region, i.e. between 10° and 70°, the colour excesses increase. The number of open clusters is greater in the region of lower dust valued, which means a selection effect is occurring in the number of open clusters known.

REFERENCES

Bok, B. 1959 *The Observatory* **79**, 58.

Feinstein, A. 1994 *Rev. Mex. Astron. Astrofis.* **29**, 141.

FitzGerald, M. P. 1968 *Astron. J.* **73**, 983.

Janes, K. A., Tilley, C. & Lyngå, G. 1988 *Astron. J.* **95**, 721.

Lyngå, G. 1987 *Publ. Astr. Inst. Czech Acad. Sciences* **69**, 121.

Elmegreen: Do you find that the gas density is in fact low where the clusters are old—as if there were no star formation there because there is no gas—or are young clusters lacking there only because it is the interarm?

Feinstein: The clusters move away after about 10 Myr from the place where they formed. These old clusters are now in an interarm region.

Chernin: The time needed for a cluster to leave a spiral arm depends, generally, on the angular velocity of the spiral pattern, the thickness of the arm, and the pitch angle. It's surprising—and perhaps instructive—that you have obtained for the combination of these figures one value of 10 Myr which is constant for the Milky Way disk.

Feinstein: It is quite clear that clusters with ages greater than 10 Myr appear to be outside the spiral arms. Your conclusion about the facts needed for the clusters to leave the spiral arms may be right.

Janes: There are two basic problems of using optical tracers to search for the spiral arms: the patchiness of the obscuration and the uncertainties in the distances.

Violent Gas Dynamics in Galactic Cosmogony: Spiral Shocks and Rotation of Star Complexes

By ARTHUR D. CHERNIN[1,2]

[1]Sternberg Astronomical Institute, Moscow University, Moscow, 119899, Russia

[2]Tuorla Observatory, University of Turku, Piikkiö, 21500, Finland

Star complexes are huge aggregates of stars and gas which are considered as the largest cells of star formation in spiral galaxies (Efremov 1988). Basic observational data on star complexes are presented with a special emphasis on their rotational properties. A possible model of the formation of star complexes and the origin of their spin momentum is discussed based on the physics of nonlinear supersonic gas dynamics effects in the interstellar medium.

1. Observations: star complexes and their rotation

Observational evidence suggests that the large majority of the young disk stars in the Galaxy are gathered into huge star complexes (Efremov 1978, 1988, 1989). The typical mass of an individual complex is 1–10 million solar masses, composed of the masses of its stars plus the masses of gas clouds in the same space volume. The typical size of these systems is about 1 kpc. Star complexes have been recognised as universal and ubiquitous entities in spiral galaxies. They seem to present the fundamental "building blocks" of the large-scale disk structures.

The data on rotational properties of star complexes are currently fairly poor. There are only a few examples of complexes with estimated rotational characteristics. The errors in these estimates remain rather large. Observational study of the rotation of star complexes present has proved to be a hard problem.

Using data on radial velocities and proper motions of the Cepheids, stellar associations and clusters in four complexes that form a chain along the Carina–Sagittarius arm, Anna Melnik of Sternberg Institute found the angular velocities of the spin rotation of these complexes (Melnik 1994):

$$\omega = 32 \pm 10 \text{ km s}^{-1} \text{ kpc}^{-1},$$
$$\omega = -89 \pm 38 \text{ km s}^{-1} \text{ kpc}^{-1},$$
$$\omega = 57 \pm 17 \text{ km s}^{-1} \text{ kpc}^{-1},$$
$$\omega = -20 \pm 23 \text{ km s}^{-1} \text{ kpc}^{-1}.$$

$$(1.1)$$

It is assumed (and there are arguments to support such an assumption) that the angular velocity vector—if it is not zero—is normal to the galactic plane, and one can define that the sign of angular velocity is positive when the rotation is parallel to the rotation of the galactic disk and negative in the contrary case, as has been adopted in the equations above.

Note that two of the complexes have retrograde rotation, and there is an alternation of the sense of the spin rotations of the systems.

2. Origin of spin of star complexes: a gas dynamic model

The nature of the spin rotation of star complexes presents a problem that may be closely related to the fundamental features of evolutionary processes in spiral galaxies. We examine here a gas-dynamics approach to this problem assuming that star complexes appear as a result of the evolution of gaseous "superclouds" with masses up to about ten million solar masses (Elmegreen & Elmegreen 1983; Efremov 1988, 1989). One of the basic gas-dynamics processes in the galactic disk is the interaction of proto-complexes (superclouds) with the spiral shock, and this will be the focus of our discussion in this section. We can demonstrate that this interaction might be responsible not only for the significant compression of the superclouds, but also for the origin of the rotation of star complexes that form out of them.

To illustrate the major features of cloud-shock interaction and estimate the vorticity introduced into the flow via this interaction, we use the results of computer simulations of such a process (Ushakov & Chernin 1983, 1984; Chernin 1993). A three-dimensional gas-dynamics model was developed in which the shock front is flat, and a non-uniform flow meets it with a constant velocity normal to the front. The flow carries a "cloud" of enhanced density which makes the flow non-uniform. The structure of vorticity–entropy wake behind the front was obtained. The vortex velocity field takes the form of two associated eddies rotating in opposite directions. The eddies are centred near the boundaries of the cloud. The cloud itself turns to be compressed along the stream direction. It is "frozen" in the down-flow and formed by entropy wave modes.

The area of the down-flow with a radius R around the centre of one of the eddies contained angular momentum (per unit mass)

$$S = aVR(\delta\rho/\rho), \quad (\delta\rho/\rho) < 1, \tag{2.2}$$

where a is a form factor depending on geometry of the up-flow, $\delta\rho/\rho$ is a measure of density non-uniformity in this flow which is the relative density perturbation due to the cloud, in our model. Each magnitude here is given in its absolute value.

It is obvious that the angular momentum vectors of the two eddies have opposite directions and the total angular momentum of the down-flow is zero, as in the up-flow.

In a particular model of a spherical cloud the form factor $a = 0.1$. Models of other geometries with clouds of various forms are topologically equivalent to this model and typically characterised by form factors of $a = 0.3 - 0.03$, typically.

Such a process can naturally develop on the spiral arm shock fronts. In this case the velocity of the flow normal to the front is

$$V = (\Omega - \Omega_0)r\sin\alpha, \tag{2.3}$$

where Ω is the angular velocity of the disk rotation (which depends on the radial distance from the centre r), Ω_0 is the angular velocity of the rotation of the spiral pattern (which is constant), and α is the pitch angle. The spin momenta vectors of the eddies are normal to the disk plane.

As is well known, there is no general consensus about the value of the angular rotation velocity of the spiral pattern in the Galaxy. If one prefers $\Omega_0 = 25$ km s^{-1} kpc^{-1} (e.g. Marochnik & Suchkov 1984; Efremov *et al.* 1989), then the solar vicinity should be considered as an area in or near the corotation zone where $\Omega - \Omega_0 \approx 0$, and so velocity V of the flows crossing the spiral shock is rather small here.

If one chooses $\Omega_0 = 13$ km s^{-1} kpc^{-1}, as in the original papers by Roberts (1969) and

Pikelner (1970), than one gets in the solar vicinity $V \approx 10 - 20$ km s^{-1}, depending on the adopted pitch angle $\alpha = 5 - 15$ degrees.

Now one can estimate the angular momentum of an eddy, assuming that it has a size of the typical supercloud, $R = 1 - 2$ kpc:

$$S \approx (5 - 10)(\delta\rho/\rho) \text{ (km s}^{-1}) \text{ kpc.} \qquad (2.4)$$

We see that the magnitudes of the momentum observed in the star complexes, $S \approx 3$ (km s^{-1}) kpc per unit mass, might be reached, if the density contrast in the non-uniform flow, $(\delta\rho/\rho)$, is near 0.3–0.5 for a gas "cloud" with the mass of about 10^7 $M_\odot$ and the initial size of 2–3 kpc.

We do not insist here on the value $\Omega_0 = 13$ km s^{-1} kpc^{-1} used for the angular velocity of the galactic rotation in the estimates above; it was taken rather as an example to demonstrate that the spiral shocks are effective enough to introduce vorticity and supply the gas mass typical for star complexes and superclouds with an adequate angular momentum. If $\Omega_0 = 25$ km s^{-1} kpc^{-1}, it would mean that the solar vicinity is located near the corotation zone, and so this gas-dynamics mechanism might be more effective outside this zone.

However in any area of the galactic disk where a gas flow crosses the spiral shock fronts, one can expect that this mechanism could work and lead to formation of couples of eddies and then, perhaps, couples of proto-complex clouds with opposite rotation—parallel and antiparallel to the rotation of the disk as a whole.

3. Conclusions

Observational studies of star complexes demonstrate their role as building blocks of the galactic structure and the origins of star formation. Studies of the rotational properties of these systems may provide a clue towards better understanding of the evolutionary processes in spiral galaxies. Treated from the gas-dynamics viewpoint, the rotation of star complexes may reveal some fundamental features of these processes.

REFERENCES

CHERNIN, A. D. 1993 *Astron. Astrophys.* **267**, 315.

EFREMOV, YU. N. 1978 *Sov. Astron. Lett.* **4**, 66.

EFREMOV, YU. N. 1988 *Stellar Complexes*. Harwood Publ. London.

EFREMOV, YU. N. 1989 *Star Formation Origins in Galaxies*. Nauka, Moscow (in Russian).

EFREMOV, YU. N., KORCHAGIN, V. I., MAROCHNIK, L. S., & SUCHKOV, A. A. 1989 *Sov. Phys. Uspekhi* **32**, 310.

ELMEGREEN, B. G. & ELMEGREEN, D. M. 1983 *Mon. Not. R. Astron. Soc.* **203**, 31.

MAROCHNIK, L. S. & SUCHKOV, A. A. 1984 *The Galaxy*. Nauka, Moscow (in Russian. English edition, Gordon and Breach, 1994).

MELNIK, A. M. 1994 in preparation.

PIKELNER, S. B. 1970 *Astron. Zh.* **47**, 752.

ROBERTS, W. W. 1969 *Astrophys. J.* **158**, 123.

USHAKOV, A. YU. & CHERNIN, A. D. 1983 *Sov. Astron.* **27**, 367.

USHAKOV A. YU. & CHERNIN, A. D. 1984 *Sov. Astron.* **28**, 11.

Shore: As I remember, a couple of giant star complexes were recently observed in the Large Magellanic Cloud.

Chernin: It would be interesting to try to find their rotation, if there is any.

Elmegreen: Have the wave-like motions of stars from the density-wave flow been removed before the rotation of the complexes was derived?

Chernin: A possible effect of the spiral wave on the value of the spin of the complexes was estimated by Melnik, and she found that uncertainties which may be due to the wave are less than the errors in her determination of the angular velocities.

Palouš: Is the evidence of complex motions coming from radial velocities or do you also use the proper motions?

Chernin: We can judge the complex motion only on the basis of the observed motions of their stars, clusters, etc. In Melnik's estimations, both radial and proper motions of stars, associations and clusters were used.

Hensler: How exact is the determination of distances and kinematics of all the cluster stars in order to faithfully conclude to which cluster they belong and assume that they are gravitationally bound?

Chernin: The errors in both distances and kinematics are rather large, and this is the major contribution to the net figure of the errors in the estimation.

Star Formation in M31

By EUGENE A. MAGNIER[1], J. VAN PARADIJS[1,2],
W. H. G. LEWIN[3], P. BATTINELLI[4]
AND YU. EFREMOV[5]

[1]University of Amsterdam, Kruislaan 403, 1098 SJ Amsterdam, The Netherlands

[2]Physics Department, University of Alabama in Huntsville, Huntsville, AL 35899, U.S.A.

[3]Massachusetts Institute of Technology, 37–627, Cambridge, MA 02139, U.S.A.

[4]Osservatorio Astronomico di Roma, Viale del Parco Mellini 84, I-00136 Roma, Italy

[5]Sternberg State Astronomical Institute, Moscow, Russia

We have begun to investigate in detail the star-formation processes in M31. We assert that a useful description of stellar groupings needs to recognise both the importance of substructure in these groups as well as the dependence of the structure on the colours, and thus ages, of the stars. We also show that the ratio of HI to blue stars shows *anti*-correlations in certain regions of M31, in addition to the expected correlations.

1. Introduction

In photographs of spiral galaxies, the spiral structure is highlighted by the clumps of bright, blue stars and the associated nebulous HII regions. Surely many generations of astronomers have been inspired by such beautiful pictures —I know I have!. As well as adding to the beauty of such pictures, the groups of bright blue stars are a useful scientific tool: because these massive stars only live for 1–10 Myr, we observe them near their birthplaces. They can thus be used to trace recent star-formation sites. We use recent observations of M31 to study the processes which contribute to star formation.

2. Observations

We have performed two major optical surveys which are relevant to the study of the star formation in M31. The first is a wide-band ($BVRI$) CCD survey of $\sim$1.7 square degrees of the disk of M31. The analysis of this survey data has yielded a catalogue of $\sim$450 000 source with rough completeness limits of $BVRI = (20.7, 21.7, 20.2, 20.5)$ (Magnier *et al.* 1992; Haiman *et al.* 1993). The second survey was an Hα, [SII] emission-line survey. The main goal of that work was the identification of new supernova remnant (SNR) candidates, and in fact we have identified 162 new SNR candidates in M31 (Magnier *et al.* 1994). However, the Hα data is quite useful for the study of the massive star populations as this data traces the locations of the gas which is ionised, primarily by the massive stars. This work is part of an ongoing project to study M31 in a global fashion at a variety of wavelengths. In addition to the two optical surveys, we have also performed an X-ray survey with the *ROSAT* PSPC (Supper *et al.* 1994) and most recently VLA radio continuum (6- and 20-centimetre) observations of many of our SNR candidates.

3. Clumps of blue stars

It is well known that young stars tend to be clustered together in groups. Ambartsumian (1947) pointed out such groups of blue (O and B) stars in the Milky Way which we now call OB associations. OB associations are groups of roughly coeval stars. In

64

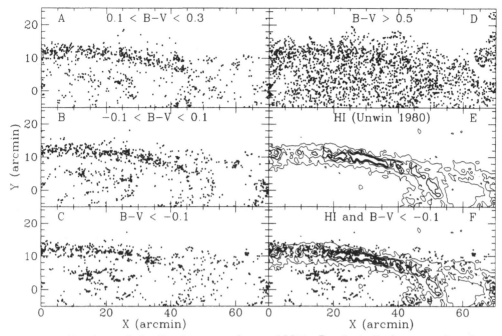

FIGURE 1. Six views of the Northwest quadrant of M31. Panels A–D show samples of stars from the Magnier *et al.* catalogue with different colour ranges. Panels E and F show the Unwin (1980) H I emission (see text for details).

the Galaxy, it is possible (though difficult for all but the nearest) to measure the proper motion of the stars in a large field and determine the association membership by tracing the stellar motions backwards to a single origination point. In M31, we do not have this luxury, and instead we must rely on the spatial distribution of stars to define the OB associations.

There have been two prior attempts to define OB associations in M31: Van den Bergh (1964) and Efremov *et al.* (1987). Both studies used visual inspection of U and V photographic plates to make the determination of the association boundaries. There has been some debate about these two studies. The van den Bergh associations were typically much larger than the associations known in our Galaxy and in the Magellanic Clouds. Hodge (1987) suggested that this difference was due to the resolution of the van den Bergh images. Efremov *et al.* used higher resolution images and identified associations more similar in size to those in the Galaxy. They also noted the larger features identified by van den Bergh and called them "complexes". We decided to use the blue stars from our CCD observations to make another attempt at identifying OB associations. One motivation was the development of correlation methods (e.g. Battinelli 1991) to make the determination objective—i.e. without the use of the potentially subjective visual inspection. Another motivation was the increased depth of our observations compared to the previous photographic studies. We applied the PLC method of Battinelli (1991) to our blue stars (Magnier *et al.* 1993). With this technique, we identified statistically significant groups of blue stars that were similar in size to the associations of Efremov *et al.* This suggests that the blue stars in M31 are grouped on scales similar to the associations in the Milky Way and Magellanic Clouds. However, as we demonstrate below, one must be careful of the rigid definition of associations or complexes.

Figure 1 shows six views of the northeast quadrant of M31. The three left-hand panels (A–C) show the distribution of three increasingly blue samples of stars from our *BVRI* survey. Panel D shows the distribution of the red ($B - V > 0.5$) stars. Panel E shows the H I contour map from the survey of Unwin (1980), while panel F shows the H I map with the bluest stars overlayed. Only stars brighter than the completeness limit of $B = 20.7$ were used in panels A–D and F, and this is reflected in the nearly uniform distribution of stars in panel D. These red stars are mostly in the foreground, and should be more or less uniformly distributed. Other than the empty areas where we have no observations, and the region near the bulge where the excess bulge light interferes with the analysis, the distribution is more or less uniform; only the faintest hint can be seen of the spiral arms, and this is probably due to the small number of evolved massive stars with the appropriate colours.

Panels A–C clearly show that, in general, the bluer stars tend to be more clustered. This is not very surprising—the bluer stars are in general younger, and are less likely to have migrated far from their birth place. However, these pictures also demonstrate that we must be very careful in our definitions of "associations". A comparison of these panels shows that the clumps of stars identified as an association on the basis of their positional clumping (Magnier *et al.* 1993) are composed of subclumps with different colours. Note, for example, the region circled at coordinates (25, 10): this region was identified by the PLC as a single clump, but closer examination shows three subclumps, two visible in panel C, and one in panel B. Clearly a simple positional coincidence is not a sufficient description. It seems clear that all of the largest groups (>100 pc) have substantial subclustering (Efremov *et al.*, 1993). We are now exploring ways to represent the range of clustering scales using the galaxy–galaxy correlation techniques as a guide, avoiding the rigid definition of a specific "association" or "complex". Furthermore, we are investigating ways of incorporating the age information implicit in the colours of the stars. Such determinations could eventually be compared with Monte Carlo simulations based on theoretical descriptions of the star formation processes on these scales.

4. H I and the star-formation efficiency

One important test of stellar models is the measurement of the star-formation efficiency, as measured by, for example, the ratio of stars produced to the initial H I mass. Such a measurement may allow for a clean distinction between stochastic and triggered star-formation models. In the stochastic models, the star-formation efficiency is roughly constant, so that observed variations in the densities of newly formed stars are due to variations in the gas mass. Models which allow for triggering mechanisms, on the other hand, predict enhanced star formation per unit gas.

Panels E and F demonstrate the connection between the blue stars and the H I gas. To first order, there is a clear correlation between the H I emission and the amount of blue stars at various points in this picture, particularly the spiral arms. In the arms, there are substantial numbers of stars and substantial H I emission. There is also some tendency for the regions of highest H I column depth to have a deficit of blue stars, something which may easily be explained by extinction effects. However, there is also evidence for an *anti*-correlation between the H I and the blue stars in some areas. The region in the inner ring—at coordinate $-5 < Y < 5$, $X < 30$, for example—contains a large number of blue stars, without any H I evident in this maps. Such an anti-correlation shows that a more complex description of the star-formation processes is needed than either the triggered or stochastic processes.

A detailed determination of the star-formation efficiency will involve comparisons between the observed stellar densities from the Magnier *et al.* survey and the H<small>I</small> column density from the Unwin (1980) or the Brinks & Shane (1984) surveys. Care must be taken to account for the effect of extinction and also the influence of the existing stars on the state of the hydrogen gas. It is well known that, once the young stars turn on, the combined actions of their winds serves to clear the surrounding regions of the remaining hydrogen gas. To estimate the real star-formation efficiency, we will need to estimate the original amount of gas from the current amounts in unaffected neighbouring areas. The H<small>I</small>, Hα, and *IRAS* data together with the *BVRI* catalogue of stars will all be needed to be used to make a believable measurement of the star-formation efficiency throughout M31.

5. Conclusions

We have begun to investigate in detail the results of the star-formation processes in M31. Our optical surveys, along with the H<small>I</small> surveys of Unwin (1980) and Brinks & Shane (1984), as well as the *IRAS* observations of M31, provide the databases necessary to study these phenomena. What is now needed are improved descriptions of the scales on which stars are clumped together for comparisons with model simulations. This work promises to provide important insight into the global process of star formation.

EAM acknowledges support by the Netherlands Foundation for Research in Astronomy (ASTRON) with financial aid from the Netherlands Organisation for Scientific Research (NWO) under contract number 782–376–011. EAM would like to thank the Leids Kerkhoven-Bosscha Fonds for travel support to attend this workshop.

REFERENCES

AMBARTSUMIAN, V. A. 1947 *Stellar Evolution and Astrophysics*, Armenian Academy of Science, 1, 33.

BATTINELLI, P. 1991 *Astron. Astrophys.* **244**, 69.

BRINKS, E. & SHANE, W. W. 1984 *Astron. Astrophys., Suppl. Ser.* **55**, 179.

EFREMOV, YU., BATTINELLI, P. & MAGNIER, E. 1993 in *La Palma Workshop on Violent Star Formation*.

EFREMOV, YU. N., IVANOV, G. R. & NIKOLOV, N. S. 1987 *Astrophys. Space Sci.* **135**, 119.

HAIMAN, Z., MAGNIER, E. A., LEWIN, W. H. G., LESTER, R. R., VAN PARADIJS, J., HASINGER, G., PIETSCH, W., SUPPER, R. & TRÜMPER, J. 1993 *Astron. Astrophys.* **286**, 725.

HODGE, P. W. 1987 *Publ. Astron. Soc. Pac.* **99**, 173.

MAGNIER, E. A., BATTINELLI, P., HAIMAN, Z., LEWIN, W. H. G., VAN PARADIJS, J., HASINGER, G., PIETSCH, W., SUPPER, R. & TRÜMPER, J. 1993 *Astron. Astrophys.* **278**, 36.

MAGNIER, E. A., LEWIN, W. H. G., VAN PARADIJS, J., HASINGER, G., JAIN, A., PIETSCH, W. & TRÜMPER, J. 1992 *Astron. Astrophys., Suppl. Ser.* **96**, 379.

MAGNIER, E. A., PRINS, S., VAN PARADIJS, J., LEWIN, W. H. G., SUPPER, R., HASINGER, G., PIETSCH, W. & TRÜMPER, J. 1994 *Astron. Astrophys., Suppl. Ser.* submitted.

SUPPER, R., HASINGER, G., PIETSCH, W., TRÜMPER, J., JAIN, A., LEWIN, W. H. G., MAGNIER, E. A. & VAN PARADIJS, J. 1994 in preparation.

UNWIN, S. C. 1980 *Mon. Not. R. Astron. Soc.* **192**, 243.

VAN DEN BERGH, S. 1964 *Astrophys. J., Suppl. Ser.* **9**, 65.

The Fine-Scale Structure and Properties of Interstellar Gas towards Globular Clusters

By BRIAN BATES

The Queen's University of Belfast, Belfast BT7 1NN, N. Ireland, U.K.

An outline is given of a programme which uses high-resolution optical spectroscopy together with data from the radio, IR and UV regions to study the structure of the gas in the foreground of globular clusters and surrounding fields. An overview is given of some programme results.

1. Introduction

In a review of absorption-line studies of the IS gas, Cowie & Songaila (1986) noted that "the extremely powerful technique of mapping the absorption line structure across full globular cluster faces has yet to be carried out." Reporting on the IS Na D lines towards stars in M92 and M15, Langer *et al.*(1990) remark that a detailed mapping programme is not easy to carry out since "the number of bright giants in a typical cluster is not large and they are not always placed the way a map maker would like." From our own experience we would agree with this comment. Nevertheless, with the recent developments in high-resolution spectroscopic instruments for 3- to 4-metre class telescopes, it is now quite routine to observe some 10 to 15 stars (V $\stackrel{\sim}{<}$ 13) across the several-arcmin field of a cluster in a reasonable observing time. Whilst such observations may not yield detailed maps of the gas properties, they are providing considerable new information on the fine-scale structure on several aspects of the ISM, such as: (i) diffuse gas and cold clouds forming the "local bubble" wall; (ii) diffuse and peculiar velocity gas in spiral arms; (iii) the fine-scale extinction variation; and (iv) halo gas and the intermediate- and high-velocity clouds. This is particularly so when the optical absorption-line studies are complemented with data from the radio, IR and UV regions.

2. Aspects of the observing programme

In the optical observing programme the brightest cluster red giants are used as background sources to probe the fine-scale properties of the foreground gas. The stellar Na D lines in such late-type stars are strong but they are Doppler shifted from the IS lines by the radial velocity of the cluster stars. Spectra have been obtained using UCLES (University College London echelle spectrograph) and UES (Utrecht echelle spectrograph) at the AAT and WHT respectively. These spectrographs provide simultaneous coverage in one CCD position (using the 79 grooves mm^{-1} grating) of the IS Na D and KI $\lambda7699$ Å lines; also simultaneously recorded is the H$_\alpha$ line, of importance for studies of mass motions in the atmospheres of cluster giants (Bates *et al.* 1993). Spectra have a velocity resolution of ≈ 6 km s^{-1} and in good observing conditions, a continuum S/N ≈ 60 is obtained for a V≈ 12.6 cluster star for a total exposure time of 4000 s.

With the long sightlines to the cluster stars the spectra usually reveal a complex, multi-component structure of the foreground gas. An interpretation of these data requires an estimate of the distance of the various gas components. This is done by recording spectra of foreground stars which lie over a range of distances and within a few degrees of the cluster. The stellar distances are determined from stellar parameters listed in the literature and also from our own LTE model atmosphere analyses of their spectra.

The most important complementary data have been the λ 21-centimetre HI profiles obtained with the 76-metre Lovell Telescope in a collaboration with the Nuffield Radio Astronomy Laboratories (NRAL), Jodrell Bank. These profiles provide HI column densities in the directions of the cluster and field stars and are used to map the gas distribution on a wider scale around the cluster. Such observations reveal the structure of the low-velocity gas and they have the sensitivity to map the weaker, intermediate- and high-velocity gas components which have a patchy distribution on the sky. Comparisons between the optical and radio data are extremely important for deriving gas properties, but the difference in the spatial sampling of these two methods must be considered. Whereas the optical spectra reveal significant variations in gas column densities across arcmin scales, the HI profiles are obtained with much larger beamwidths ($\approx$ 12 arcmin FWHM for the Lovell Telescope). Thus, towards a cluster for example, the HI profiles represent an average of gas properties over an angle which may be comparable with the angular size of the cluster. Ideally, the optical and radio data should be compared at similar spatial resolution. A major step towards this goal has been started recently using the Dominion Radio Astrophysical Observatory (DRAO) Synthesis Telescope to map the HI distribution towards M13 at a resolution $\approx$ 1 to 2 arcmin.

3. A summary of some observational data

Seven clusters have been observed in our programme to date, and the different sightlines probe the various ISM features mentioned above. An overview is given here of some programme results.

3.1. *The ω Cen sightline ($l,b \approx 309°$, $+15°$); local and spiral arm gas*

This cluster, at a distance $\approx$ 6.2 kpc, provides a good illustration of some of the programme objectives. The cluster star spectra reveal a complex structure with four distinct groups of components at LSR velocities between $\approx$ -1.5 and -40 km s^{-1} whose strength and velocity are found to vary markedly across the sampled region (Bates *et al.* 1992). Cannon (1980) had detected faint, filamentary nebulosity towards the SW of the cluster. *IRAS* 60 and 100 μm images reveal considerable similarity between the IR emission and the faint optical nebulosity, but also its larger extent in the IR. A comparison of HI column densities derived from the *IRAS* fluxes and from the Na D lines of the cluster and foreground stars indicate that the low-velocity material responsible for the nebulosity forms part of the nearby complex of gas, dust and molecular clouds (Dame *et al.* 1987) at a distance of $\approx$ 150 pc (Wood & Bates 1993). The strongest component towards the cluster stars (velocity $\approx$ -15 km s^{-1}) is produced by gas at a distance beyond $\approx$ 1 kpc and hence at least 175 pc above the galactic plane. This feature is responsible for the bulk of the *IRAS* emission detected in the cluster direction whilst a comparison of column densities of cluster and field stars suggests a rapid increase of the gas column over the distance range $\approx$ 1.3 to 1.8 kpc (Wood & Bates 1994). The gas velocity is consistent with galactic rotation according to the description of the Carina spiral arm (Grabelsky *et al.* 1987), and the gas is probably outlying material in this arm.

3.2. *The M4 cluster (l, $b \approx 351°$, $+16°$) and differential reddening*

At a distance of $\approx$ 2 kpc, M4 is the closest globular cluster to the Sun and it lies in the direction of the Ophiuchus dust complex. Na D line spectra of 16 cluster stars show that the gas foreground to M4 is dominated by two low-velocity components identified with gas in the ρ Oph dark cloud (distance $\approx$ 160 pc) and probably associated with an expanding shell around a subgroup of the Sco–Cen association (Kemp *et al.* 1993). The

Na spectra reveal a complex pattern in the column-density variation across the cluster, implying considerable differential reddening. The weaker KI lines provide accurate column densities for the gas variation across the cluster. Comparison with *IRAS* 100 μm images indicates a correlation between IR emission and gas column density. The M4 data illustrate how the KI lines might be used to study differential reddening across a cluster; such information is relevant to the determination of the parameters of the cluster stars. However, a more detailed quantitative analysis requires that the relationships between KI and HI column densities and between N(KI)–E(B–V) become more firmly established for a large sample of stars.

3.3. *The M13 cluster (l,b $\approx$ 59°, +41°) and intermediate-velocity gas*

The M13 cluster is at a distance of $\approx$ 6.3 kpc and lies in a direction close to the centre of the nearby Hercules shell, which is considered to have a supernova origin (Lilienthal *et al.* 1992). For this high-latitude sightline we detect two principal components at velocities $\approx$ +10 and −4 km s^{-1}. The positive-velocity gas has a high NaI/HIcolumn density ratio and it lies beyond the foreground stars at a distance $\overset{\sim}{>}$ 200 pc; this gas may be associated with the receding part of the Hercules shell. The negative-velocity gas is detected towards stars at a distance of $\approx$ 200 pc and is probably associated with the approaching side of the shell. Intermediate-velocity (IV) gas, previously reported in HI emission in the M13 direction and in UV spectra of the post-AGB cluster star Barnard 29, has been detected for the first time in the optical region in the Ca K absorption line spectrum of Barnard 29 at a velocity of $\approx$ −70 km s^{-1} (Bates *et al.* 1995). Using HI profiles obtained with the Lovell Telescope, the CaII/HIcolumn density ratio is consistent with values reported for intermediate- and high-velocity clouds in the lower halo. The distance to the IV gas is uncertain but it lies beyond our sample of foreground stars and thus at a distance $\overset{\sim}{>}$ 200 pc.

HI observations over a 2° square field centred on the cluster show the variability of the low-velocity gas components and the patchy distribution of the IV gas on the scale of the telescope beamwidth. The studied cluster stars also cover an area which is comparable in size with the FWHM of the telescope beam. Since the Na D profiles show clearly that gas column densities and velocities vary significantly across the cluster field, a detailed comparison of the optical and radio data is somewhat limited; such a comparison is required at a more similar spatial resolution. Recent M13 observations obtained with the DRAO Synthesis Telescope reveal the variations in HI on a scale $\approx$ 1–2 arcmin and over the velocity range of −100 to +100 km s^{-1}. It is hoped that these combined data will provide direct NaI/HI ratios over the cluster field and yield important new estimates of the fine-scale changes of the gas space density across IS clouds (Penprase 1993).

3.4. *The clusters M22 (l,b = 10°, -7.5°) and M55 (l,b = 9°, -23°) and the Riegel & Crutcher cold cloud*

M22 and M55 form part of a study of a region of sky near l = 10° and covering a range of latitudes from about −26° to +8°. The M22 (Bates & Catney 1991) and M55 (Lyons *et al.* 1994) studies together with HI observations reveal an extensive structure of nearby, diffuse gas. Close to the plane is the Riegel & Crutcher (1972, hereafter RC) cold cloud, which we have mapped in detail with the Lovell Telescope to derive some cloud properties (Montgomery *et al.* 1994). The work of Crawford (1992) and Montgomery *et al.* suggests that the RC cloud and the nearby ridge of dark molecular clouds (Dame *et al.* 1987) are denser regions contained within a much larger-scale complex of gas and dust which are moving with a similar velocity. Pöppel *et al.* (1994) have suggested a similar picture, interpreting the kinematics of the Linblad Feature A at low latitudes as "a large expanding doughnut-shaped cloud or sheet which covers a wide region of sky and which is associated with the Gould Belt." The large molecular-cloud complexes which are presumed to be related to the Gould Belt are shown to be well correlated in both position and velocity with the ridge of local HI.

4. Conclusions

In recent years there has been considerable progress concerning the description of the distribution, origin and the nature of different aspects of the IS gas; further multi-wavelength studies which identify the structure and gas properties on small spatial scales will add to this emerging picture.

It is a pleasure to acknowledge the programme contribution from QUB colleagues and in particular the role of Dr. S. Kemp. Financial support is provided by the Particle Physics & Astronomy Research Council.

REFERENCES

BATES, B. & CATNEY, M. G. 1991 *Astrophys. J., Lett.* **371**, L37.

BATES, B., KEMP, S. N. & MONTGOMERY, A. S. 1993 *Astron. Astrophys.* **97** ,937.

BATES, B., SHAW, C. R., KEMP, S. N., KEENAN, F. P. & DAVIES, R. D. 1995 *Astrophys. J.* in press.

BATES, B., WOOD, K. D., CATNEY, M. G. & GILHEANY, S. 1992 *Mon. Not. R. Astron. Soc.* **254**, 221.

CANNON, R. D. 1980 *Astron. Astrophys.* **81**, 379.

COWIE, L. L. & SONGAILA, A. 1986 *Annu. Rev. Astron. Astrophys.* **24**, 499.

CRAWFORD, I. A. 1992 *Mon. Not. R. Astron. Soc.* **254**, 264.

DAME, T. M., UNGERECHTS, M., COHEN, R. S., DE GEUS, E. J., GRANIER, I. A., MAY, J., MURPHY, D. C., NYMAN, L.-A. & THADDEUS, P. 1987 *Astrophys. J.* **322**, 706.

GRABELSKY, D. A., COHEN, R. S., BRONFMAN, L. & THADDEUS, P. 1987 *Astrophys. J.* **315**, 122.

KEMP, S. N., BATES, B. & LYONS, M. A. 1993 *Astron. Astrophys.* **278**, 542.

LANGER, G. E., PROSSER, C. F. & SNEDON, C. 1994 *Astron. J.* **100**, 216.

LILIENTHAL., D., HIRTH, W., MEBOLD, U. & DE BOER, K. S. 1992 *Astron. Astrophys.* **255**, 323.

LYONS, M. A., BATES, B. & KEMP, S. N. 1994 *Astron. Astrophys.* **286**, 535.

MONTGOMERY, A. S., BATES, B. & DAVIES, R. D. 1994 *Mon. Not. R. Astron. Soc.* in press.

PENPRASE, B. E. 1993 *Astrophys. J., Suppl. Ser.* **88**, 433.

PÖPPEL, W. G. L., MARRONETTI, P. & BENAGLIA, P. 1994 *Astron. Astrophys.* **287**, 601.

RIEGEL, K. W. & CRUTCHER, R. M. 1972 *Astron. Astrophys.* **18**, 55.

WOOD, K. D. & BATES, B. 1993 *Astrophys. J.* **417**, 572.

WOOD, K. D. & BATES, B. 1994 *Mon. Not. R. Astron. Soc.* **267**, 660.

Elmegreen: What is the smallest scale over which changes in gas properties have been detected?

Bates: There are several pairs of observed stars which have a separation of a few arcsec, but no changes have been detected yet on such scales. The smallest change so far observed is for two stars in M4 having a separation of ≈ 17 arcsec; these reveal clear differences in the nearby gas (distance ≈ 150 pc) corresponding to a spatial scales of ≈ 0.012 pc.

Excitation Conditions and Fragmentation of the Southern Orion B Molecular Cloud

By CARSTEN KRAMER[1,2], J. STUTZKI[1] AND G. WINNEWISSER[1]

[1] I. Physikalisches Institut, Universität zu Köln, Germany

[2] Present address: Instituto de Radioastronomía Milimétrica (IRAM), Granada, Spain

---

The Orion–Monoceros giant molecular clouds (GMCs) are the nearest regions with active OB-star formation. Their proximity—about 400 pc—and brightness provide an opportunity to study the interestellar matter, its morphology, kinematics and dynamics in detail. The analysis of the large-scale structure and excitation conditions of the southern part of the GMC Orion B was the aim of our multiline observations with the KOSMA (Kölner Observatorium für Sub-mm Astronomie) 3-metre radiotelescope (Kramer et al. 1994). We observed the radiation of CO and its main isotopomers in the two rotational transitions (J=3→2) and (J=2→1).

The map of velocity-integrated $^{13}CO(2\rightarrow1)$ emission (Figure 1) shows a ridge orientated north–south with two major emission regions centred on the HII region NGC 2024 to the north and the reflection nebula NGC 2023 to the south. The map shows a rather sharp drop-off to the west towards the optical nebula IC 434 and the subgroup Ib of the Orion OB association, indicating a strong influence of the OB stars on the molecular material. The southern intensity maximum protruding into IC 434 is the Horsehead Nebula B33.

At nearly all positions the line ratios are contradictory with a single temperature and density model: the line ratio of $^{12}CO(3\rightarrow2)$ over $^{12}CO(2\rightarrow1)$ is about 1. The same is valid for the ^{13}CO lines. While these ratios indicate optically thick ^{12}CO and ^{13}CO emission, the line ratio of ^{12}CO over ^{13}CO, which is about 3 in the line centres, indicates optically thin ^{13}CO emission. These are hints at temperature gradients within the clumps: the surfaces are heated by the stellar UV field whereas the inner regions are shielded from this radiation. This interpretation is in accordance with theoretical models of UV-irradiated clumps (e.g. Köster et al. 1993) which furthermore indicate that local volume densities of 10^5 cm^{-3} or more are needed to reproduce the observed line intensities and line ratios.

The large-scale observations of the southern part of Orion B reveal the fragmented structure on all scales between 0.1 pc (the resolution limit) and the extension of the observed regions (8 pc). Especially the ^{13}CO channel maps show that the molecular cloud consists of many local maxima, filaments, and cavities. The total mass of the surveyed region of Orion B is 3700 M$_\odot$. A major part of the total mass of Orion B, 44%, is found in the five most massive clumps with $M > 200$ M$_\odot$. These clumps are associated with the prominent star-forming regions NGC 2024 and NGC 2023. A Gaussian decomposition algorithm (Stutzki & Güsten 1990) identifies 244 clumps. The resulting mass spectrum is well fitted by a power law with an index of 1.74. None of the clumps found are gravitationally bound —the more massive of the well defined clumps are closer to virial equilibrium than the less massive clumps. Our data imply that external pressures between 10^4 and 10^6 Kcm^{-3} are needed to stabilise the clumps. The average clump density is 2×10^3 cm^{-3} which is significantly lower than the critical density of the CO(3→2) transition and two orders of magnitude lower than the density needed to explain the

Orion B ^{13}CO J=2−1   KOSMA 3m RT

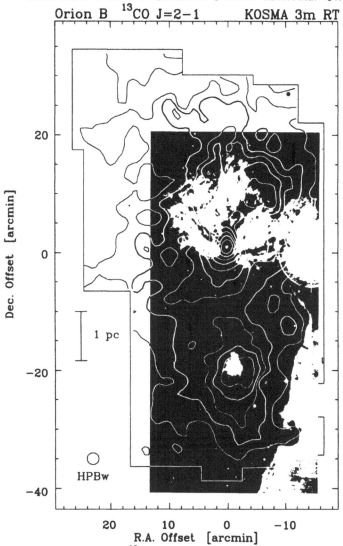

FIGURE 1. The velocity integrated ^{13}CO(2→1) emission of the southern part of Orion B over-layed on an optical photograph of the region. The (0′,0′) position is located at $5^h39^m12^s$, −1°58′ (1950). The range of integration is 0 to 20 kms^{-1} and contours are at 5.7 (=8σ), 12 (6) 60 Kkms^{-1}($\int T_R^* dv$).

temperature ratios found. These are strong indications of substructures at still smaller scales.

REFERENCES

KÖSTER, B., STÖRZER, H., STUTZKI, J. & STERNBERG, A. 1994 *Astron. Astrophys.* **284**, 545.

KRAMER, C., STUTZKI, J. & WINNEWISSER, G. 1994 *Astron. Astrophys.* in preparation.

STUTZKI, J. & GÜSTEN, R. 1990 *Astrophys. J.* **356**, 513.

High-Resolution L-Band Imaging of the Starburst Galaxy M82

By TOM W.B. MUXLOW, A. PEDLAR
AND E. M. SANDERS

N.R.A.L., Jodrell Bank, U.K.

These latest results are part of a continuing study of the starburst phenomenon in M82. The compact radio features are now established to be recent supernova remnants with ages of a few hundreds or thousands of years. A new remnant is thought to appear about every 20 years. A previous study at 5 GHz showed the M82 remnants are similar to, but younger and smaller than, the equivalent remnants in the LMC and our own Galaxy. This latest image extends this study to include larger older remnants. Suprisingly many remnants show significant low-frequency spectral turnover which is probably due to free–free absorption in the gas surrounding the remnant. The required emission measure is similar to those of giant galactic HII regions. By comparison with a VLA 5-GHz image of comparible resolution, the central 12 arcsec (180 pc) around the inner core of M82 is seen to possess a steeper radio spectral index than regions further from the core. This steeper region corresponds to that lying within the established ring of molecular gas which surrounds the nucleus of M82.

1. High resolution imaging of M82

The MERLIN array has been used to study the nuclear region of M82 at high angular resolution, complimenting the well-established VLA studies by Kronberg and Sramek (e.g. Kronberg et al. 1985). A previous study with the enhanced MERLIN at 5 GHz mapped more than 40 of the supernova remnants (SNRs) with an angular resolution of 50 mas (Muxlow et al. 1994). It was found that these SNRs follow a surface brightness-diameter relation consistent with those found in the LMC and galactic remnants. The cumulative number as a function of diameter appears to increase linearly to diameters of at least 3 pc, and indicates a supernova rate of 0.05 per year if the shells are expanding at 5000 km s^{-1}.

2. The latest images

This latest study attempts to fully image the whole of the nuclear region of M82 at L-Band. In addition to the original 1.42-GHz data (April 1993), a second run was obtained (May 1993). Furthermore, in order to enhance the density of the spatial frequency coverage, an additional run was obtained during May 1993, cycling every few minutes between 1.612, 1.658, and 1.720 GHz. This technique is known as Multi-Frequency Synthesis (MFS). In total around 50 hours of data were collected. The remaining low spatial-frequency coverage was obtained by adding a four-hour VLA A-array observation (May 1994).

The analysis of the L-Band data is still continuing, however a preliminary image at 0.3 arcsec resolution is shown in Figure 1. The dynamic range of this image is close to 2000:1 and residual image errors are close to this level. Although further improvements are expected, the present image is substantially correct. The flux densities and sizes of the more compact SNRs in Figure 1 agree well with the single frequency analysis of Sanders (1994); however, for the first time the extended emission at L-Band is fully imaged.

Figure 1   MFS MERLIN + VLA Image of M82 at L-Band

M82   IPOL   1420.000 MHZ   M82 ALL FREQ.RSTOR.4

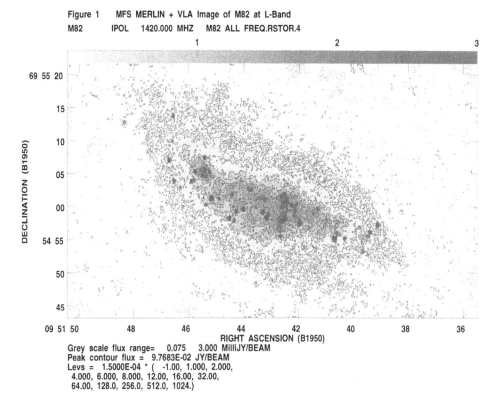

Grey scale flux range=   0.075   3.000 MilliJY/BEAM
Peak contour flux =   9.7683E-02 JY/BEAM
Levs =   1.5000E-04 * (-1.00, 1.000, 2.000,
4.000, 6.000, 8.000, 12.00, 16.00, 32.00,
64.00, 128.0, 256.0, 512.0, 1024.)

A comparison of Figure 1 with a VLA 5-GHz A-Array image of M82 (data kindly supplied by Kronberg and Sramek) observed in December 1992 with an angular resolution of 0.33 arcsec show both striking similarities and interesting detailed differences. Figure 1 is more sensitive to the extended emission around the inner 1 kpc of the nucleus of M82 than the 5GHz image (1 kpc = 66 arcsec). However, even at higher flux density levels there are signinficant differences in the extended emission between L–Band at 5 GHz which cannot be assigned to imaging errors.

Several of the SNRs show significant spectral flattening, the most extreme case being the remnant 44.01+596 which has a strongly rising spectrum between 1.4GHz and 5GHz. By including flux density measurements from VLA 8.4GHz observations (Huang *et al.* 1993, and recent (May 1994) unpublished A-Array data) it is clear that the spectra for the SNRs are steep between 5GHz and 8.4GHz. Thus we are detecting low-frequency turnovers. Around 40% of the more compact SNRs show evidence for low-frequency turnovers in their spectra. At 50-milliarsec resolution, 5-GHz MERLIN resolves virtually all these compact SNRs; the resulting sizes are thus too large to allow sychrotron self-absorption to explain the spectra. We have suggested that free–free absorption in the gas surrounding such SNRs can explain the low-frequency turnovers seen. Using the gradient of the spectrum between 5 GHz and 8.4 GHz (where we assume there is no absorption) we can extrapolate back to the value of the SNR flux density that we would expect at 1.4 GHz. By comparing this with the measured value we can calculate the optical depth and hence the emission measure (E) for a number of SNRs. Assuming an electron temperature of 10^4K we get values of E of around a few times 10^6 pc cm^{-6}, which is consistent with the emission measure in giant galactic HII regions. It is also comparable with the well-documented emission measure of the brightest SNR 41.95+575, which shows a frequency turnover below 1GHz (Wilkinson & deBruyn 1984). This strongly suggests that free–free absorption *is* the cause of the low-frequency turnover in many of the SNRs. The random distribution of the remnants exhibiting this behaviour implies that the absorption is being caused by local clumps of gas rather than one single cloud. It may also explain the differences between the 1.4-GHz and 5-GHz images of the extended structure in the nuclear region. A further point to note is that the central 12 arcsec (180 pc) in general shows an overall steeper spectral index than the surrounding regions. The reason for this is not as yet fully understood, but it is interesting to note that star formation in M82 is interpreted as moving outwards from the centre of the galaxy, and that CO observations by Nakai *et al.* (1986) show a ring of molecular gas with a radius centred on 200 pc. This is the region of current star formation and the radio SNRs appear to be situated on its inner rim. The innermost 12 arsec is thus that region lying inside the molecular gas ring.

REFERENCES

ALBERDI, A., LARA, L., MARCAIDE, J. M., ELÓSEGUI, P., SHAPIRO, I. I., COTTON, W. D., DIAMOND, P. J., ROMNEY, J. D. & PRESTON, R. A. 1993 *Astron. Astrophys.* **277**, L1.

HUANG, Z. P., THUAN, T. X., CHAVALIER, R. A., CONDON, J. J. & YIN, Q. F. 1993 *Astrophys. J.* submitted.

KRONBERG, P. P., BIERMANN, P. & SCHWAB, F.R. 1985 *Astrophys. J.* **291**, 693.

MUXLOW, T. W. B., PEDLAR, A., WILKINSON, P. N., AXON, D. J., SANDERS, E. M. & DEBRUYN, A. G. 1994 *Mon. Not. R. Astron. Soc.* **266**, 455.

NAKAI, N., HAYASHI, M., HASEGAWA, T., SOFUE, Y., HANDA, T. & SASAKI, M. 1986 in *Star Forming Regions* (ed. M. Peimbert & J. Jugaku). IAU Symposium 115, p. 614. Reidel.

SANDERS, E. M. 1994 M.Sc. thesis, University of Manchester.

WILKINSON, P. N. & DEBRUYN, A. G. 1990 *Mon. Not. R. Astron. Soc.* **242**, 529.

Galactic Warps

By EDUARDO BATTANER

Departamento de Física Teórica y del Cosmos, Universidad de Granada, E–18002, Granada, Spain

After an introductory description of the structure and dynamics of warps, the different explanations are reviewed: normal modes of oscillation, accretion and slewing, intergalactic magnetic fields, etc. Present and future observations providing constraints to current theoretical models are considered: twisting of the line of nodes, warp angle dependence or wavelength, coherent alignment of warps, etc.

1. Introduction

Excellent reviews on warps have recently been published by Binney (1991, 1992). At the University of Pittsburgh a meeting was held specifically devoted to "Warped Disks and Inclined Rings around Galaxies" (Casertano et al. 1991). These publications are taken here as a starting point. Taking them into account, we will now summarise previous models, update later results, consider some observational possibilities and emphasise one of the competing working models: that based on extragalactic magnetic fields, which may be reinforced by recent measurements in the intergalactic medium. A complementary review of galactic warps was recently presented by Combes (1994). The Milky Way will receive particular attention, but again warps of other galaxies provide a macroscopic description of those of our own, so our review will be kept as general, and include all known warps.

A disk is said to be warped (at least in the sense adopted here) if there is a smooth and moderate rise in a part of its outer region, accompanied by a descent in the azimuthally opposed region. It is important to emphasise the terms "smooth" and "moderate", because other, more dramatic, observed distortions may receive a completely different explanation. We will restrict ourselves to warps in spiral galaxies. There are certainly similar distortions in dust-lane ellipticals and polar-ring SO's. Centaurus A (Nicholson et al. 1991) is one of the best examples. However, as discussed by Galletta (1991), these structures are substantially different; they probably have an external origin, are not coeval and their kinematic and geometrical properties differ greatly with respect to the central part of the galaxy.

Most discussions on warps are based on the kinematic model by Rogstad et al. (1974) in which the disk is made up of a number of circular rings. Each ring of radius r_i has a different inclination w_i with respect to the central unwarped rings, and crosses the plane they define at two "nodes". The locus of all nodes defines the "line of nodes" which may be non-straight, but twisted. The galactic azimuth of the "ascending" node of each ring i, is called the twist angle t_i. Therefore each ring is characterised by r_i, w_i and t_i. The success of this model lies in the circularity of the rings: non-circular and even non-planar rings are more difficult to maintain and pose severe theoretical problems.

The interest in studying warps is twofold: as any other astrophysical phenomenon and as a peripherical feature, where gravity is reduced by orders of magnitude and where, therefore, some effects external to the visible galaxy have an influence. Warps can then be used to obtain information about them. This information depends on the theoretical model adopted to explain warps. Examples might include information about (i) the size of the halo core (Sparke & Casertano 1988), (ii) some geometrical parameters of the halo

(Christodoulou & Tohline 1986), (iii) the process of galaxy formation (Ostriker & Binney 1989) and (iv) intergalactic magnetic fields (Battaner *et al.* 1991).

Warps are not exceptional but ubiquitous. This fact was shown by Bosma (1991) in 21 cm and by Sánchez-Saavedra *et al.* (1990) in the visible. In the last mentioned paper, 42 out of 86 northern edge-on galaxies were observed to be warped, which would mean, according to the interpretation of the authors, a real frequency of about 83%. Barteldrees & Dettmar (1994) present photometric maps of southern edge-on spirals in which I appreciate a frequency of 12/27, in agreement with the northern results. Nearly all disks are warped.

1.1. *The warp's enemies*

The high frequency of warped disks suggests either that warps are steady-state distortions or that they are transient but easily and often excited. Probably the first possibility has received more attention. Models must identify effects counteracting the natural dissipation of warps. In the absence of external forces, warps would be transient structures because of at least two effects.

(i) If external torques are applied to a warped rotating disk, a precession motion will develop. If either the torque or the rotation period is radial dependent, a differential precession will result, thus destroying the warp. As a particular case, differential precession may be produced by the galaxy's own gravity. The warp will wind up in a characteristic time estimated by Rogstad *et al.* (1974) and by Binney (1992) to be of the order of 2 Gyr.

(ii) Autogravitation would also destroy a warp, by gradually smoothing the curvature. The characteristic time could be estimated by $\tau \sim (G\rho)^{-1/2}$ independently of the warp's curvature. At the solar galactocentric radius this time is less than 0.1 Gyr, and about 2 Gyr in the warp region.

2. Observations

2.1. *Observations in the Milky Way*

Many 21-centimetre observations have been carried out (e.g. Henderson *et al.* 1982; Burton & te Lintel Hekkert 1986), providing a precise description of the Milky Way warp. The plane of maximum warping connects galactocentric azimuths $80°$ (the north warp) and $260°$ (the south warp). Galactocentric azimuths are defined as anticlockwise when seen from the north galactic pole, and as being $180°$ for the Sun. The line of nodes, perpendicular to this direction (i.e. $170°$–$350°$, close to the direction galactic centre–Sun) is not twisted. The maximum warp angle is about $10°$ in the north warp and only $4°$ in the south warp.

HII regions (Fick & Blitz 1982), γ-rays (Mayer-Hasselwonder *et al.* 1982), dust (Sadroski *et al.* 1993), CO (Wouterloot *et al.* 1990) and other young components of the galaxy follow the HI distribution.

In the north, the warp is observable for young blue stars (Reed & Fitzgerald 1984; Miyamoto *et al.* 1988) but not for old red stars (Ichikawa & Sasaki 1984; Guibert *et al.* 1978). In the south warp, Carney & Seitzer (1993) carried out a detailed statistical analysis of more than 27 000 stars, encountering very important interpretation difficulties. No such detection has been made for red giant stars, these differences being important for theoretical reasons, as described later.

Miyamoto *et al.* (1988) found a geometrical kinematic description of the warp that (if confirmed) would require a complete reconsideration of the warping effect. In their model, there is no vertical direction in the line of nodes; the northward-bent region has a vertical motion to the north and the southward-bent region to the south; the line of nodes gyrates with the galactic rotation and orbits are not circular. The time-dependent system brings to mind an oscillatory motion. The sample used by these authors to obtain such a picture only includes stars within 3 kpc of the Sun; therefore, it may describe a local corrugation rather than the external warp itself.

2.2. *Observations in other galaxies*

HI radio observations extend to higher radii and are therefore the main technique for observing warps. Edge-on galaxies have often been investigated since the pioneering work of Sancisi (1976). Kinematic warps in face-on galaxies may also be observed, as was first demonstrated by Rogstad *et al.* (1974). There are at present detailed 21-centimetre maps of some 30 galaxies. Global samples have been extensively studied by Bosma (1991), Briggs (1990), and Christodoulou *et al.* (1993).

Warps may also be observed in the optical (van der Kruit & Searle 1981; Sasaki 1987; Hamabe *et al.* 1980, 1981; Florido *et al.* 1991a, and others). Kemp & Meaburn (1993) have recently made a study of the warp galaxy MGC 06-30-005 by using a co-added data array of APM (Automatic Plate Measurement machine) scans of Schmidt plates. This technique is highly promising for the observation of very faint features in the periphery of galaxies. Sánchez-Saavedra *et al.* (1990) presented a catalogue of all northern edge-on NGC galaxies by analysing the Palomar Observatory Sky Survey.

3. A brief account of the different theoretical models

The reader is addressed to the review by Toomre (1983) for early models, some of which are obsolete, some merely forgotten. There is a large variety of theories, their systematisation being a difficult task. We will concentrate on the three models selected by Binney (1992).

Most models require a non-spherical dark halo. The existence of massive dark halos has been questioned by Nelson (1988) and Battaner *et al.* (1993). The last model discussed here—magnetically driven warps—does not need the existence of a halo, nor is it incompatible with one.

3.1. *Discrete normal modes*

Disks may support corrugation waves as they were observed in NGC 4244 and NGC 5023 by Florido *et al.* (1991b), and local deviations from planarity in the local neighbourhood are interpreted in this way (Alfaro *et al.* 1992). One easy way to introduce discrete bending modes is as a superposition of inward and outward running waves in order to produce a steady configuration. The question is whether it is possible to find a normal mode with a shape similar to real warps. This possibility was first suggested by Lynden-Bell (1965). However, Hunter & Toomre (1969) showed that this mode cannot be present in an isolated disk with reasonably smooth edges. The idea acquired renewed interest after the suggestion by Dekel & Shlosman (1983) that a warp might be sustained if the disk were embedded in a flattened dark halo. Sparke (1984) and Sparke & Casertano (1988) have presented detailed models and the possibility is considered at present as one of the most promising.

Kruijen (1991) solved the problem in the full non-linear regime, confirming previous linear calculations and improving the agreement with actual warps. M33 has such a large warp (about 90°) that a linear model would be inappropriate. Kruijen shows that if the mass-to-light ratio in the disk is greater than a limiting value (in the case of M33, one third of the maximum disk value), the warp cannot be sustained.

The disk must initially be tilted with respect to the plane of symmetry of the oblate halo. The discrete mode would not dissipate if the disk is not too extended and the halo is not too flattened. Differential precession is eliminated by the cohesive effects of self-gravity, if the shape of the warped mode is appropriated.

One of the predictions of this model is that the line of nodes must be straight, which, as we know, is generally not true. Recently, however, Hofner & Sparke (1994) have greatly improved previous models by considering the time evolution of a disk tilted with respect to the halo. The disk settles toward the mode shape, from the centre outwards, in times similar to Hubble's time. They determined that the line of nodes is straight in the inner warp, but twisted in the outer one.

At the edge of an isolated, not sharply truncated disk, outward-running corrugation waves are not reflected; the amplitude is an increasing function of the radius, becoming non-linear and heating the outer disk. It is not impossible that warps are simply highly amplified corrugations, i.e. completely transient features, stimulated as often as are corrugations.

3.2. *Accretion and slewing*

This model has been developed by Ostriker & Binney (1989) and Binney (1991, 1992). Warps would be the result of a continuous redistribution of angular momentum in the halo, because of a permanent accretion. According to these authors, the new material— gas and any kind of cold dark matter—falls into the halo. The response of the halo is a very complicated question, and they assume that halos redistribute the angular momentum very fast, so that they spin about the same axis at all radii. There is a slewing of the angular momentum which introduces a torque in the visible galaxy, mainly in the disk. The torque would lie in the disk plane, about the warp's line of nodes, but being a function of radius. Because of self-gravitation cohesion, the inner disk is not warped, but at larger radii the mass of the rings is much lower and the disk bends. The warped zone would then be considered as left behind in the settling down of the disk into a new plane.

Ostriker & Binney (1989) considered that their model would explain how the innermost disk (≤ 3 Kpc) of our Galaxy, is tilted with respect to the outer disk about an axis, coincident with the line of nodes of the outer warp, thus connecting two phenomena with a large separation in space. The ring-ring gravitation ensures the different rings gyrate in a common plane. But if a ring is missing, self-gravity cohesion is broken there. As H_2 and HI densities have a deep minimum between 2 and 3 Kpc, they assumed that this minimum might be responsible for the misalignment observed. However, it is not at all clear that the gas depression corresponds to a global density depression. Many other galaxies, such as M31, have an annular distribution of gas with a sharp inner edge, which does not correspond to a decrease in the stellar distribution (except for OB stars). It is therefore not clear that a gap in the self-gravity cohesive force exists.

This model has much in common with an important series of papers in which, as gas settles into the equatorial plane of the dark halo, a warp develops. (Christodoulou & Tohline 1986 and references therein).

3.3. *Magnetically generated warps*

Intergalactic magnetic fields may be responsible for warps (Battaner *et al.* 1990; Battaner *et al.* 1991), if the direction of the intergalactic magnetic field does not coincide with that of the angular momentum of the galaxy and it is not contained in the plane of the disk, with the maximum efficiency to generate warps being at an angle of $45°$ between the spin and the field. The magnetic tension tends to spread the ISM plasma of the outer disk along the field lines.

If the intergalactic field vector has the same direction as the galaxy's spin, it would introduce another source of flaring. If it lies in the galactic plane it would render elliptical the outer orbits, reminding us of the isophote image of the galaxy as a "hard boiled egg" (Battaner *et al.* 1990).

As in the case of the Sun, magnetic fields become important in the periphery, when gravity has decreased by orders of magnitude. The question is whether the field is strong enough to produce warps. We estimated that a field strength of 10^{-8} G would be required. This critical value was also estimated independently by Binney (1991) with a more pessimistic result: a field strength of 10^{-8} G would only bend the disk at 40 Kpc. Certainly, four years ago the intergalactic field strength required could be considered as uncomfortably high with respect to the currently assumed values. Now, the situation has changed.

3.3.1. *The intergalactic magnetic field*

We address the reader to the excellent review by Kronberg (1994 and references therein), regarding the large amount of important data very recently acquired which demonstrates the ubiquity of very large intergalactic magnetic fields, usually of the order of $1\mu G$, being in intracluster media only slightly lower than typical galactic values. The dynamic consequences of these large fields have yet to be explored, even in the case of galaxy formation.

Such intense fields were also found in absorber systems in front of quasars (see Watson & Perry 1991), obtaining in some of them values as high as 10^{-3} G. This fact seems to indicate that galaxies have formed and evolved embedded in a highly magnetised medium (except perhaps a first generation providing first seeds, as suggested by Kronberg).

The values of the magnetic field strength seem to be of the same order of magnitude in the intercluster and extracluster media, despite the much lower density outside clusters. They are noticeably close to $3\mu G$ providing a magnetic energy density equal to the background microwave radiation. Interstellar and intergalactic magnetic fields have always saturated to this value, except in some localised systems.

Given a ubiquitous magnetic field of the order of $3\mu G$, and in particular in the medium around a galaxy, it is absolutely clear that it must exert a decisive influence on the morphology and evolution of the peripherical disk. For example, the rotation velocity becomes of the same order as the Alfvén speed in the intergalactic medium close to the galaxy.

4. Key observations

We will now discuss some observations that could provide indications to reduce the number of competing theories. Though they are not new, the limited information obtained until now requires further efforts. Among many other possibilities, we would emphasise three types of observations:

4.1. *Twist of line of nodes*

Christodoulou & Tohline (1986) gave a theoretical interpretation scheme of the twist of warps. They considered warps as transient features, resulting from differential precession of a disk tilted with respect to the symmetry plane of an ellipsoidal halo. The relative direction of precession depends on whether the frequency of vertical motions of a test particle is greater or smaller than the frequency of azimuthal motions, which in turn depends on the potential, hence on the shape of the halo. More precisely, they deduced a prograde (leading) twist if the potential well is oblate in shape, and a retrograde (trailing) twist, if it is prolate. Christodoulou *et al.* (1993) made a detailed analysis of spirals not viewed edge-on, for which good data are available. They found five "oblate" twists, two "prolate" twists and seven ambiguous cases.

Twists are otherwise interpretable. Sparke & Casertano (1988) predicted untwisted warps, but in the time-dependent model by Hofner & Sparke (1994) they would indicate different time phases before settling to the warped discrete mode. In the model by Ostriker & Binney (1989) the twist would be interpreted as a result of a change in the direction of slewing of the halo's axis. If the tilt is maintained by the halo's irregular growing, this kind of twist must necessarily be present. In the magnetic model the presence of the halo is not necessary. If it exists, a magnetically driven warp would be subject to differential precession.

But the interpretation of warp twists may be confusing, as other effects may be present. Suppose there is a radial expansion motion in the disk. The Coriolis acceleration would produce a leading twist; a contraction motion, a trailing twist. In order to quantitatively evaluate this effect, let us call R_{CD} the radius of the central unwarped disk, and R_o a radius characterising the edge. The rotation velocity V is assumed to be constant, and we take v as the radial velocity. It is easy to show that the prograde twist angle at R_o is approximately given by

$$t = \frac{(1-r)^2(1+r)}{2r}\frac{V}{v} \tag{4.1}$$

where $r = R_{CD}/R_o$. For a value $V/v \approx 10$, we obtain for $r = 0.75$ a twist angle of about $45°$. Therefore, the effect of radial motion is really important. The inclusion of these radial motions in theoretical models is of course difficult, but they are fairly common in real galaxies.

Magnetic fields are also able to produce twists. Suppose, as an example to estimate their effects, that we choose near the warp: x in the azimuthal direction; y, radial; and z, vertical. Further suppose that the galactic magnetic field is azimuthal $(B, 0, 0)$ with negligible vertical and azimuthal directions. For an extragalactic field with a non-vanishing radial component b, it is not difficult to prove that the magnetic tension produces a force in the azimuthal direction given by $-(Bb)/2(R_o - R_{CD})$. Because of this force, a twist angle is expected. If the galactic rotational field were of the order of the extragalactic field, then

$$t \approx \frac{1}{2}\left(\frac{v_{Alfven}}{V}\right)^2(1-r) \tag{4.2}$$

which produces lower but not ignorable values. It is interesting to observe that the twist would have the same direction at both extremes of the warp.

4.2. *Colour gradients within the warp*

Gravity does not make any distinction between blue and red stars, nor between stars and gas particles. However, the motion of stars is unaffected by magnetic fields, as is the motion of gas. Therefore, it should be expected that those models in which gravity is the main interaction do not predict any distinction between the gas warp and the stellar warp. Indeed, gravitational models such as those by Sparke & Casertano (1988), Ostriker & Binney (1989) and many others are mathematically formulated without specifying any distinction between gas and stars, dealing only with massive rings able to gravitationally attract and be attracted. However, the magnetic model produces a warp on the gas, not on the stars. As stars are born out of gas, a stellar warp is also expected, but less pronounced, mainly for old stars. Is the gas warp deeper than the blue-star warp, and is this in turn deeper than the red-star warp?

Certainly, magnetic fields are not the only effect causing the gaseous and stellar systems to have behave differently. Dissipation mechanisms are more efficient for the gas. Therefore, the response to gravity of gas and stars may not necessarily be the same. Also, this difference could be unimportant in the outer disk, if gas is concentrated in giant clouds with the same size as they have in the inner disk. In this case, a lower density would correspond to a lesser concentration of giant clouds, so that collisions would become infrequent. Gas clouds and stars would then behave similarly.

Observations in our own galaxy, have detected the warp, for many wavelengths characterising young objects, but a red- giant warp has been looked for and not found. Freudenreich *et al.* (1994) observed with DIRBE (Diffuse Infrared Background experiment) on *COBE* that the near-infrared ($< 5\mu$m) warp (due to old stars) has half an amplitude with respect to the far-infrared warp (due to dust).

Florido *et al.* (1991a) have shown that warps are clearly larger when observed in short-wavelength filters. They have shown that if the order from higher to lower warps is gas–blue–red, then the effect of extinction is to render the different warps closer.

The dependence of the magnitude of the warp on the age of the observed component is provided by just one image when a warped disk has a warped dust-lane. This is the case in the interesting image of NGC 4013. The effect is specially clear in the R filter (Figure 1 in Florido *et al.* 1991a). We see that in the outer parts, the dust lane is deviated from the red disk, "cutting" the northern hemisphere in the western extreme. Rather interestingly, there is no detectable warp in the near infrared (Barnaby *et al.* 1994). This effect, demonstrative in itself, is not the only case. Hamabe *et al.* (1980) reported such an inclination of the dust lane for NGC 4565.

4.3. *Coherent alignment of warps*

Intergalactic magnetic fields are not expected to have a universal common orientation, which would be in contradiction with the Cosmological Principle. But it could be random at cosmological scales, having some characteristic scales in which a coherent alignment could be present. If the size of ordered field cells is sufficient to include several warped galaxies inside, coherent alignment in groups of galaxies would be detectable. This was attempted by Battaner *et al.* (1991).

The details of the sphere statistic procedure, which can be applied to other astrophysical problems related to orientation of spins, were given by Battaner (1993). The standard deviation is small enough to suggest that warps are not randomly oriented in space. In the Virgo Supercluster, the orientation of warps is different with respect to two adjacent rich groups, being about 120° the angle formed by both magnetic field vectors. It should also be noticed that the whole sample formed by all observed warps has a non-random distribution, with the standard deviation, however, being much higher.

Therefore, at scales of about 25 Mpc, there is an appreciably ordered magnetic field. Even at scales of 100 Mpc, it is still detectable.

5. Conclusions

Some critical observations have been previously suggested. In addition, some galaxies (NGC 5866, NGC 4633+4634, NGC 4217, M102, etc.) probably have an inclined dust lane. Does NGC 4710 have a warp directed to the observer? It would be interesting to detect optical kinematic warps in face-on galaxies. Some new techniques may be of great interest to understand warps, as are those developed by Kemp & Meaburn (1993) and the use of two-dimension detectors in the near infrared (e.g. Barnaby *et al.* 1994).

The magnetically driven warp hypothesis has been reinforced now that high magnetic-field strengths have been accepted. Certainly the observations concerning intergalactic magnetic fields are not so new. For instance, the first direct measurement in the inter-cluster medium by Kim *et al.* was published in 1989, and of Faraday rotation from the Virgo cluster intergalactic medium by Vallee in 1990, to mention just two of many other measurements. But many different observations have recently been compiled, ordered and appreciated, rendering a conclusive global picture (Kronberg 1994).

Dr. Barnaby permitted me the exhibition of the near-IR image of NGC 4013 before publication.

REFERENCES

ALFARO, E., CABRERA CAÑO, J. & DELGADO, A. J. 1992 *Astrophys. J.* **399**, 576.

BARNABY, D., THRONSON JR., H. A. & ESTEP, G. M. 1994 *January Meeting American Astronomical Society.*

BARTELDREES, A. & DETTMAR, R. J. 1994 *Astron. Astrophys., Suppl. Ser.* **103**, 475.

BATTANER, E. 1993 in *II Granada Lectures in Computational Physics* (ed. P. L. Garrido & J. Marro), p. 269. World Scientific. Singapore.

BATTANER, E., FLORIDO, E. & SÁNCHEZ-SAAVEDRA, M. L. 1990a *Astron. Astrophys.* **236**, 1.

BATTANER, E., FLORIDO, E. & SÁNCHEZ-SAAVEDRA, M. L. 1990b *Astron. Astrophys.* **253**, 89.

BATTANER, E., GARRIDO, J. L., MEMBRADO, M. & FLORIDO, E. 1993 *Nature* **360**, 652.

BATTANER, E., GARRIDO, J. L., SÁNCHEZ-SAAVEDRA, M. L. & FLORIDO, E. 1991 *Astron. Astrophys.* **251**, 402.

BINNEY, J. 1991 in *Dynamics of Disc Galaxies* (ed. B. Sundelius), p. 297. Göteborg University.

BINNEY, J. 1992 *Annu. Rev. Astron. Astrophys.* **30**, 51.

BOSMA, A. 1991 *Astron. J.* **86**, 1791.

BRIGGS, F. H. 1990 *Astrophys. J.* **352**, 15.

BURTON, W. B. & TE LINTEL HEKKERT, P. 1986 *Astron. Astrophys., Suppl. Ser.* **65**, 427.

CARNEY, B. W. & SEITZER, P. 1993 *Astron. J.* **105**, 2126.

CASERTANO, S., SACKETT, P. & BRIGGS, F. (EDS.) 1991 *Warped Disks and inclined Rings around Galaxies.* Cambridge University Press.

CHRISTODOULOU, D. M. & TOHLINE, J. E. 1986 *Astrophys. J.* **307**, 449.

CHRISTODOULOU, D. M., TOHLINE, J. E. & STEIMAN-CAMERON, T. Y. 1993 *Astrophys. J.* **416**, 74.

COMBES, F. 1994 in *The Formation and Evolution of Galaxies* (ed. C. Muñoz-Tuñón & F. Sánchez). V Canary Islands Winter School, p. 317. Cambridge University Press.

DEKEL, A. & SHLOSMAN, I. 1983 in *Internal Kinematics and Dynamics of Galaxies* (ed. E. Athanassoula), p. 187. Reidel.

FICH, M. & BLITZ, L. 1982 in *Kinematics, Dynamics and Structure of the Milky Way* (ed. W. L. H. Shuter). ASSL, vol. 100, p. 151. Reidel.

FLORIDO, E., PRIETO, M., BATTANER, E., MEDIAVILLA, E. & SÁNCHEZ-SAAVEDRA, M. L. 1991a *Astron. Astrophys.* **242**, 301.

FLORIDO, E., BATTANER, E., PRIETO, M., MEDIAVILLA, E. & SÁNCHEZ-SAAVEDRA, M. L. 1991b *Mon. Not. R. Astron. Soc.* **251**, 193.

FREUDENREICH, H. T. *et al.* 1994 *Astrophys. J., Lett.* **429**, L69.

GALLETTA, G. 1991 in *Warped Disks and inclined Rings around Galaxies* (ed. S. Casertano, P. Sackett & F. Briggs). Cambridge University Press.

GUIBERT, J., LEQUEUX, J. & VIALLEFOND, F. 1978 *Astron. Astrophys.* **68**, 1.

HAMABE, M., KODAIRA, K., OKAMURA, S. & TAKASE, B. 1980 *Publ. Astron. Soc. Jpn.* **32**, 197.

HAMABE, M., OKAMURA, S., IYE, M. & NISHIMURA, S. 1981 *Publ. Astron. Soc. Jpn.* **33**, 643.

HENDERSON, A. P., JACKSON, P. D. & KERR, F. J. 1982 *Astrophys. J.* **263**, 116.

HOFNER, P. & SPARKE, L. S. 1994 *Astrophys. J.* **428**, 466.

HUNTER, C. & TOOMRE, A. 1969 *Astron. J.* **155**, 747.

ICHIKAWA, T. & SASAKI, T. 1984 in *Proc. Second Asian-Pacific Regional Meeting on Astronomy* (ed. B. Hidayat & M. W. Feast), p. 182. Tira Pustaka Pub. House. Jakarta.

KEMP, S. N. & MEAUBURN, J. 1993 *Astron. Astrophys.* **274**, 19.

KIM, K. T., KRONBERG, P. P., GIOVANINI, G. & VENTURI, T. 1989 *Nature* **341**, 720.

KRONBERG, P. P. 1994 *Rep. Progress in Physics* **57:4**, 325.

KRUIJEN, K. 1991 *Astrophys. J.* **376**, 467.

LYNDEN-BELL, D. 1965 *Mon. Not. R. Astron. Soc.* **129**, 299.

MÁRQUEZ, I. & DEL OLMO, A. 1991 in *Dynamics of Disc Galaxies* (ed. B. Sundelius), p. 43. Göteborg University.

MAYER-HASSELWONDER, H. A. ET AL. 1982 *Astron. Astrophys.* **105**, 164.

MIYAMOTO, M., YOSHIZAWA, M. & SUZUKI, S. 1988 *Astron. Astrophys.* **194**, 107.

NELSON, A. H. 1988 *Mon. Not. R. Astron. Soc.* **233**, 115.

NICHOLSON, R. A., TAYLOR, K. & BLAND, J. 1991 in *Warped Disks and inclined Rings around Galaxies* (ed. S. Casertano, P. Sackett & F. Briggs). Cambridge University Press.

OSTRIKER, E. C. & BINNEY, J. 1989 *Mon. Not. R. Astron. Soc.* **237**, 798.

REED, B. C. & FITZGERALD, M. P. 1984 *Mon. Not. R. Astron. Soc.* **211**, 235.

ROGSTAD, D. H., LOCKHART, I. A. & WRIGHT, M. C. H. 1974 *Astrophys. J.* **193**, 309.

SADROSKI, T. J. et al. 1993 in *Back to the Galaxy* (ed. F. Verter). Proc. 3rd Annual Maryland Astrophysics Conference. University of Maryland.

SÁNCHEZ-SAAVEDRA, M. L., BATTANER, E. & FLORIDO, E. 1990 *Mon. Not. R. Astron. Soc.* **246**, 458.

SANCISI, S. 1976 *Astron. Astrophys.* **53**, 159.

SASAKI, T. 1987 *Publ. Astron. Soc. Jpn.* **39**, 849.

SPARKE, L. S. 1984 *Astrophys. J.* **280**, 117.

SPARKE, L. S. & CASERTANO, S. 1988 *Mon. Not. R. Astron. Soc.* **234**, 873.

TOOMRE, A. 1983 in *Internal Kinematics and Dynamics of Galaxies* (ed. E. Athanassoula), p. 177. Reidel.

VALLEE, J. P. 1990 *Astron. J.* **99**, 459.

VAN DER KRUIT, P. C. & SEARLE, L. 1981 *Astron. Astrophys.* **95**, 105.

WATSON, A. M. & PERRY, J. J. 1991 *Mon. Not. R. Astron. Soc.* **248**, 58.

WOUTERLOOT, J. G., BRAND, J., BURTON, W. B. & KWEE, K. K. 1990 *Astron. Astrophys.* **230**, 21.

ZWEIBEL, E. G. 1988 *Astrophys. J., Lett.* **329**, L1.

Carney: I would like to make certain I understand one of your predictions. If magnetic fields produce the warp, gas will be confined to the warp but stars will not and so will diffuse away. OB stars may be found in the warp, but old stars will not. Is that correct?

Battaner: Yes. The prediction of our model is that the warp defined by the old stars would be much smaller than the warp defined by the OB stars. A weak old-star warp is to be expected, however, because the stellar warp is destroyed in characteristic times of 2 Gyr.

Franco: I was surprised to hear that the intergalactic B field is as large as 3μG. This would imply a very large intergalactic pressure. Could you please comment on this?

Battaner: This is a very large intergalactic magnetic pressure, indeed. The presence of a 3 μG magnetic field seems to be independent of the density of the intergalactic density. The magnetic pressure would even be dominant in some systems.

Palouš: Galactic spins $\vec{\Omega}$ are random. How can we get the oriented warps at 25 Mpc scale if they are due to relation of random $\vec{\Omega}$ to intergalactic $\vec{B}$?.

Battaner: But in Battaner *et al.* (1991) it was mathematically demonstrated that, with the statistical treatment used, both orientations are independent. Even if spins were perfectly oriented, this would not imply an orientation of the warps at all. On the contrary, a random distribution of spins, but an alignment of warps, was found.

Rhoads: I don't have a clear physical picture of how the intergalactic magnetic field induces a warp. Could you spend 30 seconds explaining the mechanism?. Also, what does the topology of the field lines look like where the galactic and intergalactic fields meet?

Battaner: This explanation is given by Battaner *et al.* (1990) and Binney (1991). The basic idea is that the ionised ISM has a tendency to be spread along magnetic lines as a result of the magnetic tension. The detailed topology of the field lines requires a further development of present models.

Foglizzo: According to the review by Sarrazin (1986) about clusters of galaxies, the coherence lengthscale of the intergalactic magnetic field is comparable to the size of the galaxies, which is in strong contradiction with the 25 Mpc required for your mechanism. Could you comment on the observational constraints concerning this coherence lengthscale? Moreover, from a theoretical point of view, could you comment as the possible origin of a magnetic field ordered on such a large scale?

Battaner: The prediction of a coherence lengthscale comparable to the size of galaxies was developed to explain a magnetic field which was not then measured. But other early models predict higher scales of several Mpc (e.g. Zweibel 1988). This coherence scale is not observable at all by our method, as we need sizes large enough to contain several warped galaxies. But the coherence length spectrum seems to have a (another?) maximum around 25 Mpc. There is no observational constraint. The origin of this ordered field is certainly an important problem, beyond the objective of this discussion, which we present as a possible observational fact.

García Gómez: How can the magnetic model of warp formation explain the warps in the stellar component of galaxies like the "integral sign" galaxy.

Battaner: I emphasised that the warp should be moderate. This warp is not moderate and there is no counterwarp. Probably, a tidal interaction could better explain this anomalous warp, as discussed by Márquez & del Olmo (1991).

Bystrova: If an external magnetic field produces the warps of the galactic gaseous disk, what kind of deformation from this field is it possible to expect in the gas of the Magellanic Stream?

Battaner: This is a very important question indeed. I don't know the answer and I think that this point should be investigated in detail.

Vertical Deviations of the Midplane of the Galaxy

By SANGEETA MALHOTRA AND JAMES E. RHOADS

Department of Astrophysical Sciences, Princeton University, Princeton, NJ 08544-1001, USA

Besides the integral sign warp in the outer Galaxy, the gas in the Milky Way shows small, but systematic deviations from a flat $z = 0$ plane both in the inner and the outer Galaxy. In the inner Galaxy, the tangent points have no distance ambiguity, so their distances, and hence midplane deviations, can be measured. From the tangent point analysis we find that the molecular and atomic gas layers deviate from the $z = 0$ plane with an amplitude of $\simeq 50$ pc. Whether these deviations are due to a small, smooth inner warp or are similar to the m=10 mode corrugations found in the outer Galaxy (Kulkarni, Blitz & Heiles, 1982) can be checked by looking at the two-dimensional (in Galactic radius and azimuthal angle) structure of the z deviations. For the inner Galaxy, distance ambiguity at points other than the tangent points makes the interpretation difficult, but these hypotheses can be checked in a limited way. Magnetic instabilities can cause vertical deviations of the gas, but if stars share the same deviations the origin has to be gravitational.

Disks of galaxies are flattened systems, but they do not sit flat. Most galaxies show dramatic warping of the disks in the outer parts. The theory of outer galaxy warps is reviewed by Binney (1992) and the observations summarised by Briggs (1990). In the Milky Way, where our instruments achieve high spatial resolution, a host of smaller nonplanar features—corrugations, scalloping, and perhaps a tilted bar—have been found. Evidence for the outer galaxy warp now exists for stars as well as for gas and dust. We summarise some of the reported features in Table 1.

The 3D spatial structure of nonplanar features is difficult to determine unless we use a tracer whose distance can be determined accurately. For gas interior to solar circle, a fair picture can be constructed for tangent points, where the distance is known relatively accurately (Mihalas & Binney 1981). Figure 1 shows that the midplane deviations Z_0 of the atomic and molecular gas are fairly similar (Malhotra 1994a,b). Thus we know the vertical displacement of the gas along the tangent point locus (shown in Figure 2). If, however, we want a two-dimensional picture of the midplane deviations, we have to make the more drastic assumption that the gas motion is axisymmetric. Also, along any line of sight the gas equidistant on the two sides of the tangent points has the same velocity and thus appears superimposed. We assume that since the linear resolution of the telescope is poorer at the far side, emission from the far side affects only a few points in the b-profile of the gas. The mean b estimated from the gas profile therefore represents the midplane of the near side of the gas. This is also the more conservative estimate. Extreme caution is required when comparing any resulting map with other non-axisymmetric features, however, since non-axisymmetric motions—such as streaming in spiral arms—will distort the kinematic distances.

The results of this mapping are shown in Figure 2. The effective spatial resolution along the line of sight is limited by the velocity dispersion σ_V of HI gas, which spreads the emission from a physical point over a range of apparent kinematic distance $\delta d_\odot$. For a flat rotation curve with circular speed Θ_0

$$\delta d_\odot = \left(\frac{dV_{los}}{dd_\odot}\right)^{-1} \sigma_V = \frac{\Theta_0}{R_0} \frac{\left[1 + (d_\odot/R_0)^2 - 2(d_\odot/R_0)\cos(\ell)\right]^{3/2}}{\sin(\ell)\left[\cos(\ell) - d_\odot/R_0\right]} \sigma_V$$

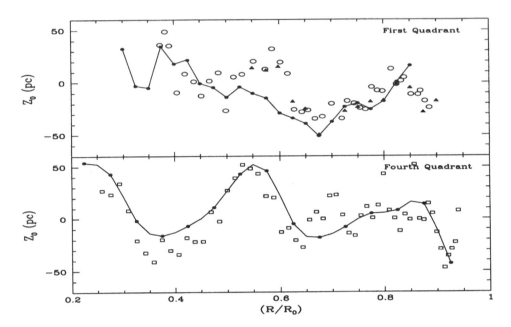

FIGURE 1. The inner Galaxy midplane elevation at the tangent points. (a) First quadrant $(0° < \ell < 90°)$; open circles: Weaver & Williams HI survey, triangles: Bania & Lockman HI survey, connected points: Knapp, Stark, & Wilson CO survey. (b) Fourth quadrant $(270° < \ell < 360°)$. open squares: Kerr *et al.* HI Parkes survey, connected lines: Bronfman CO survey.

For the assumed sun-galactic center distance $R_0 = 8.5\,\mathrm{kpc}$, $\Theta_0 = 220\,\mathrm{km\,s^{-1}}$, $\sigma_V = 9\,\mathrm{km\,s^{-1}}$, we obtain $\delta d_\odot \sim 2\,\mathrm{kpc}$ for most mapped area in Figure 2. The resolution perpendicular to the line of sight is limited only by the sampling of the survey and by the resolution of our contouring algorithm.

If stellar tracers are to yield distance information, the luminosity function must be known, and some way to constrain the luminosities of individual stars is highly desirable. This is possible with multiband photometry under some circumstances (cf. Rhoads, this volume). Non-uniform extinction is always a problem for star counts in the plane, but can be mitigated by working at infrared wavelengths. Diffuse light can be easier to model than star counts, since it does not require knowledge of the luminosity function (Hammersley *et al.* 1994).

The apparent position of the midplane can deviate due to several effects. The most trivial is an error in the observational determination of the galactic pole, or (not quite equivalently) a difference between the HI plane used to define the pole and the plane of the tracer used (e.g. stars). This will appear as a sine wave displacement of the apparent midplane from $b = 0°$, and seems to be present in stellar data. Hammersley *et al.* (1994) have quantified the effect for the *DIRBE*, *IRAS* and *TMGS* data sets. Another likely cause of midplane deviations is the Sun's location out of the galactic midplane; Carney & Seitzer (1993) and Hammersley *et al.* (1994) consider this in detail.

Finally, there are true non-planar features, such as the large scale warp, smaller scale corrugations of the disk—and, perhaps, a tilted bar. The issue of the bar is somewhat confused, as the gaseous bar is tilted with respect to the galactic plane (albeit with

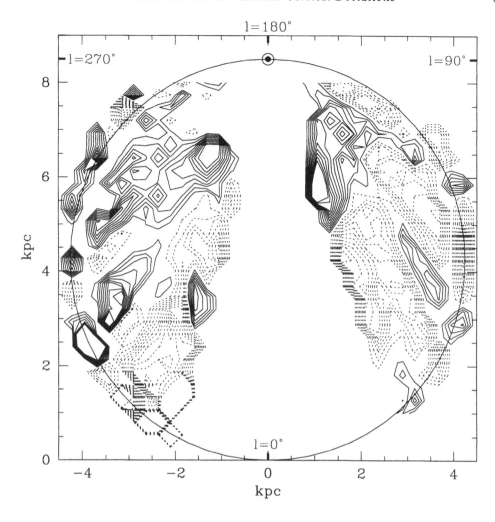

FIGURE 2. Contour map of the midplane elevation z_0 of HI gas. The solid contours correspond to $z_0 = (5, 10, 15, \dots)$ pc, and dotted contours to $z_0 = (-5, -10, -15, \dots)$ pc. The Sun is marked at $(0, 8.5)$, the galactic centre is at $(0, 0)$, and a circle marks the loci of the tangent points. The gap in the contours just inside the tangent point circle is an artifact of our mapping algorithm. The elongation of features along lines of sight is because the no-nzero velocity dispersion of gas at any physical location smears its emission over a range of apparent kinematic distances; this is the same "fingers of God" effect familiar from large-scale redshift surveys.

an uncertain inclination angle), while the stellar bar shown by the *DIRBE* data is not (Freudenreich *et al.* 1994).

A strong constraint on the relevant physical mechanism is whether or not old stars participate in the corrugations. The vertical corrugations in the gas with ~ 30 pc amplitudes and $\sim$ kpc length scales can be explained by magnetic instabilities (cf. Hanawa, this volume). If stars older than a few vertical oscillation periods (age $> 10^7$ yr) share the same midplane deviations, then these corrugations are not due to the Parker instability, since the magnetic fields affect only gas. There have been indications that young clusters (Alfaro *et al.* 1991) and HII regions (Lockman 1977) do, but these may simply reflect

Region	feature, tracer	amplitude	reference
Inner 2 kpc	HI gas on tilted barlike orbits	22°	Liszt & Burton 1980
Inner galaxy	Pop I objects	~ 20 pc	Lockman 1977, 1979
Tangent points	HI gas	~ 30 pc	Quiroga 1974
Tangent points	HI & molecular gas	~ 30 pc	Malhotra 1994a,b
Inner galaxy	HI gas	~ 30 pc	Present paper
Stellar disk	*IRAS* stars	$\sim 0.3°$	Djorgovski & Sosin 1989
Stellar disk	*COBE* diffuse NIR light	$\sim 1°$	Freudenreich *et al.* 1994
Dust disk	*COBE* diffuse FIR light	$\sim 2°$	Freudenreich *et al.* 1994
Local stellar disk	*COBE, IRAS*	$\sim 0.4°$	Hammersley *et al.* 1994
	& *TMGS* (2μ star counts)	~ 20 pc	
Local stellar disk	Open clusters	~ 100 pc	Alfaro *et al.* 1991
Outer galaxy	HI gas ("Classical warp")	~ 1 kpc	Kerr 1957; Burke 1957
Outer galaxy	Hsc i gas $m \sim 10$ scalloping	~ 0.5 kpc?	Kulkarni *et al.* 1982
Outer galaxy	Star counts	–	Carney & Seitzer 1993

TABLE 1. Observed non-coplanar features in the Galaxy

stellar birthplaces in a corrugated molecular gas layer. We are searching for old stellar counterparts to features in that map (cf. Rhoads, this volume).

J.E.R.'s work is supported by NSF grant AST91–17388 and NASA grant NAG5–2693. S.M.'s research was supported by NSF grant AST89–21700 to Princeton University. We thank the organisers of the conference and the Andalusian Regional Government for travel support for this meeting.

REFERENCES

ALFARO, E. J., CABRERA-CAÑO, J. & DELGADO, A. J. 1991 *Astrophys. J.* **378**, 106.

BINNEY, J. 1992 *Annu. Rev. Astron. Astrophys.* **30**, 51.

BRIGGS, F. H. 1990 *Astrophys. J.* **352**, 15.

BURKE, B. F. 1957 *Astron. J.* **62**, 90.

CARNEY, B. W. & SEITZER, P. 1993 *Astron. J.* **105**, 2127.

DJORGOVSKI, S. & SOSIN, C. 1989 *Astrophys. J., Lett.* **341**, L13.

FREUDENREICH, H. T., ET AL. 1994 *Astrophys. J., Lett.* **429**, L69.

HAMMERSLEY, P. L., ET AL. 1994 *Mon. Not. R. Astron. Soc.* submitted.

KERR, F. J. 1957 *Astron. J.* **62**, 93.

KULKARNI, S. R., HEILES, C. & BLITZ, L. 1982 *Astrophys. J., Lett.* **259**, L63.

LOCKMAN, F. J. 1977 *Astron. J.* **82**, 408.

MALHOTRA, S. 1994a *Astrophys. J.* **433**, 687.

MALHOTRA, S. 1994b *Astrophys. J.* submitted.

Near-Infrared Star Counts as a Probe of Asymmetries in the Galaxy's Disk

By JAMES E. RHOADS

Department of Astrophysical Sciences, Princeton University, Princeton, NJ 08544-1001, U.S.A.

By using near infrared (NIR) multiband star counts, we may try to determine if diffuse NIR light in our Galaxy traces old stars and hence mass, as has recently been suggested for other galaxies. Moreover, we can exploit the relatively low extinction in the NIR to observe giant stars at distances $\sim 17\,\text{kpc}$. We can estimate stellar distances crudely, allowing us to study the three-dimensional structure of the disk.

1. Introduction and motivation

Multiband stellar photometry can be a useful probe of galactic structure. For studies of the thin-disk component of the Milky Way, obscuration by dust is a serious difficulty, and it is best handled by working at near infrared (NIR) wavelengths. I have recently started an observing programme to study the distribution of dynamically old giant stars in the disk, looking for (i) large amplitude spiral arms and (ii) small amplitude corrugations.

The amplitude of the density contrast in spiral arms is an open question with far-reaching implications for the dynamics and structure of disk galaxies. While theoretical studies usually assume that spiral density waves are weak and linear, recent observations in red and near-IR light (Schweizer 1976; Elmegreen & Elmegreen 1984; Rix & Rieke 1993; Rix & Zaritsky 1994) suggest that they might be strong and highly non-linear. Rix and Zaritsky (1994) find that arm to interarm surface brightness ratios $S_K \geq 2$ in diffuse K band (2.2μ) light are fairly common. They estimate that at most 18% of the K-band (2.2μ) light in the galaxies they study comes from stars younger than ~ 10 rotation periods, so that the spiral pattern must also involve the dynamically old stars that constitute most of the disk mass.

In our own Galaxy, we can compare observations of diffuse NIR light and of individual stars, thereby eliminating the need for population modelling. Moreover, we can estimate distances to stars to learn about 3D structure.

Recent work by Malhotra (1994a,b) has shown that the midplane of the molecular and atomic gas differs from the nominal (galactic latitude $b = 0°$) plane by up to 50 pc in the inner Galaxy. Stars are also not distributed symmetrically about $b = 0°$. On large scales, this is apparent in the *IRAS* point-source catalogue (Djorgovski & Sosin 1989) and in the DIRBE maps of diffuse NIR light (Freudenreich *et al.* 1994). A simple tilt of the (l, b) coordinate system about the Sun–galactic centre axis can account for much but not all of the observed asymmetry. Ground-based star counts with high sensitivity and spatial resolution will allow detailed comparisons of the gas and stellar warps that are not possible with existing data.

2. Methods

We have some preliminary J (1.25μ), H (1.6μ), and K (2.2μ) band data taken during training time on the 3.5-metre telescope at Apache Point Observatory, and will get more data in November 1994. We used GRIM II, a 256^2 pixel HgCdTe NICMOS 3 array camera built by Mark Hereld and collaborators. The pixel size is 0.48″ and the field is

$(2')^2$. An approximate limiting magnitude of $K = 17$ with $s/n = 5$ can be reached in 100 s of integration. We also obtained optical images of some fields, and hope to take Gunn g- and r-band images as a standard part of our procedure in future.

The number-magnitude counts for stars at IR wavelengths have been modelled by Wainscoat *et al.* (1990). For a 2.2μ magnitude limit $K \sim 16$, we expect ~ 200 stars per $(2')^2$ field at $b = 0°$ and ~ 100 stars at $b = 2°$, of which most are G8–K3 III and F V stars. The K giants are detectable to > 10 kpc, while the dwarfs are visible to much smaller distances. We will use giants as tracers, since (i) they are common and bright; (ii) they can be identified in colour-magnitude diagrams; and (iii) their luminosity function is narrow enough ($\Delta K \approx 2$ mag FWHM, cf. Table 2 of Wainscoat *et al.* 1990) that they can to some extent be used as standard candles.

We can use interstellar reddening to separate giants from dwarfs in a colour-magnitude diagram (CMD). At fixed apparent magnitude, giant stars are further away than cool dwarfs. They will therefore suffer more reddening and will lie in a different region of the CMD (cf. Figure 1). For the method to work well, we want filters in which the typical reddening of a giant star is substantial while the extinction in the redder band is modest. The method is illustrated for a model $K, (V - K)$ CMD in Figure 1; note that $A(K)/E(V - K) \approx 0.125$ (Rieke & Lebofsky 1985).

To ensure uniform sample selection, we must determine the reddening *vs.* distance behaviour for each field independently. This should be possible using a technique akin to "fractioned two-colour diagrams" (cf. del Río & Fenkart 1987). It will not be practical to account for variations in extinction within a single field.

We get distance estimates by constructing a reddening-free magnitude based on known properties of the interstellar extinction law. Rieke & Lebofsky (1985) find $A(H)/E(H - K) = 2.8$, so that the quantity $K_H := 2.8K - 1.8H$ is independent of the extinction and gives distance estimates good to $\sim 50\%$ for spectral types near K0III. Variations in $A_H/E(H - K)$ and curvature of the reddening "vector" would introduce systematic errors in the distances. However, such errors should be small in the near IR, since the reddening law there does not vary much among different lines of sight (Mathis 1990) and the optical depth is modest.

3. Applications to galactic structure

The first question that we will address is whether K-giant counts are distributed like diffuse K-band light. We will do this by comparing CMDs for fields within and outside the prominent diffuse light feature at $l = 80°$, for which $2F_\nu(l = 80°)/[F_\nu(l = 75°) + F_\nu(l = 85°)] \approx 2$ in an extinction-corrected 3.5μ DIRBE map. Crudely, if there are αN giant stars per field at $l = 80°$ and only N at $l = 75°$, we will detect the difference in counts with significance $\frac{(\alpha-1)N}{\sqrt{\alpha N + N}} = \frac{(\alpha-1)}{\sqrt{\alpha+1}}\sqrt{N}$. If the counts trace the diffuse light ($\alpha \approx 2$), we expect a 5σ detection with only one midplane field each at $l = 80°$ and $l = 75°$.

To detect a spiral arm approximately perpendicular to the line of sight, we must use distance information, since its signature in total counts or diffuse light will be relatively weak. For a 50% overdensity and a favorable geometry (with the arm extending from 4 kpc to 6 kpc heliocentric distance), a calculation similar to the above shows that ~ 7 fields would be required for a 3σ detection of the overdensity. Recent optical CMDs of disk stars by Paczyński *et al.* (1994) and by Kałużny (1994) show distinctive main sequences that may be due to spiral arms.

Finally, we consider detection of small corrugations. Here we choose two longitudes ($l = 57°, l = 63°$) where the gas midplane deviations have opposite signs, and at each

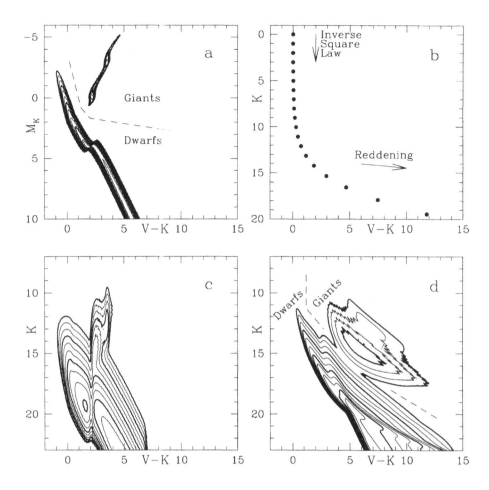

FIGURE 1. Model stellar densities in K, $(V-K)$ colour-magnitude diagrams illustrate our sample selection procedure. Contours are logarithmically spaced by factors of $10^{1/3}$; every third contour is emphasised. (a) Contours are for unreddened stars at $d = 10$ pc. (b) The reddening vector for $A(V) = 0.5$ mag is shown. The solid circles mark the location of an A0V star at distance moduli $0, 1, 2, \ldots$, assuming extinction $A_V = 0.33$ mag/kpc. (c) CMD for $l = 60°$, $b = 0°$ assuming no extinction. The lowest contour level is 1 star$/\left[(2')^2 (K\,\mathrm{mag})(V - K\,\mathrm{mag})\right]$. The noisy contours are due to finite numerical resolution. (d) As (c), but assuming $A_V = 0.33$ mag/kpc.

we take data both above and below $b = 0$. Carney & Seitzer (1993) have recently found evidence for the outer Galaxy warp in optical star counts using a similar technique. To detect a 30 pc warp in a hot stellar population with scale height $\sigma_z = 250$ pc at the 3σ level, we require $\sigma_z/\sqrt{N} = 30\,\mathrm{pc}/3$ so that $N > 600$. The actual requirements will be greater, because 50% distance errors imply 50% errors in z and because the characteristic length scale of the corrugations is relatively small.

This work will be complementary to other ongoing efforts that achieve much greater sky coverage at lower resolution and sensitivity. The Two-Micron Galactic Survey (TMGS, cf. Garzón *et al.*, this volume) in combination with scanned Schmidt-plate data should suffice to determine the contribution of rare, extremely luminous stars to the diffuse NIR light. The DENIS and 2MASS surveys, once completed, will allow a serious attempt to map the global disk structure as traced by NIR-bright stars. Studies like the one described here will then be valuable for follow-up work, since the data we will take is some 3 mag deeper at ~ 5 times the spatial resolution of DENIS and 2MASS.

I thank my advisor David Spergel, the Princeton area observing community, and Bruce Carney for discussion and advice. Also, I thank Richard Wainscoat for his model of the IR sky. Financial support for this work is provided by NSF grant AST91–17388 and by NASA grant NAG5–2693.

REFERENCES

CARNEY, B. W. & SEITZER, P. 1993 *Astron. J.* **105**, 2127.

DEL RÍO, G. & FENKART, R. 1987 *Astron. Astrophys., Suppl. Ser.* **68**, 397.

DJORGOVSKI, S. & SOSIN, C. 1989 *Astrophys. J., Lett.* **341**, L13.

ELMEGREEN, D. M. & ELMEGREEN, B. G. 1984 *Astrophys. J., Suppl. Ser.* **54**, 127.

FREUDENREICH, H. T., ET AL. 1994 *Astrophys. J., Lett.* **429**, L69.

KAŁUŻNY, J. 1994 *Acta Astron.* **44**, 247.

MALHOTRA, S. 1994a *Astrophys. J.* **433**, 687.

MALHOTRA, S. 1994b *Astrophys. J.* submitted.

MATHIS, J. S. 1990 *Annu. Rev. Astron. Astrophys.* **28**, 37.

PACZYŃSKI, B., STANEK, K. Z., UDALSKI. A., SZYMAŃSKI, M., KAŁUŻNY, J., KUBIAK, M. & MATEO, M. 1994 *Astron. J.* **107**, 2060.

RIEKE, G. H. & LEBOFSKY, M. J. 1985 *Astrophys. J.* **288**, 618.

RIX, H.-W. & RIEKE, M. J. 1993 *Astrophys. J.* **418**, 123.

RIX, H.-W. & ZARITSKY, D. X. 1994 *Astrophys. J.* submitted.

SCHWEIZER, F. 1976 *Astrophys. J., Suppl. Ser.* **31**, 313.

WAINSCOAT, R. J., COHEN, M., VOLK, K., WALKER, H. J. & SCHWARTZ, D.E. 1992 *Astrophys. J., Suppl. Ser.* **83**, 111.

Carney: How do you select your fields?

Rhoads: We look at the Palomar Sky Survey plates and choose areas of low and, more importantly, uniform extinction.

Possible Warp Tendency in the Magellanic Stream

By NATALIJA V. BYSTROVA

St. Petersburg Branch of the Special Astrophysical Observatory, Russian Academy of Sciences, 196140 St. Petersburg, Pulkovo, Russia

Just before IAU Symposium No. 108 *The Structure and Evolution of the Magellanic System* (Tübingen, Germany, September 1983), the polar-ring galaxies were detected. Neutral hydrogen observations of them (Schechter *et al.* 1984) revealed the HI density in the rings to be near that in the solar neighbourhood of our Galaxy. If the Magellanic System has some resemblance to the polar rings in other galaxies then it seemed to be worthwhile to look for some extended areas of neutral hydrogen emission around the Magellanic System's great circle.

In a paper presented during Symposium No. 108 (Bystrova, 1984) I stated that such large emission fields were possibly found at rather negative LSR velocities. This conclusion was arrived at following analysis of the data from the "Pulkovo Sky Survey in the Interstellar HI Radio Line" (Bystrova & Rachimov 1977) and from the atlas "Contour Maps to the Pulkovo Sky Survey..." (Bystrova 1980), where the two components of HI emission were given. These emission fields were found on the maps for the "structureless" component of the HI gas emission. All the maps from the galactic coordinate system had previously been regenerated into special Magellanic coordinates to avoid the distorsions in the Magellanic system images.

For the control of the existence of the above-mentioned areas of neutral hydrogen emission the RATAN–600 telescope was used (fan beam 2.5 arcmin×2.5 deg; 40 channels cover the velocity range of about 250 km s^{-1}; the bandwidth and spacing are 30 kHz).

This further series of observations was intended to be carried out as six- or seven-hour drift scans north from declination −40 degrees in five-degree intervals, and also for several special declinations. We had intended to take certain precautions during our observations in order to reveal the reality of low signals on the records, but not all of them could be carried out because the telescope was undergoing reconstruction.

On the RATAN–600 records across the Magellanic Stream zone, together with the details of the "structural" component of neutral hydrogen emission, the broad signals with halfwidths of more than two hours and amplitude of several K are also seen at least in five or six channels. This results in velocity dispersion for the broad emission areas of 30–40 km s^{-1}.

As can also be seen on the records, these broad signals for more negative velocities displace themselves in the RA direction, as does the galactic emission forming the N-warp of the neutral hydrogen gas layer. However, the displacement in the case of the Magellanic Stream is several times greater than is seen in the Galaxy. For example, at declination +22 degrees, the displacement is about 20 minutes, and at −20 degrees these Magellanic Stream observations give about one hour.

The discussion of the RATAN–600 results is currently under way.

REFERENCES

BYSTROVA, N. V. 1980 *Nauka*, Leningrad.

BYSTROVA, N. V. 1984 in *Structure and Evolution of the Magellanic Clouds* (ed S. van den Bergh & K. de Boer). IAU Symposium 108, p. 139. Reidel.

BYSTROVA, N. V., & RACHIMOV, I. A. 1977 *Nauka*, Leningrad.

SCHECHTER, P. L., SANCISI, R., VAN WOERDEN, H. & LYNDS, C. R. 1984 *Mon. Not. R. Astron. Soc.* **208**, 111.

Feedback and Regulation of the Gaseous Disk

By JOSÉ FRANCO[1], A. SANTILLÁN[1,2]
AND M. MARTOS[1]

[1]Instituto de Astronomía-UNAM, Apdo. Postal 70-264, 04510 México D. F., Mexico

[2]Dirección General de Servicios de Cómputo Académico-UNAM, Apdo. Postal 20-059, 04510 México D. F., Mexico

The main features of feedback processes associated with internal perturbations and the resulting regulation of the star-formation process are discussed. The interaction between young massive stars and their surrounding gas is probably responsible for both stimulating and shutting off the star-formation process at different scales. Both radiation and mechanical energy can play a role in the star-formation process, and a simple feedback mechanism with gravitational instabilities, assuming equilibrium between the stellar energy input and dissipation in a non-magnetic turbulent flow, is described. The presence of a magnetic field can modify the operation of these mechanisms, and the response of a magnetic disk is briefly discussed.

1. Introduction

Galaxies are open systems and their structure is continuously affected by a multitude of internal and external agents. The relative importance of the perturbations created by these agents depends on whether the host galaxy is evolving through a quiet stage in relative isolation or undergoing an interaction with a surrounding object (i.e. a nearby galaxy, an intergalactic cloud, intercluster gas, etc.). In the case of gaseous galaxies, the gas response to the resulting perturbations is controlled by the system gravity, the distribution of angular momentum, the strength and orientation of the B field, the gas and star velocity dispersions, and the abundances of heavy elements (Elmegreen 1992a, 1992b; Franco 1992). For interacting systems, the resulting effects can range from mild density and velocity perturbations to the disruption of one of the galaxies (Combes & Athanassoula 1993). Some of these interactions, particularly mergers, can collect large gas masses and drive strong bursts of star formation (see Mirabel 1992) or, as recently displayed by *HST* images of PKS 2349, could power the activity displayed by QSOs. In the case of isolated systems, on the other hand, internal torques and the energy injected from stars can produce a variety of hydrodynamical effects including radial gas flows (Habe & Wada 1993), galactic fountains (Shapiro & Field 1976), and even galactic winds (Heckman *et al.* 1990; Habe 1994). In both types of systems, isolated or interacting, the more energetic pertubations can lead to disruption, and the structure of a gaseous disk should be defined by its ability to "digest" the local energy injection rate. Thus, there should be a threshold value to the energy input above which the gaseous disk cannot maintain its integrity.

For the case of the Milky Way, a remarkable piece of information is that the star-formation rate in the solar neighbourhood has been roughly constant over the whole history of the disk (Miller & Scalo 1979; see recent review by Kennicutt 1992). In a closely related property, using the criterion for the local stability of rotating disks, the combination of the surface mass density, rotation, and stellar velocity dispersions in the solar neighbourhood indicates that the disk is marginally stable to axisymmetric perturbations (Binney & Tremaine 1987). The existence of stable and long-lived disks implies that the time-averaged energy input has been below the disruption value, and a logical

97

conclusion is that the disk of our Galaxy has feedback control mechanisms regulating its stability and the star-formation rate (Franco & Shore 1984; Shore & Ferrini 1994). Under these circumstances, the disk evolution can be followed by assuming that the overall gas structure is in dynamical equilibrium. Here we discuss the main features of feedback processes associated with internal perturbations and the resulting star-formation rates. The creation of massive gas structures and the interaction between gas and stars are the key ingredients of the process and there is a long list of recent conference proceedings discussing these and related topics (e.g. Combes & Casoli 1991; Falgarone *et al.* 1991; Ferrini *et al.* 1991; Janes 1991; Lada & Kylafis 1991; Leitherer *et al.* 1991; Tenorio-Tagle *et al.* 1992; Palouš *et al.* 1992; Franco *et al.* 1993, 1994a; Tenorio-Tagle 1994).

2. Feedback, mass exchange and star formation

The transformation of gas into stellar clusters requires, at some stage of the process, the agglomeration and compression of diffuse interstellar clouds. This can be achieved via shock fronts, thermal instabilities, magnetic instabilities, and gravitational instabilities.

One of the most obvious perturbations in a star-forming galaxy is the one due to the interaction between young stars and their surrounding gas. A quasi-steady population of star-driven bubbles can maintain the structure of the disk (Cox & Smith 1974; McKee & Ostriker 1977; Franco & Shore 1984; Dopita 1985), and the disruptive effects of the stellar energy input can illustrate how a feedback control operates.

2.1. *Stars in action*

Young massive stars stir and heat the gas in their vicinity and their collective action (via HII-region expansion) can destroy star-forming clouds and are probably responsible for both stimulating and shutting off the star-formation process (Whitworth 1979; Tenorio-Tagle 1982; Elmegreen 1983; Larson 1988, 1992; Franco *et al.* 1994c). They also create large expanding interstellar bubbles and generate the hottest gas phases (see reviews by McCray & Kafatos 1987; Tenorio-Tagle & Bodenheimer 1988; Tomisaka 1994).

When the expanding bubbles are able to break out of the disk, the gas of the hot interior escapes and the process pumps mass and energy into the galactic halo (Tenorio-Tagle & Bodenheimer 1988; Tomisaka 1994). If the thermal velocity of the hot gas remains below the escape velocity of the host galaxy, the gas cools down within the halo and simply circulates in a galactic fountain (Shapiro & Field 1976; Chevalier & Oegerle 1979; Tomisaka & Ikeuchi 1986; Ferrara & Einaudi 1992). Otherwise, when the energy injection rate into the halo reaches a value above the cooling rate, the outflow can overcome the gravitational potential and a galactic wind is created (Chevalier & Oegerle 1979; Heckman *et al.* 1990; see review by Habe 1994). Obviously, depending on the star-formation rate, the expanding bubbles can create fountains or winds at galactic scales and the associated disk heating and evaporation has a strong impact on the star-formation rate. The hot gas is unable to form stars and, moreover, any ongoing star forming activity is shut down by the density decrease. Thus, the mechanical energy injected by young stars provides a feedback control to both the star-formation process and the mass exchange between the disk and the halo.

Another interesting issue related to the evolution of superbubbles is the effect produced by the disk rotation curve. The growth along the z axis is sensitive to the strength and geometry of the general B field, and to the mass distribution and the gravitational force along the z axis. The lateral expansion, on the other hand, is affected by the general gravitational field via the rotation curve. The shear due to differential rotation produces large-scale deformations at late evolutionary times, distorting the shell into an oval shape

(Tenorio-Tagle & Palouš 1987; Silich *et al.* 1994). The epicyclic motion of the perturbed gas drives a gas flow towards the tips of the distorted shell, and large gas clouds can be created at these locations (e.g. Palouš *et al.* 1990, 1994). Eventually these clouds will form new stellar clusters; the cycle can be repeated over and over. The model evolves towards equilibrium on relatively short timescales and predicts star-formation rates similar to those observed in "normal" spirals. The end result is that, as long as the gas reservoir is not exhausted, the star-formation process maintains itself at a moderate rate: the star-formation rate is self-sustained and stable. This type of model belongs to a general class of self-propagated schemes termed "percolation" models (Seiden & Schulman 1990; Shore 1993). In contrast with earlier self-propagated models (e.g. Mueller & Arnett 1976; Statler *et al.* 1983), which are called stochastic self-propagating star-formation models, the process with sheared bubbles is not stochastic but deterministic (Palouš *et al.* 1994).

The radiation of massive stars also plays an interesting role at different times and scales. For early stages of galaxy evolution, Cox (1983, 1985) analysed the feedback control provided by the radiation field of young massive stars and pointed out that the large star-formation rates expected during galaxy formation can be controlled by photoionisation and recombination: the ionised gas is unable to form stars but recombination provides fresh material for further star formation. The star-formation process in this case, then, is limited by the rate allowed by the stellar radiation field. This mechanism cannot operate at the present time but a related scheme, associated with thermal instabilities and the general radiation field within the disk, may be now operative (Parravano 1989; Parravano & Mantilla 1991). For this case, the feedback between the radiation field and gas condensation can control the formation of star-forming clouds at large scales and, as in the case of mechanical regulation, can also maintain a moderate star-formation rate.

The efficiency of the star-forming process, which is less than 10% in average cloud complexes, is also controlled by photoionisation (Elmegreen 1983; Larson 1988, 1992; Franco *et al.* 1994b). The most efficient mechanism for cloud destruction is due to the expansion of blister H<small>II</small> regions. The expansion evaporates the mass available for further star formation, placing a severe constraint on the total number of OB stars that can be created (Franco *et al.* 1994c). The maximum number is defined by the number of expanding H<small>II</small> regions required to ionise the whole cloud, and the resulting star-forming efficiency is approximately 5%.

2.2. *Shocks and instabilities*

Aside from a direct coupling between stellar bubbles (or H<small>II</small> regions) and star formation, one can explore indirect ways in which the stellar energy is stored in a reservoir (thermal energy, turbulent motions, Alfvén waves, etc.). In this case, the energy input compensates the energy dissipated over large interstellar regions and maintains the support of the gaseous components against collapse. The main ingredients here are the loss of turbulent energy (say, via cloud collisions) and the stability conditions for the gaseous disk.

The collisions of diffuse clouds, depending on the shock velocities and cloud properties, can either lead to cloud growth or disruption (see reviews by Kwan 1988 and Elmegreen 1991). Strongly radiative shocks or gravitational instabilities can lead to cloud coalescence, but fast quasi-adiabatic shocks are basically disruptive. In the case of a magnetic cloud fluid, the energy and momentum transfer along the field lines increases the dissipation and magnetic collisions tend to agglomerate the gas more efficiently (Zweibel & Josafatsson 1983; Elmegreen 1985, 1988). In either case, however, a fraction of the gas kinetic energy is radiated away during the collision and recent models relating the cloud-collision rate to the gravitational potential of the system predict star-formation rates in agreement with observed values (Dopita & Ryder 1994).

The action of shock fronts not only dissipates energy but, given that perturbations in the velocity field of the gaseous disk generate internal torques, also results in a network of radial gas flows. Numerical simulations of a cloudy fluid with dissipation indicate that the angular momentum redistribution can drive strong gas flows into the centre of the host galaxy (Wada & Habe 1992), and these flows can trigger bursts of star formation at the centre (Lo *et al.* 1984). If the support of the gas (i.e. thermal, turbulent, or magnetic) is lost, a series of different instabilities can appear: thermal, gravitational, and Parker instabilities (see the contributions of Elmegreen and Hanawa in this volume). Any of these instabilities, alone or in combination, can operate under a variety of conditions in a sheared disk (see Elmegreen 1993).

Thermal instabilities are due to the difference in density dependence in the cooling and heating functions, and small density fluctuations can be amplified in cooling interstellar regions (Field 1965; Ferrara & Einaudi 1992). Recent numerical models with supersonic turbulent flows indicate that the compressive component of the turbulent velocity field triggers thermal instabilities, which in turn triggers gravitational instabilities and forms massive clouds (Vázquez *et al.* 1995). In the case of gravitational instabilities, the shear and gas velocity dispersions provide support against collapse, and a minimum disk column density is required to enter into the unstable regime (Safronov 1960; Toomre 1964; Quirk 1972). This minimum condition is usually called the "Toomre" criterion, and spiral waves appear whenever the stellar fluid satisfies this criterion. In the case of a plane parallel disk with a laminar magnetic field providing partial support (Parker 1966), the instabilities develop when the distorted field lines can drive and accumulate large gas masses at the magnetic valleys (see next section). This is called the "Parker" instability and it can gather giant cloud complexes at the spiral arms (Mouschovias *et al.* 1974; Shu 1974).

Obviously, shocks and instabilities can be operative for both stellar and non-stellar pertubations, and similar criteria can be applied to the interaction of high-velocity clouds with the disk of the Galaxy (Tenorio-Tagle 1981; Franco *et al.* 1988; Alfaro *et al.* 1991; Cabrera-Caño *et al.* 1995), or to a variety of other interactions involving cooling shocks. Thus, there should be a series of different channels contributing to the star-formation rate (Shore *et al.* 1987; see reviews by Ferrini 1992 and Shore & Ferrini 1994 and the contribution by Hensler in this volume).

2.3. *An example of feedback regulation: a critical rate*

In the case of stellar perturbations, one can illustrate the feedback regulation with gravitational instabilities by assuming equilibrium between the energy input and dissipation in a "turbulent" flow. On dimensional grounds, the energy dissipation rate per unit surface is

$$\dot{E}_{loss} = \frac{\sigma_g v_t^2}{2t_{diss}} \sim A_1 \rho_g v_t^3, \qquad (2.1)$$

where σ_g is the average surface gas mass density, t_{diss} is the dissipation timescale, ρ_g is the average volume gas mass density, v_t is the average turbulent velocity, and A_1 is a constant. Assuming that a collection of star-driven bubbles merging with their ambient medium regenerate the velocity field, the energy injected into the gas scales as v_t/v_{sf}, where v_{sf} is the bubble velocity at the onset of thin-shell formation (v_{sf} has a very low dependence on density and for this discussion can be regarded as roughly constant; see Franco *et al.* 1994b). For a total energy per star and per unit gas mass ϵ_*, the fraction of kinetic energy injected into the general turbulent flow per event is $\sim \epsilon_* v_t/v_{sf}$. Given that the stellar birthrate simply scales with the star-formation rate, the turbulent energy input rate per unit surface is

$$\dot{E}_{in} \sim A_2 \frac{v_t}{v_{sf}} \epsilon_* \dot{\sigma}_*,$$ (2.2)

where $\dot{\sigma}_*$ is the star-formation rate per unit surface, and A_2 is a constant. The star-formation rate required to maintain a stable disk is then given by the condition $\dot{E}_{loss} = \dot{E}_{in}$, and one obtains

$$\dot{\sigma}_* \sim C \rho_g v_t^2 = C P_t,$$ (2.3)

where $C = (A_1/A_2)(v_{sf}/\epsilon_*)$, and P_t is the gas turbulent pressure. A star-formation rate proportional to the local gas pressure has a straightforward meaning. For low-pressure regions, as is the case in low-mass irregulars or in the external parts of spirals, the system requires only a modest star-formation rate to maintain its stability and the evolution proceeds at a slow pace. At larger interstellar pressures, as in the inner regions of spiral galaxies, the gas is compressed and forced to cool faster, and a higher star-formation rate is required to compensate for the faster cooling.

The midplane rate can be easily connected with some of the basic system variables. Given that the midplane gas pressure is simply the weight of the total gas column density $P(z = 0) = \pi G \sigma_{tot} \sigma_g/2$ (where σ_{tot} is the total surface mass density), the star-formation rate can be written as

$$\dot{\sigma}_* \sim \frac{C\pi G}{2} \sigma_{tot} \sigma_g,$$ (2.4)

Obviously, the product $\sigma_{tot}\sigma_g$ varies with time and location in a given galaxy, and one can differentiate two extreme cases. At the very early stages of galaxy evolution, or at the outer parts of a gaseous disk, this product is roughly equal to σ_g^2, but for an old disk the rate tends to a simple linear dependence on the surface gas mass density. Note that this rate has the same functional form as that discussed by Dopita (1985), because he made the *explicit assumption* that the star-formation rate was proportional to the gas pressure. Here we see that such an assumption is not neccesary. Instead, it is a logical consequence of dynamical equilibrium.

An alternative way of reaching the same results is by relating—assuming hydrostatic equilibrium—the turbulent velocity field to the gas scale height. This approach leads to the same type of rates and has been used in evolutionary models of disk galaxies in hydrostatic equilibrium by Hernández (1994) and Firmani *et al.* (1995). Using the standard accretion disk formulation, they start with a viscous thin disk in rotational equilibrium, and solve its evolution as the internal torques driving radial flows and the stellar energy input compensating for the viscous dissipation. The evolution modifies the total mass distribution and the model provides, aside from adequate star-formation rates, flat rotation curves and a distribution of gas and stars that resembles actual disk galaxies. The results of these and other recent models (Firmani & Tutukov 1992; Dopita & Ryder 1994; Vázquez *et al.* 1995; Palouš *et al.* 1994; Rosen & Bregman 1995) confirm earlier suggestions that the overall structure of a non-magnetic disk could be regulated by the stellar energy injection (e.g. Cox & Smith 1974; McKee & Ostriker 1977; Franco & Shore 1984). Clearly, the presence of a B field can have a profound impact on the response of the system and the next section discusses some of the associated effects.

3. The role of magnetic fields

Despite their relatively low strength in the diffuse gas, magnetic fields can significantly alter traditional views of a variety of processes in the galactic disk. The z distributions of the "intercloud" gas (neutral and ionised), and the cosmic-ray and magnetic pressures, drop slowly compared with the cloud component. The thickness of these distributions is displayed by the galactic synchrotron emission, Faraday rotation of extragalactic sources (see references in Boulares & Cox 1990), 21-centimetre line (Lockman *et al.* 1986), trace ions such as Ti II (Edgar & Savage 1989), and pulsar dispersion measures (Reynolds 1989). These observations reveal scale heights above 1 kpc, and reach 2–3 kpc for ions like C IV, Si IV and N V (Savage & deBoer 1979; Savage & Massa 1987; see the contribution by Danly in this volume). The existence of neutral and ionised material at high latitudes above the plane, together with the observed soft X-ray background, has led to a series of different models. For the purposes of the present discussion we can divide the models in two types: magnetic and non-magnetic.

By far the most popular views of the interstellar medium belong to the first category. As described in the previous section, these are supernova-dominated models in which the shock-heated gas is pervasive, with filling factors ranging from 50 to 70%, and can circulate in a galactic fountain. They can explain the high ions and the soft X-ray background, but the existence of hot gas as a stable phase throughout the thick disk is still under debate (see Cox 1992). The soft X-ray background has a local origin (Snowden *et al.* 1990), and there is no evidence for the hot gas expected outside the local bubble (see McCammon & Sanders 1990 for a review of the subject). In an alternative, "magnetic" view, the incompressibility given to the medium by the field could isolate the evolution of supernova remnants: old remnants would be isolated bubbles of hot gas (Insertis & Rees 1991; Ferriére *et al.* 1991; see Tomisaka 1994). Similarly, in this view, the magnetic field could conceivably quench the activity of the theorised galactic fountains and the ionisation of gas high above the midplane would be due to stellar radiation (photoionisation from O stars can indeed produce the H_α surface brightness perpendicular to the galactic plane, as shown by Miller & Cox 1993). The filling factor of hot gas in this case can be as low as 10–20% (Cox 1992).

Hot halo or not, magnetic fields are certainly present and must have an impact on the overall structure of the gaseous disk. The picture of a B field providing support for material at all scales has been developed over the last decades, and MHD waves should be relevant to most interstellar processes. For instance, the heating of the halo could be due to Alfvén waves (Hartquist 1983), magnetic reconnection from cosmic-ray inflation of the disk (Parker 1990), or microflares (Raymond 1992). Obviously, the temperature of the gas at different heights will be crucial to determine the extent to which the magnetic field is important. In this respect, the relative importance of the magnetic and non-magnetic pressures can play a key role in the stability of the disk, and one has to investigate the response of the system to different perturbations and field strengths.

In fact, we are currently studying the stability of a magnetic disk under a variety of different assumptions and have performed, with the magneto-hydrodynamical code Zeus 3-D developed at NCSA-Illinois by M. Norman and associates, a series of numerical experiments in two dimensions. For instance, for disk parameters adequate to the solar neighbourhood (i.e. using the standard z distributions of gas and gravity (see Martos & Cox 1994) and a parallel magnetic field with a scaleheight of order 1 kpc, an isothermal gaseous layer with $T \sim 10^4$ K in hydrostatic equilibrium is readily Parker unstable (see Hanawa, this volume). In this case, the ratio of magnetic to thermal pressures, $\beta = P_m/P_{th}$, increases with height and a large fraction of the gas layers along the z axis can participate in the Parker instability. In 2-D, the instability forms sheets of compressed gas that extend perpendicularly to the disk (see Lerche 1967 for an analytic proof). The process evolves with a timescale of a few 10^8 yr for a 1 kpc thick disk, and the fastest growing modes are undular (as opposed to the interchange modes, which do not distort the field lines) with wavelengths of about 3 kpc or more. Given that the growth depends on the scaleheight, the onset of the instability with our assumed disk scaleheights has larger time and length scales than those obtained in previous models (see Martos & Cox 1994). An obvious consequence for an isothermal thick disk is that the instability cannot be causally related to cosmic ray leakage from the Galaxy (i.e. the cosmic ray trapping time is of the order 10^7 yr), as has previously been suggested. When the disk temperature increases with height, the ratio β is reduced and the disk can become stable (Martos 1993).

There are several irrealistic assumptions implicit in the cases just described, but these are common to most models of the Parker instability. One is the geometry of the assumed field lines: well ordered and parallel to the galactic plane. This is not sustained by observations, except near the midplane, and a significant vertical component probably exists above the plane (see Boulares & Cox 1990). Similarly, the galactic B field has an important random component, with a cell size of the order of 50 pc (Rand & Kulkarni 1989), and this disorderly character of the field will tend to stabilise the system at large scales. There are also limitations in the code, and the present version is restricted to ideal MHD (without magnetic dissipation or reconnection) of a single fluid. Additionally, the largest growth rates in 3–D correspond to the shortest wavelengths perpendicular to the field lines (the interchange mode), and there are additional ingredients, such as shear and self gravity, making the relevant wavelengths small but finite (Elmegreen 1992a). The timescales involved are only slightly longer than in 2–D (Matsumoto & Shibata 1991; Martos 1993). This behaviour in 3–D and the lack of organisation of the field, indicate that the Parker mechanism could probably be effective to form structure at relatively small scales. With these restrictions in mind, 2–D modelling of the Parker instability could still be useful to study regions with magnetic dominance at a large scale (the actual scale is yet to be clarified until the inclusion of the random component can be accomplished). It is worth noticing that the observed spacing of the associations along spiral arms, at about 1–3 kpc (see Elmegreen, this volume), matches the prediction for the wavelengths of the fastest-growing undular modes in 2–D (as first suggested by Mouschovias *et al.* 1974) for both the previous thin disk models and our thick disk models.

Given the variety of possible perturbations described in the previous sections, we performed 2-D simulations with small and large amplitude perturbations in the magnetically dominated disk model. As an example, we show in Figure 1 the results of the impact of a high-velocity cloud with an isothermal disk (clearly, similar effects should be expected for the case of collisions of globular clusters or small irregulars with a disk). The cloud is completely shocked at high latitudes but drives a strong perturbation in all directions. The perturbation excites a spectrum of different wavelengths and, eventually, the Parker mechanism is activated. Waves propagating along decreasing density gradients are accelerated, and high-velocity fields are created away from midplane. This indicates that even in the absence of classical fountains, MHD waves could maintain the observed gas velocities at high latitudes (see Danly, this volume). Also, the rarefied gas located at large z values feels a stronger gravitational pull and has a different dynamical response than the gas near the plane. As shown in Figure 1, the resulting dynamics are fairly complex and have a superposition of different modes of oscillation. The inner disk is considerably distorted at early times but, due to its large inertia and low MHD wave speeds, decouples its oscillation from the layers at high z values. The disk returns to its initial state, with a low amplitude wave propagating away from the site of impact, in timescales of a few 10^8 yr. At about these times, when the waves moving in the high-z layers can trigger the instability, the gas falls towards the magnetic valleys and a pair of density peaks (above and below the plane) begin to form. These and other experiments

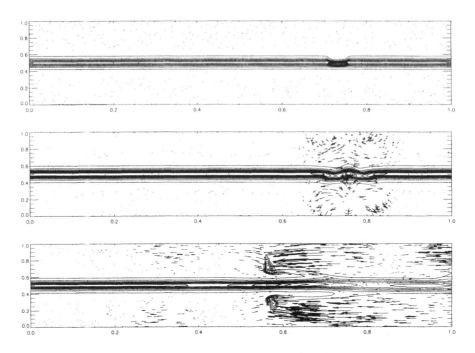

FIGURE 1. Model of the perturbations produced by an infalling cloud in a magnetised gaseous disk. The sequence shows isodensity contours and the velocity fields, indicated by arrows, at three selected times: $t = 0$ yr, 9.5×10^7 yr, and 5.4×10^8 yr. The maximum velocity values are 30, 22, and 50 km s^{-1}, respectively. The horizontal axis has a total length of 12 kpc and the vertical axis, with midplane at the centre, ranges from -1.5 to 1.5 kpc.

currently in progress indicate the strong tendency of high-β disks to release gravitational energy, sending high latitude gas back to the disk.

We are grateful to Mike Norman, Robert Fiedler, and the Laboratory for Computational Astrophysics for their invaluable help in the use of Zeus 3–D. We also acknowledge the questions and suggestions made by Stan Kurtz, Harald Lesch, Steve Shore, and Vladimir Surdin. This work has been partially supported by DGAPA–UNAM through the grant IN105894, and by an R&D grant from Cray Research Inc. The numerical simulations were performed in the CRAY–Y/MP of UNAM's Supercomputer Center.

REFERENCES

ALFARO, E. J., CABRERA-CAÑO, J. & DELGADO, A. J. 1991 *Astrophys. J.* **378**, 106.

BINNEY, J. & TREMAINE, S. 1987 *Galactic Dynamics*. Princeton University Press.

BOULARES, A. & COX, D. P. 1990 *Astrophys. J.* **365**, 544.

CABRERA-CAÑO, J., MORENO, E., FRANCO, J. & ALFARO, E. 1995 *Astrophys. J.* in press.

CHEVALIER, R. A. & OEGERLE, W. 1979 *Astrophys. J.* **227**, 39.

COMBES, F. & ATHANASSOULA, E. (EDS.) 1993 *N-Body Problems and Gravitational Dynamics*. Observatoire de Paris.

COMBES, F. & CASOLI, F. (EDS.) 1991 *Dynamics of Galaxies and Molecular Cloud Distributions*. Kluwer.

COX, D. P. 1983 *Astrophys. J., Lett.* **265**, L61.

Cox, D. P. 1985 *Astrophys. J.* **288**, 465.

Cox, D. P. 1992 in *The Astronomy and Astrophysics Encyclopedia* (ed. S. P. Maran), p. 376. Cambridge University Press.

Cox, D. P. & Smith, B. W. 1974 *Astrophys. J., Lett.* **189**, L105.

Dopita, M. A. 1985 *Astrophys. J., Lett.* **295**, L5.

Dopita, M. A. & Ryder, S. 1994 *Astrophys. J.* **430**, 163.

Edgar, R. & Savage, B. 1989 *Astrophys. J.* **340**, 762.

Elmegreen, B. G. 1983 *Mon. Not. R. Astron. Soc.* **203**, 1011.

Elmegreen, B. G. 1985 *Astrophys. J.* **299**, 196.

Elmegreen, B. G. 1988 *Astrophys. J.* **326**, 616.

Elmegreen, B. G. 1991 in *Physics of Star Formation and Early Stellar Evolution* (ed. N. Kylafis & C. Lada). NATO ASI Series C, vol. 342, p. 35. Kluwer.

Elmegreen, B. G. 1992a *Large Scale Dynamics of the ISM* 20th SAAS-FEE course. Geneva Observatory.

Elmegreen, B. G. 1992b *Star Formation in Stellar Systems* (ed. G. Tenorio-Tagle, M. Prieto & F. Sánchez). III Canary Islands Winter School, p. 381. Cambridge University Press.

Elmegreen, B. G. 1993 in *Star Formation, Galaxies and the Interstellar Medium* (ed. J. Franco, F. Ferrini & G. Tenorio-Tagle). 4th EIPC Workshop, p. 337. Cambridge University Press.

Falgarone, E., Boulanger, F. & Duvert, G. (eds.) 1991 *Fragmentation of Molecular Clouds and Star Formation*. Kluwer.

Ferrara, A. & Einaudi, G. 1992 *Astrophys. J.* **395**, 475.

Ferriére, K. M., Mac Low, M. M. & Zweibel, E. 1991 *Astrophys. J.* **375**, 239.

Ferrini, F. 1992 in *Evolution of Interstellar Matter and Dynamics of Galaxies* (ed. J. Paloŝ , W. B. Burton & P. O. Lindblad), p. 304. Cambridge University Press.

Ferrini, F., Franco, J. & Matteucci, F. (eds.) 1991 *Chemical and Dynamical Evolution of Galaxies*. ETS Editrice, Pisa.

Field, G. B. 1965 *Astrophys. J.* **142**, 531.

Firmani, C., Gallagher III, J. S. & Hernández, X. 1995 in preparation.

Firmani, C. & Tutukov, A. V. 1992 *Astron. Astrophys.* **264**, 37.

Franco, J. 1991 in *Chemical and Dynamical Evolution of Galaxies* (ed. F. Ferrini, J. Franco & F. Matteucci), p. 506. ETS Editrice, Pisa.

Franco, J. 1992 in *Star Formation in Stellar Systems* (ed. G. Tenorio-Tagle, M. Prieto, F. Sánchez). III Canary Islands Winter School, p. 515. Cambridge University Press.

Franco, J., Ferrini, F. & Tenorio-Tagle, G. (eds.) 1993 *Star Formation, Galaxies and the Interstellar Medium*. 4th EIPC Workshop. Cambridge University Press.

Franco, J., Lizano, S., Aguilar, L. & Daltabuit, E. (eds.) 1994a *Numerical Simulations in Astrophysics*. Cambridge University Press.

Franco, J., Miller, W., Arthur, S. J., Tenorio-Tagle, G. & Terlevich, R. 1994b *Astrophys. J.* **435**, 805.

Franco, J. & Shore, S. N. 1984 *Astrophys. J.* **285**, 813.

Franco, J., Shore, S. N. & Tenorio-Tagle, G. 1994c *Astrophys. J.* **436**, 795.

Franco, J., Tenorio-Tagle, G., Bodenheimer, P., Różyczka, M. & Mirabel, I. F. 1988 *Astrophys. J.* **333**, 826.

Habe, A. 1994 in *Numerical Simulations in Astrophysics* (ed. J. Franco, S. Lizano, L. Aguilar & E. Daltabuit), p. 151. Cambridge University Press.

Habe, A. & Wada, K. 1993 in *Star Formation, Galaxies and the Interstellar Medium* (ed. J. Franco, F. Ferrini & G. Tenorio-Tagle). 4th EIPC Workshop, p. 134. Cambridge University Press.

Hartquist, T. W. 1983, *Mon. Not. R. Astron. Soc.* **203**, 117.

Heckman, T. M., Armus, L. & Miley, G. K. 1990 *Astrophys. J., Suppl. Ser.* **74**, 833.

HERNÁNDEZ, X. 1994 B. Sc. thesis (unpublished), UNAM, Mexico.

JANES, K. (ED.) 1991 *The Formation and Evolution of Star Clusters.* ASP Conference Series, vol. 13. ASP.

KENNICUTT, R. C. 1992 in *Star Formation in Stellar Systems* (ed. G. Tenorio-Tagle, M. Prieto & F. Sánchez). III Canary Islands Winter School, p. 191. Cambridge University Press.

KWAN, J. 1988 in *Molecular Clouds in the Milky Way and External Galaxies* (ed. R. Dickman, R. Snell & J. Young), p. 281. Reidel.

LADA, C. & KYLAFIS, N. (EDS.) 1991 *Physics of Star Formation.* Reidel.

LARSON, R. B. 1988 in *Galactic and Extragalactic Star Formation* (ed. R. E. Pudritz & M. Fich), p. 459. Kluwer.

LARSON, R. 1992 in *Star Formation in Stellar Systems* (ed. G. Tenorio-Tagle, M. Prieto & F. Sánchez). III Canary Islands Winter School, p. 125. Cambridge University Press.

LEITHERER, C., WALBORN, N., HECKMAN, T. & NORMAN, C. (EDS.) 1991 *Massive Stars in Starbursts.* Cambridge University Press.

LERCHE, I. 1967 *Astrophys. J.* **149**, 395.

LO, K. Y., BERGE, G. L. & CLAUSSEN, M. J. 1984 *Astrophys. J., Lett.* **282**, L59.

LOCKMAN, F., HOBBS, L. & SHULL, J. 1986 *Astrophys. J.* **301**, 380.

MARTOS, M. A. 1993 PhD thesis, Univ. Wisconsin-Madison.

MARTOS, M. A. & COX, D. P. 1994 in *Numerical Simulations in Astrophysics* (ed. J. Franco, S. Lizano, L. Aguilar & E. Daltabuit). Cambridge University Press.

MATSUMOTO, R. & SHIBATA, K. 1991 in *Numerical Astrophysics in Japan 2* (ed. S. M. Miyama & M. Nagasawa), p. 177. NAO: Mitaka.

MCCAMMON, D. & SANDERS, W. T. 1984 *Astrophys. J.* **287**, 167.

MCCRAY, R. & KAFATOS, M. 1987 *Astrophys. J.* **317**, 190.

MCKEE, C. F. & OSTRIKER, J. P. 1977 *Astrophys. J.* **218**, 148.

MILLER, W. W. & COX, D. P. 1993 *Astrophys. J.* **417**, 579.

MILLER, G. E. & SCALO, J. M. 1979 *Astrophys. J., Suppl. Ser.* **41**, 513.

MIRABEL, F. I. 1992 in *Star Formation in Stellar Systems* (ed. G. Tenorio-Tagle, M. Prieto & F. Sánchez). III Canary Islands Winter School, p. 479. Cambridge University Press.

MOUSCHOVIAS, T., SHU, F. & WOODWARD, P. 1974 *Astron. Astrophys.* **33**, 73.

MUELLER, M. & ARNETT, W. 1976 *Astrophys. J.* **210**, 670.

PALOUŠ, J., BURTON, W. & LINBLAD, P. O. (EDS.) 1992 *Evolution of Interstellar Matter and Dynamics of Galaxies.* Cambridge University Press.

PALOUŠ, J., FRANCO, J. & TENORIO-TAGLE, G. 1990 *Astron. Astrophys.* **227**, 175.

PALOUŠ, J., TENORIO-TAGLE, G. & FRANCO, J. 1994 *Mon. Not. R. Astron. Soc.* **270**, 75.

PARRAVANO, A. 1989 *Astrophys. J.* **347**, 812.

PARRAVANO, A. & MANTILLA, J. 1991 *Astron. Astrophys.* **250**, 70.

PARKER, E. N. 1966 *Astrophys. J.* **145**, 811.

PARKER, E. N. 1990 in *Galactic and Intergalactic Magnetic Fields* (ed. R. Beck, P. P. Krongerb & R. Wielebenski), p. 169. Kluwer.

QUIRK, W. 1972 *Astrophys. J., Lett.* **176**, L9.

RAND, R. J. & KULKARNI, S. R. 1989 *Astrophys. J.* **343**, 760.

RAYMOND, J. 1992 *Astrophys. J.* **384**, 502.

REYNOLDS, R. 1989 *Astrophys. J., Lett.* **339**, L29.

ROSEN, A. & BREGMAN, J. N. 1995 *Astrophys. J.* **440**, 634.

SAFRONOV, V. S. 1960 *Ann. d'Astr.* **23**, 979.

SAVAGE, B. & DEBOER, K. S. 1979 *Astrophys. J., Lett.* **230**, L27.

SAVAGE, B. & MASSA, D. M. 1987 *Astrophys. J.* **314**, 380.

SEIDEN, P. & SCHULMAN, L. 1990 *Adv. Phys.* **39**, 1.

SHAPIRO, P. R. & FIELD, G. B. 1976 *Astrophys. J.* **205**, 762.

SHORE, S. N. 1993 in *Star Formation, Galaxies and the Interstellar Medium* (ed. J. Franco, F. Ferrini & G. Tenorio-Tagle). 4th EIPC Workshop, p. 367. Cambridge University Press.

SHORE, S. N. & FERRINI, F. 1995 *Fundam. Cosmic Phys.* in press.

SHORE, S. N., FERRINI, F. & PALLA, F. 1987 *Astrophys. J.* **316**, 663.

SHU, F. 1974 *Astron. Astrophys.* **33**, 55.

SILICH, S., FRANCO, J., PALOUŠ , J. & TENORIO-TAGLE , G. 1994 *Numerical Simulations in Astrophysics* (ed. J. Franco, S. Lizano, L. Aguilar & E. Daltabuit), p. 193. Cambridge University Press.

SNOWDEN, S. L., COX, D. P., McCAMMON, D. & SANDERS, W. T. 1990 *Astrophys. J.* **354**, 211.

STATLER, T., COMINGS, N. F & SMITH, B. F. 1983 *Astrophys. J.* **270**, 79.

TENORIO-TAGLE, G. 1981, *Astron. Astrophys.* **94**, 338.

TENORIO-TAGLE, G. 1982 in *Regions of Recent Star Formation* (ed. R. Roger & P. Dewdney), p. 1. Reidel.

TENORIO-TAGLE, G. (ED.) 1994 *Violent Star Formation: From QSO's to 30 Doradus.* Cambridge University Press.

TENORIO-TAGLE, G. & BODENHEIMER, P. 1988 *Annu. Rev. Astron. Astrophys.* **26**, 145.

TENORIO-TAGLE , G., BODENHEIMER , P., FRANCO, J. & RÓŻYCZKA , M. N. 1990a *Mon. Not. R. Astron. Soc.* **244**, 563.

TENORIO-TAGLE, G. & PALOUŠ , J. 1987, *Astron. Astrophys.* **186**, 287.

TENORIO-TAGLE, G., PRIETO, M. & SÁNCHEZ, F. (EDS.) 1992 *Star Formation in Stellar Systems.* Cambridge University Press.

TOMISAKA, K., HABE, H. & IKEUCHI, S. 1981 *Astrophys. Space Sci.* **78**, 273.

TOMISAKA, K. & IKEUCHI, S. 1986 *Publ. Astron. Soc. Jpn.* **38**, 697.

TOOMRE, A. 1964, *Astrophys. J.* **139**, 1217.

VÁZQUEZ, E., PASSOT, T. & POUQUET, A. 1995 *Astrophys. J.* in press.

WADA, K. & HABE, A. 1992, *Mon. Not. R. Astron. Soc.* **258**, 82.

WHITWORTH, A. 1979, *Mon. Not. R. Astron. Soc.* **186**, 59.

ZWEIBEL, E. & JOSAFATSSON, K. 1983, *Astrophys. J.* **270**, 511.

The Gaseous Halo–Disk Interaction

By LAURA DANLY

Space Telescope Science Institute, 3700 San Martin Drive, Baltimore, MD 21218, U.S.A.

Multi-wavelength observations of Milky Way halo gas are reviewed in the context of galactic fountain models. While the general characteristics of the galactic fountain theory are supported by the data, several observations suggest that both the gas distribution and physical characteristics are quite structured and patchy, suggesting that, if the galactic fountain is at work, the halo is fed by isolated "events" rather than an on-going venting of hot gas to a Spitzer-type coronal reservoir. One such halo–disk "event" is evident in the second quadrant of galactic longitude where both a preponderance of infalling high- and intermediate- velocity clouds and an enhancement in the soft X-ray emission is seen. However, estimates for the mass and energy involved in the "event" suggest that a relatively recent and significant input of disk supernova energy —on the order of that found in the 30 Dor nebula— would be required for the infalling gas to be the returning flow of a supernova-powered fountain. Observations of external galaxies suggest a variety of gaseous halo characteristics. Several systems also suggest that the gaseous halo–disk interaction is dominated by a few, active sites. However, no clear relationship is apparent between the structure and size of a galaxy's gaseous halo and the star forming characteristics of the underlying disk. A more careful analysis of the relationship between the nature of a galaxy's gaseous halo and its evolutionary state is required, especially if analyses of QSO absorption lines are employed to study galaxies at high redshift.

1. Motivation

There are three primary motivations for reviewing the nature of Milky Way halo gas as part of an investigation of the formation of our Galaxy. The first arises from the need to understand the physical processes that drive the evolution of the Galaxy. The galactic halo probably plays an important role in the evolution of the galactic ISM via the cycling of energy and enriched material through the Galaxy; the disk–halo interaction may play an important role, therefore, in regulating the evolutionary path of the Galaxy. Conversely, if the formation and maintenance of the gaseous halo of the Galaxy is driven by supernovae (as is widely believed), perhaps the star-formation history of the Galaxy is written in the structure of the Galaxy's halo gas. It is essential to understand the dominant processes at work in the disk–halo interaction, and the timescales over which they occur.

Secondly, remnants of the Galaxy's "formation" may still exist in the halo. For example, the infall of the high-velocity clouds (HVCs) may represent an ongoing process of accretion of external material on to the Galaxy.

A third motivation is to aid in the interpretation of QSO absorption lines, which provide one of the best direct means for studying galaxies at high redshift. It is thought that the material causing absorption at high redshift is in the vicinity of galaxies in the early stages of their evolution. A study of material in the vicinity of our own Galaxy, or other local galaxies, provides a means to confirm inferences on the physical conditions in halo gas from the absorption-line data, as well as the interpretation of the nature of this gas in terms of its origin and relationship to the underlying galaxy.

2. The nature and origin of halo gas

There are several excellent reviews on Milky Way halo gas (Savage 1988; Spitzer 1990; Danly 1995; see also Bloemen 1991 and articles therein), to which the reader is referred for more information. In summary, much of the theoretical and observational work of the past two decades has centred on two families of models: (i) "cool" models provide support for halo gas through magnetic-field and/or cosmic-ray pressure, while the ionisation arises from photo-ionisation; (ii) "hot" models employ thermal energy for both the ionisation and support against the gravitational potential. A significant body of work has been developed in recent years on the so-called "galactic fountain" models and their derivatives, in which disk gas is heated through supernovae and escapes to large distances above the plane, where it cools and flows back to the disk (Shapiro & Field 1976; Bregman 1980). The models commonly involve multiple correlated supernovae, where the hot gas is vented to the halo through "chimneys" (Mac Low & McCray 1988; Norman & Ikeuchi 1989). Most models provide for a vigorous fountain with numerous chimneys throughout the disk, the formation of an extended million degree "corona" (Spitzer 1956), and large mass fluxes through the halo (Norman & Ikeuchi 1989; Li & Ikeuchi 1990; Shapiro & Benjamin 1993; McKee 1993; Shull & Slavin 1994).

Several observations in the disk and halo lend support to the galactic fountain theory. Supernova-heated gas with a temperature in excess of a million degrees is indeed found in the disk where it is observed in soft X-ray emission and OVI absorption (Williamson *et al.* 1974; Jenkins & Meloy 1974; York 1974). Vertical structures in the disk HI) called "worms" are thought to mark the walls of chimneys (Heiles 1984, 1993). Evidence for hot gas in the halo has been suggested by observations of highly ionised species such as CIV, SiIV and NV at large z heights (Savage & Massa 1987; Sembach & Savage 1992), and more recently by the soft X-ray shadows seen by *ROSAT* (Burrows & Mendenhall 1991; Snowden *et al.* 1991). High-velocity clouds (HVCs) are seen at high galactic latitudes with predominantly negative velocities (Giovanelli 1980; Hulsbosch & Wakker 1988; see Wakker 1989 and references therein).

Some observations of halo gas, however, are more difficult to explain in the context of the galactic fountain theory. The distribution of the infalling clouds is far from isotropic, with most of it residing in elongated loop-like structures in the second quadrant of the northern hemisphere (Figure 1; see Danly 1992 and Kuntz & Danly 1995). If the infalling clouds are returning fountain flow, their distribution suggests that the venting of hot disk gas to the halo does not occur throughout the disk leading to the development of an extended, isotropic corona, but that hot gas is vented to the halo only in occasional isolated disk–halo "events." Given that observations indicate that interstellar "worms" appear to be numerous, with equal numbers extending to the northern and southern hemispheres, perhaps only a small fraction of worms are actually chimneys with enough energy to break out of the disk and vent hot gas to the halo.

Additional support for the idea that the galaxy is not enveloped in a smooth extended corona comes from the *ROSAT* X-ray shadow data. Observations toward the most transparent part of the galaxy (i.e the lowest HI column-density region, in Ursa Major) by Snowden *et al.* (1994) show that while the gross characteristics of the X-ray distribution are indeed dominated by a shadowing of X-rays by disk gas nearer by, the details reveal significant structure in the brightness of the X-ray emission over regions with identical values for the HI column density. Variations in X-ray brightness of over a factor of two are found despite the smooth, unvarying HI distribution. Furthermore, the value of the X-ray intensity is about a factor of seven lower than would be expected from the corona model required to explain the well-known X-ray shadow toward the Draco cloud.

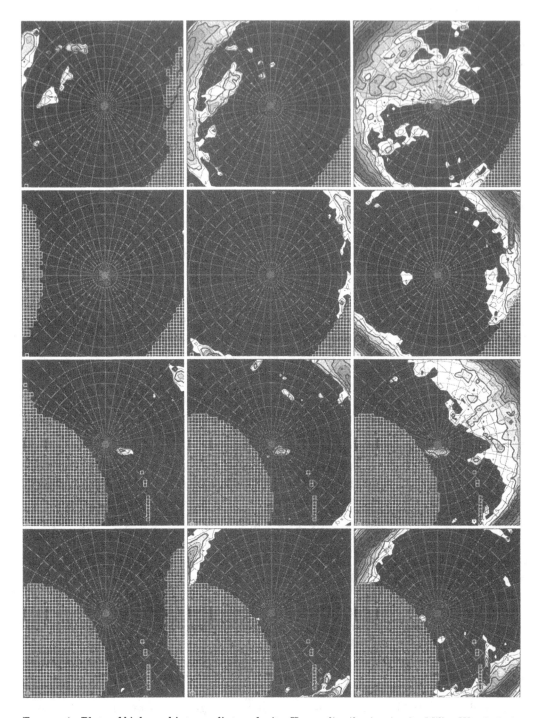

FIGURE 1. Plots of high- and intermediate-velocity H I gas distribution in the Milky Way halo in polar coordinates, with $l = 0$ at the bottom. The first two rows plot the gas distribution towards the NGP, with l increasing clockwise. The bottom two rows plot the gas distribution towards the SGP, with l increasing anticlockwise. From left to right, the velocity ranges for the columns are: $200 < |v| < 150$, $150 < |v| < 100$, $100 < |v| < 50$, with negative velocities for rows 1 and 3, and positive velocities for rows 2 and 4. The contour levels, in cm^2, are: 1.0×10^{19}, 1.8×10^{19}, 3.2×10^{19}, 5.6×10^{19}, 1.0×10^{20}, 1.8×10^{20}, 3.2×10^{20}, 5.6×10^{20}, 1.0×10^{21}, 1.8×10^{21}, 3.2×10^{21}, 5.6×10^{21}.

Finally, most of the X-ray shadow observations have been made towards clouds in the second quadrant of the northern hemisphere—just the region where the infalling clouds are found. (Indeed many of the shadows are taken against the infalling clouds since they are easy to distinguish and determine distances based on their velocities.) Variations in the distribution and temperature of a hot corona should not be surprising, again, where the nature of the halo is strongly influenced by individual disk–halo occurrences.

Another observation that is difficult to understand in the context of a quasi-steady-state fountain model is the asymmetry of the absorption high-ion characteristics toward the poles, and especially the lack of high ions with positive velocities toward the north (Danly 1987). If over-pressured hot gas were being driven from the disk through chimneys, one would expect that both instabilities in the flow and cool material entrained in the flow would give rise to regions which quickly cool from 10^6 degrees to 10^4 degrees, while still in its upward traverse (MacLow *et al.* 1989). These regions should produce observable CIV, SiIV and NV with positive velocities. The observations show that, towards the north pole, no positive velocities are seen in those species, while towards the south pole, the high ions are centred on 0 km s^{-1} and do not show any significant velocity extensions. Again, the structure in the distribution of the high ions' kinematics argue for episodic supply of material to the halo.

Our position in the galaxy is ideal for studying the nature of one of these disk–halo "events." Significant infall of material in the form of high- and intermediate-velocity clouds is found at latitudes as high as 70 degrees—on our back doorstep, so to speak. A detailed study of the physical nature of the infalling material—i.e. its temperature, density, metallicity and dynamics—can help place constraints on theories for its origin as well as on the halo environment in which the gas resides.

3. The local halo–disk "event"

The distribution of the infalling gas in the northern hemisphere as a function of velocity is shown in Figure 2, which is plotted in coordinates that centre on the HVCs and IVCs. Recent work on the distances to this gas shows that the IVCs are located at $z = 1.2 \pm 0.3$ kpc (Danly 1989; Danly & Kuntz 1995). At that distance, the mass in the IV Arch is about 6×10^5 $M_\odot$, with a corresponding energy of more than 3×10^{52} ergs. The lower limit arises because we only have one component of the motion, i.e. the radial velocity.

The dynamics of the HVCs are harder to assess, largely because their distances are largely unknown. The only distance yet determined for a HVC is for Complex M which resides at 3 ± 1.5 kpc above the disk (Danly *et al.* 1993). Lower limits for Complexes C and A are $z > 1.7$ kpc and $z > 2.9$ kpc, respectively (Danly 1989; Danly *et al.* 1995, in preparation). The total mass contained in the infalling gas shown in Figure 2 is over 10^6 solar masses in all, and the energy is over 10^{53} ergs. The energy estimate is comparable to the total energy found in the gas filaments of 30 Dor, the massive star-formation complex at the heart of the Taratula Nebula in the Large Magellanic Cloud (Chu & Kennicut 1994). A simple ballistic estimate of the timescale for the flow of the gas is of order 10^7 yr, although the actual timescale is highly dependent on the nature of the gravitational potential and hydrodynamical effects on the gas. These estimates suggest that a relatively recent and significant input of supernova energy with a highly efficient transfer of energy to the kinetic motion of the gas is required for the infalling gas to be the returning flow of a supernova-powered fountain.

Another interesting aspect of the observations is that the higher (infalling) velocity of the gas is found at larger distances above the plane, while the lower-velocity gas is nearer

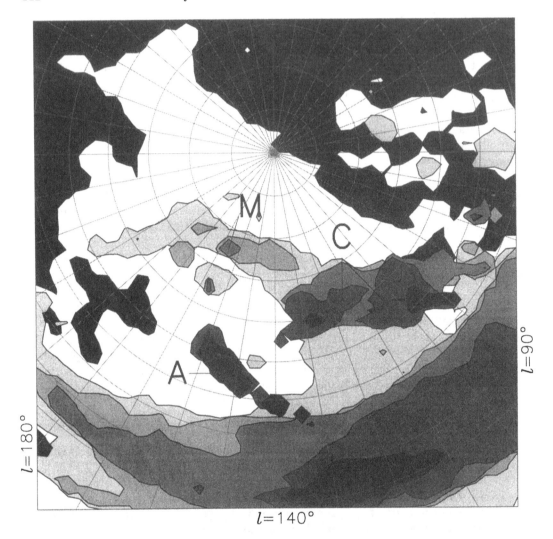

FIGURE 2. A plot of the distribution of the infalling gas in the northern hemisphere in polar coordinates, centring on the HVCs and IVCs. The letters A, C, and M denote Complex A, Complex C, and Complex M, respectively. The greyscale levels represent gas distributions of different velocity. From light grey to dark grey, the velocities in km s^{-1} are: -50, -75, -100, -125, -150, -175. The background is black.

the plane. This trend, which is seen in most of the gas in the northern hemisphere (Danly 1989) is just the opposite of what would be expected from gas which was accelerated upward from the disk and was now falling back. If significant material and/or magnetic fields exist between 1 and 3 kpc above the disk which serve to slow the infall, the energy requirements for halo "break-out" are even greater.

Ultraviolet absorption-line data toward the IV Arch show that CIV and SiIV are commonly seen, though their ratio is different at different positions along the Arch (Danly & Kuntz 1993). The IUE data provide only upper limits on the existence of NV in the Arch, which is an important discriminator between photo-ionised gas and collisionally ionised hot gas. The IUE measurements and limits can be successfully modelled with photo-ionisation alone over a reasonable range of ionisation parameter (Sokoloski &

Danly 1993). The modelling does require the depletion of refractory species from the gas phase, but the magnitude of the required depletion is in complete agreement with what would be expected from gas at the inferred density (c.f. Jenkins 1987). Higher quality *HST* data on the IV Arch are in the process of being obtained.

4. Extragalactic analogues

The unique perspective gained by studying external galaxies can offer insight on the mechanisms for the formation of galactic halo gas by providing an assessment of a galaxy's global characteristics and the observed nature of its halo. The study of edge-on galaxies permits the study of the scale height and structure of gas away from the plane, while face-on galaxies provide a means to study correlations between disk features and z motions of the gas, although the location of the emitting gas is ambiguous. Over the past few years there has been a dramatic increase in the number of observations of external systems imaged through narrow-band filters (such as H-alpha and other emission lines), radio continuum, 21-centimetre, and broad-band X-rays. The field is expanding so rapidly that any summary will undoubtedly become outdated by the time this article reaches publication. However, results from a small sample can serve to illustrate a few general points about the relationship between gaseous galactic halos and their underlying disks.

Perhaps the most important point to make is that there does not appear to be any simple trend or relationship between the star-forming nature of a galaxy and the extent or excitation of its gaseous halo. One of the best studied galaxies, NGC 891, is often discussed as a close analogue to the Milky Way, although it is in a slightly star-bursting phase. This edge-on galaxy has got a very well-developed hot gaseous halo which is observed in H-alpha, radio continuum, and soft X-rays (Rand *et al.* 1990; Dettmar 1990; Bregman & Pildis 1994). In another case, NGC 4565, soft X-rays are seen towards the centre of the galaxy with some filamentary structure, but little or no H-alpha is seen from the halo, despite a possible OIII halo (Pietsch 1993; Rand *et al.* 1994; Martin *et al.* in preparation). No soft X-ray emission is seen at all from the halo of NGC 5907 (Pietsch 1993). At H-alpha, a survey of seven edge-on galaxies reveals that only four of the seven show 1–2 H-alpha filaments extending from the disk, while none of the seven show a diffuse H-alpha halo (Pildis *et al.* 1994). In many galaxies, a few (i.e. 2–4) loops or filaments can be seen extending from the galaxy as in the case of the HI loop in NGC 4656 (Rand 1993) or the H-alpha or radio-continuum filaments seen in NGC 5775 (see Dettmar 1993). Radio-continuum measurements also show a range in halo characteristics (Hummel *et al.* 1991)

Observations of face-on galaxies also reveal a "mixed bag." Some show well-developed X-ray halos, as in the case of NGC 4258, while others show no X-rays at all, as in the case of NGC 4395, while still others show X-rays which are concentrated towards the centre of the galaxy (Pietsch 1993; Snowden & Pietsch 1994; Cui *et al.* 1995). More detailed work is required to separate contributions from nuclear emission, not superbubbles in the disk, and diffuse halo emission. HI observations of face-on galaxies do reveal "holes"—i.e. high-contrast areas of low column density—in the HI distribution, as well as superbubbles and high-velocity gas, although the relationship between these structures and star-formation or fountain activity is not yet clear (i.e. M31, Brinks 1984; M101, Van der Hulst & Sancisi 1988; Kamphuis 1993).

Certainly many more observations must be compiled before statistically meaningful conclusions can be drawn. But preliminary results from these and other galaxies suggest that there is not a simple relationship between the nature of a gaseous halo and its underlying galaxy. A wide range of halo characteristics are found without a strong connection to the global properties of the underlying disk. Most galaxies show roughly 2–4 "active" regions which have significant consequences on their gaseous halos.

5. Discussion

Comparing results from extragalactic systems to the observations of the Milky Way, one is tempted to speculate that the interstellar disk–halo interaction in spiral galaxies may be dominated by substantial, episodic "events" rather than the comparatively gentle and ongoing burbling of a galactic fountain. Kuntz & Danly (1995) estimate that if the disk–halo "event" HI features in the Milky Way reach as high as 3 kpc above the disk (typical of the scales of extragalactic spurs and loops), about one third of the disk would be observable in the Crawford–Hill data used in their analysis. They suggest that only a few such active regions may exist over the entire Milky Way, although our appreciation of the frequency of these features is limited by the view provided by our vantage point in the Galaxy's disk.

What is the nature of these "events"? The large masses, large energies, and the ionisation characteristics of the gas found in the present data may indicate that the "event" giving rise to the halo–disk activity seen in the second galactic quadrant may not be star-formation or fountain related at all. Perhaps the phenomenon is related to the collision of the HVCs with the disk, while the HVCs originate via some mechanism other than the galactic fountain. Some preliminary *HST* data show that in the region beyond the IV Arch, but less than 3 kpc distant, lies a region with very strong CIV and SiIV absorption. This is the same region where the HVC Complex M lies, and *ROSAT* data suggest significant soft X-ray emission lies in the same direction (Mebold *et al.* 1994). At present it is not known whether the X-rays originate in hot fountain gas or are the product of shocks associated with the HVCs. Further analysis and improved data are required to discriminate between the theories for its origin.

The lack of any systematic relationship between the observed properties of a galactic disk and its gaseous halo makes it difficult to draw inferences on the nature of high-redshift galaxies simply by observing absorption that may arise in their vicinities. Observations of QSOs which lie behind low-redshift galaxies but with projected distances that could intercept a gaseous halo are consistent with the picture outlined above for local galaxies. In one recent study by Bowen *et al.* (1995) two out of six galaxies with the QSO sightline impact parameter lying less than 9 kpc from the disk show no MgII in absorption. In that same study, 10 of 11 galaxy–QSO pairs with impact parameters of more than 30 kpc show no MgII absorption. In another case (Steidel *et al.* 1993), a QSO lying less than 4 kpc in projection from a blue dwarf galaxy shows no absorption from CaII down to 40 mA. The blue colour of the stars and the low gravity of the galaxy suggest that gas should have been easily driven from the disk into the halo by the latest generation of star formation.

The relative numbers of damped Lyman-alpha systems, Lyman-limit systems, and metal-line systems have led to a simple picture where the absorption line cross-sections correlate to the sizes of absorbing regions in the vicinity of protogalaxies (see excellent reviews by Steidel 1992, Lanzetta 1993, Steidel 1993, and references therein). Some have conjectured that increased star formation in the early history of a galaxy's evolution would lead to an enhanced gaseous halo as wind and/or fountain-type activity would be increased. Alternatively, a higher merger rate of galaxies or the accretion of protogalactic "shards" at the time of galaxy formation could lead to the higher cross-sections found for high-redshift systems. Certainly a more secure understanding of the nature and prevalence of gaseous haloes in nearby galaxies is required before a meaningful comparison between high- and low-redshift galaxies can be made on the basis of absorption-line data.

This work is supported under a grant from the National Aeronautics and Space Administration under grant NAG5-1509.

REFERENCES

BLOEMEN, H. (ED.) 1991 *The Interstellar Disk-Halo Connection in Galaxies.* IAU Symposium 144. Kluwer.

BOWEN, D. B., BLADES, J. C. & PETTINI, M. 1995 *Astrophys. J.* in press.

BREGMAN, J. N. 1980 *Astrophys. J.* **236**, 577.

BREGMAN, J. N. & PILDIS, R. A. 1994 *Astrophys. J.* **420**, 570.

BRINKS, E. 1984 PhD thesis, Univ. Leiden.

BURROWS, D. N. & MENDENHALL, J. A. 1991 *Nature* **351**, 629.

CHU, Y. H. & KENNICUTT, R. C. 1994 *Astrophys. J.* **425**, 720.

CUI, W. *et al.* 1995, in preprat.aion

DANLY, L. 1987 PhD thesis, University Wisconsin—Madison.

DANLY, L. 1989 *Astrophys. J.* **342**, 785.

DANLY, L. 1992 *Publ. Astron. Soc. Pac.* **104**, 819.

DANLY, L. 1995 *Annu. Rev. Astron. Astrophys.* in preparation.

DANLY, L., ALBERT, C. E. & KUNTZ, K. D. 1993 *Astrophys. J., Lett.* **416**, L29.

DANLY, L. & KUNTZ, K. D. 1993 in *Star Formation, Galaxies and the Interstellar Medium* (ed. J. Franco, F. Ferrini & G. Tenorio-Tagle). 4th EIPC Workshop, p. 86. Cambridge University Press.

DETTMAR, R-J. 1990 *Astron. Astrophys.* **232**, L15.

DETTMAR, R-J. 1993 *Star Formation, Galaxies and the Interstellar Medium* (ed. J. Franco, F. Ferrini & G. Tenorio-Tagle). 4th EIPC Workshop, p. 96. Cambridge University Press.

GIOVANELLI, R. 1980 *Astron. J.* **85**, 1155.

HEILES, C. 1984 *Astrophys. J., Suppl. Ser.* **55**, 585.

HEILES, C. 1993 *Star Formation, Galaxies and the Interstellar Medium* (ed. J. Franco, F. Ferrini & G. Tenorio-Tagle). 4th EIPC Workshop, p. 245. Cambridge University Press.

HULSBOSCH, A. N. M. & WAKKER, B. P. 1988 *Astron. Astrophys., Suppl. Ser.* **75**, 191.

HUMMEL, E., BECK, R. & DETTMAR, R-J. 1991 *Astron. Astrophys., Suppl. Ser.* **87**, 309.

JENKINS, E. B. 1987 in *Interstellar Processes* (ed. D. Hollenbach & H. Thronson). ASSL, vol 161, p. 533. Reidel.

JENKINS, E. B. & MELOY, D. A. 1974 *Astrophys. J., Lett.* **193**, L121.

KAMPHUIS, J. 1993 Neutral Hydrogen in nearby Spiral Galaxies. Holes and High Velocity Clouds. PhD thesis, Rijksuniversiteit, Groningen.

KUNTZ, K. D. & DANLY, L. 1995 *Astrophys. J., Suppl. Ser.* in press.

LANZETTA, K. M. 1993 in *The Environment and Evolution of Galaxies* (ed. M. Shull & H. Thronson). ASSL, vol 188, p. 237. Kluwer.

LI, F. & IKEUCHI, S. 1990 *Astrophys. J., Suppl. Ser.* **73**, 401.

MacLOW, M-M. & McCRAY, R. 1988 *Astrophys. J.* **324**, 776.

MacLOW, M-M., McCRAY, R. & NORMAN, M. 1989 *Astrophys. J.* **337**, 141.

McKEE, C. F. 1993 in *Back to the Galaxy* (ed. F. Verter). Proc. 3rd Annual Maryland Astrophysics Conference, p. 499. University of Maryland.

MEBOLD *et al.* 1994 in *First Symposium on the IR Cirrus and Diffuse Interstellar Clouds* (ed. R. M. Cutri & W. B. Latter). ASP Conference Series, vol. 58, p. 45. ASP.

NORMAN, C. A. & IKEUCHI, S. 1989 *Astrophys. J.* **345**, 372.

PIETSCH, W. 1993 in *Panchromatic View of Galaxies* (ed. G. Hensler *et al.*), p. 137.

PIETSCH, W., VOGLER, A., KAHABKA, P., JAIN, A. & KLEIN, U. 1994 *Astron. Astrophys.* **284**, 386.

PILDIS, R. A., BREGMAN, J. N. & SCHOMBERT, J. M. 1994 *Astrophys. J.* **427**, 160.

RAND, R. J. 1994 *Astron. Astrophys.* **285**, 833.

RAND, R. J., KULKARNI, S. R. & HESTER, J. J. 1990 *Astrophys. J., Lett.* **352**, L1.

RAND, R. J., KULKARNI, S. R. & HESTER, J. J. 1992 *Astrophys. J.* **396**, 97.

SAVAGE, B. D. 1988 in *QSO Absorption Lines: Probing the Universe* (ed. J.C. Blades *et al.*), p. 195.

SAVAGE, B. D. & MASSA, D. M. 1987 *Astrophys. J.* **314**, 380.

SEMBACH, K. & SAVAGE, B. D. 1992 *Astrophys. J., Suppl. Ser.* **83**, 147.

SHAPIRO, P. R. & BENJAMIN, R. A. 1993 *Star Formation, Galaxies and the Interstellar Medium* (ed. J. Franco, F. Ferrini & G. Tenorio-Tagle). 4th EIPC Workshop, p. 275. Cambridge University Press.

SHAPIRO, P. R. & FIELD, G. B. 1976 *Astrophys. J.* **205**, 762.

SHULL, J. M. & SLAVIN, J. D. 1994 *Astrophys. J.* **427**, 784.

SNOWDEN, S. L. *et al.* 1991 *Science* **252**, 1529.

SNOWDEN, S. L., HASINGER, G., JAHODA, K., LOCKMAN, F. J., MCCAMMON, D. & SANDERS, W. T. 1994 *Astrophys. J.* **430**, 601.

SNOWDEN, S. L. & PIETSCH, W. 1995 *Astrophys. J.* submitted.

SOKOLOWSKI, J. & DANLY, L. 1993 in *Back to the Galaxy* (ed. F. Verter). Proc. 3rd Annual Maryland Astrophysics Conference, p. 536.

SPITZER, L. 1956 *Astrophys. J.* **124**, 20.

SPITZER, L. 1992 *Annu. Rev. Astron. Astrophys.* **28**, 71.

STEIDEL, C. C. 1992 *Publ. Astron. Soc. Pac.* **104**, 843.

STEIDEL, C. C. 1993 in *The Environment and Evolution of Galaxies* (ed. M. Shull & H. Thronson). ASSL, vol. 188, p. 263. Kluwer.

STEIDEL, C. C., DICKENSON, M. & BOWEN, D. V. 1993 *Astrophys. J., Lett.* **413**, L77.

VAN DER HULST, J. M. & SANCISI, R. 1988 *Astron. J.* **95**, 1354.

WAKKER, B. P. 1989 PhD thesis, Rijksuniversiteit Groningen.

WILLIAMSON, F. O., SANDERS, W. T., KRAUSHAAR, W. L., MCCAMMON, D., BORKEN, R. & BUNNER, A.R. 1974 *Astrophys. J., Lett.* **193**, L133.

YORK, D. G. 1974 *Astrophys. J., Lett.* **193**, L127.

Superbubbles in the Milky Way

By JAN PALOUŠ[1], S. EHLEROVÁ[2]
AND M. JECHUMTÁL[2]

[1]Astronomical Institute, Academy of Sciences of the Czech Republic, Boční II 1401, 141 31 Prague 4, Czech Republic

[2]Astronomical Institute, Charles University, Švédská 8, 150 00 Prague 5, Czech Republic

Observed HI structures expanding in the interstellar medium are compared with 3D computer simulations. The z distribution of the ambient HI gas is discussed: the thick and thin disks are compared in relation to the superbubble blow-out. The upper limit of the mass that may be pushed in supershells from the HI disk to the halo of the Milky Way is $0.2 M_\odot yr^{-1}$. The criteria when the new molecular clouds may form in supershells and when the star formation may propagate in the HI disk are also discussed.

1. Observational evidence

The galactic supershells have been discovered in the HI surveys of the Milky Way by Heiles (1979). These HI expanding structures have sizes from several hundred parsecs up to ~ 2 kpc. A supershell surrounds the low density area filled with hot gas forming a superbubble. Recently, the new high sensitivity 21 cm survey has been completed with the Dwingeloo radiotelescope by Burton and Hartmann (Burton 1994; Hartmann 1994) showing many anomalous HI features including the superbubbles listed by Heiles (1979).

In nearby galaxies, the HI distribution is not smooth either: it has holes or low-density regions surrounded by the expanding envelopes. The single dish observations of LMC and SMC with the Parkes radiotelescope unravelled a wealth of HI expanding structures probably even more energetic than their analogues in the Milky Way (Brinks 1990). To resolve similar structures in other galaxies from the local group, the Westerbork Synthesis Radio Telescope and the Very Large Array of NRAO have been used. They uncovered the HI holes in M31 (Brinks & Bajaja 1986) and M33 (Deul & Hartog 1990). Beyond the Local Group, several galaxies within 10 Mpc have been examined (Kamphuis 1993; Brinks 1994): NGC 628, M 81, M101, NGC 4631, NGC 6946, IC 10, IC 2574 and HoII (Puche *et al.* 1992). All of them display the HI holes. In the near future, a high resolution HI map of the Magellanic Clouds with the Australian Compact Array may provide more detailed information eventually comparable with the optical interferometric data from the H_α studies (Brinks *et al.* 1990).

2. 3D model of superbubbles

The kinetic expansion energy of a superbubble may originate from young and massive stars of an OB association or it may be released in a collision between a high-velocity cloud (HVC) and the galactic HI plane (Tenorio-Tagle & Bodenheimer 1988). Violent star formation, creating groups of massive stars producing large energies in stellar winds, ionising photons and supernova explosions, can explain the vast majority of HI holes in nearby galaxies including many of the Milky Way superbubbles. However, in some cases the total mass and the kinetic energy of expansion are too large to be explained by star formation. For example, the total mass in superbubble No. 6 in M101 (Kamphuis *et al.* 1991; Kamphuis 1993) involves a few $\times 10^8 M_\odot$ and kinetic energy of 10^{55} erg. Such

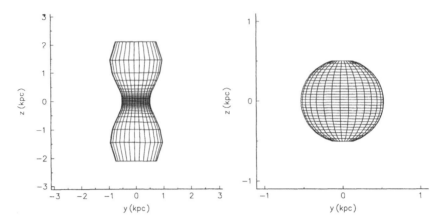

FIGURE 1. Two bubbles expanding in the disk with two different thicknesses. The HI surface density is in both cases the same, but the thickness differs 5×: the thin HI disk (left panel); the thick HI disk (right panel).

large mass and energy may originate in a collision between M101 and an HVC or a dwarf galaxy.

Expansion of the shocked interstellar medium around an OB association is an example of the astrophysical blastwave (Ostriker & McKee 1988). The thin, cold shell develops after the initial period dominated by stellar winds, radiation and supernovae. It expands supersonically accumulating the ambient medium. We assume that the thickness of the shell is much less than its diameter adopting the infinitessimally thin shell approximation invented by Sedov (1959) and developed by Kompaneets (1960) and Bisnovatyi-Kogan & Blinikov (1982).

In the 3D numerical simulations, the supershell is divided into N layers and every layer into M elements. The motion of an element is given due to the equation of motion

$$\frac{d}{dt}(Mv_S) = dS[(P - P_o) + n_o v_o (v_S - v_o)] - Mg(R, z), \qquad (2.1)$$

where M, v_S and dS are the mass, velocity of expansion and the surface; P, P_o are the inside and outside pressures; n_o, v_o are density and velocity of the ambient medium; $g(R, z)$ is the gravitational acceleration in the Galaxy and R, z are the galactocentric cylindrical coordinates. The mass of an element increases as long as the expansion velocity component normal to the supershell, $v_\perp$, exceeds the velocity of sound in the ambient medium

$$\dot{M} = v_\perp n_o dS. \qquad (2.2)$$

The 3D model using the above infinitessimally thin shell approximation was developed by Palouš (1990, 1992) and Silich (1992).

2.1. Expansion in the z direction

The z distribution of the ambient medium plays the key role: the superbubble can blow out to the galactic halo if the HI disk is sufficiently thin. With the same surface density of HI and with the same total energy in expansion, the thick disk confines a superbubble to $z < 1$ kpc, but the thin disk allows it to blow out reaching $z \sim 5 - 6$ kpc in 3×10^7 yr (Figure 1).

A supershell after 30 Myr of expansion in the thick HI disk is shown in Figure 2. The total energy put to the expansion is 10^{53} erg. Figure 2a gives the 3D shape of the

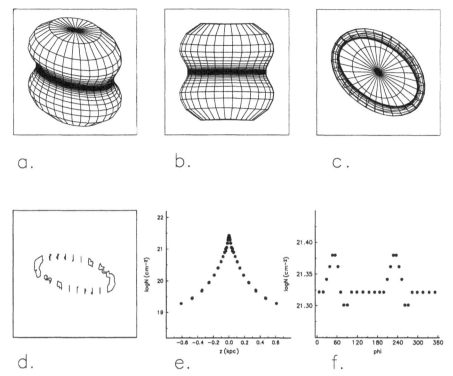

FIGURE 2. A supershell after 30 Myr of expansion in the thick HI disk: *a.* the shape of the supershell in 3D; *b.* the edge on view: the galactic differential rotation points to the right; *c.* pole on view: the galactic differential rotation points to the right, the galactic centre is towards the bottom of the page; *d.* places with HI column density $> 5 \times 10^{21} cm^{-2}$, position of the symmetry plane relative to the line of sight is the same as in 2a; *e.* the column density in the supershell as a function of the z distance from the galactic plane; *f.* the column density as the function of the position angle ϕ between the galactic centre and the element of the supershell measured anticlockwise from the centre of expansion.

supershell; Figure 2b shows its z extent, which is related to the HI distribution in the z direction; and Figure 2c shows its elliptical deformation in the plane parallel to the galactic symmetry plane, which is due to the galactic differential rotation. The surface density collected in the supershell is given in Figures 2d-f it ranges from 2×10^{19} to $3 \times 10^{21} cm^{-2}$. The highest values are reached in the galaxy symmetry plane and the lows at the caps, which are the most distant parts of the supershell from the symmetry plane. The total amount mass M_{tot} is $6 \times 10^6 M_\odot$. Most of it resides in the high-column-density region near the galactic plane. The mass of the low-column-density cups is $\leq 1\% M_{tot}$.

Only a small fraction of the mass collected in a superbubble can be pushed to high z: in the thick HI disk it is $\leq 1\%$. When the HI is concentrated to the thin disk for the same energy, the superbubble blows out; in this case the caps have even lower densities, therefore the fraction of mass at high z is even less.

We may estimate from our simulations an upper limit of the mass that may be ejected from the Milky Way in a supershell as $6 \times 10^4 M_\odot / 3 \times 10^7 yr$. Thus with ~ 100 of super-shells we get an upper limit for the total mass pushed by them to high z as $0.2 M_\odot yr^{-1}$. This upper limit may restrict the galactic fountain model of Norman & Ikeuchi (1989). We conclude that only a small fraction of the HVC can result in galactic fountains.

2.2. *Expansion in the galactic plane*

The densest and the most massive parts of the expanding supershell are concentrated near the galactic plane. The elliptical deformation of the shape, which is related to the gallactic differential rotation, is shown in Figure 2c. The influence of the differential rotation on the size, shape and column density is described by Tenorio-Tagle & Palouš (1987) and Palouš *et al.* (1990). The column density as a function of the position angle ϕ is given in Figure 2f: it has two maxima located at the tips of the elliptical supershell.

The highest values of the column density, $\sim 3 \times 10^{21}$ erg, are reached in the galactic symmetry plane at the tips of the elliptical supershell. These are the places where the column density surpasses the critical value (Franco & Cox 1986)

$$N_{crit} = 5 \times 10^{20} Z_\odot / Z \ cm^{-2}, \tag{2.3}$$

where Z is the heavy element abundance, and $Z_\odot$ is the solar value. We assume that the new molecular clouds are formed at such positions, where the opacity criterion is fullfilled.

The fragmentation of an expanding supershell has been studied by Elmegreen (1994). The stability criterion of a supershell is

$$\frac{1.5}{(Gn_o)^{1/2} v_S / c} < \frac{2.5}{\kappa}, \tag{2.4}$$

where G is the constant of gravity, c is the velocity dispersion in the supershell and κ is the epicyclic frequency. Old rings that have slowed down to v_S comparable to c are unstable and can fragment. This may be the condition of new star formation due to supershell fragmentation, which can be called the condition of the propagating star formation.

3. Conclusions

We conclude that the HI z ditribution determines whether a superbubble blows out: the thick HI disk can suppress the blow-outs. In any case the total amount of HI that may be in the connection to the clustered star formation in the Milky Way, pushed to the halo from the disk, is $< 0.2 M_\odot yr^{-1}$ providing a limit to Norman & Ikeuchi's galactic fountain model of Norman and Ikeuchi (1989).

In both the thick and the thin HI disks, the highest column densities in the expanding supershell are reached in the galactic symmetry plane near the tips of elliptical remnants. These are the places where the column density may fullfill the opacity criterium (2.3) and the new molecular clouds may be formed. If the criterium (2.4) is fullfilled, the supershell may fragment, propagating the star formation.

J. Palouš. is supported by GA ČR under grant number 0090.

REFERENCES

BISNOVATYI-KOGAN, G. S. & BLINNIKOV, S. I. 1982 *Astron. Zh.* **59**, 879.

BRINKS, E. 1990 in *The Interstellar Medium in Galaxies* (ed. H. A. Thoronson, Jr. & J. M. Shull), p. 39. Kluwer

BRINKS, E. & BAJAJA, E. 1986 *Astron. Astrophys.* **169**, 14.

BRINKS, E., BRAUN, R. & UNGER, S. W. 1990 in *Structure and Dynamics of the Interstellar Medium* (ed. G. Tenorio-Tagle, M. Moles & J. Melnick). Lecture Notes in Physics, vol. 350, p. 524. Springer.

DEUL, E. R. & HARTOG, R. H. 1990 *Astron. Astrophys.* **229**, 362.

DOVE, J. B. & SHULL, M. 1994 *Astrophys. J.* 423, 196.

ELMEGREEN, B. G. 1994 *Astrophys. J.* **427**, 384.

FRANCO, J. & COX, D. P. 1986 *Astrophys. J.* **273**, 243.

HEILES, C. 1979 *Astrophys. J.* **299**, 533.

KAMPHUIS, J. 1993 Neutral Hydrogen in nearby Spiral Galaxies. Holes and High Velocity Clouds. PhD Thesis, Rijksuniversiteit, Groningen.

KAMPHUIS, J., SANCISI, R. & VAN DER HULST, P. W. 1991 *Astron. Astrophys.* **244**, L29.

KOMPANEETS, A. S. 1960 *Soviet. Phys. Dokl.* (in Russian) **5**, 46.

NORMAN, C. A. & IKEUCHI, S. 1989 *Astrophys. J.* **345**, 372.

OSTRIKER, J. P. & MCKEE, CH. F. 1988 *Rev. Mod. Phys.* **60**, 1.

PALOUŠ, J. 1990 in *The Interstellar Disk–Halo Connectionin Galaxies — Poster Proceedings* (ed. H. Bloemen), p. 101. Space Research Leiden.

PALOUŠ, J. 1992 in *Evolution of Interstellar Matter and Dynamics of Galaxies* (ed. J. Palouš, W. B. Burton & P.O.Lindblad), p. 65. Cambridge University Press.

PALOUŠ, J., FRANCO, J. & TENORIO-TAGLE, G. 1990 *Astron. Astrophys.* **227**, 175.

PUCHE, D., WESTPFAHL, D., BRINKS, E. & ROY, J.-R. 1992 *Astron. J.* **103**, 1841.

SEDOV, L. 1959 *Similarity and Dimensional Methods in Mechanics.* Academic Press.

SILICH, S. A. 1992 *Astrophys. Space Sci.* **195**, 317.

TENORIO-TAGLE, G. & BODENHEIMER, P. 1988 *Annu. Rev. Astron. Astrophys.* **26**, 145.

TENORIO-TAGLE, G. & PALOUŠ, J. 1987 *Astron. Astrophys.* **186**, 287.

Hensler: I totally agree that superbubble models show that they are unable to sweep up enough material from the disk to account for the infall rate of the HVCs and their infall velocity. In addition, models until now overestimate the vertical expansion because they only consider the external pressure of the Lockman layer but neglect the pressure of the interstellar gas at higher stratifications like the Reynolds layer and the halo gas. Could you comment upon this?

Palouš: I agree with both of your statements. We give the upper mass limit that may be pushed from the disk to halo, which is about one order of magnitude smaller than the estimate of the total mass inflow due to HVC to the Milky Way. However, the actual value may be even smaller. To the second question: we do not consider the pressure from the Reynolds layer and from the halo gas in our simulations. It should be included, particularly if somebody is trying to justify the photoionisation due to OB associations observed in the Milky Way (Dove & Shull, 1994)

Surdin: Is there any statistical information on the real shape and orientation of superbubbles in some galaxies?

Palouš: All the information we have is related to galaxies within 10 Mpc. We may compare the observed data cubes and the information on the galactic-disk inclination and thickness with the predicted data cubes from simulations. In this way we may reach conclusions about the intrinsic shapes of superbubbles as well as about certain galactic properties, such as the z distribution of HI.

A Survey of Giant Molecular Clouds in the Perseus Arm near Cas A

By HANS UNGERECHTS[1], P. UMBANHOWAR[2]
AND P. THADDEUS[3]

[1]IRAM, Avenida Divina Pastora 7-NC, E–18012 Granada, Spain

[2]Harvard-Smithsonian Center for Astrophysics, Cambridge, MA 02138, U.S.A.

[3]Center for Nonlinear Dynamics and Dptmt. of Physics, U. of Texas, Austin, TX 78721, U.S.A.

Surveys of molecular clouds in the Perseus arm show an emission-free irregular band between -45 and $-20\,\mathrm{km\,s^{-1}}$ in the lv diagram. Where this velocity gap represents a spatial interarm gap between the Perseus and local arms, the contrast in molecular gas surface density is at least 10, and several times as high in some l ranges. However, for several large clouds photometric distances of associated H II-regions are much smaller than the kinematic distances leaving little interarm space free of CO, so that around $l = 110°$ the gap in the lv diagram may be predominantly a kinematic effect of large-scale streaming motions.

In a large-scale survey of molecular clouds in the 1–0 line of CO (Dame *et al.*, 1987) the Perseus arm appears as a string of separate giant molecular cloud (GMC) complexes at negative LSR velocities in the second quadrant; the two most prominent are associated with Cas A and W3. This survey with the 1.2-metre telescopes in New York City and in Chile has an angular resolution of only 0.5°. Using the 1.2-metre telescope, now at the Center for Astrophysics, at its full angular resolution of about 0.125°, CO surveys of the Perseus arm in the regions of Cas A (Ungerechts *et al.* 1994) and W3 (Puche *et al.* 1994) have been undertaken over the last few years. Near Cas A the survey covers 55 deg² between $l = 107°$ and 116°. A smaller area of 15 deg² between $l = 108°$ and 113° was similarly surveyed in ^{13}CO. In each pixel of angular size $A_{\mathrm{pix}} = 1/64\,\mathrm{deg^2}$ the rms noise is $T_{\mathrm{rms}} = 0.12\,\mathrm{K}$ for CO and $0.06\,\mathrm{K}$ for ^{13}CO in each channel of width $\Delta v \approx 0.65\,\mathrm{km\,s^{-1}}$.

The map of CO-line intensity integrated over velocity, $W(\mathrm{CO})$ (Figure 1), shows that most emission is concentrated in a few large clouds, several of which are associated with visible H II regions, e.g. NGC 7538. For the largest clouds and some smaller ones we calculated basic parameters (examples in Table 1); in total there are 6 GMCs with masses above $10^5 M_\odot$. With the adopted luminosity to mass conversion, the average surface density of a GMC is $\sigma = 5\,M_\odot/\mathrm{pc^2}\,(S(\mathrm{CO})/\,\mathrm{K\,km\,s^{-1}\,deg^2})(\,\mathrm{deg^2}/A)$ and values for the large clouds range from about 40 to $90 M_\odot\,\mathrm{pc^{-2}}$.

In the lv diagram emission from the Perseus arm appears very well separated from the local arm (Figure 2). The difference between the average velocity of Perseus arm clouds and local arm clouds is at least $20\,\mathrm{km\,s^{-1}}$ and an irregular band $10\,\mathrm{km\,s^{-1}}$ or wider appears nearly free of emission.

For any region with no detected CO emission in n_p adjacent pixels, a 3σ upper limit for the surface density is given by: $\sigma \leq 1.5\,M_\odot/\mathrm{pc^2}\,(T_{\mathrm{rms}}/0.12\,\mathrm{K})\,[(10/n_p)(\Delta v/0.65\,\mathrm{km\,s^{-1}})$ $(\delta v/10\,\mathrm{km\,s^{-1}})]^{0.5}$, where δv is the velocity range without emission. A comparison with Table 1 shows that typically the implied contrast in surface density between the Perseus arm GMCs and the emission free band is larger than ten, in some cases as high as 60.

This, however, does *not* prove that the gap in the lv diagram corresponds to a true spatial separation rather than a predominantly kinematic effect as, for example, in the model of Mulder & Liem (1986). It has long been known that photometric distances of objects in the Perseus arm are smaller than kinematic distances based on purely circular

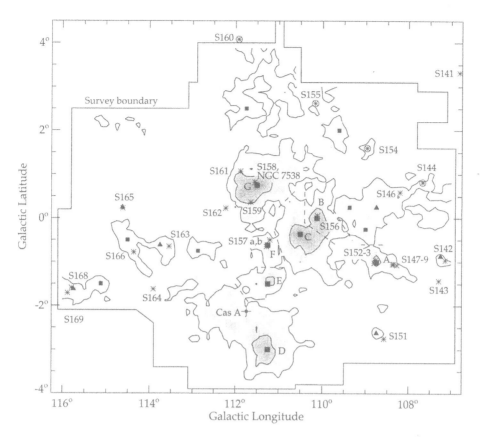

FIGURE 1. Molecular clouds in the Perseus arm toward Cassiopeia and Cepheus: CO emission integrated over the velocity range -90 to $-25\,\mathrm{km\,s^{-1}}$. The contours are at 3.5 and $22.1\,\mathrm{K\,km\,s^{-1}}$. The locations of selected Population I objects including the Cas A supernova remnant are indicated. Asterisks denote the centres of HII regions, asterisks in a circle HII regions at local arm velocities probably not associated with the molecular clouds in this map. Peak positions of molecular clouds are shown by filled squares and triangles.

motion. A recent compilation of v_{LSR} and photometric distance d_* for molecular clouds and associated HII regions is given in Brand & Blitz (1993). These distances imply that around $l = 110°$ there is *no* significant spatial separation of objects that kinematically belong to the local and Perseus arms, and that the molecular clouds in the Perseus arm may be spread out over a distance range from about 2 to 4 kpc, maybe even to 6 kpc. All GMCs and nearly all minor molecular clouds must then take part in streaming motions with velocities 20–$30\,\mathrm{km\,s^{-1}}$ to create the gap in the lv diagram.

REFERENCES

BRAND, J. & BLITZ, L. 1993 *Astron. Astrophys.* **275**, 67.

DAME, T. M., UNGERECHTS, H., COHEN, R. S., DE GEUS, E. J., GRENIER, I. A., MAY, J., MURPHY, D. C., NYMAN, L.-Å. & THADDEUS, P. 1987 *Astrophys. J.* **322**, 706.

DIGEL, S. W. 1991 PhD thesis. Harvard University, Cambridge, Massachusetts.

MULDER, W. A. & LIEM, B. T. 1986 *Astron. Astrophys.* **157**, 148.

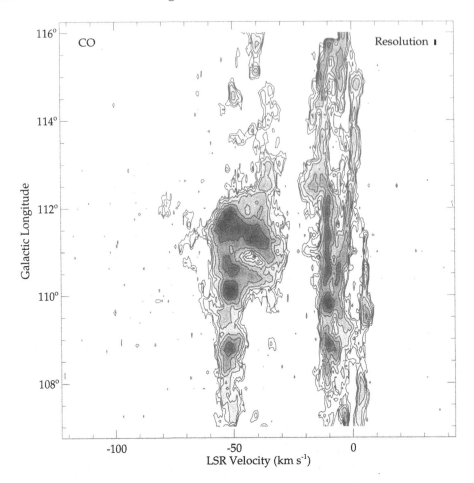

FIGURE 2. Emission of CO integrated over the observed range in latitude as a function of velocity and longitude. The lowest contour is at 0.25 K deg, and each contour is 1.58 times higher than the previous, for five contours per decade.

PUCHE, D., DIGEL, S. W., LYDER, D. A., PHILBRICK, A. J. & THADDEUS, P. 1994 in preparation.

UNGERECHTS, H., UMBANHOWAR, P. & THADDEUS, P. 1994 *Astrophys. J.* submitted.

WOUTERLOOT, J. G. A. & BRAND, J. 1989 *Astron. Astrophys.* **80**, 149.

Elmegreen: Do all the molecular cloud pieces, including those in the interarm gap, have star formation?

Ungerechts: There are Sharpless HII regions and/or *IRAS* point sources with the colour characteristics of embedded stars (Wouterloot & Brand 1989) associated with most large and strong clouds in Figure 1, but there are no HII regions and relatively few *IRAS* sources toward cloud D (Digel 1991). I am not aware of evidence for star formation in any of the small fragments that appear in the velocity gap in Figure 1.

Bania: What was your sampling coverage in b? Is it adequate so you haven't missed anything?

Ungerechts: The coverage in b (see Figure 1) was excellent for the Perseus arm and adequate for the interarm gap, while it is incomplete for local arm clouds at $b > 2.5°$.

Cloud $l \pm b$	S(CO) (K km s^{-1}deg^2)	$\langle v \rangle$ (km s^{-1})	σ_v	d_{kin} (kpc)	d	$\langle r \rangle$ (deg)	A (deg^2)	M_{CO} (10$^5 M_\odot$)	M_{vir}	σ ($M_\odot$ pc^{-2})
observed in CO and ^{13}CO, parameters from CO:										
A	5.7	−51.4	2.7	5.2	3.6	0.27	0.63	1.11	2.3	45
B	13.8	−50.8	3.2	5.1	6.4	0.31	1.08	8.53	6.9	64
C	21.2	−46.3	6.2	4.6	3.0	0.31	1.19	2.88	11.8	89
D	39.3	−39.0	5.0	3.8	3.0	0.68	3.78	5.35	16.8	52
E	5.4	−40.6	4.3	3.9	3.0	0.27	0.66	0.73	4.8	41
F	3.5	−45.6	3.7	4.5	2.5	0.24	0.50	0.33	2.7	35
G	32.1	−53.2	3.8	5.2	2.2	0.46	2.39	2.35	4.8	67
examples of clouds with known H II-regions, observed only in CO:										
108.8+0.3	7.6	−52.2	2.6	5.3	3.0	0.35	1.03	1.03	2.3	37
114.6+0.3	0.3	−34.1	1.5	3.2	1.6	0.15	0.09	0.01	0.2	19

S(CO): CO luminosity integrated over velocity and solid angle. $\langle v \rangle$: mean velocity. σ_v: line width. d_{kin}: kinetic distance assuming purely circular rotation. d: adopted distance: photometric where available, otherwise 3 kpc assumed. $\langle r \rangle$: intensity-weighted mean radius. A: area above W(CO) = 1.8 K km s^{-1}. M_{CO}: mass, using a W(CO) to N(H$_2$) conversion factor $X_{12} = 2.3 \times 10^{20}$ cm^{-2}(K km s^{-1})$^{-1}$. M_{vir}: virial mass from σ_v and $\langle r \rangle$. σ: cloud surface density.

TABLE 1. Parameters of selected clouds in the Perseus arm

Elmegreen: Is the whole GMC complex near Cas A virialised to within a factor of two or are the pieces clearly strung out on the line of sight?

Ungerechts: While most of the large GMCs are virialized to within a factor of a few, the complex as a whole is not; in fact there are large velocity differences from GMC to GMC. If the available photometric distances are correct, the GMCs are strung out along the line of sight over several kiloparsecs.

Chernin: Could you give please just three rough figures: the length of the Perseus arm, its thickness, and the number of GMCs in it?

Ungerechts: In the large-scale survey (Dame *et al.* 1987) the molecular Perseus arm is a distinct feature at least from about $l = 95°$ to 155° corresponding to a length of 3–4 kpc. Given the distance uncertainties around Cas A, the thickness is easier to estimate toward W3, where Puche *et al.* (1987) find a scale height of ≈ 150 pc assuming $d = 4$ kpc. We see 16 clouds with masses of at least $10^4 M_\odot$ in the range $l = 107°$–116° and Puche *et al.*, 1994, have 20 for $l = 131°$–144°; but of course there are more in the other longitude ranges.

Towards a Complete Description of the Galactic Fountain

By MIGUEL A. AVILLEZ[1], D. L. BERRY[1]
AND F. D. KAHN[2]

[1]Department of Physics, University of Evora, Largo dos Colegiais 2, 7000 Evora, Portugal

[2]Department of Physics and Astronomy, The University, Manchester, M13 9PL, U.K.

Several models have been developed which aim to describe the important features of the galactic fountain. They concentrate on a one- or two-dimensional description of the fountain flow with or without cosmic ray and magnetohydrodynamic contributions. Evidence for the existence of descending flows comes from UV and 21-centimetre absorption lines. Here, we review these models and propose a more complete description, aiming to predict those features as yet not predicted.

1. Introduction

The major fraction of interstellar space within the galactic disk is occupied by the hot intercloud medium (ICM), which has a low density (10^{-26} g cm^{-3}) and a high temperature (10^6 K), and derives its thermal energy from shocks in supernova remnants. This energy eventually transforms into kinetic energy which is sufficient for the hot intercloud medium to escape from the galactic disk, but is not sufficient for escape from the Galaxy as a whole. The rate of flow of mass of the hot intercloud medium from the disk is approximately 1.5 $M_\odot$yr^{-1} (Kahn 1993), which is much larger than the Galaxy can sustain in a Hubble timescale. Also, it has been shown that the infall rate of HI at intermediate and high velocities in the solar vicinity is 5 $M_\odot$yr^{-1} (Wakker 1990), a value ten times larger than the estimated accretion rate of extragalactic high-velocity HI (Mirabel 1989). Thus, most of the observed neutral gas that is falling towards the disk must have originated from the disk. These arguments suggest that there is an intensive disk–halo circulation, with the hot intercloud medium flowing upwards into the halo in a continuous process followed by a returning flow set up under the gravitational pull, resembling a fountain, in addition to a moderate inflow of extragalactic HI clouds.

2. Observational evidence

Systematic surveys by means of ultraviolet and 21-centimetre measurements in both the northern and southern galactic hemispheres have revealed the widespread presence of diffuse clouds in the halo with line-of-sight velocities relative to the local standard of rest from 20 km s^{-1} up to about 300 km s^{-1}, with a preference for negative over positive velocities at high latitudes. These clouds have been classified according to their velocities as very-high- (VHVC), high- (HVC) and intermediate- (IVC) velocity clouds (see review by van Woerden et al. 1985; and Wakker 1991); the VHVCs are very rare.

The high-velocity clouds (HVCs) have been found at high latitudes and are concentrated in structures located primarily in the second quadrant of galactic longitude, forming large complexes like C I, C III and A, according to the classification by Wakker (1990). The clouds are distributed up to distances of several kiloparsecs from the galactic disk, the bulk of the gas being located between 250 pc and 1 kpc, but up to one third of the total material may be beyond 1 kpc (Albert et al. 1993; Lockman 1991).

The IVCs detected in the northern hemisphere in these surveys are also primarily

126

located in the second quadrant and form part of the large complexes C II and M II with velocities between -50 and -90 km s^{-1} (Danly & Blades 1989; Wakker 1990). They show no correlation with the low-velocity gas in the disk; the complexes are concentrated in directions where emission from the local gas is weakest. This circumstance indicates a removal or displacement of the gas from the disk. Further, their morphological structure suggests that the complexes may have been related to the disk (Danly 1991).

Surveys have shown the abundance of gaseous Ti in the halo to be nine times that in the disk. This result is consistent with the existence of a thick extended disk intermixed at great heights with material from a galactic fountain (Albert *et al.* 1993). The enhancement is present at all velocities and increases with distance from the galactic plane, possibly due to motions in the halo, but more probably as a result of grain disruption in the hot interior of the supernova remnants where the halo gas originated (Seab 1988). The evidence for this is provided by the observation that most of the IVCs (presumably recently shocked) have approximately solar elemental abundances.

It is expected that interstellar gas of a large range of temperatures should be observed, with abundances roughly proportional to the respective cooling times. Detections have been made of hot ionised gas with a temperature of 10^6 K and also partially recombined "warm" or already cold gas, with temperatures between 10^2 and 10^4 K, by means of the absorption lines of SiII, SiIV, CIII, CIV and NIII in the ultraviolet (Danly 1991). When cooling of the hot gas takes place, recombination will occur, forming neutral hydrogen clouds. This effect would be partially counterbalanced by photoionization, with radiation coming from below. The neutral hydrogen in the descending clouds will be observed by its 21-centimetre line emission.

3. Flows on the galactic scale

It is generally accepted that the presence of gas in the halo is due to certain dynamic phenomena occurring in the galactic disk. The hot gas injected into the halo will eventually cool and condense into clouds that rain back onto the disk. Two distinct classes of models have been proposed for the outflow. In one class, referred to as galactic fountains, it is assumed that the flow consists of the amalgamated hot remains of neighbouring supernova remnants whose inner parts were heated by the blast wave from the supernova explosion, and which take a very long time to lose their thermal energy by radiation, forming an outflow on the galactic scale (Shapiro & Field 1976; Bregman 1980; Houck & Bregman 1990; Kahn 1981, 1991, 1993). In the second class, referred to as the "chimney models", it is assumed that the flow of gas has its origin in clusters of supernovae of type II that form a superbubble through which the hot gas flows ballistically into the halo (Norman & Ikeuchi 1989; Tomisaka 1991). The number of supernovae needed for such a multiple explosion varies between 50 and 100. Events like this must be regarded as unusual, but have been observed in other galaxies (cf. Meaburn 1980).

Cosmic rays have also been proposed as an important additional energy source, able to set up the motion. Cosmic rays are coupled to the thermal plasma by Alfvén waves, so their pressure provides support against gravity and may even drive the thermal plasma out of the Galaxy in a galactic wind (Breitschwerdt *et al.* 1991; Kahn 1994).

4. The galactic fountain model

Models of the flow on the galactic scale were first introduced by Shapiro & Field (1976) and subsequently developed by Bregman (1980), Kahn (1981) and others. Hydrodynamic models exist in one dimension (Shapiro & Field 1976; Houck & Bregman 1990) and two dimensions, with symmetry about the z axis through the galactic centre (Bregman 1980). Calculations have been performed assuming an injection of matter from the galactic disk, starting in a thin layer at the galactic plane, with a temperature of 10^6 K and a certain rate of injection of mass. According to these models, outward winds are generated, which expand into the halo, where the pressure and density are low. The gas will condense into clouds whenever the gas temperature falls below 10^4 K (Shapiro & Field 1976; Bregman 1980; Houck & Bregman 1990; Wakker 1990) largely as a result of radiative cooling.

The height reached by the hot gas and the expected velocities of the condensations in the cooling gas depend on the temperature of the gas at the base of the fountain and on the rate of cooling of the upflowing gas, regardless of whether there are heating processes occurring at large distances above the galactic plane. The falling fountain gas cools and condenses before reaching hydrostatic equilibrium (Shapiro & Field 1976; Kahn 1981).

Kahn (1981, 1991, 1994) proposed that the interesting fountain characteristics are associated with a critical point where the upward flowing gas goes from subsonic to supersonic, at a distance of about 500 pc above the galactic disk. In Bregman's (1980) quasi-static models, the upward flow remains subsonic everywhere. The details depend on the assumed temperature; in some cases the flow approaches but does not pass through the sonic point (Houck & Bregman 1990).

After the clouds form, they can be assumed to move ballistically, independent of the ambient pressure and density, eventually falling back towards the galactic plane. With the values of initial temperature and rate of mass injection cited above, the clouds form at a height of a few kiloparsecs, with each fluid element at a distance ϖ from the galactic axis of symmetry rather larger than its initial value. This outward motion comes partly from a radial pressure gradient, assumed to have been present in the hot gas from which the clouds later condensed. It also results from the decrease of g_ϖ with increasing distance from the galactic plane; g_ϖ is the component of gravitational acceleration perpendicular to the axis of symmetry. A cloud on a ballistic orbit conserves its angular momentum, and its distance from the axis must increase when g_ϖ drops, because there is a change in balance between gravity and centrifugal acceleration. It therefore performs a large-scale Lindblad oscillation in which its distance from the axis first increases and later decreases. In a typical case, the gas will drift about 850 pc from its initial axial distance by the time it returns to the disk (Kahn & Brett 1993).

The model predictions should be compared with the 21-centimetre data for IVCs and HVCs. Velocity estimates show that the fountains driven by hot gas (1 to 2×10^6 K), can reproduce the pattern of motions of the HVCs and IVCs, in particular the predominance of negative velocities around the galactic poles and throughout the range $30° < l < 210°$ of longitude (Bregman 1980) and for $l > 330°$ (Wakker 1990). The fountain models in which the flow is driven by cooler gas (2 to 3×10^5K) predict clouds with velocities consistent with the existing ultraviolet data for gas in the lower halo (Houck & Bregman 1990). The IUE observations of NV absorption lines can be explained by the cooling gas of a fountain with a flow rate of 4 $M_\odot \text{yr}^{-1}$ on each side of the galactic plane, corresponding to a local flux of 6×10^{-9} $M_\odot \text{yr}^{-1}$ pc^{-2} (Edgar & Chevalier 1986). This value is in good agreement with the estimated observed mass flux in IVCs, 3×10^{-8} $M_\odot \text{yr}^{-1}$ pc^{-2} (Mirabel 1989) and in HVCs, 10^{-9} $M_\odot \text{yr}^{-1}$ pc^{-2} (Kaelble *et al.* 1985).

5. Conclusions

It is clear that substantial progress has been made in modelling galactic fountains since they were introduced by Shapiro. It seems that a complete description of the support and ionisation of the halo gas may require many of the features involved in the previous models and such a "complete model" is yet to be realised. The model has to include the effects of supernovae in creating and heating the hot intercloud medium that feeds the fountain flows. The kinematic asymmetry in the halo gas between the galactic hemispheres may be explained by the presence of flows resulting from fountains or other disturbances in the disk; the flow of a fountain may not be symmetric with the galactic plane; this may result in a difference of the heights where condensation occurs and thus, in different timescales for the flows above and below the plane. Other phenomena are important such as magnetic fields and cosmic rays. There may be regions where thermal energy determines the motion of the gas and results in the fountain flow. Elsewhere cosmic rays and magnetic fields may provide the energy density resulting in the loss of some of the gas in a galactic wind.

M.A.A. and D.L.B. are grateful to the Department of Astronomy, University of Manchester for the facilities provided during the writing of this paper. M.A.A. is grateful to the LOC for the travelling grant that made possible his participation at this meeting. This work is supported by a JNICT grant under contract number BD/2679/93.

REFERENCES

ALBERT, C. E., BLADES, J. C., MORTON, D. C., LOCKMAN, F. J., PROULX, M. & FERRARESE, L. 1993 *Astrophys. J., Suppl. Ser.* **88**, 81.

BREGMAN, J. N. 1980 *Astrophys. J.* **236**, 577.

BREITSCHWERDT, D., McKENZIE, J. F. & VÖLK, H. J. 1991 *Astron. Astrophys.* **245**, 79.

DANLY, L. 1991 in *The Interstellar Disk–Halo Connection in Galaxies* (ed. H. Bloemen). IAU Symposium 144, p. 53. Kluwer.

DANLY, L. & BLADES, J. C. 1989 in *Structure and Dynamics of the Interstellar Medium* (ed. G. Tenorio-Tagle, M. Moles & J. Melnick), p. 408. Springer.

EDGAR, R. J. & CHEVALIER, R. A 1986 *Astrophys. J., Lett.* **310**, L27.

HOUCK, J. C. & BREGMAN, J. N. 1990 *Astrophys. J.* **352**, 506.

KAELBLE, A., DE BOER, K. S. & GREWING, M. 1985 *Astron. Astrophys.* **143**, 408.

KAHN, F. D. 1981 in *Investigating the Universe* (ed. F. D. Kahn), p. 1. Reidel.

KAHN, F. D. 1991 in *The Interstellar Disk–Halo Connection in Galaxies* (ed. H. Bloemen). IAU Symposium 144, p. 1. Kluwer.

KAHN, F. D. 1994 *Astrophys. Space Sci.* **216**, 325.

KAHN, F. D. & BRETT, L. 1993 *Mon. Not. R. Astron. Soc.* **263**, 37.

LOCKMAN, F. J. 1991 in *The Interstellar Disk–Halo Connection in Galaxies* (ed. H. Bloemen). IAU Symposium 144, p. 15. Kluwer.

MEABURN, J. 1980 *Mon. Not. R. Astron. Soc.* **92**, 365.

MIRABEL, I. F. 1989 in *Structure and Dynamics of the Interstellar Medium* (ed. G. Tenorio-Tagle, M. Moles & J. Melnick), p. 396. Springer.

NORMAN, C. A. & IKEUCHI, S. 1989 *Astrophys. J.* **345**, 372.

SEAB, C. G. 1988 in *Dust in Universe* (ed. M. E. Bailey, & D. A. Williams), p. 303. Cambridge University Press.

SHAPIRO, P. R. & FIELD, G. B. 1976 *Astrophys. J.* **205**, 762.

TOMISAKA, K. 1991 in *The Interstellar Disk–Halo Connection in Galaxies* (ed. H. Bloemen). IAU Symposium 144, p. 407. Kluwer.

VAN WOERDEN, H., SCHWARZ, U. J. & HULSBOSCH, A. N. M. 1985 in *The Milky Way Galaxy* (ed. H. van Woerden, R. J. Allen & W. B. Burton). IAU Symposium 106, p. 387. Reidel.

WAKKER, B. J. 1990 Interstellar Neutral Hydrogen at High Velocities. PhD thesis, University of Gröningen.

WAKKER, B. J. 1991 in *The Interstellar Disk–Halo Connection in Galaxies* (ed. H. Bloemen). IAU Symposium 144, p. 27. Kluwer.

Star Formation and High–Velocity Clouds

By FERNANDO COMERÓN AND JORDI TORRA

Departament d'Astronomia i Meteorologia, Universitat de Barcelona, Av. Diagonal, 647,
E–08028 Barcelona, Spain

In recent decades, the interaction of high-velocity clouds with the galactic disk has been invoked to account for the formation of large shell-like structures and as a mechanism of star formation in galactic disks. Here we consider some aspects of the feedback implied by star formation, with its subsequent energetic output into the interstellar medium, and the formation of high-velocity clouds in the galactic halo gas, which in turn can trigger further star formation. We also pay attention to the spatial arrangement of star-forming complexes and some kinematic consequences of star formation triggered by high-velocity cloud impacts.

1. Introduction

The circulation of gas between the galactic disk and the halo is a major element in modelling the evolution of the gaseous disks of spiral galaxies. Hot gas produced by the activity of massive stars is expelled to the halo (MacLow et al. 1989; Norman & Ikeuchi 1989; Tenorio-Tagle et al. 1990; Minisighe et al. 1993), and large amounts of neutral gas exist in cloud complexes in the halo of our and other galaxies falling towards the disk (van Woerden et al. 1985; Wakker 1991; Wakker & van Woerden 1991). Observations of other galaxies indicate that star-forming activity is correlated to the existence of high-velocity neutral gas in the halo (Schulman et al.1994). The impacts of high-velocity clouds (HVCs) can be important in shaping the galactic HI layer and in locally triggering star formation (Heiles 1984; Tenorio-Tagle & Bodenheimer 1988; Franco et al. 1988; Alfaro et al. 1991; Comerón & Torra 1992, 1994; Lepine & Duvert 1994).

Evidence has accumulated in recent years that at least some of the largest complexes of HVCs lie at distances of several kiloparsecs from the galactic plane (Haud 1990, Danly et al. 1993), implying masses of order 10^5 M$_\odot$ and more. HVC surveys suggest a mass spectrum of the form $N(M)\,dM \propto M^{-1.5}\,dM$ (Wakker & van Woerden 1991), implying that most of the infalling mass is in large clouds. Since the impacts of the largest clouds can be expected to have the most noticeable observational consequences, it is interesting to wonder about their potential contribution to the total star-formation rate in the galactic disk, as well as about the observational traces implying evidence for such a contribution.

In this work we use simple arguments suggesting that HVC-induced star formation may contribute to about 10% of the total star-formation rate in our Galaxy. We also discuss how an HVC-triggered population may reveal itself in observations, and suggest that the local system of early-type stars may be dominated by such a population.

2. Infall-induced star formation

As a schematic approach, we decompose the net OB-star-formation rate in the Galaxy as:

$$\frac{dN_{OB}}{dt} = \left(\frac{dN_{OB}}{dt}\right)_{GD} + \left(\frac{dN_{OB}}{dt}\right)_{infall} - \left(\frac{dN_{OB}}{dt}\right)_{SN}$$

where the three terms in the right-hand side respectively represent the stars formed per unit time as a consequence of processes taking place in the galactic disk (unrelated to gas infall), the stars formed per unit time as a consequence of the infall, and the supernova rate among OB stars. We will consider $\left(\frac{dN}{dt}\right)_{GD}$ to be a constant. The infall term can be expected to reflect the energetic activity in the galactic disk at a time in the past of order of the recycling time t_R of the disk gas fuelled into the halo and then returned to the disk:

$$\left(\frac{dN_{OB}}{dt}\right)_{infall} = \alpha N_{OB}(t_R) \tag{1}$$

Finally, $\left(\frac{dN}{dt}\right)_{SN}$ is set equal to the total star-formation rate at a time in the past equal to the lifetime of an OB star, t_{OB}, to keep stable the number of OB stars in the disk. Most uncertainties in this scenario are related to $\left(\frac{dN}{dt}\right)_{infall}$. However, we can obtain an estimate of its value which may ultimately allow comparison to $\left(\frac{dN}{dt}\right)_{GD}$. A cloud with mass M_{cl} falling on the galactic disk with a highly supersonic vertical component of the velocity, v_z, produces a strong compression of gas which in the end triggers star formation. If the compression ends when the shocked material becomes subsonic in the galactic disk, the mass of gas M_{aff} affected by the shock is, from momentum conservation:

$$M_{aff} = M_{cl} \frac{v_z}{c_0}$$

where c_0 is the ambient sound speed in the galactic disk. If halo gas infall proceeds at a rate $\dot{M}_{infall}$ at a typical vertical speed v_z at the top of the HI disk layer, and assuming that most or all clouds are massive enough to create OB-type stars (so that the initial mass function generated in an impact is not truncated at the massive end),

$$\left(\frac{dN_{OB}}{dt}\right)_{infall} = SFE \times f \times \frac{v_z}{c_0} \times \dot{M}_{infall}$$

where SFE is the star-forming efficiency and f is the number of OB stars formed per each unit of mass converted to stars. As we said above, $\dot{M}_{infall}$ is considered to reflect the past stellar activity in the disk as expressed by α, and can therefore be considered to be proportional to $N_{OB}(t_R)$; this defines the value of the proportionality coefficient α appearing in Equation (1).

To assess the relative importance of gas infall with respect to other star-forming processes in the galactic disk, we have assumed a supernova rate of 1/60 SN yr^{-1} (types Ib+II) (Weiler & Sramek 1988), a star-forming efficiency of 2% in the material shocked by the cloud collisions, comparable to that of giant molecular clouds (Blitz 1993), and an infall rate $\dot{M}_{infall} \sim 1$ M$_\odot$ yr^{-1}. This value is very uncertain: modelling of the galactic fountain has provided different estimates (Bregman 1980, Norman & Ikeuchi 1989, Houck & Bregman 1990) ranging between 0.3 and 3 M$_\odot$ yr^{-1}. Column densities of CIV, SiIV and NV in cooling gas in the halo suggest higher values, between 4 and 8 M$_\odot$ yr^{-1} (Spitzer 1990). We thus take our assumed value as a rather conservative estimate. The infall velocity is taken as 150 km s^{-1}, which is characteristic of material condensed at a height of about 5 kpc from the galactic plane. The ambient sound speed is set to 10 km s^{-1}. Finally, a Miller & Scalo (1979) initial mass function is used both for the supernova-generating population of the disk and for the stellar complexes formed by gas infall.

With these ingredients, we find that $\sim 12\%$ of the galactic population of OB stars may be a consequence of halo gas infall in the form of high-velocity clouds.

3. Anatomy of oblique HVC impacts

The general case of an oblique HVC impact has some differences with respect to the head-on collision studied by Tenorio-Tagle (1981) and Tenorio-Tagle *et al.* (1986, 1987). Oblique collisions have been discussed by Comerón & Torra (1992, 1994). After the HVC encounters the boundary of the HI layer, a double shock structure develops that tends to align parallel to the galactic plane due to the density gradient of the galactic ISM. The shocked layer contains gas coming both from the HVC and from the galactic disk. The latter enters the shock with a certain parallel component of the velocity with respect to it, which is conserved. This causes a flow of dense gas out from the dense layer which moves perpendicular to the galactic disk, accreting more gas as it moves through the galactic ISM. If the accretion stops when the layer becomes subsonic by momentum conservation, the final surface density of this dense gas is

$$\sigma_f = \rho_g(z) \frac{v_s^2}{v_t c_0} l$$

where $\rho_g(z)$ is the density of the unperturbed galactic ISM at a distance z from the plane, v_s is the vertical velocity of the shocked layer at this height (a function of v_z and the density ratios between the cloud and the galactic disk), v_t is the velocity component parallel to the galactic plane, c_0 is the effective sound speed in the diffuse galactic HI and l is the linear size of the cloud. If star formation in the shocked layer starts when its size is of order of the fastest growing unstable perturbation, the height of the shocked layer where star formation begins is given by

$$\rho_g(z_i) = \frac{c_s^2 v_t c_0}{2 G l^2 v_s^2}$$

where c_s is the effective sound speed in the shocked layer and G is the gravitational constant. This unstable portion of the escaped gas acquires a subsonic vertical speed at a height z_f implicitly given by

$$\int_z^z \rho_g(z)\, dz = \rho_g(z_i) \frac{v_s l}{v_t} \left(\frac{v_s}{c_0} - 1 \right)$$

Adopting the Dickey & Lockman (1990) expression for $\rho_g(z)$, and unless a very dense cloud is assumed ($n_H \sim 1$ cm^{-3} or greater), z_f does not exceed a value of order 100 pc, similar to the scale height of Population I objects. On the other hand, the fastest growing length generally becomes of order the size of the cloud only when the vertical velocity of the accreting layer approaches c_0.

As a conclusion, HVC impacts are not likely to form stars either with high velocity or very far from the galactic plane: HVC-induced star formation is not likely to reveal itself in stellar kinematics. However, the positions of the stars formed after the impact may be expected to have a spatial arrangement reflecting the path followed by the material escaping from the shocked layer. The tilted plane containing most of the extreme Population I objects in the solar neighbourhood, known as the Gould Belt, may reflect such an impact (Comerón & Torra 1994). Recent HVC impacts in the solar vicinity have also been proposed by Lepine & Duvert (1994), and have been invoked to account for some features in the spatial distribution of young open clusters (Alfaro *et al.* 1991; Phelps 1993; Edvardsson *et al.* 1994).

This work has been supported by the DGICYT under contract PB91-0857 and by the CICYT under contract ESP94–1311-E.

REFERENCES

ALFARO, E. J., CABRERA-CAÑO, J. & DELGADO, A. J. 1991 *Astrophys. J.* **378**, 106.

BLITZ, L. 1993 in *Protostars and Planets III* (ed. E. H. Levy & J. I. Lunine). University of Arizona Press.

BREGMAN, J. N. 1980 *Astrophys. J.* **236**, 577.

COMERÓN, F. & TORRA, J. 1992 *Astron. Astrophys.* **261**, 94.

COMERÓN, F. & TORRA, J. 1994 *Astron. Astrophys.* **281**, 35.

DANLY, L., ALBERT, C. E. & KUNTZ, K. D. 1993 *Astrophys. J., Lett.* **416**, L29.

DICKEY, J. M. & LOCKMAN, F. J. 1990 *Annu. Rev. Astron. Astrophys.* **28**, 215.

EDVARDSSON, B., PETTERSSON, B., KHARRAZI, M. & WESTERLUND, B. 1995 *Astron. Astrophys.* **293**, 75.

FRANCO, J., TENORIO-TAGLE, G., BODENHEIMER, P., RÓZYCZKA, M. & MIRABEL, I. F. 1988 *Astrophys. J.* **333**, 826.

HAUD, U. 1990 *Astron. Astrophys.* **230**, 145.

HEILES, C. 1984 *Astrophys. J., Suppl. Ser.* **55**, 585.

HOUCK, J. C. & BREGMAN, J. N. 1990 *Astrophys. J.* **352**, 506.

LEPINE, J. R. D. & DUVERT, G. 1994 *Astron. Astrophys.* **286**, 60.

MacLOW, M. M., McCRAY, R. & NORMAN, M. L. 1989 *Astrophys. J.* **337**, 141.

MILLER, G. E. & SCALO, J. M. 1979 *Astrophys. J., Suppl. Ser.* **41**, 513.

MINESIGHE, S., SHIBATA, K. & SHAPIRO, P. R. 1993 *Astrophys. J.* **409**, 663.

NORMAN, C. A. & IKEUCHI, S. 1989 *Astrophys. J.* **345**, 372.

PHELPS, R. 1993 PhD thesis, Boston University.

SCHULMAN, E., BREGMAN, J. N. & ROBERTS, M. S. 1994 *Astrophys. J.* **423**, 180.

SPITZER, L. 1990 *Annu. Rev. Astron. Astrophys.* **28**, 71.

TENORIO-TAGLE, G. 1981 *Astron. Astrophys.* **94**, 338.

TENORIO-TAGLE, G. & BODENHEIMER, P. 1988 *Annu. Rev. Astron. Astrophys.* **26**, 145.

TENORIO-TAGLE, G., BODENHEIMER, P., RÓZYCZKA, M. & FRANCO, J. 1986 *Astron. Astrophys.* **170**, 107.

TENORIO-TAGLE, G., FRANCO, J., BODENHEIMER, P. & RÓZYCZKA, M. 1987 *Astron. Astrophys.* **179**, 219.

TENORIO-TAGLE, G., RÓZYCZKA, M. & BODENHEIMER, P. 1990 *Astron. Astrophys.* **237**, 207.

VAN WOERDEN, H., SCHWARZ, W. J. & HULSBOSCH, A. N. M. 1985 in *The Milky Way Galaxy* (ed. H. van Woerden, R. J. Allen & W. B. Burton). IAU Symposium 106, p. 387. Reidel.

WAKKER, B. P. 1991 *Astron. Astrophys.* **250**, 499.

WAKKER, B. P. & VAN WOERDEN, H. 1991 *Astron. Astrophys.* **250**, 509.

WEILER, K. W. & SRAMEK, R. A. 1988 *Annu. Rev. Astron. Astrophys.* **26**, 295.

Large-Scale Structure of Galaxies: Who Ordered That?

By STEVEN N. SHORE

Department of Physics and Astronomy, Indiana University South Bend, South Bend, IN 46634-7111, U.S.A.

This paper summarises the second open discussion section on the origin and maintenance of large-scale structure in galaxies. Topics include the halo–disk separation, spiral structure and its characterisation, the high-velocity clouds and large-scale structures in the magnetic field, non-axisymmetric structures such as bars, and the feedback between star formation on parsec scale and the structure of the Galaxy on kiloparsec length scales. Magnetic fields are newly implicated in structuring the halo.

1. Introductory remarks and a historical prelude

The title for this session comes from a casual remark that summarised a crisis in physics about 50 years ago. During upper atmosphere observations, a particle appeared that had the charge of an electron but considerably more mass. Yet it was expected. The Yukawa theory of nuclear structure predicted that nuclear matter was cohesive through the exchange of a particle with a mass greater than an electron but smaller than a nucleon, and it seemed at first that this particle had indeed been observed. Its mass was, however, wrong—too light to explain the strong interaction. For the first time, something truly unexpected had appeared in the microworld and I. I. Rabi reacted with "Who ordered that?" We face the same problem when dealing with the large-scale structure of the Milky Way. It isn't a spherically symmetric, isotropic, homogeneous system. Why? We'll see shortly just how far it is from this idealised state.

Large-scale galactic structure was discovered accidentally, and so early in the history of research on galaxies that it wasn't even realised that they were extragalactic. In Herschel's Index Catalogue, and later in Dreyer's revision in the New General Catalogue, nebulae were divided according to their morphology. Some were noted as being disk-like, some even were distinguished by their degree of central concentration, but the apertures used to observe them were typically small and details were lacking. The first clear evidence of spiral structure was reported for M51 by Parsons in 1850 and for M101 by Hunter in 1851 (see Berendzen *et al.* 1976). By the end of the century, other nebulae had been resolved, sometimes erroneously (observers, it seems, sometimes see what theorists say they ought to see).

Although Herschel and Wolf had employed taxonomic criteria for "nebulae" that depended on large-scale structure, the first comprehensive morphological scheme of galaxies based on physically motivated guidelines was Hubble's (1936). Hubble's taxonomy is still the most successful. He used a minimum parameter set, partly guided by Jeans's theoretical work, to distinguish galaxies: the degree of central flattening concentration of the spheroidal bulge and the openness of the spiral arms. In his broad classification, the ellipticals were grouped according to the ratio of minor to major axes, a measure of the eccentricity of the spheroid assuming that it was cylindrically symmetric. Spirals had the added feature of the presence of a bar and the flatness of the spheroid in addition to the resolution of the arms, in effect using the pitch angle.

Dynamical theories for the origin of the arms assumed that the arms were perturbations on the overall stellar distribution. Lindblad† assumed that they were a stationary pattern supported by the differential rotation of the galaxy. Although the disk is self-gravitating, the orbits were initially treated as Keplerian circulation about a central point source. The characteristic timescale for the perturbation was related to the epicyclic frequency in such a way that there were two points in the galaxy where the wave would be in resonance—the inner and outer Lindblad resonances at $\Omega \pm \kappa$ (ILR, OLR hereafter), where κ is the epicyclic frequency and Ω is the local orbital frequency. Once you know the rotation curve, the resonances are immediately fixed.

In a sense, besides the social upheavals of the day, the 1960s were a "Golden Age" of instabilities. Parker (1966) computed the response to gravity of a gaseous disk supported by cosmic rays and magnetic fields, modifying and extending the Kruskal–Schwarzschild criterion for plasmas to galactic scales and including a relativistic fluid (i.e. cosmic rays) (see Parker 1979 and Zweibel 1987 for especially good discussions of the physical development of the instability). About the same time, Field (1965) performed the first systematic examination of the cooling instability in an optically thin gas. It is important to note, however, that these were all local phenomena. In particular, the thermal instability has no characteristic length scale. Recognising the importance of rotation in structuring the system, Lin & Shu (1964) examined global quasi-stationary density waves as unstable modes in thin self-gravitating disks. This was later expanded to finite disks by Shu (1968, PhD thesis). Toomre (1964)‡ established the importance of the $Q = c_s^2/2\pi G\kappa^2$ as the dimensionless parameter measuring the importance of disk self-gravitation, the extension of the Jeans criterion to differentially rotating (i.e. sheared) slabs. It is the controlling physical variable for gravitational instability in a differentially rotating disk. A major advance came with hydrodynamic modelling by Roberts (1969), who provided the first non-linear calculation of flow in a spiral potential. This work established the paradigm that we still debate, that star formation observed along the arms is triggered by the passage of a large-scale disturbance through the gas at a pattern speed different from the local circulation velocity (see Roberts 1975; Roberts & Shu 1975; Elmegreen, this volume).

This catalogue of local and global instabilities highlights the complexity intrinsic to the study of the origins of the structures seen in galaxies. All of the mechanisms operate on similar timescales and act in concert, as we shall soon see.

2. Guidelines

As a road map for the discussions that follow, let me provide a list of questions that encapsulate the essential features of the physics.

• *Would we recognise an unstructured system?* Wittgenstein remarked in the *Tractatus Philosophicus* that we would not be able to recognise an illogical world for what it is. This applies as well to large-scale structure. To ensure that we don't prejudice our interpretation of galactic structure, we require objective analytical methods. The qualitative classification system introduced by Hubble is fine as far as it goes for distinguishing types of galaxies. But once we seek to move past taxonomy and towards a deeper understanding of mechanism, it is useless. It lacks the unambiguous physical foundation of the MK classification system for stellar spectra. The inverse problem is

---

† See IAU Symposium 10 and Lindblad (1963) for thorough discussion of the dynamical background to single particle density wave approaches.
    ‡ For all references on early work on spiral structure, see Toomre 1977.

also important to solve. Without an adequate physical understanding of the large-scale structure, we can't create trustworthy taxonomic criteria.

• *What do we mean by "large"?* Christine Jones, who was unable to attend this discussion, left a little bombshell behind when she asked what "large" means. This is a profound question because it focuses attention on the mechanisms. I think we usually use the term to mean long-range, like the length scale of the visible disk in a spiral. But this is far too great a length for the scale over which self-gravitation acts.

• *What do we mean by "structure"?* First, we see that galaxies have complex symmetries. Triaxial structures are quite important—for instance, bars and ellipsoidal halos. This is followed by the observation of disks and bulges. Ring systems come in many varieties. Some are generated by bars, the Θ-ring galaxies, where the structuring is strictly planar. Another class resembles the Cartwheel, the off-centred nucleus elliptical rings first discussed by Lynds & Toomre (1976) and Theys & Spiegel (1976). Finally, we see polar rings, those galaxies that have extended structure orthogonal to the plane of symmetry. Some disk galaxies show warps in the symmetry plane, especially noticeable in maps of the HI gas distribution. But the most compelling is the spiral structure observed in disk galaxies, separated into the flocculent and grand-design subtypes (e.g. Elmegreen & Elmegreen 1990) in an extension of the Hubble and RC2 sequences. Notice that all of these are big, extending more or less coherently over substantial fractions of the size of the system.

• *The "greening" of galactic structure: act locally, think globally.* Star formation is a very local phenomenon, on the scale of a few parsecs. Buried deep within molecular clouds, there are knots that reach densities high enough and temperatures low enough for them to collapse. The energetic phenomena that feed back into the large-scale structure of cloud complexes begin on scales of only few stellar radii, far below the smallest observable extragalactic scale. As Ferrini discussed in his panel, the feedback to galactic gas and stellar dynamics works its way up from this tiny size to kiloparsec scales through the effects of winds and supernovae.

• *Is there a horizon in the fossil record?* Catastrophism, as Tom Bania noted in his talk, is becoming more popular these days among geophysicists. Maybe we can take a leaf out of their book. As in the extinction of the dinosaurs being marked on this planet with a horizon, the Cretaceous–Tertiary boundary, we should ask whether there is evidence for such horizons in the galaxy. Does the metallicity history contain any discontinuities? Have the population mixtures been altered over time? In short, is there any reason to believe that the Galaxy has evolved as an isolated system over the past 10 Gyr?

• *Does the Milky Way suffer from dynamical amnesia?* Freud might have asked, Does the system remember its possibly traumatic past? Dynamical relaxation is slow in a system as sparse as the disk of the Milky Way. Once a strong perturbation has acted on the stellar velocity distribution, its record should be preserved. Is there any indication that violent accretionary events have occurred over galactic history?

• *Is the mirror providing a correct representation?* In *Julius Caesar*, Shakespeare posed an important observational question. Cassius asks, "Tell me, Brutus, can you see your face?" to which his companion replies, "No, for the eye sees not itself but by reflection, by some other thing." Thus, stated in our particular language, we can ask whether the Milky Way is *the* galaxy or rather *a* galaxy? For instance, the first suggestion of spiral structure for the Galaxy was by Easton (1900) based primarily on an analogy with M51, a colliding system. Is our view of processes in our system thus biased by our perceptions of external galaxies? We all know that it is singularly difficult to determine the structure of the system in which we're embedded.

Finally, since we are in Granada, we should recall an experience from the visit last evening to the Alhambra. When viewed at close range, the symmetries in the tiling pattern are different from the large-scale structure that they create as one steps back from the plane. The first discussion concentrated on the individual tiles. This panel concentrates on the big picture. The pattern is, unfortunately, a dynamic one that we see statically merely because of our finite lifetimes.

3. Scales of length and time

The length scales are well known. On the smallest scale—parsecs to tens of parsecs—single stars and strongly dissipative hydrodynamic flows predominate. Turbulence, whether generated by internal sources like protostellar winds or external sources like shear flows, finds its source scale here. On the intermediate scale of a few hundred parsecs, collective mechanical and radiative input from groups of stars, such as associations, plays the major role. OB associations produce supershells. These are ordered by the next larger scale, but for the most part they act locally. We should also not forget the omnipresent mechanism of cloud–cloud collisions for inducing star formation and cloud dispersal. So the return of gas and large-scale formation of stars occurs on a length scale that is small compared with the next higher length scale, kiloparsecs, on which we encounter the action of magnetic fields, cosmic rays, hydrodynamic diffusive flows, and any dynamics that depend on the gradient in the gravitational potential of the disk–halo system. Put another way, on the scale of tidal forces—the Oort constants for instance—the minimal degree of ordering is introduced into the system. There is a scale that is even larger, involving the entire galaxy, namely galactic collisions and mergers. These can't be neglected, especially in light of the horizon question.

4. Determination of structure

How can we determine whether a system possesses large-scale structure or not? One way is qualitative, following the statement of an American judge who declared of pornography that he couldn't define it but "I know it when I see it." The other is to use impersonal statistics, modal analysis, correlation functions, and moments of the luminosity distribution in rings around the centre. As we shall see, this topic was also discussed during the session and in posters at the meeting.

The existence of large-scale structure requires large-scale ordering. A trivial statement, perhaps, but one that focuses attention on the basic feature of spiral galaxies: they rotate. More specifically, they rotate differentially. This means that there exists in all of them a characteristic timescale at any position—the inverse of the epicyclic frequency, $t = \kappa^{-1}$, where $\kappa^2 \sim \partial^2\Phi/\partial r^2$. Here $\Phi(r, z)$ is the unperturbed gravitational potential. The basic feature of the disk, its mass distribution, determines the length and timescale most typical for any flows. It is important to remember that the vertical timescale for a stellar orbit is $\omega_z^2 \sim \partial^2\Phi/\partial z^2$, which is comparable to κ, while the orbit timescale is given by $\Omega^2 \sim \partial\Phi/\partial r$.

Galaxies are extended mass distributions and their typical scale lengths are kiloparsecs. Thus there is a scale imposed on any flows by potential gradients both within the plane and in the vertical direction. One of the most interesting features of spiral galaxies is that they are remarkably fine-tuned. Nearly all of the relevant dynamical timescales are the same. This makes it extraordinarily hard to distinguish the triggering mechanisms for star formation. It may also contribute to a systemic chaos, in that the selection of a particular agent may be extremely sensitive to noise. For instance, the gravitational collapse time for a molecular cloud is about the same as the epicyclic timescale through the Q parameter. The critical local input here is the local sound speed. Because of the expansion velocities of supernova and wind-generated bubbles, the mixing time for the interstellar medium is also about the same as κ^{-1}. Of course, the density wave timescale, determined by the pattern speed at a given galactocentric radius, depends on the differential rotation timescale. Finally, the evolution timescale for the massive stars that power much of the motion of the gas phase is about the same as the interval for large-scale motion.

In the face of these coincidences, the question of local triggering *vs.* large-scale ordering becomes very difficult to answer.

The overall systematics of structure and dynamics are pretty well solved, largely because of the heroic efforts of previous generations of optical observers who have determined the distribution of massive stars in the solar vicinity, and through the data returned by the *IRAS* and *COBE* satellites. These, however, are relatively low resolution views, lacking a crucial datum—the distance to the different components. The three dimensional structure must be recovered using inversion methods that require considerable tinkering and many assumptions.

One of the chief discussions in this session focused on how to separate and characterise stellar groups in the halo. More of this has been summarised in Carney's discussion section, and in talks by Majewski, Alfaro, Agostinho and Ojha at this meeting. When we separate the stars into population components distinguished by their scale heights, we must be mindful that we require the three-dimensional velocity distribution in order to ascertain whether the population is really isothermal. This is the hidden assumption underlying this parameterisation. Here we face the problem of the history of the system, in particular whether the current state of the galaxy is representative of conditions in its distant past. Recent work has also argued that the Galaxy has accreted at least one comparatively low-mass system, and we have heard of this during our meeting as well. Almost annually, the Local Group grows more populous and the immediate vicinity of the Milky Way more crowded. In fact, just after the meeting, a massive galaxy—Dwingaloo 1—was detected in the "zone of avoidance" in the galactic plane and described in the literature (Kraan-Korteweg *et al.* 1994). Thus collisions and even mergers are likely, and this means that the population concept becomes progressively fuzzier. In addition, as Levine stressed, the dynamical models for the local group must be re-thought.

5. Characterising spiral structure

The recognition of large-scale structure in other galaxies has been discussed by Elmegreen (1991, this volume), who employed log-spiral decomposition. Levine emphasised that the advantage of choosing this template is that the eye is a wonderful pattern-matching device, capable of spotting symmetries and also noting where they are broken. However, more quantitative measures are required to identify the role of the dynamical agent for this structure. These were described by del Rio & Cepa. They use Fourier decomposition in radial annuli to obtain the azimuthal power spectrum of the light distribution in various filters. They find that there is a change in the modes at the corotation radius, with low orders (like $m = 2$, simple two-arm spirals) changing to the next even order as the corotation radius is crossed. They observe that the pitch angles of the arms change at radii that are identified, from the rotation curve, as resonances and corotation. The arms also split at the corotation radius, at which they also note decreases in the rate of star formation. Asymmetry profiles across the arms also change; specifically the third moment of the distribution changes sign.

The Fourier techniques assume, implicitly, that the rings are dynamically separate. However, if there is non-circular motion of the gas, or if the stellar orbits have appreciable eccentricities, it may be better to use wavelet methods for the global analysis of various size scales of structure. Radial flows along arms may change the skew. One way of checking this is to conformally map the galaxy into a logarithmic spiral which, as Levine pointed out, is precisely what the Elmegreen & Elmegreen decomposition does. Given a rotation curve, the sheet that can be overlaid on the galaxy is "rubbery"—i.e. once you have one of the radii located (such as the 4:1 radius, or OLR), then it's easy to iteratively match all of the other critical radii.

The idea that there should be trailing modes has been uppermost in our minds since the first resolution of the extragalactic spiral structure. What we require is a more precise objective measure. Lesch pointed out that the N-body simulations by Sellwood & Carlberg (1984) produced transient unstable modes that change on a rotation timescale. In order to stabilise the disks, they required dynamical cooling to prevent the stellar velocity dispersion from becoming too large.† Therefore, Lesch argued, the individual critical radii may be meaningless since the modes come and go. So what use is it to identify them? He continued that large asymmetries in the potential—like a bar—mechanically maintain these waves. The pattern must exist, even in these transient spiral modes, and with them must also exist the ILR. Its interaction with the particles causes the pattern to move. Thus as long as there is a driving force, the structure will exist but the modes don't couple to this.

Rhodes noted that there were other ways, not dependent only on morphology, of studying the location of the resonances and corotation. Tremaine & Weinberg, Westphfal and others have developed methods of determining the corotation radius using the continuity equation. However, if there are warps in the Galaxy, there may be uncertainty in these procedures.

Palouš remarked on the connection between the intermediate and large scales. The large scale—spiral arms and the like—influences the small scale. For our Galaxy, the question is whether there is a bar and what its properties are. Also, is the bulge spheroidal or triaxial? The centre is important as an influence on the large-scale structure. Cusp orbits near the ILR are important because they increase the star formation, as Vietri also noted. This confuses the issue of identification of increased star formation with the corotation radius. Both speakers emphasised the crucial role a bar may play in the production of large-scale structure.

Ferrini previewed his session's discussions by introducing the ecological problem. The spiral arms, whether transient or permanent, are important for the gas just at the start of star formation. They seem to last long enough for the system to be organised non-homogeneously. However, the scale for star formation is much smaller than the giant molecular clouds (GMCs), and even smaller than the molecular clouds. On the real scale of star formation, there is an enormous structural complexity. These smallest scales are a long way from the large-scale structure. The small scale looks completely stochastic, and the smallest structures are transient. They require some help from existing stars to form new stars. The large-scale structure organises the compartments. Is there something operating on an intermediate scale? Large-scale gradients, intermediate organisation, smallest scale for the formation of the stars themselves.

This brings up the question of stochastic *vs.* deterministic mechanisms for generating spirals, a question that has been examined recently by Palouš and his collaborators and Shore & Ferrini (1995). Zasov discussed the formation of coherent structure from

† Levine added that "this is Toomre's Q hitting you smack in the face."

GMC generation in the tips of sheared supershells. He argued that large-scale structure can also be the product of these hydrodynamic features in the galaxy amplifying the star formation. The expansion timescale for the supershells is very close to the vertical oscillation timescale for the galaxy. This is about the same as the timescale for the differential rotation.

Hensler noted that Gerola and Seiden (see Seiden & Schulman 1990) had many problems with forming grand-design spirals and had to choose specific timescales to obtain these. The stochastic modes of self-regulated spiral structures should lead to smoother systems. What are the differences between the Sc and Sa galaxies? Is it something about their interstellar media? Perhaps they have the same internal properties but what differentiates them is the propagation length for star-formation activity, which is equivalent to changing the critical probability p_c in the stochastic disk models.

Elmegreen clarified the differences. Grand-design spirals (GDSs) are bigger, perhaps by 40%, than flocculent systems. Otherwise the colours and other properties are the same. But GDSs tend to have bars or companions. You should make a distinction between galaxies with irregular looking long arms and those with many shorter arms. The former are truly waves, the short arms aren't; irregularity doesn't mean a lack of waves. He added that it is a mystery why the interstellar medium should be different between these two broad classes of galaxies. The disk-to-halo mass ratio, through Q, governs everything. If this parameter is low, you produce spirals; if not, any of a variety of processes act. He added that Q is known from observations, but it has not been compared between the different types of disk systems.

Franco made the point that Q for spirals increases with time as the disks heat, but Majewski said that the highest latitude stars may belong to another separate population. He also noted that it is very hard to find mechanisms for heating the stars in the high halo. You can't heat the stars to speeds of 45 km s^{-1} or higher through secular processes.

What about violent relaxation. How do we know that the collapse was dissipational? When did the halo form stars. What if the collapse was dissipationless? Are there any clues? Why should the halo be isothermal. Carney pointed to the metallicity gradients as signatures of the dissipational collapse. He also emphasised that the halo is only a trace component of the mass distribution.

What about dark matter? While not physically equivalent, massive dark halos serve *functional* roles similar to cooling mechanisms. They delay the onset of mode damping by raising κ to large values rather than reducing c_s. We trace out a gravitational potential but what's its origin? This is crucial for arguments of star formation based on Q.

6. Magnetic fields as originators of structure

Whenever astrophysicists gather, magnetic fields will eventually be invoked to provide conversation and dynamics, and this meeting was no exception. In the context of high latitude gas and the high-velocity clouds (HVCs), some participants conjectured that magnetism might be required for any eventual explanation. Foglizzo added a few comments about whether the Parker instability explains the origin of the HVCs. He noted that it is hard to understand the observed velocities by such a mechanism, which are of the order of 100 km s^{-1}. The clouds are isolated. They don't seem to be connected with the kind of large arc-like structures you would expect from the Parker instability.

Continuing the discussion of what role magnetic field might play in the structuring of halo gas, Lesch described plausible solar analogues. For instance, arches are configurations of cylindrical magnetic fields and more compact flux tubes. The confined gas "sleeps on a magnetic bed." Cooling of the gas leads to non-gravitational trajectories in the flows, but the relevant speeds and field strengths have to be calculated. The cooling of the material causes non-gravitational trajectories as the stuff returns to the photosphere in the sun. Perhaps, he remarked, this analogy works as well for the halo.

To this, Danly was encouraging, mentioning that it's been about 20 years since anyone tried to model the structure of the high-latitude clouds employing a solar analogy (Sturrock & Stern 1980). She noted that it is important to distinguish between the large and small HVCs. For the large loops, Zeeman splitting measurements have not been obtained yet—there isn't enough signal detected so far to permit this. But for the masses and energies involved, she thought that the magnetic fields required for their structuring must be quite large. Just how large is an open question. Danly added a comment about the driver for the HVC dynamics. In the abstract of her talk, she described the engine driving HVC and halo gas dynamics as being star formation. Danly said she is now moving toward a picture in which the driver of the HVCs is a galactic dynamo. Star formation may be completely unrelated to their origin.

Notice how substantially the introduction of magnetic fields changes the debate concerning large-scale features. There is a natural scale length in the system if a dynamo is at all important because resistive terms and inductive terms conspire to create a natural length; with gravitation alone, there is no characteristic length. Thus opens a broad new class of models.

Lesch added that there is new evidence for the role played by magnetic fields in HVCs. Zeeman measurements for one HVC—in the N complex—were 35 microgauss at 6-sigma (just detected with Güsten) at the edge of an HVC. Franco pointed to the north polar spur as a possible example of a magnetic arch and asked about possible magnetic field strengths. Battaner responded that intergalactic magnetic fields have been measured that are of order 3 μG on a scale of hundreds of megaparsecs! This is really large scale! The field is assumed to be generated by the motion of galaxies in the intergalactic medium that generates this. If the fields are so high then this only works for very weak fields, of order 10^{-8} to 10^{-10} Gauss. Now, he said, the problem is to explain much stronger fields. Maybe they are primordial or due to an early generation of galaxies that produced seed fields.

The determination of such high fields is, however, based on Faraday rotation measures towards radio lobes of active galaxies. How well do we understand the fine structure of the fields and the gas? Do we understand the distribution of magnetic fields and gas at high latitude? What about the holes in the polarisation maps? Are these Faraday-rotated ordered fields by intervening galactic high-latitude clouds? What about Fornax A, for instance, which covers nearly a degree on the sky? Removal of the foreground screen is very hard. Lesch explained, in response, that in the holes, there is sometimes a change in the sign of the RM measure on a scale of 1.5–2kpc. Thus the field structures must be large loops.

7. Intrinsic and extrinsic origins of structure

A question arose as to whether the Milky Way is a polar ring galaxy, referring specifically to the Magellanic Stream. Here we see another kind of large-scale structure, extrinsically generated and supplied. Where, though, are the stars from the stream? Is it the remnant of the Clouds as they are now? Were there other galaxies present sometime in the past that have now been tidally disrupted? These are important questions. Observations of IC 342 and other modelling of the local Local Group suggest that at least some of the systems near ours could be very massive. Levine pointed out that stability of accreted flows consequently becomes a much more complex problem. The rings are very stable in *some* potentials. But how stable would the rings be in the face of strongly interacting complex systems? Some of the dwarf spheroidals may line up in planes, perhaps related to a common origin. This remains an open question. Levine added that there seems to be plenty of evidence that we've been hit sometime in the relatively recent past.

Majewski also noted that work on more distant clusters in our galaxy suggests that these could be good dynamical tracers of past events. Scl moving toward Fornax. The ones we have left may not fairly represent the original population. Majewski, Carney, and Schuster noted that stellar streams and inhomogeneity in the halo may trace the history of interactions within the Local Group. However, Schuster noted that single-orbit modelling shows clumping of stellar orbits in an otherwise stochastic sea. Since all of the relevant timescales are comparable, perhaps this is affecting the orbits and producing chaos. This will be a fruitful area for further investigation.

8. Conclusions

As you can see from the discussions, the problem of large-scale structure cannot be divorced from questions about local instabilities, star formation and chemical evolution, or the disk–halo connection. We are left with more questions than solutions, topics for those clever graduate students to come who we hope will find the answers where we have failed. Collisions and mergers are now playing important roles in our thinking. The Galaxy is clearly a barred spiral with a massive halo, and this combination is central to the generation and stabilisation of spiral modes. As in the early 1960s, magnetic fields are again pivotal agents because of the obvious needs to couple phenomena in the gas and stellar phases. We did not hear much at the meeting about the galactic scale dynamo, but it was a *leitmotif* of the meeting after this panel. What happens in the fluid is once again being linked to stellar dynamics. The seeds planted here will certainly flower in the years ahead.

I wish to thank all of the discussion participants for making this such a lively and productive session. I hope they will feel content with the way their views have been recorded for posterity. My deepest thanks for E. Alfaro and the LOC for their superb work in arranging the technical details of the panel and for arranging to obtain clandestine recordings (in typical "talk-show" fashion). I also want to especially thank J. Cepa, A. D. Chernin, L. Danly, B. Elmegreen, F. Ferrini, J. Franco, A. Kolesnik, H. Lesch, S. Levine, J. Palouš, G. Tenorio-Tagle, M. Tosi, and M. Vietri for extended discussions on some of the topics raised during the meeting. Participation in this meeting was partly supported by NASA.

REFERENCES

BERENDZEN, R., HART, R. & SEELEY, D. 1976 *Man Discovers the Galaxies*. Science History Publications.

EASTON, C. 1900 *Astrophys. J.* **12**, 190.

ELMEGREEN, B. G. & ELMEGREEN, D. M. 1990 *Astrophys. J.* **355**, 52.

ELMEGREEN, B. G. & ELMEGREEN, D. M. 1991 in *Dynamics of Galaxies and their Molecular Cloud Distributions* (ed. F. Combes & F. Casoli). IAU Symposium 146, p. 113. Kluwer.

ELMEGREEN, D. M. & ELMEGREEN, B. G. 1984 *Astrophys. J., Suppl. Ser.* **54**, 127.

FIELD, G. 1965 *Astrophys. J.* **142**, 531.

HUBBLE, E. P. 1936 *The Realm of the Nebulae*. Yale University Press.

KRAAN-KORTEWEG, R. C., LOAN, A. J., BURTON, W. B., LAHAR, O., FERGUSON, H. C., HENNING, P. A. & LYNDEN-BELL, D. 1994 *Nature* **372**, 77.

LIN, C. C. & SHU, F. H. 1964 *Astrophys. J.* **140**, 646

LINDBLAD, B. 1963 *Stockholm Obs. Ann.* **22**, 55.

LYNDS, R. & TOOMRE, A. 1976 *Astrophys. J.* **209**, 382.

PARKER, E. N. 1979 *Cosmical Magnetic Fields.* Oxford University Press.

ROBERTS, W. W. 1969 *Astrophys. J.* **158**, 123.

ROBERTS, W. W., ROBERTS, M. S. & SHU, F. H. 1975 *Astrophys. J.* **196**, 381.

SEIDEN, P. E. & SCHULMAN, L. S. 1990 *Adv. Phys.* **39**, 1.

SELLWOOD, J. A. & CARLBERG, R. 1984 *Astrophys. J.* **282**, 61.

SHORE, S. N. & FERRINI, F. 1995 *Fundam. Cosmic. Phys.* in press.

STURROCK, P. A. & STERN, R. 1980 *Astrophys. J.* **238**, 98.

THEYS, J. & SPIEGEL, E. A. 1976 *Astrophys. J.* **208**, 650.

TOOMRE, A. 1964 *Astrophys. J.* **139**, 1217.

TOOMRE, A. 1977 *Annu. Rev. Astron. Astrophys.* **15**, 437.

ZWEIBEL, E. G. 1987 in *Interstellar Processes* (ed. D. Hollenbach & H. A. Thronson). ASSL, vol 161, p. 202. Reidel.

Metallicity Distributions in the Galactic Disk

By KENNETH JANES

Astronomy Department, Boston University, 725 Commonwealth Ave., Boston,
MA 02215, U.S.A.

A great deal of effort has been expended to study the compositions of stars and gas across the galactic disk. Although a few general trends are evident in all the data, some apparently dramatic discrepancies are present as well. New studies of open cluster stars together with a fresh look at some of the old data suggest an entirely new picture of galactic chemical evolution. On a global scale, an abundance gradient radially along the galactic disk is well established, but there is little or no solid evidence for either a significant metallicity gradient perpendicular to the plane or an age-metallicity correlation among disk stars. On the other hand, the data now clearly show local irregularities in the global patterns, in the solar neighbourhood and elsewhere in the disk. Both the large-scale distributions and local irregularities are consistent with a picture of galactic evolution in which the Galaxy has been heavily influenced by interactions with external systems.

1. Introduction

The study of the metallicities of disk stars would probably be considered by most people to be a "mature" field: some general trends have been agreed upon for many years, and few would expect any major surprises. Building on the general ideas of the well-known Eggen, Lynden-Bell and Sandage (1962) paper, most models for the chemical evolution of the Galaxy have predicted (i) a distinct correlation between stellar ages and compositions; (ii) a statistical correlation between the distance of a star from the galactic plane (or its z velocity near the plane) and its composition; and (iii) a correlation between a star's galactocentric distance and its composition. A number of observational studies (e.g. Twarog's (1980) study of the ages and metallicities of the F stars; the Hartkopf & Yoss (1982) survey of G giants towards the galactic poles; and the Shaver et al. (1983) review of radial abundance gradient results) have apparently confirmed these predictions, in some detail. Nevertheless, there have long been some apparently inconsistent observational data, most notably the so-called G-dwarf problem (Pagel & Patchett 1975). In just a few years, many of the older conclusions have been called into question for various reasons, so it now appears that only the radial metallicity gradient can be confirmed. Furthermore, there is strong evidence that even among stars of similar ages located in the same part of the Galaxy there is a substantial dispersion in their metallicities. A complete re-thinking of galactic chemical evolution is in order.

To learn how such a complete change in our view of galactic evolution has occurred, it is necessary to understand something about the nature of the observational data and what the limitations are.

2. A brief survey of metallicity surveys

The number of studies and surveys of stellar metallicities is truly enormous, far beyond the scope of this brief review. A recent general review of the field of galactic chemical evolution can be found in Rana (1991), but it is useful to highlight here a few studies that have in one way or another been particularly influential in shaping our view of

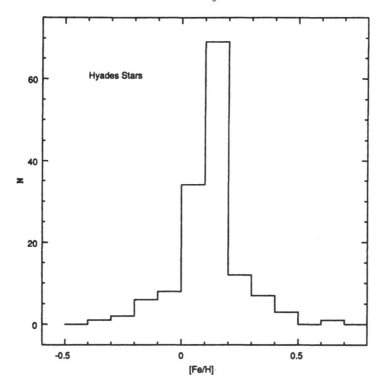

FIGURE 1. Histogram of measured metallicities for Hyades stars, taken from the catalogue of
Cayrel de Strobel *et al.* (1992)

galactic chemical evolution. There are two issues to evaluate: the global, galactic-scale
trends and the local irregularities. For the latter question, it is necessary to separate the
observational dispersion in a data set from any intrinsic dispersion that may exist. It is
convenient to divide metallicity surveys into several categories depending on how fine a
spectral analysis was done.

2.1. *High-resolution spectroscopy*

Most recent high-resolution studies of stellar abundances have been included in the Cayrel
de Strobel *et al.* (1992) compilation of [Fe/H] studies—the latest edition of this comprehensive literature survey includes 3252 measures for 1676 stars. The Cayrel catalogue
is a heterogeneous sampling of the astronomical literature, and it is difficult to make
a systematic characterisation of the errors of a typical entry in the catalogue. As a
demonstration of the typical quality of the data that makes up the Cayrel catalogue,
Figure 1 shows a histogram of metallicity measures for stars of the Hyades cluster; presumably all (or nearly all) of these stars actually have the same composition. Known
or suspected Am stars were excluded from the figure. The mean of the distribution is
[Fe/H]= 0.12 dex, close to the "accepted" value for the cluster (e.g. Boesgaard & Friel
1990), and the standard error is ±0.12 dex. Notice, however that the full range of values
for the metallicities of Hyades stars is over one order of magnitude. Since the Hyades
stars are likely to represent a best-case situation, the implication of Figure 1 is that
the errors of high-resolution abundances are probably no better than about 10%, and
possibly much worse.

Keeping Figure 1 in mind, it is interesting to examine Figures 3a and 3b of Cayrel *et al.* These two figures show the metallicity distributions of all stars in the catalogue with normal MK spectral types (their Figure 3a), and those with B, A, or F spectral types (their Figure 3b). In both cases, the distributions are substantially broader than that of Figure 1 (but note that the bin size is also three times larger), consistent with the existence of a real spread in composition among typical solar-neighbourhood disk-stars. The standard deviation of the distributions in the Cayrel *et al.* Figures 3a and 3b are perhaps ±0.5 or somewhat less. The surprising thing about these two figures is that they are, except for the number of stars, *virtually identical.* Figure 3b, containing only B, A and F stars, should be a sample of relatively young stars, whereas Figure 3a should contain both young and old stars. Thus, if there were a progressive increase in metallicity with time in the galactic disk, one would expect the distribution in Figure 3a to be broader than that of Figure 3b and biased towards lower metallicities. The data in the Cayrel catalogue show no evidence of an age–metallicity relation.

Edvardsson *et al.* (1993; see also the paper by Andersen in this volume) have recently derived abundances of many elements for 189 nearby F and G stars from high-resolution spectroscopy. Although there are far fewer stars than in the Cayrel catalogue, the Edvardsson *et al.* study has the advantage of being a uniform, systematic survey, so the quoted relative errors of their abundances are only 0.05 dex. As their Figure 14 shows, for disk stars at a given galactocentric distance, there is little or no systematic change in metallicity with age. There is, however, a substantial dispersion in metallicity at all ages.

Friel & Boesgaard (1992) obtained high-resolution spectroscopy of stars in several relatively young open clusters and moving groups. The typical errors of their mean cluster values are of the order of ±0.05 dex. They find no correlation of age with metallicity. Yet, even though all of their clusters are in the immediate solar neighbourhood, there is a significant dispersion in the metallicities.

2.2. *Low- and moderate-resolution spectra*

Friel and Janes (1993) used moderate-resolution (4 Å) spectra of red giant stars in 24 clusters to derive the cluster metallicities, and elsewhere in this volume Friel describes preliminary results for several additional clusters. This work will be discussed in the following sections, but in summary the open cluster data confirm the existence of a dispersion in metallicities at all ages at a given location in the Galaxy, and demonstrate the lack of an age–metallicity correlation.

Carney *et al.* (1990) have combined spectroscopically determined compositions and space motions for nearby subdwarf and dwarf stars to derive the relationship between stellar kinematics and composition. Although they used high-dispersion spectra, their abundances are derived from the cross-correlation of their spectra with template synthetic spectra, and so they could use very low signal-to-noise ratio samples, which were therefore effectively of moderately low spectral resolution. The power of the cross-correlation method is that even spectra that would ordinarily be considered underexposed will yield precise overall metallicities. Their primary aim was to study the galactic halo, but their sample includes a large number of disk stars. They found a clear separation between the halo and disk composition–velocity distributions, with little overlap between the two populations. Furthermore, within their disk star population, there is little or no correlation between kinematics and composition.

2.3. *Photometric surveys*

At the lowest spectral resolution characterised by intermediate- and broad-band photometry, there are some discordant results. It is unlikely that metallicities can be estimated with uncertainties smaller than about 0.2 dex from photometric techniques but, on the other hand, it is possible to measure a great many stars with relative ease. There is first the influential work of Twarog (1980) who used *uvby* photometry for a large sample of F-stars to derive their compositions and ages. He found a distinct correlation between stellar ages and compositions, a correlation that is easily reproduced by theoretical chemical evolution models. A major complication of analysing a sample similar to that of Twarog is that one must account for the effects of stellar evolution and composition on the probability of finding a star at a given stage of evolution. Unfortunately, it appears that it may not be possible to correct fully for these effects, which may account for the differences between Twarog's results and the above studies.

Cameron (1985) compiled UBV photometry of open clusters with a wide range in ages. His analysis clearly shows the radial abundance gradient, but because he had only a very few really old clusters, his data do not give any real information about the age–metallicity relation. There is certainly a large dispersion in metallicity at each age, but because of the heterogeneity of the source data, much of that dispersion is observational.

Geisler *et al.* (1992) obtained Washington photometry of open cluster giants in nine old disk clusters in the third galactic quadrant. They found several clusters with extraordinarily low metallicities, and at least two —NGC 2324 and NGC 2660— with [Fe/H] less than -1. Although all of the above studies are consistent with a significant range in composition among stars of a fixed location and age, none of other surveys found disk stars at such low metallicities. Furthermore, there is a considerable disagreement between the Geisler metallicities of a few of the clusters and the values found by Cameron (1985) and Friel & Janes (1993); in all such cases the Geisler values are substantially lower than the others. Nevertheless, it is difficult to reject the conclusion that some of their clusters have rather low metallicities.

An indirect corroboration of the low metallicities that Geisler *et al.* found among clusters in the third galactic quadrant comes from Alfaro *et al.* (1991), who showed that a distortion in the disk they call the "Big Dent" exists in the disk in this direction. The Big Dent is a region where the young clusters and other young objects lie systematically below the galactic plane. A possible explanation for the Big Dent is that it was created in the impact of a high-velocity hydrogen cloud with the galactic disk. If the Big Dent and other similar structures were created by the impacts of clouds from outside the disk, there is the possibility that the impacting cloud could be metal-poor, thereby diluting the local gas and leading to the formation of stars and clusters with low metallicity.

There is no direct evidence that the low-metallicity clusters found by Geisler *et al.* have anything to do with the Alfaro *et al.* Big Dent, except that they are located in the same general part of the Galaxy. The Big Dent is defined by very young objects, whereas the Geisler *et al.* clusters are all substantially older. Nevertheless, both observations point up the need to account for local irregularities and disturbances when trying to understand either the dynamical or chemical evolution of the galactic disk.

3. The galactic system of old open clusters

Two classic problems have long confounded efforts to probe the chemical history of the galactic disk: finding ways to select stars by age and selecting unbiased samples of stars in distant regions of the galactic disk. One way to at least minimise these difficulties is to select stars in open clusters. The open clusters are easily recognisable structures, spread widely through the disk, and their ages and distances can be estimated with some degree of reliability. Although in principle there is still a strong selection effect, in that the more distant clusters will necessarily be found further from the galactic plane on the average than the nearby ones, as will be shown in the following section, that is not a serious problem in practice, at least not for the older clusters. The problem with the open clusters is that they are almost all very young, compared to the age of the galactic disk.

The potential importance of the old open clusters for galactic structure studies has led our group at Boston University (which consists of myself, Eileen Friel, Randy Phelps and Kent Montgomery) to conduct a search for previously unrecognised old clusters. Several lists of potential old clusters have been published (see Janes & Phelps 1990) and these lists are being used by several groups as target lists of potentially interesting clusters (e.g. Kaluzny, this volume).

As a result of this effort by a number of people, the number of known old clusters has increased dramatically in the past several years. A summary of this work, and a discussion of the implications of the old open cluster properties for the formation and evolution of the galactic disk are given in Phelps *et al.* (1994) and Janes & Phelps (1994). The following discussion is abstracted from those two papers. Consider first the basic observational properties of the old cluster sample:

3.1. *Number of old clusters*

Through a combination of compiling data from the literature and obtaining our own photometry, we identified 73 clusters with ages $\geq$ Hyades ($\sim$ 700 Myr) (see Table 1 of Janes & Phelps 1994). A small number of other clusters may fall within this age range, but for various reasons they could not be included in the table. These clusters represent the sample of clusters available for probing the history of the galactic disk.

3.2. *Cluster ages*

The oldest known open cluster is now Be 17 which has an age of $\sim$ 12 Gyr. The age distribution is roughly exponential with an $\sim$ 5-Gyr age scale. At least 19 clusters are approximately the age of M67 (5 Gyr) or greater.

3.3. *Galactic distribution*

The dominant young open-cluster population is distributed very differently in the Galaxy, compared to the old clusters. As Figures 2 and 3 show, all the old clusters are more than 7 kpc from the galactic centre, whereas in contrast, the young clusters are distributed almost symmetrically about the solar position in the Galaxy. Although the observed distribution of clusters of any age must be affected by dust clouds and other selection effects, it is difficult to imagine a selection effect that could cause Figures 2 and 3 to look so very different. Perpendicular to the galactic plane, the young and old clusters also have different distributions. The scale height of the old clusters perpendicular to the galactic plane is $\sim$ 375 pc, whereas the scale height of the young clusters is $\sim$ 55 pc.

3.4. *Radial velocities*

The radial velocities, which are primarily a reflection of motions in the galactic plane, are typical of the old disk population. The only real exception is Be 17, which has a substantially larger velocity than the rest of the open clusters.

3.5. *Chemical compositions*

The chemical compositions of the old open-cluster stars are completely typical of the disk population at the same location projected onto the galactic plane. In agreement with other studies, a distinct radial composition gradient exists. However, there is no correlation of cluster composition with distance from the galactic plane, nor is there a correlation of cluster composition with age (see Friel & Janes 1993).

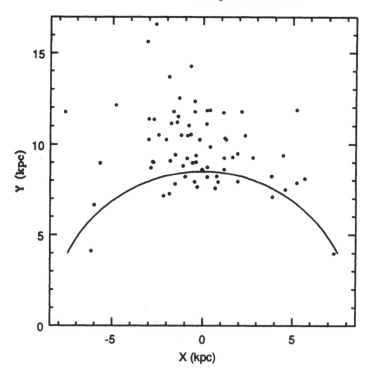

FIGURE 2. Distribution of old open clusters, projected onto the galactic plane. The Sun is located at $Y = 8.5$ kpc, $X = 0$ kpc; the arc has a radius of 8.5 kpc centred at the galactic centre. From Janes & Phelps (1994).

4. Global trends in the disk

Although these new results may seem surprising, it is useful to compare them with some previous studies of the global characteristics of the galactic disk.

4.1. *Radial composition gradient*

A distinct gradient certainly exists, although it may be discontinuous. For example some recent studies find the following:

Friel & Janes (1993)—$d[Fe/H]/dR = -0.09 \pm 0.02$ dex kpc^{-1}

Cameron (1985)—$d[Fe/H]/dR = -0.11 \pm 0.02$

Panagia & Tosi (1981)—$d[Fe/H]/dR = -0.095 \pm 0.034$

See also Shaver *et al.* (1983) for a review of earlier studies, most of which give results consistent with the above values. However, Neese & Yoss (1988) and Lewis & Freeman (1989) found no gradient.

4.2. *The z-distribution*

There is little or no correlation of [Fe/H] with z for true thin-disk objects (Friel & Janes 1993). This is consistent in a general way with the results of Carney *et al.* (1990) who find distinct halo and disk samples but no correlation of metallicity with kinematics within each sample. Hartkopf & Yoss (1982) did find a correlation, but their data set included old disk stars, thick disk stars and halo stars, with no easy way to separate them into groups.

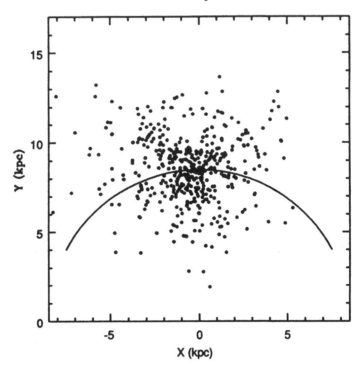

FIGURE 3. Distribution of young open clusters, projected onto the galactic plane. The Sun is located at $Y = 8.5$ kpc, $X = 0$ kpc the arc has a radius of 8.5 kpc centred at the galactic centre. From Janes & Phelps (1994).

4.3. *The age–metallicity relation*

Although Twarog (1980) found a gradual but distinct trend of increasing metallicity with time, Edvardsson *et al.* (1993) find little or no trend of [Fe/H] with age. Furthermore, Friel & Janes (1993) find no correlation between [Fe/H] and age. One is forced to conclude that there is little evidence for a systematic increase in metallicity in the galalactic disk in the past 10 Gyr.

5. A new view of galactic evolution

These new results, together with other information both about our Galaxy and from extragalactic studies, force a considerably new picture of the way our Galaxy formed and evolved. Although there are still some considerable uncertainties, the following scenario is suggested by recent observations and theory.

5.1. *The pre-galactic or chaotic era (ca. 15 Gyr b.p.)*

The "true halo" globular clusters and other halo stars formed during this period from more-or-less independent fragments over a large region centred about the developing central bulge of the protogalaxy.

5.2. *The collapse era (ca. 15 Gyr b.p)*

Nearly simultaneously, a general collapse began, with a rapid coalescence of the nearby objects. There was a (thick) disk at the end of this period.

5.3. *The accretion era (15–10 Gyr b.p.)*

The continued coalescence and mergers with small galaxies sparked starbursts promoting much of the chemical enrichment of the interstellar medium. The thin disk began to condense out of the halo and/or thick disk. The central bulge continued to develop. By the end of the era, the merger events occurred much less frequently and were generally less disruptive.

5.4. *Galactic maturity (10 Gyr b.p–present)*

A near-equilibrium state exists between the star-formation rate and the accretion–infall events. The abundance gradient is maintained through radial inflows driven by low-angular-momentum mergers and the age–metallicity correlation is suppressed by the same inflows. Occasional small merger events produce substantial local deviations from the mean distributions. These occasional disturbances also create clusters with orbits advantageous for long-term survival.

REFERENCES

ALFARO, E. J., CABRERA-CAÑO, J. & DELGADO, A. J. 1991 *Astrophys. J.* **378**, 106.

BOESGAARD, A. M. & FRIEL, E. D. 1990 *Astrophys. J.* **351**, 467.

CAMERON, L. M. 1985 *Astron. Astrophys.* **147**, 47.

CARNEY, B. W., LATHAM, D. W. & LAIRD, J. B. 1990 *Astron. J.* **99**, 572.

CAYREL DE STROBEL, G., HAUCK, B., FRANCOIS, P., THEVENIN, F., FRIEL, E., MERMILLIOD, M. & BORDE, S. 1992 *Astron. Astrophys., Suppl. Ser.* **95** 723.

EDVARDSSON, B., ANDERSEN, J., GUSTAFSSON, B., LAMBERT, D. L., NISSEN, P. E. & TOMKIN, J. 1993 *Astron. Astrophys.* **275**, 101.

EGGEN, O. J., LYNDEN-BELL, D. & SANDAGE, A. 1962 *Astrophys. J.* **136**, 748.

FRIEL, E. D. & BOESGAARD, A. M. 1992 *Astrophys. J.* **387**, 170.

FRIEL, E. D. & JANES, K. A. 1993 *Astron. Astrophys.* **267**, 75.

GEISLER, D., CLARIA, J.J. & MINNITI, D. 1992 *Astron. J.* **104**, 1892.

HARTKOPF, W. I. & YOSS, K. M. 1982 *Astron. J.* **87**, 1679.

JANES, K. A. & PHELPS R. L. 1990 in *CCDs in Astronomy II* (ed. A.G.D. Philip), p. 117. L. Davis Press.

JANES, K. A. & PHELPS R. L. 1994 *Astron. J.* **108**, 1773.

LEWIS, J. R. & FREEMAN, K. C. 1989 *Astron. J.* **97**, 139.

NEESE, C. L. & YOSS, K. M. 1988 *Astron. J.* **95**, 463.

PAGEL, B. E. J. & PATCHETT, B. E. 1975 *Mon. Not. R. Astron. Soc.* **172** 13.

PANAGIA, N. & TOSI, M. 1981 *Astron. Astrophys.* **96**, 306.

PHELPS, R. L., JANES, K. A. & MONTGOMERY, K. A. 1994 *Astron. J.* **107**, 1079.

RANA, N. C. 1991 *Annu. Rev. Astron. Astrophys.* **29**, 129.

SHAVER, P. A., MCGEE, R. X., NEWTON, L. M., DANKS, A. C. & POTTASCH, S. R. 1983 *Mon. Not. R. Astron. Soc.* **204**, 53.

TWAROG, B. A. 1980 *Astrophys. J.* **242**, 242.

Delgado: How would the [Fe/H]-*vs.*-R diagram look for clusters with low reddening, including only those with E(B–V) ≤ 0.3?

Janes: I don't think that would have much effect. There is of course a tendency for the most distant clusters to be more heavily reddened, which increases the uncertainty in the metallicity. But this effect is more or less independent of galactic longitude, so there are heavily reddened clusters with R similar to the Sun, and other ones at very large distances that are not very reddened.

Carney: I have a comment and a question. Anne Fry and I have been doing abundance analyses of the Cepheids in clusters and associations that calibrate the P–L relation, and even for these very young stars we find a fairly large dispersion in [Fe/H], independent of the radial metallicity gradient. The question is: Your data suggest that the oldest open clusters are in the outer parts of the Galaxy. Many models of disk formation are "inside-out", which your data seem to contradict. Have the searches for old clusters in the inner disk been thorough enough to rule out the presence of such old clusters? If so, could tidal disruption—via, say, the bar—be a possible explanation?

Janes: I am not surprised to hear about the dispersion in the abundances of the Cepheids. All the evidence points to local irregularities in the composition of both old stars and young stars and gas. As for the lack of old clusters towards the galactic centre, there certainly have been searches and surveys, but few have been found. The star density is higher of course in the direction of the galactic centre and there is many more dark, obscuring clouds, so some clusters might have been missed. But many *young* clusters are known towards the galactic centre, so that cannot be the entire story. Tidal destruction is the likely answer, not necessarily by the bar, but more likely by encounters between clusters and giant molecular clouds (GMCs). The GMCs are much more numerous toward the galactic centre than in our part of the Galaxy, and they are extremely effective in disrupting an open star cluster.

Franco: You indicated that the [Fe/H] gradient with galactocentric radius may well be discrete, with at least two separate groups at about 10 kpc. The number of data points however, seems to be small at that same radius. This paucity could well create an apparent jump in an otherwise smooth gradient.

Janes: I have not been able to convince myself either that the discontinuity is real or that it is an illusion. As you say, we need more data points. I might note however, that I first noticed the effect some years ago. Our recent data has, if anything, strengthened the apparent effect, but we still need more data, and we also need some rational explanation for the phenomenon, if it is even real.

Hensler: I have three comments: (i) We should withdraw from the simple picture of a homogeneous temporal evolution of metallicity gradients in the galactic disk. Dynamical models reveal that the gradient can alter—e.g. steepen and then flatten again. Therefore we should not expect a clear trend of a gradient if you mix clusters of different age. (ii) The dilution due to PN ejecta should be taken into account in addition to the aspect of mass infall in order to reduce the metal content of gas. (iii) Toth & Ostriker (1992, *Astrophys. J.* **389**, 5) concluded from heating and adiabatic cooling arguments that not more than 4% of the disk mass within the solar circle could be accreted by our Galaxy during the last 5 Gyr.

Janes: These are all things that must be included in any proper model of galactic chemical evolution. I should stress that the radial metallicity gradient has not changed noticeably over the course of galactic history. There certainly could have been short-term changes which would not be detectable with the present data, but over the broad sweep of galactic history, the gradient has been remarkably steady.

Chemical Evolution of the Galaxy

By MONICA TOSI

Osservatorio Astronomico di Bologna, Via Zamboni 33, I–40126, Bologna, Italy

Standard models for the chemical evolution of the Galaxy are reviewed with particular emphasis on the history of the abundance gradients in the disk. The effects on the disk structure and metallicity of gas accretion are discussed, showing that a significant fraction of the current disk mass has been accreted in the last Gyrs and that the chemical abundances of the infalling gas can be non primordial but should not exceed $\sim 0.3\ Z_\odot$. The distributions with time and with galactocentric distance of chemical elements are discussed, comparing the observational data with the corresponding theoretical predictions by *standard* models, which reproduce very well the ISM abundances at various epochs, but not equally well all the features derived from observations of old stellar objects.

1. Introduction

In the last few years a new generation of models for the chemical evolution of the Galaxy has started to appear, in which the dynamics of the system is also taken into account (e.g. Sommer-Larsen & Yoshii 1990; Chamcham & Tayler 1994; Hensler, this volume, and references therein). This new class of models can provide a complete scenario for the evolution of the Milky Way but is still in a rather preliminary phase. The aim of this presentation is therefore to review the current state of *standard* models for the chemical evolution of the galactic disk, with particular emphasis on the effect of gas accretion on the element abundances and gradients.

These models are quite successful in accounting for the large-scale long-term phenomena taking place in the Galaxy, and reproduce its main observed features, such as the age–metallicity relation and the G-dwarf distribution in the solar neighbourhood, the elemental and isotopic abundances and ratios, the present star-formation rate, gas and total mass densities. To avoid misleading conclusions, however, it is necessary to test the models by comparing their predictions not only with the observational constraints derived for the solar neighbourhood but also with the data relative to other galactic regions; first of all because the solar neighbourhood is not representative of the whole disk, and secondly because the distribution with galactocentric distance of several quantities provides important information about the history of the Milky Way.

The current star-formation rate (SFR) represents an excellent example of the importance of modelling the whole disk. One of the most popular approximations for the SFR is a linear proportionality with the gas density, and some authors consider it not only a simple and intuitive law but also a realistic one, as it can reproduce several properties observed in the solar neighbourhood objects. Since the observed radial distribution of the gas in the disk is rather flat (see the shaded area in the bottom panel of Figure 2), this approximation inevitably implies a flat radial distribution of the present SFR. However, Lacey & Fall (1985) have shown that the current SFR in the disk, derived from a large sample of young objects, pulsars, O stars, etc.), is actually a steeply decreasing function of the galactocentric distance. Thus, the radial distribution predicted by models assuming a SFR linearly proportional to the gas density is totally inconsistent with the observed one. On the contrary, models assuming the SFR proportional to both the gas and the total mass densities (e.g. Tosi 1982, 1988; Matteucci & François 1989, hereafter MF) are in agreement with the observed trend, thanks to the steep decrease of the total mass density with galactic radius.

2. Gas infall and abundance gradients

The idea of a long-lasting infall of metal-poor gas on the galactic disk was first suggested by Larson (1972) and Lynden-Bell (1975) to solve several inconsistencies of the first simple models: a too-rapid gas consumption that prevented the reproduction of the amount of gas currently observed in the disk, the unlikelihood of a complete collapse of the whole protogalactic halo in a few 10^8 yr, and the existence of very few metal-poor, long-living stars in the solar neighbourhood, compared to the relatively large predicted percentage (the so called G-dwarf problem). Since then, gas accretion has turned out to be necessary to explain most of the characteristics of our disk, and all the chemical evolution models (with or without dynamics) in better agreement with the largest set of observational constraints assume a significant amount of gas infall throughout the disk's lifetime.

2.1. *Interstellar medium abundances*

One of the most evident effects of gas infall on galactic evolution concerns the absolute value and radial distribution of the element abundances. If no gas accretion is assumed after the disk formation, all chemical evolution models with reasonable SFR and initial mass function (IMF) predict too-large present abundances and/or inconsistent radial distributions. This problem is apparent in the top panel of Figure 1, where the distribution of the current oxygen abundances predicted by models with exponentially decreasing SFR and Tinsley's (1980) IMF is compared with that derived by Peimbert (1979) and Shaver et al. (1983) from HII region observations. The models divide the disk into concentric rings, 1 kpc wide, and assume the sun at 8 kpc. Gas motion can be allowed between consecutive rings, whereas stars are assumed to die in the same region where they were born. If the galactocentric distances of the HII regions are properly rescaled assuming the Sun at R=8 kpc, the observational oxygen gradient is $\Delta \log(\mathrm{O/H})/\Delta R =$ -0.103 dex kpc^{-1}. The long-dashed line corresponds to a model with no infall after the disk formation 13 Gyr ago: it is too flat and lies above the observed range of oxygen abundances. The solid line, instead, fits very well the data and corresponds to a model with an SFR e-folding time of 15 Gyr and constant infall of primordial gas with a density rate F = 4 10^{-3} M$_\odot$kpc^{-2}yr^{-1} all over the disk. This uniform density rate implies the infall of a larger mass of metal-free gas in the outer than in the inner rings and favours the development of negative metallicity gradients as steep as those observed. This model reproduces the most important features of the Milky Way and from now on it will be referred to as the *reference* model. It must be emphasised, however, that other combinations of the SFR and infall parameters may lead to a similarly good agreement with the data, as shown in Figure 1 by the short-dashed line, corresponding to a model assuming a shorter e-folding time for the SFR (5 Gyr) and a lower infall density rate (F = 2 10^{-3} M$_\odot$kpc^{-2}yr^{-1}). What is generally found is that models in better agreement with the observational constraints assume an SFR e-folding times in the range 5–15 Gyr and e-folding times for the gas accretion rate longer than for the SFR. This requirement is not unrealistic, if we consider that according to Sofue's (1995) models the Magellanic Stream has regularly been supplying the Milky Way with gas for 10 Gyr.

If the mass of infalling gas is assumed to increase inwards rather than outwards, the model predictions are much less satisfactory. The dotted line corresponds to a model with an infall rate proportional to the total mass of each ring, which is then increasing toward the centre. As a consequence, there is a larger dilution of the inner interstellar medium (ISM) resulting in a flat abundance distribution inconsistent with that derived from HII regions.

The above results refer to infalling gas with primordial chemical composition, which

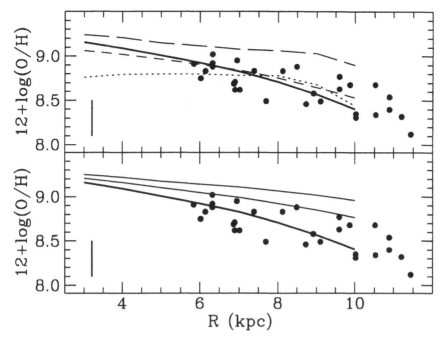

FIGURE 1. Present distribution of the oxygen abundance with galactocentric distance, as derived from HII region observations (dots) and theoretical models. The average observational uncertainty is shown in the bottom left corner. Top panel: models with primordial infall (see text for details). Bottom panel: models with metal-enriched infall.

would be available at best in the intergalactic medium or in the early halo. If the gas originates from regions already polluted by stellar nucleosynthesis, such as the current halo or the Magellanic Stream, it has most probably a non negligible metal content. The intermediate solid line in the bottom panel of Figure 1 shows that if the metallicity of the infalling gas is 0.5 $Z_\odot$, the predictions of the *reference* model are at the upper edge of the observed distribution. The top solid line shows that if the infall metallicity is solar, the resulting oxygen abundance is definitely outside the observational range. From a large variety of models, Tosi (1988b) has found that to allow for a good agreement with the data the infall metallicity should not exceed 0.3 $Z_\odot$. The same result has been obtained by MF with different models and different assumptions on the relative abundances of the infalling elements. This limit is perfectly consistent with the metal content attributable to both the galactic halo and the Magellanic Clouds.

Depending on the model parameters, the present infall rate for the whole disk ranges between 0.3 and 1.8 $M_\odot yr^{-1}$. The lower limit of this range is in agreement with the amount inferred from very high-velocity clouds which, with the Magellanic Stream, are the most reliable observational evidence for this phenomenon. If a fraction of high-velocity Clouds could be considered of non-disk origin as well (see Danly, this volume), the amount of infalling gas observationally detected would cover all the theoretical range.

The metallicity gradients predicted by chemical evolution models depend on the ratio between the SFR and the interstellar and infall gas predicted at each epoch and at each galactocentric distance. The top panel of Figure 2 shows the radial distribution of the SFR and infall rate resulting from the *reference* model at three different epochs: the dotted line corresponds to the epoch of disk formation (assumed to be 13 Gyr ago), the dashed line to the epoch of Sun formation (8.5 Gyr later), and the solid line to the present.

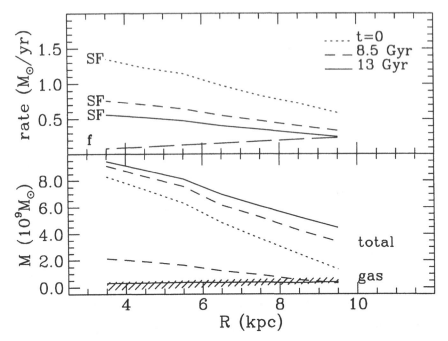

FIGURE 2. Radial distribution at three different epochs of quantities predicted by the *reference* model. Top panel: SF and infall rates; bottom panel: gas and total mass. The shaded area corresponds to the observed gas mass range as published by Lacey & Fall (1985)

Since the infall rate is assumed to be constant, only one line appears in the figure. The bottom panel of Figure 2 shows the radial distributions of the gas and total mass at the same three epochs. The disk is supposed to evolve from an initial configuration of pure gas with radially decreasing mass (dotted line). The initial SFR is radially decreasing as well, so that in inner regions there is more astration and therefore larger stellar production of metals. However, the amount of ISM gas which must be polluted by these metals is much larger in the inner regions and therefore the efficiency of the ISM enrichment is quite modest. Thus, at early epochs the predicted abundance gradients are either flat or even positive, depending on the model parameters (see also Mollá *et al.* 1990). After several gigayears the situation changes significantly, because the larger astration of the inner regions leads to a higher gas consumption which is not totally compensated by infall since the gas accretion is assumed to be increasing outwards. Thus, at a certain time (see for instance the dashed line corresponding to the situation 4.5 Gyr ago) the larger SFR in the inner regions corresponds to a higher efficiency in the metal enrichment of the medium and negative abundance gradients start to develop. Since then, and as long as the star-formation activity remains a decreasing function of the galactocentric distance, the slope of the gradients keeps steepening, because the gas radial distribution becomes increasingly flat, the infall dilution is more efficient outside, and the metal enrichment inside. As shown in Figure 2, the disk of the Galaxy is currently in this phase with a very flat radial distribution of the gas mass, a radially decreasing SFR and an SFR/infall rate ratio progressively smaller for increasing R and equal to 1 at 9–10 kpc (see also Wilson & Matteucci 1992).

A steepening with time of the abundance gradients is predicted by most of the models (with or without dynamics) which are able to reproduce the observational features of the

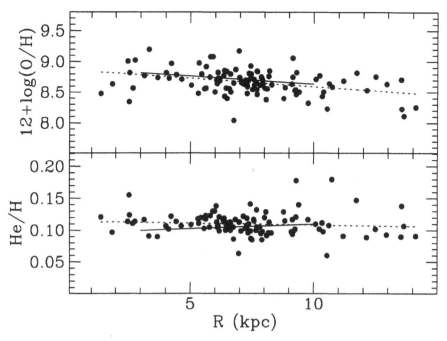

FIGURE 3. Top panel: Radial distribution of the oxygen abundances 3 Gyr ago derived from the *reference* model (solid line) and from observations of PNeII (dots). The data are from PP and the dotted line represents their best fit. Bottom panel: same, but for the He abundance.

Galaxy (e.g. MF; Chamcham & Tayler 1994; Koppen 1994), despite the rather different model characteristics (see, however, Ferrini *et al.* 1994 for different predictions). It is then important to verify if this predicted trend is indeed consistent with the available observational constraints. Since the gradient derived from data on HII regions is representative of the situation in the current ISM, to test the model predictions older objects must be examined as well.

Planetary nebulae, specially of Peimbert's (Peimbert & Torres-Peimbert 1983) type II (PNeII), are also very good indicators of the ISM metallicity. PNeII have stellar progenitors with lifetimes in the 1–5 Gyr range, and they therefore represent the ISM conditions around 3 Gyr ago. Two recent and extensive studies (Pasquali & Perinotto 1993, hereinafter PP; Maciel & Koppen 1994) show that the abundance gradients derived for several elements in PNeII are systematically flatter than the corresponding gradients derived from HII regions. For instance, the oxygen gradient derived by PP is $\Delta\log(O/H)/\Delta R =$ -0.03±0.01 dex kpc^{-1} and that derived by Maciel & Koppen is -0.07 ± 0.01. The latter authors have also found hints of increasing slopes of the gradients with decreasing age of the PNe (i.e. from type III to type I), in agreement with the model predictions. Figure 3 shows the helium (bottom) and oxygen (top) abundances as derived by PP from PNeII and the corresponding predictions of the *reference* model. The agreement between the model solid line and the empirical best fit to the data (dotted line) is excellent. This confirms that in the last few billion years the slope of the abundance gradients in the ISM has actually steepened.

No other gaseous indicators are available to check whether the gradients were increasingly flatter at earlier epochs. Stars and stellar clusters of whatever age are instead visible in a fairly large range of distances and can therefore indicate the earlier scenario.

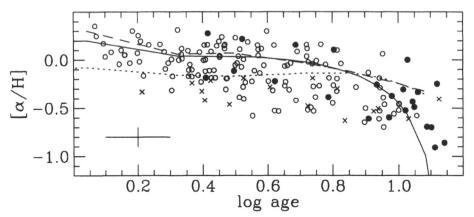

FIGURE 4. Age–metallicity distribution for α elements as derived from the *reference* model and from Edvardsson *et al.* (1993) data. Filled circles and dashed curve refer to stars in rings with $4 \leq R \leq 7$ kpc, open circles and solid curve to stars with $7 \leq R \leq 9$ kpc, crosses and dotted line to stars with $9 \leq R \leq 11$ kpc. Age is in Gyr.

2.2. *Stellar abundances*

As far as single stars are concerned, the situation is unfortunately rather confused. Lewis & Freeman (1989) found no significant metallicity gradient in a sample of 600 old K giants, but more recently Edvardsson *et al.* (1993) have argued that the radial metallicity distribution derived from a sample of 189 F and G dwarfs is similar to that derived from HII regions. The major result of this accurate and extensive work is the scatter on the derived abundances which turns out to be much larger than the observational uncertainties and should therefore be considered an intrinsic feature of the analysed stellar population. Edvardsson *et al.* therefore avoided formally deriving the slopes of the abundance gradients, but one can obtain them from their Table 14, where the analysed stars are divided into different groups according to their age and galactocentric distance and the average [Fe/H] of each group is given. Despite the poorness of the sample in the older and more distant bins, and the corresponding weakness of the statistics, it is interesting to note that the resulting formal slopes get flatter for increasing age (i.e. towards earlier epochs) and that the oldest bin even shows a positive gradient (derived from two single points, however!), thus giving some further support to the predictions of the chemical evolution models.

Another interesting feature of the Edvardsson *et al.* data is the different distribution of metallicity with age for stars at different galactic locations. As pointed out by Pagel (1994) and shown in Figure 4, if one divides their stars into three groups according to their mean galactocentric distances (inner objects, solar-ring objects, and outer ones), one finds that the outer stars show a much flatter age–metallicity distribution, with average abundances in the last ten billion years (i.e. over most of the disk's lifetime) systematically lower that those of the other objects. As already mentioned above, the data show a large intrinsic scatter in the derived metallicities of stars of any age and this scatter cannot be directly reproduced by standard chemical evolution models such as the *reference* one, which assumes both the SFR and the gas accretion in a sort of steady-state. However, the age–metallicity relations predicted for the three ranges of galactocentric distances are consistent with each of the corresponding average empirical distributions. Notice that the relations predicted for the solar (solid line) and the outer (dotted line) rings flatten off at recent epochs, whereas the relation predicted for the

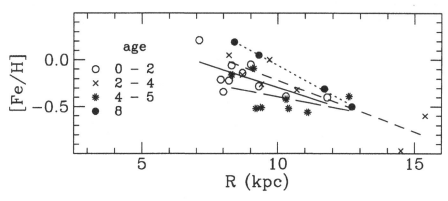

FIGURE 5. Radial distribution of open cluster metallicity as derived from Friel & Janes (1993) and Thogersen *et al.* (1993) samples. The clusters have been divided into age bins and the best linear fit for each bin is shown (dotted line for the oldest bin, long-dashed for the 4–5 Gyr bin, short-dashed for the 2–4 Gyr bin, and solid line for the youngest one.

inner ring (dashed line) keeps increasing up to the present time, as already shown by MF. Besides, François & Matteucci (1993) have argued that even the spread of the observed age–metallicity distribution can be accounted for by *standard* models, once the different birthplaces of the sample stars are considered.

The major problem of single-star analyses is the uncertainty in the derived R, age and metallicity of objects beyond a limited distance from the Sun, as confirmed by the Edvardsson's *et al.* survey. From this point of view, open clusters are, in principle, safer indicators. There are, however, several problems also affecting the determination of the cluster parameters, such as the non-homogeneity of most age estimates and the uncertainty about the cluster's birthplace, the cluster disruption due to disk friction which can alter the original distributions, etc. (see, however, Carraro & Chiosi 1994).

Several years ago, it was suggested that young open clusters indicate steeper abundance gradients than old open clusters (Mayor 1976; Panagia & Tosi 1981). However, more recent and extensive studies (Friel & Janes 1993; Thogersen *et al.* 1993) do not seem to support this hypothesis. The metallicity gradient derived by Janes and collaborators is $\Delta[\text{Fe/H}]/\Delta R = -0.09\pm0.02$ dex kpc^{-1} for the whole sample of clusters of any age, and does not seem to depend on the cluster age. Bearing in mind that for field stars in the disk oxygen has been empirically found to follow the relation $[\text{O/Fe}] \simeq -0.3[\text{Fe/H}]$ (e.g. Edvardsson *et al.* 1993), and assuming that this relation applies to open clusters as well, the iron gradient corresponds to an oxygen gradient $\Delta\log(\text{O/H})/\Delta R = -0.06$, flatter indeed than that derived from HII regions and more similar to that of objects a few gigayears old—such as the PNeII discussed above.

The most striking feature of the Friel & Janes sample is that at each galactic radius the oldest clusters are also the most metal-rich (see Figure 5). Whilst it is true that there are only four published clusters in the oldest age bin (≥ 8 Gyr) and the corresponding statistics are therefore too poor, two additional clusters of roughly the same age have been found (Friel 1994, private communication) with metal abundances which are less extreme but still higher than average. It is of crucial importance to verify this result with a larger sample of old clusters and with more accurate and homogeneous methods to derive their ages, chemical abundances and galactic original locations. It might well be, in fact, that this anomaly is fictitious, resulting results from the uncertainty in the metallicity and/or, more probably, age determination. However, if confirmed, this phenomenon would have remarkable implications for our understanding of the Galaxy's evolution, because it is opposite to any intuitive age–metallicity relation derivable from "steady-state" scenarios, where old stars are inevitably more metal-poor than young objects and may result from short, intense phenomena not considered in our models.

Another characteristic of old open clusters is that they are all located beyond 7–7.5 kpc from the galactic centre, contrary to younger clusters which are likely to be equally distributed everywhere in the disk (Janes & Phelps 1994). On the one hand, the external location of the older clusters in the observed sample might not reflect an odd distribution of all the clusters formed several gigayears ago and be the result of a more efficient disruption in the inner than in the outer regions. On the other hand, it may instead correspond to a non-homogeneous star formation activity, perhaps related to external phenomena like the first impact on the disk of the Magellanic Stream (Sofue 1995). The latter scenario might also provide an explanation to the large metallicity of the oldest clusters, in terms of a transitory metal enhancement of the ISM due to the larger SFR triggered by the sudden event and later smeared out during the following "steady-state" evolution.

3. Summary

In conclusion, the comparison between the abundance distributions in the galactic disk predicted by *standard* chemical evolution models and derived from observational data can be summarised as follows:

(i) Abundance distributions derived from observations of gaseous objects of various ages (HII regions and PNe) are very well reproduced by "steady-state" models with slowly decreasing SFR and large, long-lasting infall of metal-poor gas.

(ii) Average abundance distributions derived from stars are also reproduced. These *standard* models, however, do not reproduce the observed spread in the metallicity distribution of field stars and the anomalously high [Fe/H] of the oldest open clusters.

We must bear in mind, however, that old stars can have quite eccentric orbits and may therefore have formed in galactic regions other than where they are observed now. According to François & Matteucci this might explain most of the abundance spread in the Edvardsson *et al.* sample. On the other hand, Carraro & Chiosi (1994) have suggested that the same argument cannot apply in the case of old open clusters, for which other explanations are thus needed, unless all the inconsistencies with the model predictions can be attributed to observational errors.

A possible reason for the different agreement found for gas and stellar objects is that the gas mixes rapidly, compared to the timescales for galactic evolution, and therefore "forgets" local perturbations occurred in the past and follows the "steady-state" scenario. Stars, however, "remember" the local perturbations occurring at, or just before, their birth and therefore deviate more from that scenario, showing intrinsic large scatter and anomalous behaviour. To interpret their observed features in detail, more sophisticated models which also take into account also the small-scale, short-term phenomena are therefore required.

I wish to thank Francesca Matteucci for always being ready to discuss and compare the results and the different approaches of our models: a praiseworthy attitude rather unusual among theoreticians.

REFERENCES

CARRARO, G. & CHIOSI, C. 1994 *Astron. Astrophys.* **287**, 761.

CHAMCHAM, K. & TAYLER, R. J. 1994 *Mon. Not. R. Astron. Soc.* **266** 282.

EDVARDSSON, B., ANDERSEN, J., GUSTAFSSON, B., LAMBERT, D. L., NISSEN, P. E. & TOMKIN, J. 1993 *Astron. Astrophys.* **275**, 101.

FERRINI, F., MOLLÁ, M., PARDI, C. & DÍAZ, A. I. 1994 *Astrophys. J.* **427**, 745.

FRANÇOIS, P. & MATTEUCCI, F. 1993 *Astron. Astrophys.* **280**, 136.

FRIEL, E. D. & JANES, K. A. 1993 *Astron. Astrophys.* **267**, 75.

JANES, K. A. & PHELPS, R. L. 1994 *Astron. J.* **108**, 1773.

KOPPEN, W. E. 1994 *Astron. Astrophys.* **281**, 26.

LACEY, C. G. & FALL, S. M. 1985 *Astrophys. J.* **290**, 154.

LARSON, R. B. 1972 *Nature* **236**, 21.

LEWIS, J. R. & FREEMAN, K. C. 1989 *Astron. J.* **97**, 139.

LYNDEN-BELL, D. 1975 *Vistas Astron.* **97**, 139.

MACIEL, W. J. & KOPPEN, W. E. 1994 *Astron. Astrophys.* **282**, 436.

MATTEUCCI, F. & FRANÇOIS, P. 1989 *Mon. Not. R. Astron. Soc.* **239**, 885. (MF)

MAYOR, M. 1976 *Astron. Astrophys.* **48**, 301.

MOLLÁ, M., DÍAZ, A.I. & TOSI, M. 1990 in *Chemical and Dynamical Evolution of Galaxies* (ed. F. Ferrini, J. Franco & F. Matteucci), p. 577. ETS Pisa.

PAGEL, B. E. J. 1994 in *The Formation and Evolution of Galaxies* (ed C. Muñoz-Tuñón & F. Sánchez). V Canary Islands Winter School, p. 149. Cambridge University Press.

PANAGIA, P. & TOSI, M. 1981 *Astron. Astrophys.* **96**, 306.

PASQUALI, A. & PERINOTTO, M. 1993 *Astron. Astrophys.* **280**, 581.

PEIMBERT, M. 1979 in *The Large Scale Characteristics of the Galaxy* (ed W. B. Burton), p. 307. Reidel.

PEIMBERT, M. & TORRES-PEIMBERT, S. 1983 in *Planetary Nebulae* , IAU Symp 103, p. 233. Reidel.

SHAVER, P. A., MCGEE, R. X., NEWTON, L. M., DANKS, A. C. & POTTASCH, S. R. 1983 *Mon. Not. R. Astron. Soc.* **204**, 53.

SOFUE, Y. 1995 *Publ. Astron. Soc. Jpn.* in press.

SOMMER-LARSEN, J. & YOSHII, J. 1990 *Mon. Not. R. Astron. Soc.* **243**, 468.

THOGERSEN, E. N., FRIEL, E. N. & FALLON, B. V. 1993 *Publ. Astron. Soc. Pac.* **105**, 1253.

TINSLEY, B. M. 1980 *Fundam. Cosmic Phys.* **5**, 287.

TOSI, M. 1982 *Astrophys. J.* **254**, 699.

TOSI, M. 1988a *Astron. Astrophys.* **197**, 33.

TOSI, M. 1988b *Astron. Astrophys.* **197**, 47.

WILSON, T. L. & MATTEUCCI, F. 1992 *Astron. Astrophys. Rev.* **4**, 1.

Palouš: As you have shown, SFR(R) is a function of both σ_{gas} and σ_{tot}: SFR $= \psi(\sigma_{gas}, \sigma_{tot})\, e^{-t/\tau}$. In the model of propagating SF, recently published by Palouš *et al.* (1994, M.N.R.A.S.) the SFR $\propto [\sigma_{gas}, shear(R)]$. Shear follows the total mass distribution and decreases with R. It seems then that the propagating SF model reaches conclusions similar to yours.

Chernin: Do I understand you correctly that there might be two characteristic timescales of the chemical evolution of the galactic disk—15 and 5 Gyr—and that the available observational data do not enable us to make a choice between them? And, by the way, is there any room for dark matter in your models, whatever its physical nature may be?

Tosi: Yes, with a proper combination of the other parameters, both an SFR with an e-folding time of 15 Gyr and one with 5 Gyr provide results consistent with the disk observational data. I haven't included dark matter in my models.

Tenorio-Tagle: Would you comment about outflows?

Tosi: Outflows haven't been introduced yet in *standard* chemical evolution models, because we don't have enough observational constraints on their chemical composition, on the significance of the phenomenon or on the final fate of the ejected gas—does it eventually escape from the system or fall back—and where? In the absence of such information, too many free parameters should be assumed in the models to allow for a safe interpretation of the results.

Serrano: What would be the effect on the standard model of outflows larger at small R and smaller at large R?

Tosi: If the ejected gas is lost for ever, it should lead to a reduction of the inner abundances and therefore to a flattening of the radial gradient of the outflowing chemical elements. If the wind is made only of the gas ejected by SNe, the effect will be restricted to the elements—like oxygen—synthesized by their progenitors, otherwise it will be on all the metals.

Martín: In your models you need a total amount of infalling gas mass comparable to the mass of the disk. So much infall cannot come from the Magellanic Clouds: where do you think it's coming from?

Tosi: The infalling gas has presumably a composite origin: at early epochs it must have been mostly collapsing halo material, with a small component of extragalactic gas; nowadays it is most probably due to external gas. The sum of these components can easily account for the required $\sim 10^{10} M_\odot$.

Andersen: I am still somewhat concerned about the amount and age of data on which some crucial features of the models depend rather critically. Edvardsson *et al.* did not discuss possible radial gradients from their data, simply because we did not have a fair sample of the stars originally inside or outside the solar circle.

Tosi: I mentioned indeed that the statistics on the older and more distant bins are quite poor. With this caveat in mind, I do however think that it is useful to check whether or not the model predictions are consistent with such a large and good sample of data.

Chemical Abundances in the Outer Milky Way

By JOSÉ M. VÍLCHEZ AND CÉSAR ESTEBAN

Instituto de Astrofísica de Canarias, E–38200 La Laguna, Tenerife, Spain

New spectroscopic observations have been obtained for a sample of HII regions located toward the galactic anticentre region. The sample includes HII regions with known galactocentric distances, having their ionising stars clearly identified from previous work. The spectra correspond to low- to intermediate-excitation nebulae, ionised by one (few) late O–early B star(s). The global parameters of the HII regions as well as their physical parameters have been derived. Abundances have been computed using direct determinations of the electron temperature for some of the objects or alternatively, by adopting an electron temperature from radio data or model fitting. Overall, the data appear to show a nearly flat gradient of oxygen and nitrogen abundances up to 18 Kpc of galactocentric distance.

1. Introduction.

The chemical abundances of the interstellar medium (ISM) in gas-rich galaxies are observed to be different depending of the precise location within a given galaxy, and also from one galaxy to another. Since the nucleosynthesis of the different chemical elements occurs in stars of different mass, the study of the chemical abundances in the disks of spiral galaxies is a powerful way to understand the history of star formation and evolution in these galaxies. These ISM abundances can be derived by making use of the spectra of ionised nebulae such as planetary nebulae and/or HII regions. The existence of gradients of the chemical composition of the ISM in galaxies has been known since the pioneering work of Searle (1971). For the Milky Way, the chemical abundance gradient has been derived using planetary nebulae (e.g. Maciel 1990). However, in order to derive abundances representative of the current ISM values, the HII regions are preferred, since the ionised gas in planetary nebulae may be affected by self-enrichment produced by their central stars.

With respect to the abundances of galactic HII regions, the work by Shaver et al. (1983) remains a key reference. They obtained optical spectra for a sample of HII regions located between galactocentric radii from 5 to 12 kpc approximately and, with the help of convenient radio observations, the physical properties and abundances were derived. This work established the existence of a negative gradient for the abundance of the elements heavier than helium in the disk of our galaxy. Subsequent studies of the chemical evolution of the galaxy have greatly bennefit of this work (e.g. Tosi & Diaz 1985; Matteucci & Francois 1989; Ferrini et al. 1994).

Nonetheless, the range of galactocentric radii of the HII regions included in Shaver et al. (1983) limits this work to the study of HII regions with galactocentric radii within some ± 5 Kpc of the solar neighbourhood. Therefore, the sampling of the whole galactic disk is still poor, in particular towards the galactic anticentre. This fact has limited the study of the Milky Way, as compared to the nearby external galaxies, and so reduced the application of models of galaxy evolution—especially those aspects related to the evolution of the outermost parts of the disk, represented by the anticentre, which, at least from the point of view of the chemical evolution, might be considered much closer to the pre-galactic initial conditions.

However, the determination of distances to the HII regions in the galactic anticentre has been very scarce and with large uncertainties. Following the work of Fich & Blitz (1984) and Blitz *et al.* (1989), the distances to a substantial set of regions in the anticentre have been derived. In addition, Chini & Wink (1984) carried out a study of the outer rotation curve of our galaxy using photometry of ionising stars and spectroscopy of the HII regios. Also, Hunter & Massey (1990) studied the properties of the central stars of a set of HII regions including their distances. All these studies are consistent with the existence of an end of the star formation in the optical disc at some 18–20 Kpc of galactocentric radius (see also Moffat *et al.* 1979). But recent data appear to favour massive star formation even beyond a galactocentric radius of 20 Kpc (De Geus *et al.* 1993).

Recently, Fich & Silkey (1991, hereafter FS) have performed a spectroscopic study of a sample of HII regions located towards the galactic anticentre. They suggested the existence of a rather flat N/H gradient when going to the outermost parts of the Galaxy. Unfortunately, the lack of measurements of the [OII]λ 3727 Å lines together with those of [OIII]$\lambda\lambda$ 4959, 5007 Å (both only available for one object), combined with the rather low signal-to-noise ratio of the data, renders any derivation of the O/H abundances unreliable. In any case, the possibility of a rather flat N/H gradient in the outermost Galaxy is very exciting, and indeed, our own work on the distant HII region S 266 (Manchado *et al.* 1989) pointed in the same direction.

On the other hand, recent work on abundances in external galaxies (e.g. Garnett & Shields 1987; Vílchez *et al.* 1988a,b; Díaz *et al.* 1991; Zaritsky *et al.* 1989) has demonstrated the importance of having a complete sampling of the disks, in order to interpret the results in terms of the nucleosynthesis and star-formation history of the galaxies. A global fit to the gradient may be dangerously driven by the richness of observational points in the regions of intermediate/innermost radii. Therefore, the study of the abundances in the distant anticentre HII regions is of great interest.

2. Observations and Data Reduction

The observations were carried out in several runs between 1990 and 1992 at the Observatorio del Roque de los Muchachos (ORM), La Palma, using the 4.2-metre WHT. The ISIS double arm spectrograph was used at the Cassegrain focus. The slit centre and position angle were chosen accordingly to the morphology and surface brightness of the nebulae as shown in the POSS plates and/or from recent available work (Hunter 1992, hereafter H92; Mampaso 1991). An EEV detector with 22.5 μ pixels was used for each arm. Typical seeing values were about 1–1.5 arcsec throughout the observations, and the slit width was set to 1.5 arcsec, having a total length of some 3 arcmin in the spatial direction. A dichroic was set at an effective wavelength of λ 5500 Å in order to separate the spectral ranges of blue and red arms. Two gratings—316R and 300B, giving 316 g/mm and 300 g/mm for red and blue arms, respectively—were used, with corresponding reciprocal dispersions of 60.4 Å/mm and 62.2 Å/mm .

This set-up allowed us a wide spatial coverage including from [OII] $\lambda\lambda$ 3727 Å in the blue up to the [SIII] λ 9532 Å and P8 λ 9546 Å lines in the near infrared. A total of eight anticentre HII regions were observed for this project: S 98, S 127, S 128, S 209, S 219, S 266, S 283 and BFS 31.

The data reduction was performed at the IAC using the standard software package FIGARO (Shortridge 1990), following the standard procedure of bias substraction, flat-fielding, wavelength calibration, sky substraction and flux calibration. The correction for atmospheric extinction was performed using an average curve for the continuous atmospheric extinction at the ORM observatory (King 1985) and the flux calibration was achieved by the observation of flux standards during the same nights (Oke 1974).

3. Results

Electron densities have been derived for all the observed regions using the [SII]λ λ 6717/6731 Å ratio following the standard procedure (eg. Castañeda *et al.* 1992). Density values are found to be in the low density limit (≤ 100 cm^{-3}) only for S 98 and S 285. For most of the regions the density appears to be typically of several hundreds, with the highest measurements being 597 cm^{-3} and 548 cm^{-3} for S 127 and S 209, respectively. The values of the density are consistent with previous determinations for the regions in common with FS and H92, although with a large dispersion, which can be somehow expected given the extended nature of the objects and different slit placements.

Electron temperatures can be derived directly from the spectra of some of the observed regions. For S 127 and S 128, the value of t([OII]), the electron temperature of the singly ionised zone, can be calculated from the ratio of the [OII] doublets $\lambda\lambda$ 7320,30/3727 Å. This is also the case of S 266, for which this line ratio can be determined from the spectra of the two zones of the nebula. For this last nebula, a value of the t([NII]) electron temperature can also be derived using the [NII] $\lambda\lambda$ 5755/6584 Åline ratio.

The values observed for the line ratio [NII]/[OII] are quite homogeneous, clustering around Log [NII]/[OII] $\approx$ -0.7, no matter the range of galactocentric radius. A similar situation holds for the abundance parameter R$_{23}$; for this parameter all the values are around Log R$_{23}$ $\approx$ 0.6 and show no correlation with galactocentric distance. In contrast, the ratio [NII]/[SII] does show a clear increase from inner to outer regions up to a radius of about 16 Kpc. Beyond this radius, the three outermost HII regions do not follow a similar behaviour. Other spectroscopic parameters derived in our study include the radiation softness parameter η' (Vílchez & Pagel 1988), the excitation measured by the [OIII]/[OII] ratio, and the S$_{23}$ parameter, which measures the total emission in the optical sulphur lines with respect to Hβ. A deeper study of the data for the HII regions referred to here, including some reference objects taken from FS, whith information about their ionising stars, will be published elsewhere (Vílchez & Esteban, in preparation).

For those objects for which a value of η' can be derived, it is apparent a clear anticorrelation of NLy$_{CNeb}$, the total number of Lyman continuum photons, with the value of η', reflecting the expected trend between the effective temperature of MS ionising stars (as measured by η') and the total number of ionising photons.

Preliminary analysis of the data presents some very interesting properties. In particular, a nearly flat oxygen abundance gradient is revealed in the outer parts of the Galaxy. In addition, the observed behaviour of the nitrogen-to-oxygen ratio (N/O) also appear to show an abundance distribution which is independent of galactocentric distance. All these results favour a differentiated chemical evolutionary scenario for the outer Galaxy with respect to the solar neighbourhood.

JMV thanks the organisers for an enjoyable and very constructive meeting.

This work has been partially financed by DGICYT grants PB91–0531 (GEFE), PB90–0570 and by IAC P14/86.

REFERENCES

CASTAÑEDA, H. O., VÍLCHEZ, J. M. & COPETTI, M. V. F. 1992 *Astron. Astrophys.* **260**, 370.

CHINI, R. & WINK, J. E. 1984 *Astron. Astrophys.* **139**, L5.

DE GEUS, E. J., VOGEL, S. N., DIGEL, S. W. & GRUENDL, R. A. 1993 *Astrophys. J., Lett.* **413**, L97.

DÍAZ, A. I., TERLEVICH, E., VÍLCHEZ, J. M., PAGEL, B. E. J. & EDMUNDS, M. G. 1991 *Mon. Not. R. Astron. Soc.* **253**, 245.

FERRINI, F., MOLLÁ, M., PARDI, M. C. & DÍAZ, A.I. 1994 *Astrophys. J.* **427**, 745.

FICH, M. & BLITZ, L. 1984 *Astrophys. J.* **279**, 125.

FICH, M., BLITZ, L. & STARK, A. A. 1989 *Astrophys. J.* **342**, 272.

FICH, M. & SILKEY, M. 1991 *Astrophys. J.* **366**, 107.

HUNTER, D. A. 1992 *Astrophys. J., Suppl. Ser.* **79**, 469.

HUNTER, D. A. & MASSEY, P. 1990 *Astron. J.* **99**, 846.

KING, D. L. 1985 *La Palma Technical Notes*, **15**.

MACIEL, W. J. 1992 in *Elements and the Cosmos*, p. 210. Cambridge University Press.

MAMPASO, A. 1991 PhD thesis. University of La Laguna.

MATTEUCCI, F. & FRANCOIS, P. 1989 *Mon. Not. R. Astron. Soc.* **239**, 885.

MOFFAT, A. F. J., FITZGERALD, M. P. & JACKSON, P. D. 1979 *Astron. Astrophys., Suppl. Ser.* **38**, 197.

OKE, J. B. 1974 *Astrophys. J., Suppl. Ser.* **27**, 21.

PANAGIA, N. 1973 *Astron. J.* **78**, 929.

SEARLE, L. 1971 *Astrophys. J.* **168**, 327.

SHAVER, P. A., McGEE, R. X., NEWTON, L. M., DANKS, A. C. & POTTASCH, S. R. 1983 *Mon. Not. R. Astron. Soc.* **204**, 53.

TOSI, M. & DÍAZ, A. I. 1985 *Mon. Not. R. Astron. Soc.* **217**, 571.

VÍLCHEZ, J. M., EDMUNDS, M. G. & PAGEL, B. E. J. 1988b *Publ. Astron. Soc. Pac.* **100**, 1428.

VÍLCHEZ, J. M. & PAGEL, B. E. J. 1988 *Mon. Not. R. Astron. Soc.* **231**, 257.

VÍLCHEZ, J. M., PAGEL, B. E. J., DÍAZ, A. I., TERLEVICH, E. & EDMUNDS, M. G. 1988a *Mon. Not. R. Astron. Soc.* **235**, 633.

Bania: I would like to know your opinion about the quality of the distances to your sample of HII regions, especially the most distant sources.

Vílchez: The distances to the most distant HII regions are uncertain, with estimated errors of the order of 25% in some cases, such as for S 266. Nonetheless, the adopted distances for the sample HII regions result from compilation of distance determinations, including spectrophotometric studies of their stellar content. This would reduce the effect of the systematic errors. In any case, it seems unlikely that errors in the distance determination could be affecting the overall shape of the abundance gradient.

Torrelles: Do you find any dependence between the spectral type of the exciting stars of the HII regions and metallicity?

Vílchez: I do not find any clear trend, but our sample is still not large enough to perform this test. What we can clearly say is that the presence of the hottest O stars within typical anticentre HII regions, with the lowest metallicity, does not seem to be favoured by the observations.

Non-Standard Abundance Patterns in Disk Stars

By JOHANNES ANDERSEN
AND BIRGITTA NORDSTRÖM

Niels Bohr Institute for Astronomy, Physics and Geophysics, Astronomical Observatory,
Brorfeldevej 23, DK–4340 Tollose, Denmark

1. Introduction

The evolution of the galactic disk is recorded in the distribution, motions, chemical composition, and ages of its stars. Observed correlations among these parameters form some of the basic observational constraints on physical models of the evolution of the disk. There is, however, much scatter in these correlations, traditionally ascribed mostly to observational error. In addition, it is not always possible to observe each of the above parameters independently of the others. Most importantly, the way in which stars are selected for such studies unavoidably introduces biases in the sample which may, in unfortunate cases, determine the outcome of the investigation even before the first observation is actually made. As a result, although we conveniently classify the stars into a number of populations, their exact definition remains diffuse, and it is rare that an individual star can be unambiguously assigned to any one of them.

The first step towards understanding the Galaxy is an inventory of the solar neighbourhood. In the belief that the most informative tests are based on the most accurate data for the most complete and well-defined samples, we are engaged in a programme to provide accurate and independent ages, metallicities, and kinematics for several thousand nearby F and G dwarfs (see overview by Strömgren 1987). While the bulk of the data is still being analysed, the phase of the project which consists of a detailed chemical analysis of 189 nearby F dwarfs has recently been completed by Edvardsson et al. (1993, hereafter EAGLNT). Perhaps the most noteworthy result of the paper is the new questions it raised rather than the definitive answers it did *not* provide to the classical questions it was designed to address. Without repeating the extensive discussion of EAGLNT, we address in the following a few points which were not raised there, and list several issues that will be the focus of our continued work.

2. Mean relations in the local sample

The title of this section is a deliberate pun: A main result of EAGLNT was indeed that the "local sample" of stars is far from representative of the local neighbourhood, and that average relations between the main parameters are if not misleading, then at least less informative than the deviations from them: There is in reality no such thing as "the" age–metallicity or age–velocity relation for solar-neighbourhood stars.

Figure 1 shows minimum and maximum galactocentric distances (R_a, R_p) *vs.* [Fe/H] for the EAGLNT sample (all now within 60 pc of the Sun), computed from galactic orbits based on the observed space motions. Clearly, these "local" stars have travelled widely in the Galaxy. $R_m = (R_a + R_p)/2$ is our best indicator of their origin (see discussion in EAGLNT), and in Figures 2 and 3 we have divided the stars roughly into groups belonging "inside", "at", and "outside" the solar circle.

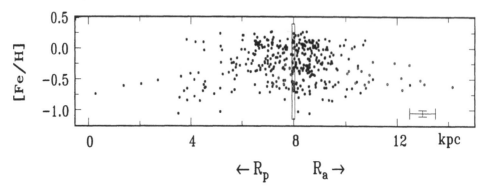

FIGURE 1. Peri- and apogalactic distances *vs.* [Fe/H] for the EAGLNT stars (presently all located inside the box).

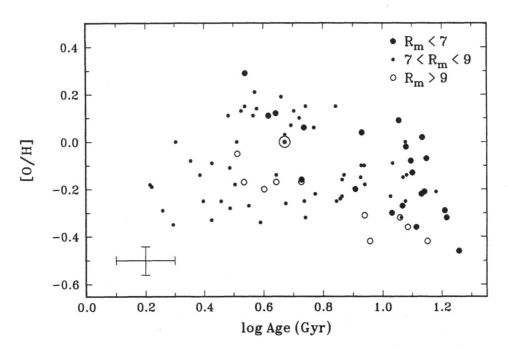

FIGURE 2. Age–metallicity diagram for oxygen for the EAGLNT stars (and the Sun), sorted by galactic origin

Combined with the large and real scatter in the age–[Fe/H] relation, the change in mean [O/Fe] with [Fe/H], and the result that the relative α-element content also correlates with R_m (all from EAGLNT), Figure 2 signals the demise of the concept of a uniquely defined age–metallicity relation for the disk, overall as well as for single elements. Moreover, since EAGLNT found effectively zero scatter in the proportions of all elements heavier than hydrogen in stars of equal age *and* R_m, hydrogen appears to be the element occurring in locally variable proportions in the ISM during the history of the disk. Infall of metal-poor gas might explain this as well as the weak increase of the average [Fe/H] at the same time (but see Discussion).

Finally, Figure 3 shows the age–vertical velocity diagram for the sample, showing some abrupt increases in velocity dispersion near 2.5 and 10 Gyr which should become better

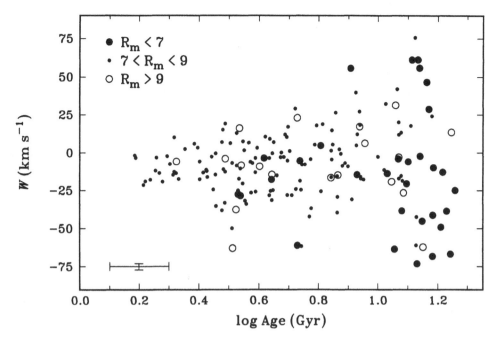

FIGURE 3. Age *vs.* vertical velocity W for the EAGLNT stars, sorted as before

defined when based on the several thousand stars in our main program. These features appear much less clear-cut in an [Fe/H]–W diagram, which is understandable when no tight correlation exists between age and [Fe/H]. It is noteworthy also that the abrupt increase in velocity dispersion at 10 Gyr, which might be interpreted as a signature of the thick disk, seems to be caused mainly by stars originating inside the solar circle. This may, however, mostly reflect the density gradient in the disk.

The basic messages of the EAGLNT paper seem to be that (i) samples of local stars should be defined from galactic orbits, not present positions; (ii) even the best high-resolution spectroscopic analyses of (disk) stars cannot be interpreted without knowing their ages and galactic orbits; (iii) the evolution of the overall metallicity in the disk cannot be modelled without knowing the detailed element ratios; and (iv) better understanding, not just better data, is needed to improve the usefulness of the classical correlation diagrams between age, [Fe/H], and kinematics.

3. What's next

Based on these lessons, we have drawn up a list of new questions to address over the next few years with our full sample of 10 000 stars. Like EAGLNT, it is kinematically unbiased, but avoiding the strong metallicity bias of EAGLNT allows us to ask, for example:

- What is the *real* [Fe/H] distribution for old stars (the G-dwarf problem)?
- What is the *real* age–metallicity relation for stars on the solar circle?
- What are the *real* age–velocity relations (U, V, W) for stars on the solar circle?
- Is the build-up of oxygen really very different in the inner and outer disk (Figure 2)?
- What happens near $[Fe/H] = -0.4$ (Figure 1; and EAGLNT Figures 15 and 18)?

- Is there a real bimodality in [Fe/H] for stars inside the solar circle (Figure 1)?
- Do "all" old metal-poor stars originate inside the solar circle?
- Do "all" stars with large vertical velocities originate inside the solar circle?
- How precisely constant are the detailed abundance ratios for given age and R_m?
- To how low [Fe/H] do the detailed abundance ratios remain precisely constant?

It is sobering to recall that at the outset of this project, the default expectation was a well-defined age–metallicity relation for the solar circle, tight ratios between the various elements as functions of [Fe/H], but increasing scatter towards the low-metallicity end where fewer stellar generations went before and incomplete mixing could be expected. What EAGLNT found from the most accurate abundance data currently obtainable was the exact opposite on all accounts. As shown by Nissen *et al.* (1994) and Sneden *et al.* (1994), these results for disk stars remain valid well into the halo domain, but break down for the most very-metal-poor stars known. Theorists and observers alike should have much fun giving each other a hard time for another good many years.

We thank the organisers for a pleasant and inspiring meeting and the editors for much patience. This project is a collaboration with several colleagues, notably E. H. Olsen, M. Mayor, and the EAGLNT group, and supported by observing time awarded by ESO and the Danish Board for Astronomical Research, and financial support from the Danish Science Research Council and the Carlsberg Foundation.

REFERENCES

EDVARDSSON, B., ANDERSEN, J., GUSTAFSSON, B., LAMBERT, D. L., NISSEN, P. E. & TOMKIN J. 1993 *Astron. Astrophys.* **275**, 101. (EAGLNT).

NISSEN, P. E., GUSTAFSSON, B., EDVARDSSON, B. & GILMORE, G. 1994 *Astron. Astrophys.* **285**, 440.

SNEDEN, C., PRESTON, G. W., McWILLIAM, A. & SEARLE, L. 1994 *Astrophys. J., Lett.* **431**, L27.

STRÖMGREN, B. 1987 in *The Galaxy* (ed. G. Gilmore & R. Carswell), p. 229. Reidel.

Carney: You suggested that the constant [α/Fe] amidst a wide range in [Fe/H] indicates variable amounts of hydrogen. But it would take a lot of hydrogen to change n_{Fe}/n_H by factors of two or three, which would imply the ISM is not at all mixed. Could it be, instead, since [X/Fe] is independent of distance from the sites of nucleosynthesis, whereas [Fe/H] is distance-sensitive, that what you're seeing is a spotty distribution of supernovae relative to the sites of star formation?

Andersen: You are absolutely right, a lot of hydrogen is required, but at the same time the average metallicity of the disk seems to vary little despite the enrichment we know goes on all the time. Moreover, your picture would require that all supernovae produce the same heavy-element mix (and we know it changes with R_m, at least), but perhaps the variations are below even the detection threshold of EAGLNT.

The Evolution of Radial Abundance Gradients in the Galaxy

By CRISTINA CHIAPPINI AND WALTER J. MACIEL

Instituto Astronômico e Geofísico, Universidade de São Paulo, São Paulo–SP, Brasil

One of the most important constraints of chemical evolution models (CEMs) of the Galaxy is the observed radial abundance gradients in the disk. To explain these gradients, different models have been proposed, and a large number of them are successful in accounting for the observed values, although they make use of very different assumptions. The investigation of the temporal behaviour of these metallicity gradients can contribute to clarify the problem of the non-uniqueness of CEMs for the Galaxy. As an example we have the opposite results obtained by the models of Tosi (this volume) and Ferrini *et al.* (1994). With a simplified chemical evolution model we investigate some of the fundamental processes responsible for the formation and evolution of the radial abundance gradients in the galactic disk. Preliminary numerical results for the radial variation of the metallicity at three different epochs show that these gradients depend both on time and position in the disk. They seem to be steeper in the inner regions than in the outer parts and tend to increase with time for galactocentric distances smaller than solar, being constant for greater distances.

1. Physical mechanisms for the origin and evolution of the gradients

The model chosen is similar to that presented by Götz & Köppen (1992) for the galactic disk, where the main physical mechanisms for the origin and evolution of the metallicity gradients are: (i) radial variation of the yield ($y = y(r)$); (ii) radial variation of the star-formation timescale ($\Psi(r,t)/g(r/t)$); (iii) non-linear dependence of the star formation on gas density ($\Psi(r,t) = C_n(r)g^n(r,t)$); (iv) radial variations of the infall parameters; (v) radial gas flows. The fundamental hypotheses are the radial dependence of the radial flow velocity field, azimuthal symmetry and instantaneous recycling approximation. We also assume a primordial infall and adopt an expression for the infall rate, $f(r,t)$, given by Köppen (1994). From the two fundamental equations of CEMs, $\partial g/\partial t$ and $\partial gZ/\partial t$, where Z refers to the metallicity of the system, we can obtain an analytical expression for the temporal evolution of the metallicity gradients:

$$\frac{\partial}{\partial t}\left(\frac{\partial lnZ}{\partial r}\right) = \frac{\alpha y \Psi}{gZ}\left[-\frac{\partial lnZ}{\partial r} + \frac{dln(\alpha y)}{dr} + \frac{\partial ln(\Psi/g)}{\partial r}\right] - \frac{f}{g}\frac{\partial ln(f/g)}{\partial r} - \frac{dv}{dr}\frac{\partial lnZ}{\partial r} - v\frac{\partial^2 lnZ}{\partial r^2}$$

$$(1.1)$$

From this equation we can observe: (i) that one of the terms capable of forming gradients vanishes if $n = 1$; (ii) that for $n > 1$ and $C = constant$, the metallicity gradient will be proportional to the gas density; (iii) radial flows can only modify an existing gradient but never generate a new one, so they cannot explain their origin, i.e. they can only modify an existing gradient; and (iv) that the temporal variation of abundance gradients due to infalling gas on the disk depends on the relative properties of the gas density and infall rate.

2. Preliminary results and discussion

Solving numerically the equation for the temporal variation of the metallicity—$\partial gZ/\partial t$— we obtain the radial variation of the metallicity gradients at different epochs of galactic history (Figure 1). This was made for a simple case, where we assumed: (i) that the

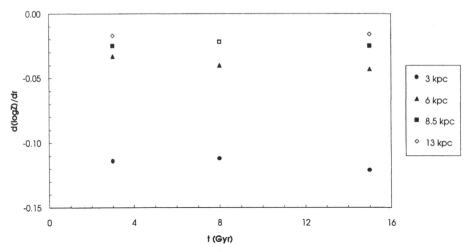

FIGURE 1. Radial gradients at different galactocentric distances for three different epochs

radial variation of the yield is that given by Peimbert & Serrano (1982); (ii) a linear star-formation rate; (iii) $v = -0.5r/r\odot$ for the radial flows velocity field; and (iv) that the functions $g(r,t)$ and $Z(r,t)$ are separable, adopting the observed radial gas density distribution and an exponential temporal gas density distribution for the galactic disk.

The model discussed in this paper is only a preliminary attempt at solving the problem of CEMs and radial gradients. But, even with this simplified model we can see, from Figure 1, that the radial gradients seem to be steeper in inner regions than in the outer parts of the galactic disk, in accordance with previous suggestions (Vílchez, this volume). The figure shows that for $r = 3$ and $r = 6 kpc$ (inner part of the disk) the gradients tend to increase with time, while for $r > r\odot$ they seem to be constant. We shall improve the present model and investigate other star-formation rates (with $n \neq 1$), velocity fields and yield variations in order to place constraints on these parameters. Another possibility is to use an updated sample of planetary nebulae of types I, II and III, which correspond to different times in galactic history, to infer the metallicity gradients at different epochs observationally (Maciel & Köppen 1994).

Work partially supported by FAPESP and CNPq.

REFERENCES

FERRINI, F., MOLLÁ, M., PARDI, M. C. & DÍAZ, A. I. 1994 *Astrophys. J.* **427**, 745.

GÖTZ, M. & KÖPPEN, J. 1992 *Astron. Astrophys.* **262**, 455.

KÖPPEN, J. 1994 *Astron. Astrophys.* **281**, 26.

MACIEL, W. J. & KÖPPEN, J. 1994 *Astron. Astrophys.* **282**, 436.

PEIMBERT, M. & SERRANO, A. 1982 *Mon. Not. R. Astron. Soc.* **198**, 563.

Oxygen Abundance in the Milky Way

By JOSÉ RODRÍGUEZ-QUINTERO
AND MANUEL LOZANO

Departamento de Física Atómica, Molecular y Nuclear, Universidad de Sevilla, Apdo. 1065,
E–41080 Sevilla, Spain

1. Description and Results

The radial distribution of metallicity (as other related quantities), especially that of oxygen, presents a negative gradient that has been widely discussed. In order to study the influence on this radial distribution, which arises from the processes which diffuse the material enriched in heavy elements produced inside stars, a model based on the original hypotheses of the "Simple Model" (SM) (Schmidt 1963) but including diffusive and edge effects has been developed. The starting point was a model used by Wang & Silk (1993) to discriminate different IMFs through the connection between the star-formation rate and the oxygen abundance. We have generalised this model by considering a local metallicity Z and introducing diffusive effects produced by supernova bursts. Thus, local equations for the description of the galactic chemical evolution are derived. The crucial point is the spatial range of these effects causing three different cases to be established, one of which is the above-mentioned model (Rodríguez-Quintero & Lozano 1994). The hypotheses concerning the chemical evolution for the galactic disk that give theoretical support to the model are described in detail by these authors; in summary:

(i) The system to be described (i.e. galactic disk formed mainly by stellar and interstellar gas distribution) is two-dimensional. We define σ_g and M_g as the local and total gas distribution, respectively. The local stellar distribution is directly related to the IMF and the local distribution of stellar formation can be obtained from the latter.

(ii) The instantaneous recycling approximation is assumed. This is the main reason for considering the case of oxygen, to which this approximation can best apply (Wang & Silk 1993; Rodríguez-Quintero & Lozano 1994; Pardi & Ferrini 1994).

(iii) The heavy-element yield, in particular for oxygen, is introduced in our calculations following the work of Wang & Silk (1993). The diffusion mechanism of materials ejected by supernovae is incorporated into our approach by considering a uniform distribution in a circle centred at the supernova event. The radius of this circle, R_g, is a parameter of the model. The physical arguments for that assumption can be found in Rodríguez-Quintero & Lozano (1994).

(iv) Local balance equations to describe the chemical evolution of the system are derived:

$$\frac{\partial}{\partial t}(Z_0(\vec{r},t)\sigma_g(\vec{r},t)) = -Z_0(\vec{r},t)\phi(\vec{r},t) + \int_{S\,(\vec{r})} d\vec{r}'^2 \frac{Z_0(\vec{r}',t)}{A_f(\vec{r}')}\langle m\rangle(\vec{r}',t) +$$

$$\int_{S\,(\vec{r})} d\vec{r}'^2 \frac{1 - Z_0(\vec{r}',t)}{A_f(\vec{r}')}\langle o\rangle(\vec{r}',t) \qquad (1.1)$$

while σ_g satisfies: $\frac{\partial}{\partial t}\sigma_g(\vec{r},t) = -\phi(\vec{r},t) + \int_{S\,(\vec{r})} d\vec{r}'^2 \frac{\langle m\rangle(\vec{r},t)}{A\,(\vec{r})}$.

We have defined: $\phi(\vec{r}) = \int_{m_0}^{\infty} dm \mathcal{F}(m,\vec{r})m$, $\langle m\rangle(\vec{r}) = \int_{m_0}^{\infty} dm \mathcal{F}(m,\vec{r})(m - m_w)$ and

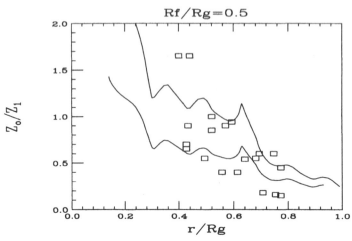

FIGURE 1. Oxygen abundance Z_0 normalised to the solar Z_1. The curves are obtained from Equation (1.2). R_f is taken as $0.5R_g$. Data are from Wang & Silk (1993).

$\langle o \rangle(\vec{r}) = \int_m^\infty dm \mathcal{F}(m, \vec{r})(m - m_w)\delta_0(m)$. Here m_0 and m_l are the lowest mass limits for becoming a star and ending up as a supernova, respectively; m_w is the mass fraction locked up for ever after the supernova event; and $\delta_0(m)$ the above-mentioned oxygen yield. The upper mass limit is implicitly taken in such a way that the stellar distribution vanishes since a certain value. The values of these physical quantities are discussed by Wang & Silk (1993). In Equation (1.1), $A_f(\vec{r})$ is the region where the material ejected from $\vec{r}$ is distributed, and $S_f(\vec{r})$ is the region of points $\vec{r}'$ under the influence of the material ejected from the point $\vec{r}$. $\mathcal{F}(m, \vec{r})$ is the local distribution of the star deaths (or formations).

Assuming that the IMF consists of two factorising terms depending on the mass and position, it can be shown that the best choice for the mass-dependent term is Scalo's IMF (Wang & Silk 1993; Rodríguez-Quintero & Lozano 1994). Introducing a Heavyside function $\theta(R_g - r)$ for the position term, although the results are not very sensitive to the global spatial inhomogeneities (Rodríguez-Quintero & Lozano 1994), and defining $\mathcal{H}(r) = \frac{\langle m \rangle T(r)}{\phi - \langle m \rangle T(r)}$, we obtain the two following limits:

$$Z_0^{lower} = \frac{\langle o \rangle}{\langle m \rangle}(1 - \mu^{\mathcal{H}(r)}), \quad Z_0^{upper} = -\frac{\langle o \rangle}{\langle m \rangle}\mathcal{H}(r)Ln\mu, \qquad (1.2)$$

where μ is the gas fraction; zero initial oxygen abundance is assumed.

2. Conclusion

The diffusion of materials ejected by supernova events is an important effect and with appropriate geometrical considerations it can help to explain such observations as the negative gradient of abundances, even in the framework of the "Simple Model."

REFERENCES

PARDI, M.C. & FERRINI, F. 1994 *Astrophys. J.* **421**, 491.

RODRÍGUEZ-QUINTERO, J. & LOZANO, M. 1994 preprint CERN-TH.7476/94, FAMNSE-94-08.

SCHMIDT, M. 1963 *Astrophys. J.* **137**, 758.

WANG, B. & SILK, J. 1993 *Astrophys. J.* **406**, 580.

Age Distribution of Old Open Clusters

By GIOVANNI CARRARO AND CESARE CHIOSI

Department of Astronomy, Padova University, Vicolo dell'Osservatorio 5, 35122 Padova, Italy

A new and homogeneous age ranking for the family of old open clusters in the galactic disk is presented. The implications for the age–metallicity relation (AMR) and the cluster-formation history in the disk are also discussed. Finally, NGC 6791, the supposedly oldest open cluster so far detected, offers us the opportunity to clarify a point regarding the mass-loss rate along the red giant branch (RGB) and its dependence on the metallicity.

1. Ranking the old open clusters

In a recent paper (Carraro & Chiosi 1994) we proposed a new and homogeneous compilation of ages for the old open cluster population in the galactic disk based on the new stellar models and isochrones of the Padova group (see Bertelli *et al.* 1994, and references therein).

Our idea is that for a correct understanding of both the chemo-dynamical evolution and the star-formation history in the galactic disk, a preliminary step is to arrange the objects one wants to study—the star clusters, for example—along a homogeneous age scale. This is far from being an easy task, since it requires the comparison with complete sets of stellar models including the most updated physics inputs, covering large intervals both in metallicity and in mass (or age).

The detailed procedure to obtain reliable ages is amply described in the quoted paper. We selected 10 well-studied clusters (Table 1) and derived a relation between the logarithm of the ages in gigayears and the magnitude difference between the turn-off (TO) and the clump (the so called ΔV index). With this relation it is possible to get an estimate of the age for any cluster whose ΔV index is measurable. This calibration has been used to obtain an updated AMR (see Figure 1), homogeneous both in ages and in metallicities. The resulting trend shows the lack of any correlation between age and metallicity, suggesting that the galactic disk was already metal-enriched at the epoch of its formation (Friel & Janes 1993).

Recently, Phelps *et al.* (1994) completed a survey of old open clusters in the disk and found a significant population of very old clusters (7–9 Gyr, according to our calibration). The existence of this group of clusters suggests a probably constant past star-cluster formation rate in the galactic disk (see Janes, this volume).

2. Mass loss *vs.* metallicity

The oldest open cluster so far identified is NGC 6791 (Figure 2). Its age has been established at around 7–9 Gyr, depending on the exact metallicity. The most recently published value is [Fe/H] = 0.19 (Friel & Janes 1993), confirming previous suggestions that the cluster has a super-solar metal abundance. NGC 6791 contains a handful of very hot stars recently studied by Liebert *et al.* (1994). The authors conclude that three of these are likely extended horizontal branch (EHB) stars with a mass of around 0.5 $M_\odot$.

In addition, assuming that the clump-star (red horizontal branch) mass is $M = 0.7 M_\odot$ as in globular clusters, they suggest that the mass loss the stars undergo climbing the RGB is quite great (0.4-0.5 $M_\odot$, for a TO mass of 1.1 $M_\odot$). Taking into account the high

Cluster	l	b	ΔV	[Fe/H]	Z_{der}	Z_{ad}	E_{B-V}	DM_0	Age (Gyr)
NGC752	137.2	-23.4	1.00	-0.16±0.05	0.014	0.014	0.02	8.20	1.5
IC4651	340.0	-8.0	1.30	-0.16±0.10	0.014	0.014	0.10	9.90	1.6
NGC3680	286.8	+16.9	1.70	-0.16±0.05	0.014	0.014	0.04	9.90	1.8
NGC2506	230.6	+9.9	1.75	-0.52±0.07	0.006	0.008	0.08	12.50	1.9
NGC2420	198.1	+19.7	1.90	-0.42±0.07	0.007	0.008	0.08	11.80	2.1
NGC2243	239.5	-18.0	2.15	-0.56±0.17	0.005	0.004	0.02	12.90	4.5
M67	215.6	+31.7	2.25	-0.09±0.07	0.016	0.020	0.01	9.47	4.8
Berkeley39	223.5	+10.1	2.50	-0.31±0.08	0.009	0.008	0.10	13.20	6.5
NGC188	123.0	+22.0	2.70	-0.06±0.10	0.017	0.020	0.03	11.16	7.5
NGC6791	70.0	+11.0	2.95	+0.19±0.19	0.028	0.028	0.15	13.50	9.0

TABLE 1. Basic parameters of the open cluster sample

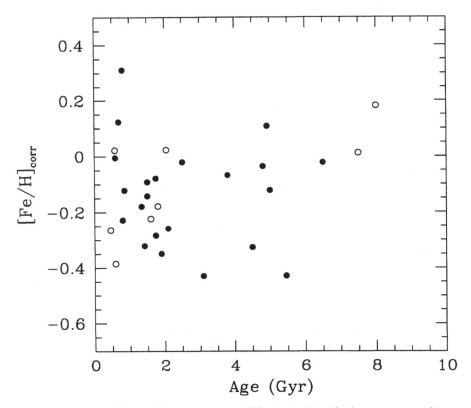

FIGURE 1. The final homogeneous AMR obtained with the new age ranking

metal content of NGC 6791, the conclusion of a strong correlation between mass loss and metallicity is straightforward. This has important consequences; on the interpretation of the UV output from elliptical galaxies (Bressan *et al.* 1994), for instance.

The result by Liebert *et al.* (1994) completely relies on the assumption that the clump-star mass is around 0.7 $M_\odot$ as in globular cluster.

It is easy to verify (see Figure 2) that it is possible to obtain a pretty good global fit of the NGC 6791 colour-magnitude diagram (CMD), assuming that the standard mass

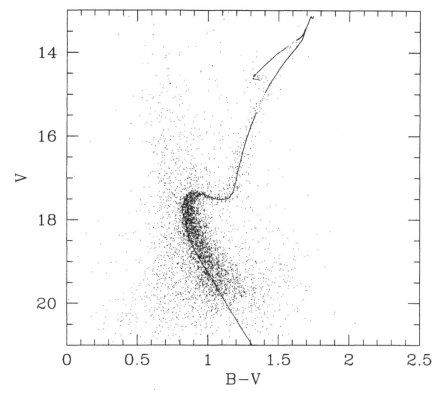

FIGURE 2. CMD of NGC 6791 by Kaluzny & Udalsky (1992). Overimposed is an isochrone for Z=0.028 by Bertelli *et al.* (1994): the η value adopted is 0.35. Distance modulus and colour excess are as in Table 1.

loss parameter η has a value between 0.35 and 0.50 as in the case of the galactic globular clusters. In this case the clump mass turns out to be between 0.85 and 0.90 $M_\odot$.

In conclusion our suggestion is that looking at NGC 6791 no hints can be found to support the idea of their being any strong relation between the mass-loss rate and the metallicity.

The authors have benefitted greatly from lengthy discussions with G. Bertelli, A. Bressan and F. Fagotto.

REFERENCES

BERTELLI, G., BRESSAN, A., CHIOSI, C., FAGOTTO, F. & NASI, E. 1994 *Astron. Astrophys., Suppl. Ser.* **106**, 275.

BRESSAN, A., CHIOSI, C. & FAGOTTO, F. 1994 *Astrophys. J., Suppl. Ser.* **94**, 63.

CARRARO, G. & CHIOSI, C. 1994 *Astron. Astrophys.* **287**, 761.

FRIEL, E. D. & JANES, K. A. 1993 *Astron. Astrophys.* **267**, 75.

KALUZNY, J. & UDALSKY, A. 1992 *Acta Astron.* **42**, 29.

LIEBERT, J., SAFFER, R. A. & GREEN, E. M. 1994 *Astron. J.* **107**, 1048.

PHELPS, R. L., JANES, K. A. & MONTGOMERY, K. A. 1994 *Astron. J.* **107**, 1079.

Andersen: Comparing a set of 2-Gyr isochrones (solar abundance) from various groups, one thinks that those from your group are about 0.05 mag. redder in $(B - -V)$ than the others, for the same T_{eff}. Would you comment on this difference?

Carraro: The reason for this difference is not yet understood. Anyway, our isochrones match the Sun, and this is the main requirement.

Kaluzny: NGC 6791 is an exceptionally rich object. Its giant branch is very well populated. Did you try to use the luminosity function for the age determination?

Carraro: No, I did not. We computed the ratio between the evolved stars above and below a given magnitude level along the RGB (*A&AS* **103**, 375), and compared this ratio with the theoretical predictions at varying ages. We found that this ratio is fairly well reproduced with the age derived from the isochrone fit.

New Results on the White Dwarf Luminosity Function

By JORDI ISERN[1],
M. HERNANZ[1], R. MOCHKOVITCH[2],
A. BURKERT[3] AND E. GARCÍA-BERRO[4]

[1]Centre d'Estudis Avançats de Blanes (CSIC), Spain

[2]Institut d'Astrophysique de Paris, France

[3]Max Planck Institut für Astronomie, Germany

[4]Departament de Física Aplicada (UPC), Spain

1. Introduction

Understanding the evolution of the Galaxy is a formidable task that is still far from being concluded. One of the best methods of determining the dominant physical processes and of identifying the general properties of our Galaxy is to examine the characteristics of its oldest stars. Often, this type of study is limited to the low-mass long-lived stars which were formed during the early stages of the Galaxy. However, it is also possible to use white dwarfs as tracers of galactic evolution. White dwarfs are the final outcome of stars with masses $M \leq 8M_\odot$ and their lifetimes can be as long as that of the Galaxy. Their luminosity function, combined with their kinematic properties, can provide valuable information about different aspects of galactic evolution, such as the star-formation rate as a function of time (Noh & Scalo 1990; Díaz-Pinto et al. 1994), the age of the disk (Winget et al. 1987; García-Berro et al. 1988a,b) or the properties of the halo (Mochkovitch et al. 1990). Due to the intrinsically low luminosity of the objects, the properties deduced in this way are strictly valid for the solar neighbourhood.

2. Model and results

The luminosity function can be written as:

$$n(l) = \int_{M_l}^{M_u} \Phi(M)\psi[T_{disk} - t_{cool}(l, M) - t_{MS}(M)]\tau_{cool}\,dM \tag{1}$$

where l is the logarithm of the luminosity in solar units, $l = \log(L/L_\odot)$, M is the mass of the parent star (for convenience all white dwarfs are labelled with the mass of their main-sequence progenitor), $\tau_{cool} = (dl/dt)^{-1}$ is the cooling rate, M_u is the maximum mass of the main-sequence stars able to produce a white dwarf, and M_l is the minimum mass of a main-sequence star currently able to produce a white dwarf with luminosity l. Therefore, M_l satisfies the condition: $t_{cool}(l, M_l) + t_{MS}(M_l) = T_{disk}$, where t_{cool} is the time necessary to cool down to a luminosity l, t_{MS} is the lifetime in the main sequence and T_{disk} is the age of the disk. $\Phi(M)$ is the initial mass function and $\Psi(t)$ is the star-formation rate per unit volume. All these values can be obtained from Hernanz et al. (1994) and Ségretain et al. (1994).

There are two ways of obtaining the star-formation rate from this equation: one is to solve the inverse problem; the other is to restrict the luminosity function to the massive white dwarfs only. In this case, the lifetime in the main sequence of the parent stars,

179

($M \geq 2M_\odot$), is much shorter than the age of the disk and so can be neglected. Since t_{cool} is weakly dependent on the mass of the white dwarf, it is possible to remove Ψ from the integral. This method has been described in detail by Hernanz *et al.* (1993) and Díaz-Pinto *et al.* (1994), and we will not consider it here. Instead, let us discuss the solution of the inverse problem.

In order to solve the inverse problem it is convenient to change the variable of integration, M, of Equation (1) by the time, t_b, at which the progenitors of the white dwarfs were born: $t_b = T_{disk} - t_{cool}(l, M) - t_{MS}(M)$. This implies that it is possible to define the mass, $M = M(t_b)$, of the stars that, being born at the time t_b, provides a white dwarf of luminosity l at the present time. Equation (1) can be written as:

$$n(l) = \int_0^t K(l, t_b)\Psi(t_b)dt_b$$

with

$$K(l, t_b) = \Phi[M(t_b)]\tau_{cool}[l, M(t_b)]\frac{dM}{dt_b}$$

where $t_b^{up} = T_{disk} - t_{cool}(l, M_u)$ is the time at which stars with the maximum possible mass must be born if a white dwarf of luminosity l is to be produced. The function $K(l, t_b)$ is the kernel of the transformation. It is not symmetric in l and t_b, and shows quite complicated behaviour. Therefore, according to Piccard's theorem, Ψ cannot be directly obtained and a single solution is not guaranteed.

One way of tackling tackle the problem is to define a trial function and adjust its parameters using maximum-likelihood techniques. The last part can be easiliy achieved by using the MINUIT subroutine, from the CERN Library, while the trial function can be obtained in the following way:

The models of chemical evolution of the Galaxy assume that the star-formation rate per unit of disk surface, $\dot{\Sigma}$, is proportional to the surface density of gas. The solution thus takes the form $\Sigma = \Sigma_0 \exp(-t/\tau_s)$ in $M_\odot/\mathrm{Gyr}/\mathrm{pc}^2$. Since the gas has been gradually settling onto the disk, it seems reasonable to assume that its scale height is of the form $h = h_1 + h_0 \exp(-t/\tau_h)$, where h_1 is the present value. Therefore, the star-formation rate per unit volume can adopt the form

$$\Psi(t) = \frac{\Sigma(t)}{h(t)} \propto \frac{e^{-t/\tau}}{He^{-t/\tau} + 1}$$

where the constant of proportionality is obtained by normalising the luminosity function to $l = -3$ in order to avoid the uncertainties linked to the total density of white dwarfs.

Figure 1 shows the star-formation rate that provides the best fit to the observations. The values of the parameters are: $\tau_s = 24$ Gyr, $\tau_h = 0.7$ Gyr, $H = 485$ and $T_{disk} = 12$ Gyr. If this result is taken at face value it turns out that after 2 Gyr of low activity— the coolest stars on the luminosity function were born at this epoch—the star-formation rate suddenly increased, later reaching a maximum value after 4 Gyr. After that it remained virtually constant for 6 Gyr (the star-formation activity has merely decreased by a 20% during this time). The tail is very sensitive to the binning of data, thus implying that the present distribution of white dwarfs is compatible with the existence of a long period of low activity, which, in turn, would imply that is not possible to determine a precise age for the disk.

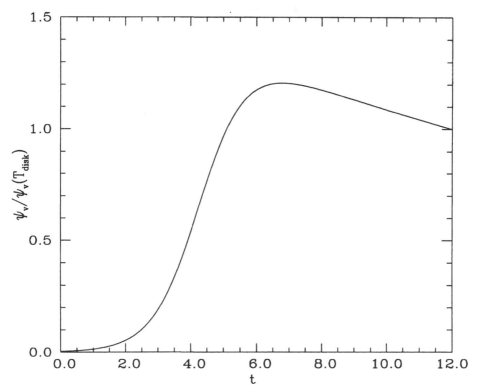

FIGURE 1. SFR derived from the white-dwarf luminosity function

3. Conclusions

The star-formation rate that best fits the white-dwarf luminosity function shows that in the solar neighborhood there was a period of low activity at the beginning of the life of the Galaxy. After 2 Gyr, the activity suddenly increased and reached a maximum after 4 Gyr. Since then the star-formation rate has decreased very slowly, 20%. The time elapsed from the maximum to now is 6 Gyr, so the estimated age of the disk is therefore 12 Gyr.

This work has been supported by DGICYT grants PB91–060 and PB93–0820–C02–02, by the CESCA consortium and by an AIHF. A. Burkert gratefully acknowledges a grant from the EC Action "Access to Large-Scale Supercomputing Facilities in Europe" within the Human Capital and Mobility Programme.

REFERENCES

DÍAZ-PINTO, A., GARCÍA-BERRO, E., HERNANZ, M., ISERN, J. & MOCHKOVITCH, R. 1994 *Astron. Astrophys.* **282**, 86.

GARCÍA-BERRO, E., HERNANZ, M., ISERN, J. & MOCHKOVITCH, R. 1988a *Nature* **333**, 642.

GARCÍA-BERRO, E., HERNANZ, M., MOCHKOVITCH, R. & ISERN J. 1988b *Astron. Astrophys.* **193**, 141.

HERNANZ, M., DÍAZ-PINTO, A., ISERN, J., GARCÍA-BERRO E. & MOCHKOVITCH, R. 1993 in *White Dwarfs: Advances in Observation and Theory* (ed. M. Barstow), p. 15. Kluwer.

HERNANZ, M., GARCÍA-BERRO, E., ISERN, J., MOCHKOVITCH, R., SÉGRETAIN, L. & CHABRIER, G. 1994 *Astrophys. J.* **434**, 652.

MAZZITELLI, I. & D'ANTONNA, F. 1986 *Astrophys. J.* **308**, 706.

MOCHKOVITCH, R., GARCÍA-BERRO, E., HERNANZ, M., ISERN, J. & PANIS, J. F. 1990 *Astron. Astrophys.* **233**, 456.

SÉGRETAIN, L., CHABRIER, G., HERNANZ, M., GARCÍA-BERRO, E., ISERN, J. & MOCHKOVITCH, R. 1994 *Astrophys. J.* **434**, 643.

NOH, H. R. & SCALO, J. 1990 *Astrophys. J.* **352**, 605.

WINGET, D. E., HANSEN, C. J., LIEBERT, J., VAN HORN, H. M., FONTAINE, G., NATHER, R., KEPLER, S. O. & LAMB D. K. 1987 *Astrophys. J., Lett.* **315**, L77.

The Very Young Cluster Westerlund 2: Optical and Near-IR Photometry

By MARIA D. GUARNIERI[1]†, M. G. LATTANZI[2,3],
G. MASSONE[2], U. MUNARI[4] AND A. MONETI[5]

[1]Universitá degli Studi di Torino, Italy

[2]Osservatorio Astronomico di Torino, Torino, Italy

[3]European Space Agency at STScI, Baltimore, USA

[4]Osservatorio Astrofisico di Padova, Padova, Italy

[5]European Southern Observatory, La Silla, Chile

Galactic open clusters are excellent tools for the understanding of the formation and evolution of our Galaxy from both the chemical and structural points of view. Most of the open clusters present large optical reddenings, so infrared observations become very important when studying their stellar content. In particular, the estimated reddening of Westerlund 2 is E(B–V)=1.68 (Janes & Adler 1982; Lang 1991). In this frame, we present optical and near-infrared observations of the young open cluster Westerlund 2, with emphasis on both reduction techniques and absolute calibrations. The visual observations were carried out with the 1-metre telescope at the Sutherland station of the South African Astronomical Observatory using the RCA 512×320 CCD camera in UBVRI filters. IR data were obtained at the 2.2-metre MPI/ESO telescope at La Silla (Chile), using the new NICMOS3 256×256 IR camera IRAC2 in JHK filters. This multi-band data set allows us to derive accurate values for age and distance to the cluster, and to study in detail the cluster's stellar content providing mass and temperature of individual stars.

1. The colour-magnitude diagrams

Westerlund 2 (W2) [C 1022-575, $\alpha = 10^h 22^m 6^s$; $\delta = -57°30'$ (1950.0), mean $E(B-V) = 1.68$ (Janes & Adler 1982; Lang 1991)] is a very young, not very well-studied open cluster in the optical band. The first photoelectric photometry observations of nine stars are from Moffat & Vogt (1975), while the only available CMD is the one by Moffat et al. (1991). They obtained UBV photometry for ~220 stars over a region of approximately 4'×4', and found 82 cluster members. W2 lies in the galactic plane (Carina region), where the extinction is quite high in the optical bands. Prior to this work, no R, I or infrared photometry was available, so we felt it was worthwhile attempting to re-observe and extend this search to higher wavelengths in the same region.

Figure 1 shows the (K, J–K), and (V, B–V) diagrams. The optical data are comparable with, or better than, those presented by Moffat et al. (1991) and, according to Moffat et al. (1991), the dispersion —especially in the (V, B–V) plot— implies variable extinction.

The absolute calibrations in both the optical and the near infrared are quite good and will enable us to carry out a thorough investigation of W2, to include a photometric membership classification based on the reddening of each individual star, age and distance through isochrone fitting, luminosity function and initial mass function (including correction for the contribution of pre-main-sequence stars).

† Present address: Osservatorio Astronomico di Torino, Strada Osservatorio 20, 10025 Pino Torinese, Torino, Italy

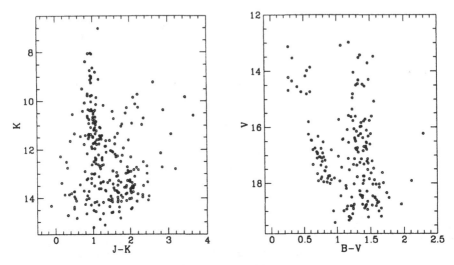

FIGURE 1. (K, J–K), and (V, B–V) CMD's of Westerlund 2

A preliminary discussion of the reduction techniques and the presentation of the absolute calibrations has already been presented in Guarnieri *et al* (1994a), while all the details concerning data acquisition and reduction, and a comprehensive astrophysical discussion of the comparison with the Moffat *et al.* data, will be published in a forthcoming paper (Guarnieri *et al.* 1994b).

REFERENCES

GUARNIERI, M. D., LATTANZI, M. G., MUNARI, U., MONETI, A., GAI, M. & MASSONE, G. 1994a in *XXIInd IAU General Assembly, Session 167: New Developments in Array Technology and Applications* in press.

GUARNIERI, M. D., LATTANZI, M. G., MUNARI, U., MONETI, A., GAI, M.& MASSONE, G. 1994b in preparation.

JANES, K. & ADLER, D. 1982 *Astrophys. J., Suppl. Ser.* **49**, 425.

LANG, K. R. 1991 *Astrophysical Data: Planets and Stars.* Springer.

MOFFAT, A. F. J., SHARA, M. M & POTTER, M. 1991 *Astron. J.* **102**, 642.

MOFFAT, A. F. J. & VOGT, N. 1975 *Astron. Astrophys., Suppl. Ser.* **20**, 125.

Open Clusters as Tracers of the Evolution of the Galaxy

By JANUSZ KALUZNY[1], W. KRZEMINSKI[2]
AND B. MAZUR[3]

[1]Warsaw University Observatory, Al. Ujazdowskie 4, 00–478 Warsaw, Poland

[2]Las Campanas Observatory, Casilla 601, La Serena, Chile

[3]Copernicus Astronomical Center, Bartycka 18, 00–716 Warsaw, Poland

1. Introduction

Over the past few years we have conducted a photometric survey of a large sample of rich and distant open clusters. Our goals included: (i) the identification of the oldest open clusters, which could set an interesting lower limit for the age of the galactic disk; (ii) the study of the chemical evolution of the Galaxy (metallicity relation *vs.* age and galactocentric distance). Multicolour CCD photometry was collected for about two dozen clusters. The observational data were obtained at KPNO and LCO observatories. In this contribution we present results for some most interesting objects from our sample.

2. The metal-rich clusters: Cr 261, BH 176 and Be 54

It is known that a substantial fraction of stars from the galactic bulge are objects of metallicity significantly higher than solar (Frogel & Whitford 1987). Recently, a few globular clusters were identified with [Fe/H] likely to be solar or even higher then solar (eg. Bica *et al.* 1994). A signature of high metallicity is the presence of extremely red giants with $V - I > 3$ on the extension of the red giant branch. Such red stars were recently discovered in the extremely old open cluster NGC 6791 by Garnavich *et al.* (1994). In Figure 1, we present colour-magnitude diagrams (hereafter CMD) for three open clusters: Cr 261, BH 176 and Be 54. These objects belong to the group of the oldest open clusters known. Their ages can be estimated at 5–8 Gyr. Extremely red stars which are likely to be red giants are present in central parts of these clusters. CMDs of Cr 261 and Be 54 were published recently by Phelps *et al.* (1994). Our photometry of these objects is however deeper and more extended in comparison with their data. BH 176 is usually listed with globular clusters. Our photometry shows that its age is close to the age of NGC 6791 (7–9 Gyr). BH 176 can be classified either as a young globular cluster or as an old open cluster.

3. Very old open clusters Be 18 and Tr 5

In Figure 2, we present the CMDs for two poorly known open clusters, Be 18 and Tr 5, whose ages can be estimated at about 4 Gyr. A paper with detailed analysis of photometric data for these and several other clusters will be submitted by the end of 1994 to *Astronomy & Astrophysics*. Photometry for old clusters Be 17, Be 22, Be 29 and Be 54 was published recently by Kaluzny (1994a, 1994b). Photometry for Cr 261 will be published soon by Mazur *et al.* (1995).

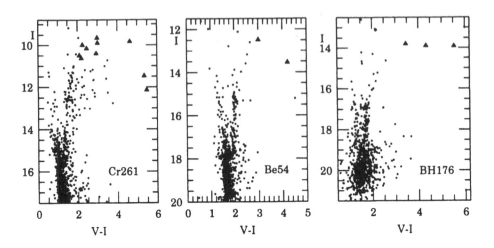

FIGURE 1. The I *vs.* $V - I$ CMDs of Cr 261, Be 54 and BH 176. The large symbols mark extreme RGB candidates.

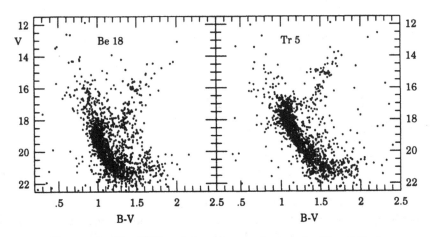

FIGURE 2. The CMDs of the old open clusters Be 18 and Tr 5.

REFERENCES

BICA, E., ORTOLANI, S. & BARBUY B. 1994 *Astron. Astrophys., Suppl. Ser.* **106**, 161.

FROGEL, J. & WHITFORD, A. E. 1987 *Astrophys. J.* **320**, 199.

GARNAVICH, P. M., VANDENBERG, D. A., ZUREK, D. R. & HESSER, J. E. 1994 *Astron. J.* **107**, 1097.

KALUZNY, J. 1994a *Acta Astron.* **44**, 247.

KALUZNY, J. 1994b *Astron. Astrophys., Suppl. Ser.* **108**, 151.

MAZUR, B., KRZEMINSKI, W. & KALUZNY, J. 1995 *Mon. Not. R. Astron. Soc.* in press.

PHELPS, R. L., JANES, K. A. & MONTGOMERY, K. A. 1994 *Astron. J.* **107**, 1079.

Berkeley 17: The Oldest Open Cluster?

By RANDY L. PHELPS[1]†, K. A. JANES[2],
E. D. FRIEL[3] AND K. A. MONTGOMERY[2]

[1]Phillips Laboratory / GPOB, 29 Randolph Road, Hanscom AFB, MA 01731-3010, U.S.A.

[2]Department of Astronomy, Boston University, 725 Commonwealth Avenue, Boston, MA 02215, U.S.A.

[3]Maria Mitchell Observatory, 3 Vestal Street, Nantucket, MA 02554, U.S.A.

In a recent survey of candidate old open clusters, Phelps *et al.* (1994a) found 72 clusters with ages as great, or greater, than the Hyades (age about 0.8 Gyr) and 19 clusters with ages as great, or greater, than M67 (age about 5 Gyr). Among the oldest open clusters we found, and perhaps the oldest open cluster yet discovered, is Berkeley 17 (Be 17). Based upon a "morphological age index" (MAI), Janes & Phelps (1994) determined that Be 17 may in fact be as old as the youngest globular clusters. If Be 17 is indeed that old, it becomes a useful probe of the transition from the halo to the disk.

In this paper we present improved photometry of Be 17, taken with the Kitt Peak National Observatory (KPNO) 2.1-metre telescope. With the addition of spectroscopic data obtained with KPNO 4-metre telescope and the HYDRA multi-fibre spectrograph, we are able to improve our understanding of this important cluster and obtain the reddening, distance, metallicity, radial velocity, and age of the cluster. These data indicate that Be 17 is indeed likely to be the oldest open cluster yet discovered and is as old as, or older than, the youngest globular clusters.

1. Discussion

In an extensive literature and CCD photometric survey of potential old open clusters, Phelps *et al.* (1994a, hereafter PJM94) brought the total number of open clusters known to be as old or older than the Hyades to 72, with 19 of the clusters being as old or older than the ∼5-Gyr-old M67. Using parameters based on the luminosity difference between the main sequence turnoff and the horizontal branch and on the colour difference between the turnoff and the giant branch, a "morphological age index" (MAI) was established for the clusters in the PJM94 list and for a sample of globular clusters (Janes & Phelps, 1994, hereafter JP94).

The age distribution of the open clusters found by JP94 overlaps that of the globular clusters, indicating that the galactic disk began to develop toward the end of the period of star formation in the galactic halo. This assertion by JP94 was based primarily on the MAI of the oldest open cluster in the sample, Berkeley 17, whose faintness made interpration of the 0.9-metre telescope photometry difficult.

We have susequently observed Be 17 with the Kitt Peak National Observatory (KPNO) 2.1-metre telescope with the TEK1024 CCD and standard UBVI filters. Point spread function (PSF) photometry was undertaken using the Stellar Photometry Software (SPS) as described in Janes & Heasley (1992).

This paper reports the early results of our subsequent photometric and spectroscopic study of this important cluster and confirms the PJM94 and JP94 result that Berkeley 17 is indeed the oldest open cluster yet discovered (Figure 1), and has an age as great as, or greater than, the youngest globular clusters. The adopted parameters for Be 17

† Present address: The Observatories of the Carnegie Institution of Washington, 813 Santa Barbara Street, Pasadena, CA 91101-1292, USA

Age:	12.0 ± 2 Gyr	E(B-V):	0.72 ± 0.05	[Fe/H]:	0.00 to -0.30
$RV^{1)}$:	-84 ± 10 km s^{-1}	(m-M)o:	11.90 ± 0.20	Distance:	2.5 ± 0.2 kpc
$R_{gc}^{2)}$:	11.0 ± 0.2 kpc	Z:	-160 ± 20 pc	Diam:	7 ± 2.5 pc

TABLE 1. Be 17: Adopted Values
1) From Scott et al. 1994. 2) Assumes $R_o = 8.5$ kpc.

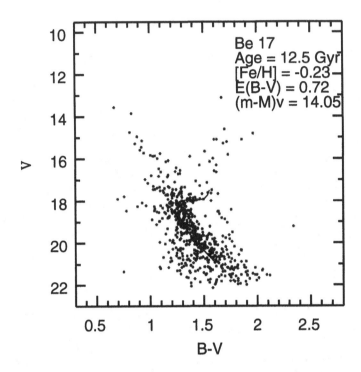

FIGURE 1. Fit of the VandenBerg 1985 [Fe/H] $= -0.23$ isochrone to the CMD of Be 17

are listed in Table 1. Further discussion of the derivation of these parameters and their implications will be presented in a forthcoming paper (Phelps *et al.* 1994a).

REFERENCES

JANES, K. A. & HEASLEY, J. 1992 *Publ. Astron. Soc. Pac.* **105**, 527.

JANES, K. A. & PHELPS, R. L. 1994 *Astron. J.* **108**, 1773. (JP94)

PHELPS, R. L., JANES, K. A., FRIEL, E. D. & MONTGOMERY, K. A. 1994b in preparation.

PHELPS, R. L., JANES, K. A. & MONTGOMERY, K. A. 1994a *Astron. J.* **107**, 1079. (PJM94)

SCOTT, J. E., FRIEL, E. D. & JANES, K. A. 1994 *Astron. J.* submitted.

VANDENBERG, D. A. 1985 *Astrophys. J., Suppl. Ser.* **58**, 711.

New Results for the Oldest Open Clusters: Kinematics and Metallicities of the Old Disk

By EILEEN D. FRIEL[1]†, K. A. JANES[2], L. HONG[1], J. LOTZ[1] AND M. TAVAREZ[1]

[1]Maria Mitchell Observatory, 3 Vestal Street, Nantucket, MA 02554, U.S.A.

[2]Boston University, Dept. of Astronomy, 725 Commonwealth Ave, Boston, MA 02215, U.S.A.

We present results from an ongoing spectroscopic survey of the oldest and most distant open clusters in the galactic disk, including six newly discovered clusters with ages greater than about 5 Gyr. Radial velocities support the conclusion that the clusters rotate like the thin, old disk. New determinations of metallicities further refine the slope and intrinsic dispersion in the disk radial-abundance gradient. We investigate the chemical homogeneity in CN strength among cluster giants, and the correlation of Mg strength with cluster age with the new, enlarged sample. These observational results have significant implications for theories of galactic-disk formation and chemical evolution.

1. Introduction

Photometric surveys have revealed a significant number of open clusters with ages greater than or equal to the Hyades (Phelps *et al.* 1994), but for many of them little is known about their detailed properties. These clusters are excellent tracers of the structural and chemical evolution of the outer disk, and detailed studies of their kinematics and intrinsic properties offer clues to how they survived to such great ages. As part of our effort to use these clusters as probes of the evolution of the galactic disk, we have obtained spectroscopic data, from which we derive metallicities and radial velocities for a number of the oldest clusters.

Our data consist of moderate resolution spectroscopy acquired with the multi-object spectrographs of the CTIO and KPNO 4-metre telescopes. Overall metallicities accurate to about 0.1 dex are derived using spectroscopic indices (Friel & Janes 1993), and radial velocities are measured to an accuracy of about 15 km s^{-1}.

2. Results and Discussion

Results for the radial velocities of clusters observed at KPNO are presented by Scott *et al.* (1994). We find, for a sample of 36 clusters from Phelps *et al.* (1994) which have radial velocities from our data or the literature, that the clusters have kinematics consistent with rotation with the old, thin disk, or assuming a constant velocity solution, that $V_{rot} = 212 \pm 8$ km s^{-1} with a line-of-sight dispersion of 28 km s^{-1}.

Preliminary mean cluster abundances from these new data are presented in Table 1. Combining these metallicities with those for old clusters published or compiled in Friel & Janes (1993), we find a radial abundance gradient of Δ ([Fe/H]/R$_{GC}$) $= -0.091 \pm 0.014$ dex kpc^{-1}, which shows no dependence on age within the sample.

With the addition of six clusters with ages greater than about 5 Gyr (twice the number in previous solutions), we see no significant correlation of cluster metallicity with age. The observed dispersion of 0.17 dex in mean cluster [Fe/H] at any position in the disk

† Present address: Maria Mitchell Observatory, 3 Vestal Street, Nantucket, MA, 02554, USA

Name	l	b	MAI[a]	R_{gc} (kpc)	[Fe/H]	s.d.	N stars	V (km s^{-1})[b]
NGC188	122.8	22.5	7.2	9.31	-0.05	0.11	22	
King5	143.7	-4.3	0.9	10.35	-0.38	0.20	12	-52
Be17	175.6	-3.7	12.6	11.12	-0.29	0.13	13	-84
Be20	203.5	-17.3	4.9	16.29	-0.75	0.21	5	
NGC2158	186.6	1.8	2.2	12.35	-0.23	0.07	6	22
Be31	206.3	5.1	6.3	11.92	-0.50	0.16	17	61
Be32	207.9	4.4	7.2	11.31	-0.58	0.10	11	101
To2	232.9	-6.8	2.5	13.10	-0.35	0.08	15	
NGC2324	213.5	3.3	0.9	11.28	-0.31	0.14	8	
Cr261	301.7	-5.6	9.5	7.48	-0.14	0.14	24	
King11	117.2	6.5	6.3	9.70	-0.36	0.14	12	-34

TABLE 1. Cluster properties, abundances and radial velocities
(a) From Janes & Phelps 1994. (b) From Scott et al 1994.

or at any age suggests a significant intrinsic dispersion in [Fe/H], in agreement with field-star studies.

These new data continue to show the same trend of weaker Mg in younger clusters relative to older clusters at a given [Fe/H], though with greater scatter than was found originally (Friel & Janes 1993). Theoretical modelling of the indices (McQuitty *et al.* 1994) suggests that both gravity and [Mg/Fe] variations can explain the magnitude of the effect, although we continue to investigate inadequacies in the empirical calibrations.

Cluster mean CN strength scales with overall metallicity, as expected for Population I stars, and most clusters show star-to-star variations in CN strength consistent with observational error. Several clusters show a larger dispersion, which is most likely due to variable reddening, and there are several instances of strong-CN stars.

The combination of radial abundance gradient, lack of age–metallicity relationship, and modest chemical inhomogeneities in the old open cluster population suggest a complex picture of galactic chemical evolution which probably involves prompt (localised) enrichment of the disk, coupled with continued infall and/or radial flows over the lifetime of the disk. These new data, particularly for the oldest disk clusters, call for detailed models of chemical evolution that can link the evolution of the galactic halo with the oldest disk population.

We thank Rocio Patino and Kendra Barkocy for help in the data reduction. This research is supported by grants from the Perkin Fund and the NSF (AST–9300391).

REFERENCES

FRIEL, E. D. & JANES, K. A. 1993 *Astron. Astrophys.* **267**, 75.

JANES, K. A. & PHELPS, R. L. 1994 *Astron. J.* **108**, 1773.

McQUITTY, R. G., JAFFE, T., FRIEL, E. D. & DALLE ORE, C. M. 1994 *Astron. J.* **107**, 359.

PHELPS, R. L., JANES, K. A. & MONTGOMERY, K. A. 1994 *Astron. J.* **107**, 1079.

SCOTT, J. E., FRIEL, E. D. & JANES, K. A. 1994 *Astron. J.* submitted.

Element Abundances: Pollution in an Ecosystem

By FEDERICO FERRINI

Dipartimento di Fisica, Sezione di Astronomia e Astrofisica
Piazza Torricelli 2, I–56126 Pisa, Italy

1. Introduction

The subject of element abundances is interrelated with all the other subjects in the Conference. From the presented papers it became evident just how difficult it is to even recognise clear trends in the observational scenario. Consider that the history of our Galaxy is written in its element abundances, in the sense that the formation of the global structure, the laws of conversion of gas into stars and the nucleosynthesis internal to stars produce, rather like a symphony, the present abundance status. The elements— these tiny little things—contain in the mystery of their quantity and distribution one possible key to understanding the history of the complex, self-regulated system that is the Milky Way.

At least three different aspects are involved in determining the present observed status: star formation, dynamics and nucleosynthesis. This unfair situation stimulates a series of questions, which are of a sufficiently simple and general nature to be very appropriate for a panel discussion.

As general references on the situation of our knowledge concerning elements, their distribution and origin, there have been several workshops in the last few years and various reviews papers (e.g. Gilmore *et al.* 1989; Rana 1991; Franco *et al.* 1993; Shore & Ferrini 1995).

2. The Questions

2.1. *The "zero" question: Are we talking about something solid or not?*

Are the element abundances homogeneously derived from the observations? Is there a reasonable agreement between photometric and spectroscopic calibrations? We have a very large amount of data: can they be collected together (different groups of stars, open and galactic clusters) to give a really rich and more stringent picture of the 3D situation of our Galaxy? What are the uncertainties, both systematic and due to dispersion on the empirical data? Once, we believed in relations such as the age–metallicity relation (AMR); now, and evidence has been shown presented at this Workshop, there are distributions, diagrams, non-injective functions. What are the errors in the abundances and in the ages? This is crucial also for a deterministic or stocastich-accretion scenario.

Are we dealing with well-defined quantities or ill-defined ones? "Metallicity:" does everybody mean the same thing? I quote three examples of rather common assumptions, taken from papers of very accurate authors, to illustrate the situation:

• often authors introduce "theoretical global metallicity" such as: $\log Z = \alpha[\text{Fe/H}] - \beta$, extrapolating from the fit for a selected sample of data; this obviously generates a lot of confusion in the interpretation of different sets of data.

• knowing [Fe/H] directly from spectroscopical data, it is often assumed that the oxygen abundance for low iron-abundance stars ([Fe/H] < -1) can be deduced from relationships such as: [O/H] = 0.4[Fe/H], which is a traditional but rough fit to objects of possibly different origin and situation. This assumption gives [O/Fe] = −0.6[Fe/H], which is an overestimate by a factor of two to the data of the same authors, when referring to stars for which they have accurate observations of both elements at their disposal (Edvardsson *et al.* 1993).

• Another typical assumption concerns the abundance relative to iron of heavy unobserved elements. It is generally considered to be the same as in the Sun, while they may refer to something completely different: why the hell must a Population I star have the same abundance ratios as a Population II star?

This last point introduces another crucial question concerning the stellar evolution results, which are used for different populations, with a certain ease. Z is used as the quoted parameter but Fe/O, for example, changes. What does Z mean, and how would different abundance mixtures affect ages from models? Furthermore, stars produce elements according to their initial composition, and this is not only a question of Z. It is easy to imagine that there could have been differences in the nucleosynthesis and ages.

A thorough comment by Andersen has been addressed to this question: his answer to the first two parts of Question "zero" is clearly a resounding *No!* One only has to look at the various photometric and spectroscopic techniques used to study from one to a collection of elements in different types of stars and gas clouds to see why—especially in view of what we now know about the chemical diversity even in thin-disk stars.

Yet one should consider the intended use of the data. If the goal is *absolute* abundances (the number of atoms of a given species per 10^{12} H atoms), e.g. to calculate stellar models to determine the absolute ages of globular clusters, then many unsolved problems remain. If, however, the goal is to look for the degree of homogeneity (or lack thereof) in a well-defined set of stars, then it is possible to obtain—*not* compile!—a set of carefully controlled data that can discriminate abundance variations *differentially* at the ± 10% level. This was the approach taken by Edvardsson *et al.*. O is a special problem here, both because of its different nucleosynthesis history from that of, for example, Fe (because it is the most commonly studied element in gaseous nebulae), and because it is observationally difficult even in stars. It is well known that the OI-triplet near 7770 Å gives appreciably higher abundances than the 6300 Å[OI] line, and many of the currently quoted O determinations for halo stars are derived from the triplet, which is affected not only by NLTE effects, but probably even more severely by atmospheric inhomogeneities (granulation, etc.—cf. Edvardsson *et al.* 1993).

Finally, a word about stellar ages. There remains considerable confusion in the field of stellar models regarding opacity tables, convection prescriptions, and temperature–colour conversions, but any recent set of models can probably be used to assign *relative* ages to an accuracy of perhaps 15–25%. *None* of the current models, however, incorporates our present knowledge of the chemical composition of metal-poor stars. The van den Berg and Bergbusch models include O enhancement, but not the other elements known to have non-solar abundance ratios. Since the detailed abundance ratios depend on both [Fe/H] and galactic position, one should ideally base age determinations on a similarly detailed set of models; interesting systematic effects could arise from the neglect of this.

2.2. *Question 1: abundances and populations*

• Are there any laws in the distributions?

• How do you separate the components of the galaxy? Remember the "old" argument that classified Halo and Disk stars via a "chemical" distinction between the two populations, the threshold being posed at [Fe/H] = −1. Many models of galactic evolution based their structure on this criterium, calling "halo" everything with [Fe/H] = < −1 and "disk" the subsequent matter; it is evident that this argument cannot give much idea of the possible evolution. This criterium has been used and abused for long time: it is indeed very easy to use, but perhaps devoid of meaning (cf. the poster by Laird *et al.*, this volume). Kinematical separation is much more significant and important.

- Observational mixing *vs.* real mixing: which sort of mixing happens in nature? We may envisage the possibilities of no mixing at all (what is newly produced, remains forever where it is deposited), a continuous mixing (a gentle source like "uno Zefiro Gentile" will allow for a continuous diffusion process) or a stochastic mixing (setting up supernovae or high-velocity clouds (HVC); fireworks to animate the *siesta* and turn it into a *movida*.

According to Tenorio-Tagle, there is a full chapter of Astrophysics that has yet to be written: the mixing of elements. We do not really know how metals mix; we only seem to recognize that the material ejected from supernovae mixes on a very rapid timescale relative to evolutionary times. But we should also consider that the time and length scales of mixing may be various, i.e. a part of the metals can stream out of the Galaxy.

Tosi presented a critical case for mixing: for IZw18, the most metal-poor irregular galaxy ever known, the abundances derived from HII regions are $Z \sim 1/30 Z_\odot$, while the recent observations from *HST* concerning HII gas result in an abundance $1/1000$ lower. The mixing was not very efficient there, indeed, but the idea of Kunth, Sandage and collaborators is that within at most 1 Gyr everything will be mixed.

Shore noticed that supernovae are very hard to watch for mix because their ejecta could take too long to slow down and mix—too long for the lifetime of a PhD student. But novae and massive stars very often throw out very rich material in dense environments where we can actually watch them in real-time mix. η Carinae is a perfect subject; it's got everything: dust, nucleosynthesis, radiation pressure, gas. In at least in a couple of novae the abundance productions are particular markers: Al, Na are real trace elements that have very few seeds in the galaxy. You can really watch them forming in small amounts and have the hope of seeing how they are getting diluted in the background.

- Why do we expect to see an age–metallicity relation (AMR)? I have a couple of arguments, one in favour of establishing an AMR and the other against.

The formation of stars takes place inside molecular clouds: numerous independently evolving clouds formed the halo stars. From a conceptual point of view we expect fluctuations in the large-scale distributions of elements: clouds behave ballistically; in a viscous medium there is orbital mixing, spread of birth places. Clouds born together from gas (same Z) are spread around, so the birth places of stars are mixed in space; the instability conditions for clouds are dependent on the position not on age: collisions, differential rotation, tidal stresses, etc. will fragment clouds. The new-born stars will have same Z but different ages! Or we can find also stars with same age and different metallicity, when they are born in different zones (halo, thick and thin disk).

Suppose the sampling occurs as a point process at random intervals in space and time (i.e. star formation) and that the system has very small fluctuations in, say, Z. Then Z for a single cluster $\simeq <Z>$. But if $<\delta Z^2>^{\frac{1}{2}}$ is large then the same process for any one cluster can have a large deviation from $<Z>$. So is the galaxy irregular on short times and only smooth on long times?

At the same time we expect a tendency to saturation —i.e. non-exploding behaviour— in the evolution: galaxies are complex systems, where many physical ingredients work together like the instruments in an orchestra, and hence, in this highly non-linear evolving structure we wait for stable solutions on long timescales.

The evolution does not have to go to saturation, as suggested by Shore. In our particular neck of the woods, it looks as though it does go to saturation: this tells us that the system is non-linear as we know it is. Here is a strong constraint: we do not see chaos—we see dispersion. In our neighbourhood, all the timescales being about the same, when we look out either into the past or away from us the scales start to separate. For example, when the halo is collapsing, the evolutionary times for the individual stars that were formed and the return time for the gas were probably very different from the formation and dynamical times. When the system (or a part of it) settles down, again we get back to the dissipation—we wind up in a system where all the timescales are suddenly equal. So maybe one of the reasons why we do not see a single age–metallicity relation is because everything has been mixed up by the competing timescales.

2.3. *Question 2: kinematics and metallicity*

- Is memory a function of scale?
- From what scale does star formation begin?
- At what level does the physical state no longer matter for the final large-scale answer?
- At what scale do the physical state and all its consequences—star formation, cloud properties, abundance inhomogeneities, i.e. the coarse-grain statistical quantities—average out?

Shore has synthesized this ecological problem in galactic evolution well: *The "Greening" of Galactic Structure: Act Locally, Think Globally*. Feedback and self–regulation turn out to be more important than the numerous sorts of interactions: it is as if on some intermediate scale between the single cloud and the small star association size and the few kpc complexes, a company were organising the economy of the system.

The convivence of very different timescales has been underlined by Serrano: the self-regulation processes intervening in clouds and concerning star formation act on 10^7 yr, while the overall evolution for large-scale star formation and enrichment takes about 5 Gyr.

Franco finds a problem for the range of possible abundances: we easily see very low abundances (10^{-2} solar) but very rarely do we find values higher than two or three times solar. Just with quasars, Hamann & Ferland (1991) have found an abundance of Nitrogenum as high as 2–9 times solar. QSOs are supposed to be early stages in galaxy evolution so we are a long way from the present age of galaxies. This seems to contradict the objection that in our neighbourhood starting from zero or a value very close to zero, it took 10 Gyr to reach solar abundances, and so it would come as no surprise if in another 5 Gyr this value had doubled (Laird). The choice of the IMF, with a given nucleosynthesis scheme, fixes, for Serrano, the maximum production of elements which is therefore not a function of time. This can be considered as an example of saturation as long as you have a certain combination of IMF and nucleosynthesis; of course, this happens because a lot of elements are in dust and stellar remnants (neutron stars, white dwarfs, black holes). You will never increase the abundances much more than we see.

The analysis of Carraro and coworkers reveals the existence of different dynamical families in the system of galactic globular clusters, important markers for the evolution and the formation of the Milky Way. The dynamical interactions with the disk eroded the inner halo population and modified the initial abundance distribution in clusters.

Very high values of N and O are seen in the inner part of external galaxies, up to 2–3 times solar, according to Vílchez. Also in some parts of the Galaxy high values are reached; abundances of N higher than ten times solar are observed in WR stars, LBVs and blue and red supergiants. These cases are anyway different from the rest of the Galaxy, since their ejecta are not mixed. The presence of gradients (see Question 3) is connected with the presence of high abundances in the central part of spirals where they are very high, and this is why it is so difficult to measure gradients. We are somehow biased to measure abundances from solar to one-tenth solar.

A technical but fundamental point has been raised by Shore about the determination of abundances in HII regions via O and WR stars. People who are using stellar models have not matched the abundances of stellar models and opacities of modern stellar atmospheres to the HII regions in which the stars themselves are embedded. If you have a metal-rich HII region you may very well have a metal-rich star sitting inside: this is a real problem, if what you are using is a solar abundance atmosphere and also one which does not have the proper line blanketing. Hence the determination of abundances (e.g. N^+, O^+) may be extremely difficult for early O stars (e.g. OIV). Vílchez agreed with Shore, adding that only recently have stellar atmosphere models begun to treat NLTE and non-solar abundances (in the sense of $Z << Z_\odot$ but always with solar ratios among the elements).

Vietri made a few general remarks about the importance of dynamics:

• There is a quantity in collisionless systems—i.e. globular cluster, halo, bulge—conserved from the very beginning of the Universe: the distribution of mass as a function of specific angular momentum. Disk viscous stresses and collisions with giant molecular clouds can alter the distribution of angular momentum for the other components of the Galaxy.

• Alfaro found a number of globular clusters streaming one way from the centre of the Galaxy and the same number streaming the opposite way. If you believe in the presence of a triaxial bulge in the centre of our Galaxy, boxy orbits—those that plunge toward the center of the Galaxy—do not exist, while the only allowed orbits are streaming and counterstreaming orbits. Certain objects that for some reason get stuck close to the centre cannot move easily in and out, but will necessarily corotate with the rotating bulge or counter-rotate. This local explication is an alternative to the division into two sub-populations.

• Another example of how dynamics is important was raised in the question of Franco on why we do not see more metal-rich stars. Perhaps we are not looking in the right places: the right to look for objects that are very metal-rich and close to the centre of the Galaxy, but only with very radial orbits—i.e. high, eccentric orbits—starts here!

• A final example of the importance of dynamics is related to Ferrini's question about whether chemical properties alone can determine the fate of galactic evolution. He recalled, for example, the paper by Eggen, Lynden-Bell and Sandage (1962) based on coupling dynamics and chemical properties.

Tosi pointed out that it is not true that we do not look in the right places to find metal-rich objects: after all we do look at elliptical galaxies which are indeed metal-rich but not that rich—five times solar, no more. She believes in a conspiracy of nature, with nature in some way trying to saturate. Elliptical galaxies have a lot of star formation and so make a lot of metals, but then at a certain point they run out of gas and stop making metals, because stars cannot form anymore. The galaxies where more metals are made are the first to stop making them. In other galaxies there is a lot of gas but the star-formation rate is lower or the outflow is much more efficient as in irregular galaxies, or infall is present. With the exception of Quasars, are they really metal-rich or is there a problem of determination of abundances? Galaxies of all morphological types have the tendency to saturate and perhaps we do not see anything above five times solar because they are unable to make it.

2.4. Question 3: abundance gradients

• Disk: do element abundance gradients vary with time, and how?

I will start by quoting a perentorious declaration by J. Vílchez. He stated that he does believe in gradients and that the gradients vary from galaxy to galaxy. It is indeed evident from the observations of external galaxies (see, for example, Oey & Kennicutt, 1993) that gradients are different in the various Hubble types. The gradients are flatter in early-type spiral galaxies: Sa has a gradient which is virtually zero, while as the sequence towards Sd the negative gradients steepen. The distinctive property throughout the Hubble spiral types is the star-formation history of the disk, as theoretical models (e.g. Larson & Tinsley, 1978; Ferrini & Galli, 1988) and phenomenological arguments (e.g. Sandage, 1986) demonstrate. Galaxies which had a more active star formation in their early evolution (Sa) consumed most of their gas in a short time, and the abundances, as a result of efficient stellar production and hence nucleosynthesis, soon reached the maximum possible value, and hence the evolution was completed. This situation of long-term equilibrium is apparent in the absence of abundance gradients. Galaxies with an early low star formation (Sd, Sm) for a long time kept enough gas as a reservoir for recurrent star formation; their enrichment is not yet completed in the external regions, where star-formation efficiency is relatively lower. Consequently, the abundance gradients are rather steep and will flatten with time, since the internal regions evolved more than the external ones. These arguments conflict, as discussed in the contribution by Tosi, with all the standard models except for the multiphase approach by Ferrini, Shore and coworkers. The situation for the Milky Way seems to be more controversial with regard to the observations (see Tosi's contribution in this volume and Ferrini *et al.*, 1994), but we could expect that the situation for the Galaxy would be analogous to other galaxies.

• Halo: Can gradients be cancelled out? Can the timescale for disk formation and/or bulge formation be constrained by halo gradients?

It seems quite reasonable that the argument on the ballistic behaviour of clouds, reinforced by the presence of the thick-disk zone and the time delay between the ages of stars pertaining to thin and thick disk, suggests clearly that an ordered distribution of stellar abundances in the halo—especially in its lower part—can be excluded (see Pardi, Ferrini & Matteucci 1995).

• Infalls and outflows: what is their origin and nature?

Infall has been assumed as a prop for models in difficulties (see below), but now it seems clear that the accumulation phase of the internal galactic regions is real, as is the possibility of matter removal: galactic regions and the whole Galaxy are open systems. Naturally, any assumption about infall rate must satisfy two dynamical constraints: the surface mass distribution in the disk and the observed rotation curve.

Similar arguments have been presented by Serrano: from 4 kpc onwards chemical abundances do not yet reach saturation, and this is why we observe gradients in the outer part of galaxies. Also the results derived by Isern from white dwarfs point in the direction of a disk taking a long time to form, expecially in the outer parts.

There is a great deal of discussion about stellar dynamics, according to Palouš and its influence on abundance distributions, concerning not only large scale motions, but also bars (see, for example Friedli *et al.* 1994) and radial streaming.

2.5. *Question 4: the models*

We are immediately confronted with a dilemma: simplicity or complexity? I mean, is it permissible to reduce the description of the evolution of galactic subsystems to a set of laws, based on phenomenological deductions, or should we put together an impressive array of machinery to describe the interactions in some detail? The advantages and drawbacks of the two competitive approaches are evident.

It seems to me that a couple of arguments force us to take this unpleasant decision: the phenomenological laws are established on systems *after* their evolution (see the Schmidt law for star formation, $\Psi \propto \rho^n$, where ρ is volume or surface density of HI or H_2 or both). So feedbacks must be present if not dominating the evolution (Franco, this volume).

How well do we know the origin of individual elements? The abundances produced from a stellar generation are determined by the nucleosynthesis processes of single stars and by the initial mass function (IMF) of stars, and this is another kind of "dark matter" in models: there is clear evidence that the Salpeter mass function is unrealistic for any portion of the Universe, but it assumed to hold in 95% of galactic-evolution models!

Theoretical studies of chemical evolution from the seventies to mid-eighties look like a *Paraiso cerrado para muchos* [a closed paradise for many] (Pedro Sato de Rojas, Granada, 1584–1658)—they are hermetic in their assumptions/equations, but diffuse in their results. A school of uniform and homogeneous thinking was established. Excellent work has been done; by focusing on the nucleosynthetic aspects of star formation they took the first fundamental steps, but they formed a club which paid scant attention to a rapidly growing field: the connection of gas to stars. They insisted on adopting $\Psi \propto \rho^n$, while there were no indications of such a "law", neither in nature (where we see the consequences of the star formation, not really the process), nor in simulations, which do not produce stars, or even clouds, i.e. self-gravitating structures!

The driving question behind many of the attempts to determine the law for star formation, and hence the rules for galactic evolution, is the G-dwarf problem (Audouze & Tinsley, 1976; Pagel & Edmunds, 1981; Rana, 1991). The stars of spectral type G or later and main-sequence luminosity are low-mass stars, about 0.8 $M_\odot$, with an evolutionary timescale of about 15 Gyr, similar to the estimated age of the Galaxy. They represent a monotonously increasing population; thus, a complete sample of these stars in the solar neighbourhood preserves the record of local star-forming activity. The G-dwarf problem can be stated as follows: in a closed system undergoing star formation at an exponentially decreasing rate from the gas that is being polluted with metals at a rate dependent on the star-formation rate, how can there be so few stars with very low metal abundances? Note that this question contains several testable conditions. The first is that the system is closed. The suggestions of infall by Larson (1972), Ostriker & Thuan (1975) and many subsequent studies have centred on this part of the problem. The second assumption is that the star formation must necessarily have been higher in the past. In a self-regulated system this is not necessarily so. Finally, the rate of gas depletion is very different for the non-linear models than when the star-formation rate is assumed. A multipopulation–multizone approach can naturally achieve this situation (see, for example, Shore, Ferrini & Palla, 1987; Ferrini *et al.*, 1992, Shore & Ferrini, 1995).

Franco noticed that there is an ingredient often missing in the discussion of galactic evolution: dust. We have to consider it for reddening corrections, but also because it holds half of the elements that are present in the interstellar medium. Radiation effects on dust can accelerate grains out of the galaxy, and we may easily pollute the intergalactic medium with the heavy elements.

Vílchez stressed the problem of the high dust level around quasars: How were such large amounts of C produced at such an early epoch? Quoting some results for the determination of abundances considering NLTE effects and how they teach us that we do not need to be quite so keen to try to make everything fit: systematic effects can be present.

Binaries are also absent from the present schemes of galactic evolution, as mentioned by Shore.

2.6. *Question 5: Dei ex machina or real physical effects?*

- Variable IMF—do we really need it?

- Does star formation proceed in a stochastic way?
- Non-isolation of the Milky Way—how much material, enriched or not, do we need?

Hensler comments that with the simple model we could not understand the evolution of our Galaxy, so the need for open-box models quickly emerged, although there was no physical reason to introduce the infall term. Now we know that material is falling on the disk in the form of HVCs, or from the Magellanic Stream, giving support to open models and excluding the closed ones.

With regard to the variations of IMF, most evidence suggests that if it varies, it does not vary too much: if you make the slope of the high-mass end of the IMF flatter, you will get a rate of oxygen production that is too high. If you change the slope at the low-mass end, you will have problems with the production of nitrogen, as Tosi suggested.

2.7. *Question 6*

Can the element distributions unequivocally constrain the possible scenarios for the formation and evolution of the Milky Way?

3. The answers

This section has yet to be written by an international community of galactic-evolution fans, who I believe have every chance of becoming a winning team.

I wish to thank all the discussion participants; Emilio Alfaro and the LOC, for their perfect organisation; and Steve Shore, for years of discussions.

REFERENCES

AUDOUZE, J. & TINSLEY, B. M. 1976 *Annu. Rev. Astron. Astrophys.* **14**, 43.

EDVARDSSON, B., ANDERSEN, J., GUSTAFSSON, B., LAMBERT, D. L., NISSEN, P. E. & TOMKIN, J. 1993 *Astron. Astrophys.* **275**, 101.

EGGEN, O. J., LYNDEN–BELL, D. & SANDAGE, A. R. 1962 *Astrophys. J.* **136**, 748.

FERRINI, F. & GALLI, D. 1988 *Astron. Astrophys.* **195**, 27.

FERRINI, F., MATTEUCCI, F., PARDI, M. C. & PENCO, U. 1992 *Astrophys. J.* **387**, 138.

FERRINI, F., MOLLÁ, M., PARDI, M. C. & DÍAZ, A. I. 1994 *Astrophys. J.* **427**, 745.

FRANCO, J., FERRINI, F. & TENORIO–TAGLE, G. (EDS.) 1993 *Star Formation, Galaxies and the Interstellar Medium*. 4th EIPC Astrophysical Workshop. Cambridge University Press.

FRIEDLI, D., BENZ, W. & KENNICUTT, R. 1994 *Astrophys. J., Lett.* **430**, L105.

GILMORE, G., WYSE, R. F. G. & KUIJKEN, K. 1989 *Annu. Rev. Astron. Astrophys.* **27**, 555.

HAMANN, F. & FERLAND, G. J. 1991 *Astrophys. J., Lett.* **391**, L53.

LARSON, R. B. 1972 *Nature* **236**, 7.

LARSON, R. B. & TINSLEY, B. M. 1978 *Astrophys. J.* **219**, 46.

OEY, M. S. & KENNICUTT, R. C. 1993 *Astrophys. J.* **411**, 137.

OSTRIKER, J. & THUAN, T. X. 1975 *Astrophys. J.* **202**, 353.

PAGEL, B. E. J. & EDMUNDS, M. G. 1981 *Annu. Rev. Astron. Astrophys.* **19**, 77.

PARDI, M. C., FERRINI, F. & MATTEUCCI, F. 1995 *Astrophys. J.* in press.

RANA, N. C. 1991 *Annu. Rev. Astron. Astrophys.* **29**, 129.

SANDAGE, A. 1986 *Astron. Astrophys.* **161**, 89.

SHORE, S. N. & FERRINI, F. 1995 *Fundam. Cosmic Phys.* in press.

SHORE, S. N., FERRINI, F. & PALLA, F. 1987 *Astrophys. J.* **316**, 663.

The Intermediate Population II—
Extended, High-Velocity, Metal-Weak,
Thick Disk—Inner, Flattened Spheroid—
Lower, Contracted Halo

By STEVEN R. MAJEWSKI

The Observatories of the Carnegie Institution of Washington, Pasadena, CA, U.S.A.

The spatial, kinematical, chemical and age properties of the intermediate Population II (iPII) are discussed in the context of constraining formation models. It is suggested that all the flattened populations named above are the same, or parts of the same, ubiquitous galactic component having broad kinematical and chemical distributions. The formation of the iPII was likely independent from (although simultaneous with) that of the halo, along the lines of recent "dual halo models". However, it is still not clear whether the iPII is a distinct population from, or an extension of, the thin disk, nor whether the iPII formed in a dissipative collapse (distinct from the formation of the halo) or via the dynamical heating of an existing thin disk by the accretion of a galactic satellite into the disk.

1. Introduction

Are there any real differences between the variously named populations in the above title? It is my contention that the differences may be only semantic and that the various terms all refer to the same, or parts of the same, component of the Galaxy.

There has long been a recognition of the need for some stellar population "intermediate" between the old disk population and the halo. The "intermediate Population II" (iPII) component was introduced at the 1957 Vatican Conference on Stellar Populations (O'Connell 1958). Since then, understanding the nature and importance of this component in the formation of the Galaxy has become an ever more central theme, although it was not pursued with vigour until the very existence of this component was debated in the early 1980s. Some of the properties ascribed to the iPII at the Vatican Conference are similar to modern determinations: a scale height of 700 pc, an age reaching that of the extreme halo population (though the age scale from contemporary stellar evolution theory gave substantially lower *absolute* age values), and the inclusion of "high-velocity stars" with $w > 30$ km s^{-1}. Because various other names for this component have taken on certain connotations of formation, for this discussion I will adopt the name "iPII".

Unfortunately, the iPII is elusive and its study is frustrated by selection biases and preconceptions. Simply defining what this component is presents a major problem: names attached to it are derived from the direction of approach or predilections toward delineation by location, age, kinematics, metallicity, interface with other components, or origin. For example, if a flattened component is found in a metal-poor sample, we get a "flattened spheroid". If a high-velocity component is found in a metal-rich sample, the name "high-velocity disk" obtains.

It is the "intermediateness" of the iPII that makes it so elusive. It significantly overlaps other components in every possible defining property. Figure 1 demonstrates the difficulty in simply attempting to delineate members of this component from the mix of (typically blue) halo stars and (typically red) disk stars at a magnitude range where the dominant contribution from each component is main-sequence stars, $V = 20 - 21$.

199

2. Origin scenarios

Of course, our objective is to determine "The Formation of the Milky Way". In spite of the difficulties, it is worthwhile to study the iPII because it plays a key role in recent formation models. iPII formation scenarios are summarized in Gilmore *et al.* (1989) and Majewski (1993). They break down generally into "top-down" models, whereby the iPII is a transitional phase between the formation of the halo and the thin disk, and "bottom-up" models, where the formation of the iPII occurs after that of the thin disk.

The bottom-up models typically involve some action on, or in response to, a previously formed, thin disk. For example, the original proposal to explain "thick disks" in external galaxies (van der Kruit & Searle 1981), and suggested by Gilmore & Reid (1983) to explain our own, is that they represent the gravitational response of the halo to the disk potential. This scenario is no longer regarded as viable due to the low surface mass density of the disk (Gilmore *et al.* 1990). Alternatively, thin disk stars may have been kinematically heated into a more extended iPII configuration. This heating may have been violent, perhaps from the accretion of a satellite galaxy (Carney *et al.* 1989; Quinn *et al.* 1992). In this case, any preexisting gradients in the thin disk are expanded and the asymmetric drift of the iPII is expected to be modest (Quinn *et al.* 1992). The thin disk regenerated after the violent event, making the iPII disjoint from both the halo and the thin disk (which at its oldest is younger than the time since the formation of the iPII). A variant of this model is that the iPII represents the debris from the accreted satellite; however, Statler (1988) points out that for this scenario to work, special conditions regarding the satellite's orbit (low eccentricity) and/or the need for the dark matter halo to be truncated at $R_{GC} = 30$ kpc must apply.

More gradual heating may also have "puffed up" the thin disk. Among the proposed secular heating agents, (i) the density of molecular clouds is apparently too low to account even for kinematically hot, thin disk stars (Lacey 1984; Villumsen 1985; Binney & Lacey 1988); (ii) spiral arms or wavelets are relatively ineffective at imparting kinetic energy vertically, as they predominantly couple to planar motions (Carlberg 1987); (iii) $10^6 M_o$ black holes appear to be incapable of producing an iPII as massive as that observed (Lacey & Ostriker 1985), which is $\approx 10\%$ of the total disk mass (see Section 3). In general, studies of the age–velocity relationship (cf. Wielen *et al.* 1992 and references therein) are in general agreement that vertical velocity dispersions of at most $\sigma_w \approx 15-25$ km s^{-1} can be achieved in 10–15 Gyr, while the observed σ_w's for the iPII is near 45 km s^{-1}.

In the top-down models, the kinematical and chemical properties of the present iPII are the product of the interplay of the histories of star formation and dissipation as the proto-disk gas collapsed. Gradual formation scenarios (Sandage 1990) give relatively smooth transitions between the halo, iPII and thin disk, and leave kinematical, chemical and age gradients within the iPII. Here the iPII represents the first phases of pressure support as the halo collapses. However, Larson (1976) and Gilmore (1984) suggested that star formation must pause after halo formation to allow the collapsing gas to settle into a disk, after which gradual, pressure-supported collapse of the disk ensues. Alternatively, the disk may have collapsed rather quickly due to an increase in the dissipation rate from radiative line cooling when the metallicity reached [Fe/H] ≈ -1 (Wyse & Gilmore 1988; Burkert *et al.* 1992). The latter two scenarios may be distinguished today by a metallicity gradient in the former, and the lack of a metallicity gradient and a small range of ages in the latter. Unlike the smoothly evolving Galaxy in the first top-down model, the latter two scenarios would have an iPII with properties disjointed from those of the halo.

In summary, of the bottom-up scenarios, only formation by violent heating, presumably from satellite accretion, has theoretical support. Our first-order goal then is to distinguish between the accretion origin and formation through one of the dissipative schemes. The abruptness of the accretion origin implies a kinematical, chemical and age disjointedness from the thin disk, which would have to reform after the event. In contrast, the top-down models would be expected to show relatively smooth kinematic gradients (the "spin down" described by the collapse model of Eggen *et al.* 1962). Chemical gradients might also be expected in the top-down models, except perhaps in the "rapid line cooling" model. On the other hand, it is becoming increasingly clear that metallicity is not a good chronometer. Studies of the disk age–metallicity relationship show very broad metallicity ranges at all ages and only very modest mean increases in metallicity with age (Carlberg *et al.* 1985; Geisler 1987; Marsakov *et al.* 1990; Edvardsson *et al.* 1993).

The division of the models discussed thus far has been based on the relative ordering of the thin disk and the iPII, under the assumption that the formation of the halo preceded both. However, formation of the iPII may have begun before (Jones & Wyse 1983), or simultaneously with (Norris & Ryan 1991; Majewski 1993; see also the discussion of "dual halo" models in Section 6), the halo. In this "independent evolution" case, the formation of the iPII may have proceeded along the various lines already described, with the exception that it may have had little connection with the formation of the halo.

3. Physical dimensions

How "thick" is the "thick disk"? Figure 2 shows how dramatically the relative fraction of halo, iPII and thin-disk changes at the NGP under several different combinations of exponential scale height, h_z, and local relative density, as have been reported. The importance of knowing the density law for attempts at surveying and characterising the iPII cannot be understated, especially when the appreciable kinematical and chemical overlap with other galactic populations is considered. As one example of the problem, take the rotational velocities derived for the iPII compared to the respective density characterisations (h_z, local fraction) from *in situ* NGP surveys by Spaenhauer (1989; $V_{rot} = 140$-160 km s^{-1} with a 1.3 kpc, 2% iPII) and Ojha (this volume; $V_{rot} = 182 \pm 3$ km s^{-1}, 0.8 kpc, 5.6%), and also the V_{rot} *gradient* suggested by Majewski (1992; V_{rot} from 200 to 100 km s^{-1}, 1.4 kpc, 3.8%). On the other hand, these surveys are in very good agreement about the kinematics of *all* stars at the NGP when viewed without regard to population definitions (see Figure 1 of Majewski 1994). Clearly, what you find for the iPII is defined by how you "slice up" the Galaxy.

Figure 3 is an attempt to cull iPII characterisations from the literature. They are grouped into studies which make use of presumed iPII tracers, and those which use star-count analyses. Note that the tracer studies tend to find a broader range of iPII density parameters than do the star-count techniques, although the most recent star-count surveys show a distressing divergence. One limitation of the latter type of work is that with brighter magnitude limits it is more difficult to break the degeneracy of satisfactory solutions possible by adjusting the inverse relationship of h_Z and local normalisation. Appreciable leverage on h_Z can be gained by probing to deep magnitudes (V> 21), but the number of reliable star catalogues at magnitudes where star/galaxy/QSO separation is critical (by V$\approx$ 21 galaxies outnumber stars by an order of magnitude and QSOs make up 30-50% of faint blue "stars"; Reid & Majewski 1993) is embarrassingly small. Indeed, the Kron (1980) star-count catalogues remain the most widely used deep data over any appreciable area, and these were obtained as a byproduct of a galaxy count survey!

In an attempt to redress this deficiency, several new, faint star-count programmes have

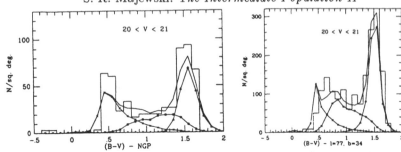

FIGURE 1. B–V colour distributions (histogram) and those predicted (curves) in two directions for magnitudes V=20-21 by the "interim model" (with an iPII scale height of 1.4 kpc and 2.5% local normalisation, shown by boxes) of Reid & Majewski (1993). By colour and magnitude criteria it is easier to define relatively more pure halo (crosses) and disk (triangles) samples than iPII samples. Note that photometric calibration is not yet completed in the (l=77, b=34) field, which results in some mismatch between the model and data for blue stars.

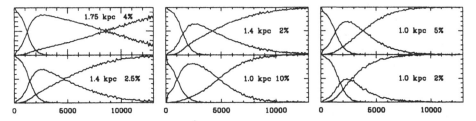

FIGURE 2. The relative fraction of thin-disk, iPII and halo stars as a function of Z (pc) for an observing cone at the NGP for various iPII density laws.

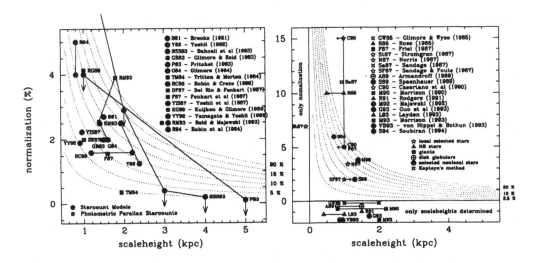

FIGURE 3. Scale heights and local normalisations for the iPII from various studies. Dotted lines show loci of relative mass density to a disk with $h_Z = 325$ pc.

been initiated. Reid & Majewski (1994) describe an extension of the Kron (1980) KPNO 4-metre photographic survey. In our initial analysis, there is good agreement (Figure 1) for the colour/count data in three new areas with the "interim model" proposed by our initial study (Reid & Majewski 1993) at both the NGP and SGP. The latter analysis found that "standard" parameterisations of the Galaxy (including an $h_Z = 1.2$ kpc,

2% iPII component) tend to over predict the number of faint halo stars by a factor of two at the expense of the disk stars, which are generally under predicted by a factor of two. Significantly better fits to the NGP data were obtained with the "interim model" having *both* a higher local iPII normalisation ($\approx 2.5\%$) and a higher h_Z (≈ 1.4 kpc). In addition, we hope to incorporate the Sandage (1983) 4-metre plate survey in eight fields along the $l = 0, 180$ meridional plane. Finally, Reid, Majewski & Thompson have begun a long-term programme of star counts in Kapteyn Selected Areas with a 2048^2 CCD on the Swope 40" telescope at Las Campanas Observatory. To date, we have surveyed approximately 13 $\deg^2$ in the B, V and I bands with better than 5% photometry to V=21.5.

In summary, galactic astronomers have yet to converge on a common density law for the iPII, and this inadequacy complicates the task of deriving the chemical, kinematical and age distributions of this population. At best, what may be said (Figure 3) is that the relative mass fraction of the iPII to the thin disk is likely to be in the range 5–15%.

4. Age of the iPII

Possibly the least controversial aspect of the iPII is its age, as various studies are converging to agreement on its antiquity. The question of the true identity of Rose's (1985) iPII "red horizontal branch" (RHB) stars has perhaps not been satisfactorily settled; if they really are RHBs, an age similar to that of the disk globulars is suggested, but if they are clump giants (Norris 1987) they are probably significantly younger. Rose & Agostinho (1991) argued that the difference in colour of their RHBs with those of 47 Tuc is due to the iPII being several 0.1 dex more metal poor than the cluster, but the unexpectedly high density of these RHBs (Norris & Green 1989) is still unexplained. But the turn-off colours of intermediate metallicity stars (Carney *et al.* 1989, 1995; Rose & Agostinho 1991; von Hippel & Bothun 1993) support an age at least as old as 47 Tuc for the bulk of the iPII; this is reinforced by the small number of $B - V \leq 0.5$ stars of iPII-like metallicity and at iPII-dominated z-distances (say $1 \leq z \leq 5$—Croswell *et al.* 1991; Majewski 1992). Ages derived from Strömgren photometry for $-0.5 > [Fe/H] > -1.2$ stars suggest an age equal to (18.5 ± 3.2 Gyr—Nissen & Schuster 1991) or only 1–2 Gyr younger than that of the halo field population (Márquez & Schuster 1994). This is consistent with the work by Carney *et al.* (1995) using the tidal circularisation of binary star orbits. The combination of the Zinn "old halo" plus the disk globular clusters may be tracers of the iPII population (see below). If so, this would lend further support to the notion that the iPII was among the first populations to have formed in the Galaxy.

A question of importance to formation models is whether there is an age gap between the iPII and the thin disk. If the oldest thin-disk stars are $\approx$11–12 Gyr old (Barry 1988) or younger (cf. Butcher 1987; Fowler 1987; Wood 1992), then an age gap is implied. However, at least one apparently open cluster—Berkeley 17—appears to be older than some young halo globulars (see this volume: Kaluzny; Phelps *et al.*; also, evidence for a large iPII age scatter in Agostinho *et al.*, this volume).

5. Kinematics

The question of vertical kinematic gradients in the iPII is central to discriminating formation scenarios. As mentioned in Section 3, there is disagreement about the rotational characteristics of the iPII. It has been common to describe the iPII with a single-valued asymmetric drift, with the estimates ranging from 20 km s^{-1} (Norris 1987) to 100 km

s^{-1} (Wyse & Gilmore 1986). However, Carney *et al.* (1990b) have explored the [m/H]-$|Z_{max}|$ distribution of stars in their sample and found that a metallicity gradient persists for $|Z_{max}| < 3$ kpc until an asymmetric drift limit of $V \leq -150$ km s^{-1} is imposed; this implies the existence of [Fe/H] ≥ -1.45 (presumably iPII) stars with $70 \leq V_{rot} \leq 120$ km s^{-1}. This V_{rot} is near the apparent "break" between halo and iPII clumps seen in various [Fe/H]-V diagrams (Yoshii & Saio 1979; Carney *et al.* 1990a; Majewski 1992; Schuster *et al.* 1993), a feature that has been cited as strong evidence of the disjointedness of the halo and the iPII. Eggen (1990) also finds a discontinuity between "old disk" and halo stars at $V_{rot} = 100$ km s^{-1}. The point is that the iPII seems to show a broad range of kinematics with a significant number of members rotating as slowly as 100 km s^{-1}, or more than one σ from many cited mean V_{rot}'s for iPII.

Among the various surveys, there is some correlation between the mean measured iPII asymmetric drift and the median $|Z|$ distance of the iPII stars: more distant samples tend to give larger drifts than more local surveys (Figure 6 of Majewski 1993). This trend may represent either an intrinsic iPII Z-gradient in V_{rot} or the effects of iPII sample contamination by thin-disk stars at low $|Z|$ and by halo stars at high $|Z|$. There is good agreement (Figure 1 of Majewski 1994) among the various galactic pole proper motion surveys (including the new study by Ojha *et al.* 1994, and this volume) on the overall kinematical Z-gradients exhibited by *all* stars as a function of $|Z|$. However, contrast Ojha et al's population deconvolution (with an $h_Z = 0.8$ kpc iPII), which yields a single 182 km s^{-1} rotational velocity, with Majewski's (1992) deconvolution, which attributes a large fraction of $1 \leq Z \leq 5$ kpc stars to the iPII and results in a V_{rot} Z-gradient *within* the iPII. A possible explanation for the contrast may be the difference in magnitude (i.e., distance) limits, so that a thicker iPII with an intrinsic gradient would be less obvious in the more local (often limited to ≈ 3 kpc) surveys.

Preliminary analysis of the red portion of the Majewski (1992) NGP sample lends further support to the notion of an iPII gradient; Figure 4a shows an apparent V gradient to $Z \approx 5$ kpc in the expanded NGP sample that includes K–M dwarfs to V< 21.5. Figure 4b makes use of complete space velocity data (with radial velocities obtained in the Majewski *et al.* 1994 programme) and shows the distribution of peculiar velocities ($V_{pec}^2 = U^2 + V^2 + W^2$) as a function of Z for 236 stars from the expanded NGP sample. The iPII population is apparent out to $Z \approx 5$ kpc and is superimposed on a halo population of stars with a large spread in peculiar velocities and predominantly at $Z > 3$ kpc. The locus of V_{pec} for the iPII stars from $Z = 2$ to 5 kpc shows a gradient.

What is not evident in Figure 4b is whether there is kinematical continuity between the thin disk and the iPII. Overall, this still remains a matter of considerable uncertainty. Norris & Ryan (1991) found that either a continuous or a discrete thin/thick disk model could account for the kinematical properties as a function of metallicity in the Carney *et al.* sample. Ratnatunga & Yoss (1991), however, argue for a discrete thin/thick disk model based on their modelling of the Yoss *et al.* (1987) data. Von Hippel & Bothun (1993) conclude that a sample with at least 2500 iPII stars is needed before discrimination between discrete and continuous models is possible.

6. Metallicity Distribution and "Dual Halo" Models

The mean metallicity of the iPII is typically given as $-0.6 \leq$[Fe/H]≤ -0.4, but it is clear (Norris *et al.* 1985; Norris 1986; Morrison *et al.* 1990; Majewski 1992; Morrison 1993; Rodgers & Roberts 1993) that it also contains members as metal-poor as [Fe/H]=-1.6. Morrison *et al.* (1990; Morrison 1993) claim that 80% of $-1.0 \leq$ [Fe/H] ≤ -1.6 locally are members of the iPII. It should be noted that the DDO metallicity scale used

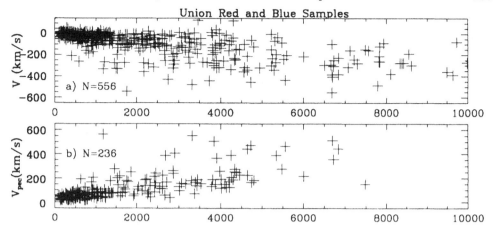

FIGURE 4. a. Asymmetric drift velocities as a function of Z (pc) for both red and blue samples of the Majewski (1992) survey. b. Peculiar velocities, $(U^2 + V^2 + W^2)^{1/2}$, as a function of Z for those NGP stars with radial velocities by Majewski *et al.* (1994).

in several of these studies, and in turn the magnitude of the metal-weak contribution of the iPII, has recently been questioned (Twarog & Anthony-Twarog 1994; Ryan & Lambert 1994).

Any vertical metallicity gradient in the iPII appears to be small (< -0.10 dex kpc^{-1}; Yoshii *et al.* 1987; Majewski 1992; Armandroff 1989 if the disk globulars are considered tracers) or nonexistent (Yoss *et al.* 1987; Sandage & Fouts 1987; Carney *et al.* 1989). Given the apparent unreliability of [Fe/H] as a chronometer (stars with a broad metallicity range were being formed over a large fraction of the life of the Galaxy), it is unlikely that the metallicity gradient issue can be used to discriminate between formation models yet.

It has become increasingly evident in the last few years that a very metal-poor population exists in the Galaxy with a flattened distribution, and this has given rise to the notion of a "dual halo". Hartwick (1987) showed that the spatial distribution of metal-poor RR Lyrae stars requires, in addition to a spherical component, a component with $c/a = 0.6$ (equivalent to an exponential with $h_Z = 1.6$ kpc). This flattened component dominates the spherical component locally. Sommer-Larsen & Zhen (1990; see also Sommer-Larsen *et al.* 1993) studied the orbits of a local sample of [Fe/H] ≤ -1.5 "halo" stars (with mean [Fe/H]=-2.0) and found that 40% of this sample populated a flattened component with $|Z| < 3$ kpc. A similar analysis of an [Fe/H] ≤ -2.0 sample by Allen *et al.* (1991) also showed two rather distinct populations: one with Z_{max} extending to at least 50 kpc and with a net retrograde rotation, and another with a net prograde rotation but which is confined to $Z_{max} < 4$ kpc. In their analysis of the Carney *et al.* sample, Ryan & Norris (1991) found that a gradient in σ_w exists to the lowest [Fe/H]. They proposed the existence of a component (the "contracted halo" in the more recent discussion of Norris 1994) with low σ_w but with a metallicity distribution that overlaps that of the more spherically distributed component and which extends to the most extreme, metal-poor limits. More recently, Carney *et al.* (1995, this volume) have found evidence for such an arrangement and propose the existence of "high" and "low" halo populations, dominating at $Z > 5$ kpc and $Z < 5$ kpc, respectively. In the same vein, the recent study of field blue horizontal branch (BHB) stars by Kinman *et al.* (1994) finds two halo populations with approximately the same metallicity distribution: a spheroidal population following an $R^{-3.5}$ density law dominating at $Z > 5$ kpc, and a "flat halo" component dominating at $Z < 5$ kpc and with an equivalent exponential h_Z of 2.2 kpc. The ratio of "flat halo" to "spheroidal halo" stars locally is 80:20.

On the assumption that age is the second parameter of horizontal branch (HB) morphology, Zinn (1993) has divided the halo globular clusters into "young" and "old" groups by their position in the [Fe/H]-HB Type diagram. The young halo globulars are distributed spherically, have no metallicity gradients, and appear to have a net retrograde rotation. The old halo globulars are in an oblate configuration, do exhibit radial and Z metallicity gradients, and rotate at 70 ± 22 km s^{-1}. Zinn suggests that the combination of the flattened populations of disk and old halo globulars may represent one dissipational component, as together they show smooth kinematical and metallicity gradients. Recent work to obtain full cluster space velocities supports this notion (see summary in Majewski 1994, Table 2). A number of relatively metal-poor, old halo globulars—M107 ([Fe/H]= -0.99), M4 (-1.28), M28 (-1.44), M12 (-1.61), M22 (-1.75) and NGC 6397 (-1.91)—have disk-like orbits (see also Alfaro *et al.*, this volume), with Z_{max} smaller than that of the archetypal disk globular 47 Tuc (3.5 kpc). The disk globulars are often considered as tracers of the iPII. If true, then Zinn's hypothesis that the disk population is merely the metal-rich portion of a single, flattened system with members having metallicities spanning the entire range of halo globulars suggests the existence of a corresponding iPII field population. The existence of such a very metal-poor iPII field-star population was suggested in Majewski (1992, 1993) in order to reconcile the dilemma of the halo rotational velocity: measurements using local star samples—even those constrained to be very metal-poor (say [Fe/H]≤ -2.0)—find $V_{rot} \approx +30$ km s^{-1}, whereas samples with $Z >\approx 5$ kpc tend to find retrograde V_{rot} (Reid 1990; Allen *et al.* 1991; Majewski 1992; Schuster *et al.* 1993; Carney *et al.* 1990b). If the iPII stretches to extremely poor metallicities, and in addition, has kinematics as broad as that of Zinn's disk+old halo globulars (which seems likely, as discussed in Section 5), then the local "halo" samples may be contaminated by iPII stars, which accounts for the differences in measured halo V_{rot}.

7. Zero population growth and an appeal to Occam's razor

The theme of this section may be more appropriate for another international conference being held today 3500 km east of here!† It is proposed (cf. Majewski 1993) that the very metal-weak extension of the iPII suggested above by both the Zinn hypothesis (a single disk+old halo globular system) and the halo rotation dilemma, and the various metal-poor, "lower" or "flattened" halo components in the "dual halo" models discussed above, are one and the same population. The proliferation of flattened subpopulations in the Galaxy, as suggested by the title of this paper, seems to me to place an undue burden on us; the implied divisions may only be semantic and even untoward if they are simply references to the same, or parts of the same, population. No evidence has yet arisen that the various groups are not simply artificially slicing up a single, continuously and broadly distributed population by selection or other biases. On the other hand there is great economy of hypothesis as well as utility in the paradigm of a ubiquitous iPII/thick disk, with (i) a scaleheight and normalisation such that it dominates the halo to at least $Z = 5$ kpc and perhaps even higher; (ii) a broad metallicity distribution reaching to the most metal-poor halo stars; and (3) a wide kinematical distribution, with a significant

† The U.N. (Cairo) International Conference on Population Growth.

number of members having $V_{rot} \leq 100$ km s^{-1}. Such an iPII would overlap the halo age, kinematic, and metallicity distributions, yet may have evolved very differently.

An extreme of this position would be that the iPII is merely a tail of the thin disk and that referring to a separate iPII is still one "population" beyond need. For now, however, this may be a reasonable compromise – at least what have been called here the "iPII" and the "thin disk" span very different ages, spatial distributions, metallicities and kinematical ranges and, if not separate populations, then at least they represent the extremes of one continuous distribution. In contrast, it is not obvious that the "metal-weak thick disk" is distributed any differently from the "thick disk" spatially, kinematically, or in terms of age, or that it is distinguishable in any way from the "lower" or "contracted halo". Our difficult task, as always in the population decomposition game, is to distinguish between continuously distributed groups of stars exhibiting correlations in properties, and truly distinct populations that overlap in some properties.

8. Concluding Remarks

The iPII appears to have no, or only a very modest, metallicity gradient, but has a V_{rot} gradient with Z, reminiscent of expectations from an Eggen *et al.* (1962) "spin-down". All indications are that the iPII is distinct from the halo, at least kinematically. Their disjointedness is particularly evident if the halo has a retrograde rotation: a single collapse could not account for the majority of both populations if they have opposite angular momentum vectors. However, the halo and iPII contain stars of a similar age and they overlap in metallicity distributions. It is probable that they evolved simultaneously, but perhaps relatively independently. The connection between the thin disk and the iPII has yet to be satisfactorily untangled. The solution to the question of one or two distinct populations may require substantial modelling efforts and large samples of stars. It is clear that the iPII is ancient, whereas the bulk of the thin disk may be younger than 12 Gyr or so, with the implication of an age gap and therefore disjointedness. However, it may be that the disk has been built up by a series of bursts separated by low activity pauses (Larson 1976, Berman & Suchkov 1991, Burkert *et al.* 1992, Katz 1992), in which case the iPII may simply represent the oldest major star formation event in the disk. The existence of other peaks and lulls in the disk star-formation rate are evident in the work of Barry (1988), Noh & Scalo (1990) and others.

Of the many models that have been proposed for the origin of the iPII (Section 2; Majewski 1993, Table 1), the evidence at present seems to favour either of two models: (i) rapid dynamical heating of an early disk by satellite accretion into the disk, or (ii) a model in which the iPII formed through a collapse, probably dissipatively, and probably relatively separate from the formation processes which resulted in the present halo. The latter represents a merging of the Searle & Zinn (1978) picture (for the halo) with the Eggen *et al.* (1962) model (for the iPII), à la suggestion by Sandage (1990). A limitation of the accretion model is that the iPII is as old as the halo; this places a rather tight constraint on the timing of the accretion event (though it is not unreasonable to expect that such events were more common in the early universe). Moreover, if the iPII asymmetric drift gradient is real, and it extends to very low V_{rot}, this would be in conflict with the predictions of Quinn *et al.* (1992). While together this would seem to favor a collapse origin, apparently not all disk galaxies have discernible "thick disks" (van der Kruit & Searle 1981; Morrison *et al.* 1994), which suggests that formation of these components may not be a requisite for the formation of all disk galaxies, and that more stochastic processes—such as satellite accretion—may be at work.

REFERENCES

ALLEN, C. SCHUSTER, W. J. & POVEDA, A. 1991 *Astron. Astrophys.* **244**, 280.

ARMANDROFF, T. E. 1989 *Astron. J.* **97**, 375.

BARRY, D. C. 1988 *Astrophys. J.* **334**, 436.

BERMAN, B. G. & SUCHKOV, A. A. 1991 *Astrophys. Space Sci.* **184**, 169.

BINNEY, J. & LACEY, C. 1988 *Mon. Not. R. Astron. Soc.* **230**, 597.

BURKERT, A., TRURAN, J. W. & HENSLER, G. 1992 *Astrophys. J.* **391**, 651.

BUTCHER, H. R. 1987 *Nature* **328**, 127.

CARLBERG, R. G 1987 *Astrophys. J.* **322**, 59.

CARLBERG, R. G, DAWSON, P. C., HSU, T. & VANDENBERG, D. A. 1985 *Astrophys. J.* **294**, 674.

CARNEY, B. W., AGUILAR, L., LATHAM, D. W. & LAIRD, J. B. 1990b *Astron. J.* **99**, 201.

CARNEY, B. W., LAIRD, J. B., LATHAM, D. W. & AGUILAR, L. 1995 in preparation.

CARNEY, B. W., LATHAM, D. W. & LAIRD, J. B. 1989 *Astron. J.* **97**, 423.

CARNEY, B. W., LATHAM, D. W. & LAIRD, J. B. 1990a *Astron. J.* **99**, 572.

CROSWELL, K., LATHAM, D. W., CARNEY, B. W., SCHUSTER, W. & AGUILAR, L. 1991 *Astron. J.* **101**, 2078.

EDVARDSSON, B., ANDERSON, J., GUSTAFSSON, B., LAMBERT, D. L., NISSEN, P. E. & TOMKIN, J. 1993 *Astron. Astrophys.* **275**, 101.

EGGEN, O. J. 1990 *Astron. J.* **100**, 1159.

EGGEN, O. J., LYNDEN-BELL, D. & SANDAGE, A. R. 1962 *Astrophys. J.* **136**, 748.

FOWLER, W. A. 1987 *Q.J.R. Astron. Soc.* **28**, 87.

GEISLER, D. 1987 *Astron. J.* **94**, 84.

GILMORE, G. 1984 *Mon. Not. R. Astron. Soc.* **207**, 223.

GILMORE, G. F., KING, I. R. & VAN DER KRUIT, P. C. 1990 in *The Milky Way as a Galaxy*. Univ. Sci. Books.

GILMORE, G. & REID, N. 1983 *Mon. Not. R. Astron. Soc.* **202**, 1025.

GILMORE, G., WYSE, R. F. G. & KUIJKEN, K. 1989 *Annu. Rev. Astron. Astrophys.* **27**, 555.

HARTWICK, F. D. A. 1987 in *The Galaxy* (ed. G. Gilmore & B. Carswell), p. 281. Reidel.

JONES, B. J. T. & WYSE, R. F. G. 1983 *Astron. Astrophys.* **120**, 165.

KATZ, N. 1992 *Astrophys. J.* **391**, 502.

KINMAN, T. D., SUNTZEFF, N. B. & KRAFT, R. P. 1994, preprint.

KRON, R. G. 1980 *Astrophys. J., Suppl. Ser.* **43**, 305.

LACEY, C. G. 1984 *Mon. Not. R. Astron. Soc.* **208**, 687.

LACEY, C. G. & OSTRIKER, J. P. 1985 *Astrophys. J.* **299**, 633.

LARSON, R. B. 1976 *Mon. Not. R. Astron. Soc.* **170**, 31.

MAJEWSKI, S. R. 1992 *Astrophys. J., Suppl. Ser.* **78**, 87.

MAJEWSKI, S. R. 1993 *Annu. Rev. Astron. Astrophys.* **31**, 575.

MAJEWSKI, S. R. 1994 in *Astronomy from Wide-Field Imaging* (ed. H. T. MacGillivray, E. B. Thomson, B. M. Lasker, I. N. Reid, D. F. Malin, R. M. West & H. Lorenz). IAU Symposium 161, p. 425. Kluwer.

MAJEWSKI, S. R., MUNN, J. A. & HAWLEY, S. L. 1994 *Astrophys. J., Lett.* **427**, L37.

MÁRQUEZ, A. & SCHUSTER, W. J. 1994 *Astron. Astrophys., Suppl. Ser.* **108**, 341.

MARSAKOV, V. A., SHEVELEV, YU. G. & SUCHKOV, A. A.1990 *Astrophys. Space Sci.* **172**, 51.

MORRISON, H. L. 1993 *Astron. J.* **105**, 539.

MORRISON, H. L., BOROSON, T. A. & HARDING, P. 1994 *Astron. J.* **108**, 1191.

MORRISON, H. L., FLYNN, C. & FREEMAN, K. C. 1990 *Astron. J.* **100**, 1191.

NISSEN, P. A. & SCHUSTER, W. J. 1991 *Astron. Astrophys.* **251**, 457.

NOH, H.-R. & SCALO, J. 1990 *Astrophys. J.* **352**, 605.

NORRIS, J. 1986 *Astrophys. J., Suppl. Ser.* **61**, 667.

NORRIS, J. 1987 *Astron. J.* **93**, 616.

NORRIS, J. 1994 *Astrophys. J.* **431**, 645.

NORRIS, J., BESSEL, M. S. & PICKLES, A.J. 1985 *Astrophys. J., Suppl. Ser.* **58**, 463.

NORRIS, J. & GREEN, E. M. 1989 *Astrophys. J.* **337**, 272.

NORRIS, J. E. & RYAN, S. G. 1991 *Astrophys. J.* **380**, 403.

O'CONNELL, D. J. K. (ED.) 1958 *Specola Vaticana.*

OJHA, D. K., BIENAYMÉ, O., ROBIN, A. C. & MOHAN, V. 1994 *Astron. Astrophys.* **290**, 771.

QUINN, P. J., HERNQUIST, L. & FULLAGER, D. P. 1992 *Astrophys. J.* **403**, 74.

RATNATUNGA, K. U. & YOSS, K. M. 1991 *Astrophys. J.* **377**, 442.

REID, N. & MAJEWSKI, S. R. 1993 *Astrophys. J.* **409**, 635.

REID, N. & MAJEWSKI, S. R. 1994 in *Astronomy from Wide-Field Imaging* (ed. H. T. MacGillivray, E. B. Thomson, B. M. Lasker, I. N. Reid, D. F. Malin, R. M. West & H. Lorenz). IAU Symposium 161, p. 423. Kluwer.

RODGERS, A. W. & ROBERTS, W. H. 1993 *Astron. J.* **106**, 2294.

ROSE, J. A. 1985 *Astron. J.* **90**, 787.

ROSE, J. A. & AGONSTINHO, R. 1991 *Astron. J.* **101**, 950.

RYAN, S. G. & LAMBERT, D. L. 1994 *Astron. J.* submitted.

RYAN, S. G. & NORRIS, J. E. 1991 *Astron. J.* **101**, 1835.

SANDAGE, A. 1983 in *Kinematics, Dynamics and Structure of the Milky Way* (ed. W. L. H. Shuter), p. 315. Reidel.

SANDAGE, A. 1990 *J. R. Astron. Soc. Can.* **84**, 70.

SANDAGE, A. & FOUTS, G. 1987 *Astron. J.* **93**, 74.

SCHUSTER, W. J., PARRAO, L. & CONTRERAS MARTÍNEZ 1993 *Astron. Astrophys., Suppl. Ser.* **97**, 951.

SEARLE, L. & ZINN, R. 1978 *Astrophys. J.* **225**, 357.

SOMMER-LARSEN, J., BEERS, T. C. & ALVAREZ, J. 1992 *Bull. Am. Astron. Soc.* **24**, 1177.

SOMMER-LARSEN, J. & ZHEN, C. 1990 *Mon. Not. R. Astron. Soc.* **242**, 10.

SPAENHAUER, A. 1989 *Contrib. Van Vleck Obs.* **9**, 45.

STATLER, T. S. 1988 *Astrophys. J.* **331**, 71.

TWAROG, B. A. & ANTHONY-TWAROG, B. J. 1994 *Astron. J.* **107**, 1371.

VAN DER KRUIT, P. C. & SEARLE, L. 1981 *Astron. Astrophys.* **95**, 105.

VILLUMSEN, J. V. 1985 *Astrophys. J.* **290**, 75.

VON HIPPEL, T. & BOTHUN, G. D. 1993 *Astrophys. J.* **407**, 115.

WIELEN, R., DETTBARN, C. FUCHS, B. JAHREISS, H. & RADONS, G. 1992 in *The Stellar Populations of Galaxies* (ed. B. Barbuy & A. Renzini). IAU Symposium 149, p. 81. Kluwer.

WOOD, M. A. 1992 *Astrophys. J.* **386**, 539.

WYSE, R. F. G. & GILMORE, G. 1986 *Astron. J.* **91**, 855.

WYSE, R. F. G. & GILMORE, G. 1988 in *Chemical and Dynamical Evolution of Galaxies* (ed. F. Ferrini, J. Franco & F. Matteucci), p. 19. ETS Editrice.

YOSHII, Y., ISHIDA, K. & STOBIE, R. S. 1987 *Astron. J.* **93**, 323.

YOSHII, Y. & SAIO, H. 1979 *Publ. Astron. Soc. Jpn.* **31**, 339.

YOSS, K. M., NEESE, C. L. & HARTKOPF, W. I. 1987 *Astron. J.* **94**, 1600.

ZINN, R. 1993 in *The Globular Cluster–Galaxy Connection* (ed. G. H. Smith & J. P. Brodie). ASP Conference Series, vol. 48, p. 38. ASP.

Delgado: How are the lowest-metallicity thick-disk stars spatially distributed? Are they more or less randomly distributed, or preferentially located in certain regions?

Majewski: It is not yet clear that the lowest-metallicity iPII stars are distributed any differently from their more metal-rich counterparts. There is some evidence from Morrison (1993) that the scale height of the metal-weak members is higher ($\approx$ 2 kpc, but for R_{GC} =4-7 kpc) and this may be consistent with very modest metallicity gradients measured within the iPII.

Surdin: Is there any difference in metallicity, HB, mass, etc. between globulars with prograde and retrograde orbits?

Majewski: Zinn's (1993) "young halo" globular clusters, defined by HB Type and metallicity, have mean retrograde rotation, whereas the "old halo" clusters have a net prograde rotation. Rodgers & Paltoglou (1984, *ApJ* **283**, L5) showed that the group of clusters with $-1.3 \leq$[Fe/H]≤ -1.7 was in net retrograde rotation, and more recently, van den Bergh (1993, *AJ* **103**, 971; *AJ*, **106**, 2145) showed that (i) 8 of 10 retrograde clusters are in this metallicity range, and 5 are in the even tighter range $-1.5 \leq$[Fe/H]≤ -1.6; (ii) retrograde clusters tend to have smaller half-light radii; and (iii) retrograde clusters tend to be of Oosterhoff class I or have red HBs, and prograde clusters tend to be of Oosterhoff class II or have blue HBs.

UVW Components, Proper Motions, Age and Chemical Composition of Disk Stars at the South Galactic Pole

By RUI J. AGOSTINHO[1], S. ANTÓN[1], D. MAIA[1],
J. A. ROSE[2] AND J. STOCK[3]

[1]Department of Physics, University of Lisbon
Av. Prof. Gama Pinto n⁰ 2, 1699-Lisboa-CODEX, Portugal

[2]Department of Physics & Astronomy, Univ. of North Carolina at Chapel Hill, U.S.A.

[3] Centro de Investigaciones de Astronomía, Mérida, Venezuela

The spatial distribution of stars towards the south galactic pole (SGP) has been studied using the sample by Rose & Agostinho (1991) in two fields covering about 55 °, which is complete down to B = 12.5. Through the use of digitised objective prism spectra, their data yielded information on each star's magnitude, spectral type, luminosity class, an estimate of [Fe/H], radial velocity and proper motions. Maximum likelihood computations showed that the disk-like component(s) in the galactic z direction is best described by a two-exponential model with scale heights of 145±10 pc and 287±13 pc. The relative normalisations of these two components vary from nearly zero at F5 to about 3.6 for G0–G3 stars. From $uvby\beta$ photometry of 250 late-F to early-G dwarf stars we conclude that the intermediate population (iPII) group shows a chemical distribution compatible with that of the thick disk, and its age shows a large variation with limits of 6 and 14 Gyr. There is a shift towards older ages in Population I when the metallicity decreases, but one cannot safely determine an age–metallicity relation.

1. Introduction

In order to understand the formation and evolution of the Galaxy we need a comprehensive inventory of the kinematics, chemical composition, and evolutionary states of its stellar populations. Objective prism spectroscopy offers the most efficient way of obtaining a large and kinematically unbiased sample of stars. Thus, a programme with the 10° prism on the Curtis Schmidt is being carried out at CTIO. From the digitised spectra, quantitative information is obtained for a complete sample of stars in the magnitude range B < 13. Information on T_{eff}, log g and [Fe/H] is obtained from a system of spectral indicators from digitised objective prism plates, developed by Rose (1984, 1991). A discussion on the radial velocity determination methods—based on the technique of opposed dispersion plate pairs by Stock (1992)—and the first results for the velocity dispersions can be found in Agostinho (1992) and Stock et al. (1994). Absolute magnitudes, metallicities and ages are also obtained from intrinsic colours in the Strömgren system, for F and early-G dwarf stars. The data are already complete for two fields near the SGP, and some results will be briefly presented in this paper.

2. Kinematics

Using BVRI and $ubvy\beta$ photometry we were able to establish an absolute magnitude (Crawford 1975) calibration for the dwarfs, i.e. Hδ/FeI vs. M_v, to determine distances to individual stars, using the prism B magnitude. The Astrographic Catalog served as a first epoch for the determination of proper motions for all stars in common with our survey. The final accuracy was ≈ 0.003–4 "/yr (Figures 1a and b). The V component distribution for all dwarfs is assymetric and cannot be well described by a single Gaussian (Figure 1c)

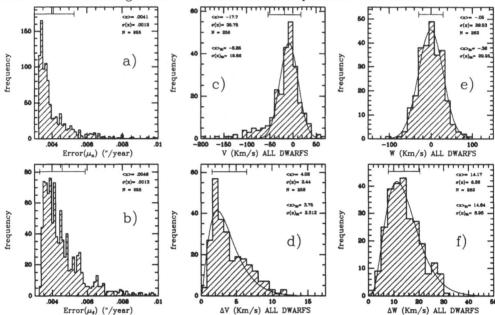

FIGURE 1. a, b: distribution of the proper motion errors in RA and Dec, respectively; c, d: distribution of the V velocity and its error (respectively) for all dwarfs (note the assymetry in c and the small bump around -35 km s^{-1}, typical of the iP$_{II}$); e, f: distribution of the W velocity and its error for all dwarfs

but is a better fit with two Gaussians—one corresponding to the iP$_{II}$ with an assymetric drift of ≈ -35 km s^{-1}. The W component (almost equal to the radial velocity) suffers from the larger error in V_{rad} (Figure 1f). The velocity dispersion $\sigma_W = 25.2 \pm 3.1$ km s^{-1} is computed subtracting in quadrature the errors (Agostinho 1992; Stock *et al.* 1994), and is larger than the typical ≈ 12 km s^{-1} for Population I (Knude 1994), basically due to the two components (one kinematically hotter) one finds.

3. Scale heights of the disk population(s)

We compare the measured distribution of stars with the predictions from a Bahcall-Soneira-like model, based on a Wielen (1994) local luminosity function, a relative fraction of dwarfs to giants and scaleheights as advised by Bahcall (1986); the halo contribution was ignored. In Figure 2a we can see that the model overestimates the number of early-type stars and fails to reproduce the sharp increase near Hδ/Fe$_I$ ~ 0.55, i.e. a model of one exponential with a smoothly varying scale height gives a poor fit to the vertical distribution of stars. Checking the model parameters to look for possible causes for the discrepancies found, the stars were then divided into Hδ/Fe$_I$ groups, and n exponentials were fitted in each group using the maximum likelihood technique (e.g. Figure 2c). The results and fitting procedures were tested using Monte Carlo simulations, to look for biases associated with metallicity variations and errors, and proved to be consistent.

We have found that the sum of *two* exponentials gives much better results. The scale heights were essentially the same, regardless of the Hδ/Fe$_I$ interval considered; namely, $h_1 = 145 \pm 10$ pc and $h_2 = 287 \pm 13$ pc (internal accuracy). The relative normalisation of these two components (in the sense hot/cooler component) varies from nearly zero at F5 to about 3.6 for G0–G3 stars (see Figure 2b). The two-exponential model density

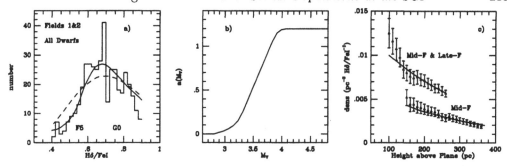

FIGURE 2. a. observations plotted against the Bahcall–Soneira Model (dashed line) and the two-exponential model (full line); b. relative normalisation factor $a(M_v)$ as a function of M_v; c. the density of F stars $(pc^{-3}[H\delta/FeI]^{-1})$ (the full line is the $h = 287$ pc component only)

$D(z, M_v)$ is of the form:

$$D(z, M_v) = \left[\frac{1}{2.1\,[a(M_v) + 1]}e^{-\frac{z}{145}} + \frac{1.45 \times a(M_v)}{a(M_v) + 1}e^{-\frac{z}{287}} \right] \times \Phi(M_v) \qquad (3.1)$$

where z is the height above the plane in parsecs, $\Phi(M_v)$ the Wielen luminosity function, and $a(M_v)$ is one third of the relative normalisation (Figure 2b); a comparison of the two exponentials model with observations is given in Figure 2a.

4. Ages and Metallicities

Since Strömgren photometry is particularly apropriate for the determination of reddening and of intrinsic stellar parameters in the spectral range F0–G2, we have determined the intrinsic colours of 250 stars. Using the chemical criterium of Strömgren (1984), one can divide the stars into Population I and iPII. The metallicity (Nissen 1989) distribution of these groups is shown in Figures 3a and b, respectively. The values of iPII are roughly concentrated between -0.7 and -0.3, whereas Population I shows a broad distribution. The corresponding mean values and standard deviations are: $[Fe/H]_{Pop-I} = -0.10 \pm 0.29$, and $[Fe/H]_{iPII} = -0.49 \pm 0.27$. The age is estimated from isochrones adequated to the metallicities of the stars ($[Fe/H]_{uvby}$ in the plots). The Population I stars were divided into three metallicity groups to account for the dispersion shown by this population. The corresponding $\log T_{eff} - M_v$ diagrams are presented in Figures 3c–e. The isochrones with $Z=0.02$ and $Z=0.008$ are from Schaller *et al.*(1992); those with $[Fe/H]= -0.47$ are from Bergbursh & van den Bergh (1992). One can see that the age limits for the Population I group are 1 and 12 Gyr, while iPII candidates (cf. Figure 3f) have ages of between 5 and 14 Gyr. The high metallicity tail of iPII indicates that some iPII candidates might actually belong to the old thin disk, which also explain the younger age limit.

5. Conclusions

We have shown that the density of later-type stars in the z direction includes not only the classical $h \approx 300$ pc disk component, but also that a significant fraction of these stars remains close to the plane, in a component that falls off by a factor of $1/e$ at ≈ 150 pc, the latter being the only disk-component in the case of the early-type stars. The results demonstrate that there is no significant difference in age between Population I and iPII. The separation between iPII and Population I needs a kinematical criterium, namely the assymetric-drift V component, which is the one with the smallest error in our case.

This work was partially supported by grants number 2543/91–RM, 2189/91–RM and PESO/PRO/23/92 to S. Antón, D. Maia and R. J. Agostinho, respectively, from Junta

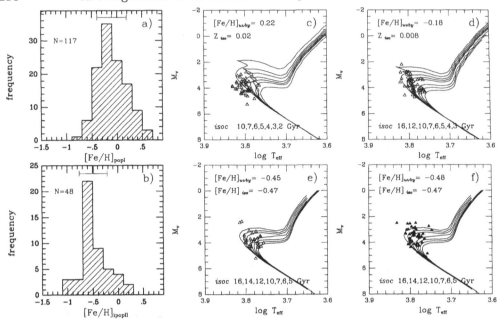

FIGURE 3. Metallicity distribution for: a. Population I stars; b. iPII stars; c–e. HR diagrams for Population I stars, and f. for iPII stars

Nacional de Investigação Científica e Tecnológica, Portugal, and NSF grant AST–8919455 to the University of North Carolina at Chapel Hill.

REFERENCES

AGOSTINHO, R. J. 1992 PhD thesis, Univ. North Carolina at Chapel Hill, U.S.A.

BAHCALL, J. 1986 *Annu. Rev. Astron. Astrophys.* **24**, 577.

BERGBUSH, J. & VAN DEN BERG, S. 1992 *Astrophys. J., Suppl. Ser.* **81**, 163.

CRAWFORD, D. 1975 *Astron. J.* **80**, 955.

KNUDE, J. 1994 in *Galactic and Solar System Optical Astrometry* (ed. L. V. Morrison & G. F. Gilmore), p. 174. Cambridge University Press.

NISSEN, J. 1989 *Astron. Astrophys.* **221**, 65.

ROSE, J. A. 1984 *Astron. J.* **89**, 1238.

ROSE, J. A. 1991 *Astron. J.* **101**, 937.

ROSE, J. A. & AGOSTINHO, R. J. 1991 *Astron. J.* **101**, 950.

SCHALLER, G., SCHAERER, D., MEYNET, G. & MAEDER, A. 1992 *Astron. Astrophys., Suppl. Ser.* **96**, 269.

STOCK, J. 1992 *Rev. Mex. Astron. Astrofis.* **24**, 45.

STOCK, J., AGOSTINHO, R. J., ROSE, J. A. & UPGREN, A. R. 1994 in *Galactic and Solar System Optical Astrometry* (ed. L. V. Morrison & G. F. Gilmore), p. 96. Cambridge University Press.

STRÖMGREN, B. 1984 in *Proc. Nordic Astronomy Meeting*, p. 7.

WIELEN, R. 1974 in *Highlights of Astronomy* (ed. G. Contopoulos). Vol. 3, p. 395. Reidel.

Kinematical Properties of the Thick Disk of the Galaxy

By DEVENDRA OJHA

Observatoire de Besançon, BP 1615, F–25010, France

We try here to find new constraints on the thick-disk population using samples at intermediate latitude and the north galactic pole (NGP), which include photometry and proper motions. The algorithm SEM (Stochastic–Estimation–Maximization; Celeux & Deibolt 1986) was used to deconvolve the stellar components up to large distances above the plane, facilitating the study of their statistical properties independently. Multivariate discriminant analysis (MDA) is used to qualify the thick disk using observations in multidimensional space (V,B-V,U-B, μ_l & μ_b).

1. Samples

The data sets from the following six surveys have been used to derive the kinematical and structural parameters of the thick-disk population :

- field in the direction of galactic anticentre ($l = 167°$, $b = 47°$ (Ojha *et al.* 1994b); hereafter GAC.
- field in the direction of galactic centre ($l = 3°$, $b = 47°$ Ojha *et al.* 1994c); hereafter GC.
- field in the direction of galactic antirotation ($l = 278°$, $b = 47°$ Ojha *et al.* 1994a); hereafter ELLIPSOID.
- field in the direction of NGP-I ($l = 58°$, $b = 80°$ Soubiran 1993).
- field in the direction of NGP-II ($l = 42°$, $b = 79°$ Kharchenko *et al.* 1994).
- field in the direction of NGP-III ($l = 124°$, $b = 87°$ Kharchenko *et al.* 1994).

2. Methods

2.1. *SEM*

The aim of the SEM algorithm is to resolve the finite mixture density estimation problem under the maximum likelihood approach using a probabilistic teacher step. Through SEM one can obtain the number of components of the Gaussian mixture, its mean values, dispersions and the percentage of each component with respect to the whole sample. Applying the SEM algorithm on the real data sets from 6 *in situ* surveys, a thick disk population has been identified in different distance bins (up to $z \sim 3.5$ kpc). Figure 1 shows the asymmetric drift measurements of thick disk population from selected studies as a function of z distances. We find a similar value of the asymmetric drift of thick disk from the NGP and intermediate-latitude data sets. As can be seen in Figure 1, no clear dependence with z distance is found in the asymmetric-drift measurements of the thick-disk population. However, a gradient can be seen in V velocity if *no separation* was made between the three populations (thin disk, thick disk and halo). Vertical bars indicate the error $\frac{\sigma}{\sqrt{N}}$ in V velocity, where N is the number of stars in each distance bin.

2.2. *MDA*

We have done the multivariate discriminant analysis using the two data sets and the Besançon model of population synthesis (Robin & Crézé 1986; Bienaymé *et al.* 1987). The observed data in five-dimensional space (V, B–V, U–B, μ_l, μ_b) have been used for

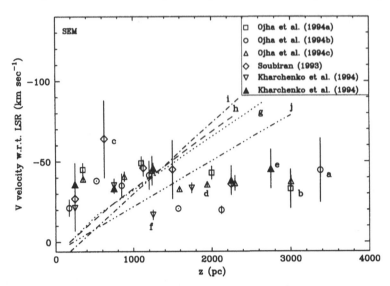

FIGURE 1. The measured asymmetric drift of the thick-disk population plotted as a function of z distances from the proper motion selected samples. Open circle = a. Ojha *et al.* (1994c), open square = b. Ojha *et al.* (1994b), open diamond = c. Soubiran (1993), open triangle up = d. Ojha *et al.* (1994a), filled triangle up = e. Kharchenko *et al.* (1994), (open triangle down = f. Kharchenko *et al.* (1994). The dotted (g), dashed (h), dashed-dotted (i) and dashed-dotted-dotted (j) lines represent GC, GAC, NGP (Soubiran 1993) and ELLIPSOID fields, respectively, with *no separation* between the three populations.

this analysis. The two data sets we used are the GC field ($l = 3°$, $b = 47°$) and GAC field ($l = 167°$, $b = 47°$). The U–B and μ_l parameters are necessary to make a good discrimination between the three populations, because U–B is sensitive to the metallicity and μ_l is parallel to the V velocity and discriminate the populations by their asymmetric drift.

We used the model simulations to find the best discriminant axes, able to separate the thick-disk population from the thin disk and halo. To avoid too much Poisson noise in the Monto Carlo simulations, we computed at least 10 simulations in 100 square degrees for each of the models tested in our analysis. The best discriminant axis for the circular velocity of 180 km s^{-1} of thick disk in the direction of GC is given by
$x = 0.024(B - V) + 0.139(U - B) - 0.079V - 0.310\mu_l - 0.069\mu_b$

To quantitatively estimate the adequacy of the models with various circular velocities, we applied a χ^2 test to compare the distribution of the sample on the discriminant axis with a set of model predicted distribution assuming different circular velocities of thick disk. Table 1 shows the values of the probability (in sigmas) of each model to come from the same distribution as the observed sample. Where sigma is an approximation to the distribution of χ^2 (Kendall & Stuart 1969). The most probable value for the circular velocity of thick disk comes out to be 180 km s^{-1}, corresponding to a lag or asymmetric drift of thick disk of the order of 40±10 km s^{-1}.

The distribution over the discriminant axis of observed stars (solid points) towards GAC with best model predictions (with three populations) is shown in Figure 2.

3. Conclusion

We have deduced the new estimates of the kinematical parameters of thick disk population. The thick disk population is found to have a rotational velocity of $V_{rot} = 177$

V_{thd} (km s^{-1})	Lag (km s^{-1})	χ^2 (in sigmas)	
		GC	GAC
150	70	8.7	6.8
165	55	6.8	5.7
175	45	4.1	3.8
180	40	4.3	3.4
185	35	4.9	3.5
190	30	5.5	4.0
215	5	6.7	5.1

TABLE 1. χ^2 test for models with different circular velocities of thick disk. χ^2 is given in number of sigmas. Lag or asymmetric drift is w.r.t. the LSR, assuming $V_{LSR}=220$ km s^{-1}.

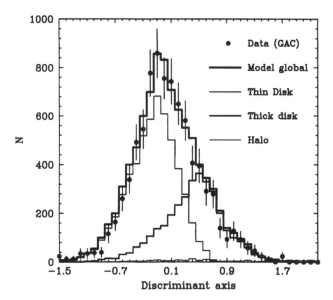

FIGURE 2. Distribution over the discriminant axis of observed stars (solid points) towards the GAC direction; best model predictions (solid full thick line) and model predicted stars according to their populations (thick line = thin disk; full thick line = thick disk; and thin line = halo)

km s^{-1}, and with velocity dispersions $(\sigma_U, \sigma_V, \sigma_W) = (67,51,46)$ km s^{-1}. Our data are consistent with no dependence of the thick disk's asymmetric drift with distance above the galactic plane (up to $z = 3$ kpc). We obtain a unique value for the asymmetric drift of thick disk in two directions (galactic centre and anticentre), showing that no radial gradient seems to occur on the asymmetric drift of thick disk on a base of 3 kpc around the Sun. From the number ratio of the thick disk stars in a pair of directions (towards the galactic centre and anticentre), we deduce the scale length h_R of the thick disk, which is found to be 3.6±0.5 kpc. The density laws for stars with $3.5 \leq M_V \leq 5$ as a function of distance above the plane, follow a single exponential with scale height of ~ 260 pc for $150 \leq z \leq 1200$ pc, and a second exponential with scale height ~ 770 pc for z distances from ~ 1200 pc to at least 3000 pc. We identify the 260 pc scale height component as a thin disk, and the 770 pc scale height component as a thick disk (Ojha *et al.* 1995).

This research work was partially supported by the Indo–French Centre for the Promotion of Advanced Research (IFCPAR), New-Delhi (India). We thank Dr. E. Schilbach for letting us use their data.

REFERENCES

BIENAYME, O., ROBIN, A. C. & CREZE M. 1987 *Astron. Astrophys.* **180**, 94.

CELEUX, G. & DIEBOLT, J. 1986 *Rev. Statistique Appliquée* **34**, 35.

KENDALL, M. G. & STUART, A. 1969 *The Advanced Theory of Statistics*, vol. 1, p. 371.

KHARCHENKO, N., SCHILBACH, E. & SCHOLZ, R.-D. 1994 *Astron. Nachr.* **315**, 291.

OJHA, D.K. & BIENAYME, O. 1995 in *IAU Symposium 166* (ed. E. Hoeg & J. Kovalevsky), in press.

OJHA, D. K., BIENAYME, O., MOHAN, V. & ROBIN, A. C. 1994a *Astron. Astrophys.* in preparation.

OJHA, D. K., BIENAYME, O., ROBIN, A. C. & MOHAN, V. 1994b *Astron. Astrophys.* **284**, 810.

OJHA, D. K., BIENAYME, O., ROBIN, A. C. & MOHAN, V. 1994c *Astron. Astrophys.* **290**, 771.

ROBIN, A. C. & CREZE, M. 1986 *Astron. Astrophys.* **157**, 71.

SOUBIRAN, C. 1993 *Astron. Astrophys.* **274**, 181.

Carney: Since your kinematics depend solely on tangential velocities, your absolute magnitudes are very important. Did you use a metallicity sensitivity? if not, how might that affect your results?

Ojha: Since the local value of the vertical metallicity gradient is not well determined, we assumed a model of vertical metallicity gradient as in Kuijken & Gilmore (1989). This model gives an indicative rather then conclusive estimate of the true gradient. The relation of the sensitivity of M_V to [Fe/H] was taken from Laird *et al.* (1988) and the UV excess–metallicity ralation was taken from Carney (1979). In our distance determination, we did not consider the radial metallicity gradient. The overall error is about 10–20% in the velocity, which is mainly due to the photometric uncertainties.

Majewski: Do you allow for kinematical gradients in your second analysis—the multivariate discriminant analysis, not the SEM—or do you assume discrete components only?

Ojha: In the case of MDA, we have selected a subsample of stars—where the majority of thick disk stars exits—with the help of the Besançon model. So we assume the thick disk as a discrete component and did not allow the kinematical gradients in this analysis.

Carney: Did you try to derive the velocity ellipsoid for the thin disk ? Your thick disk velocity ellipsoid suggests near-equality of the θ and z components, which is different from the thin-disk results. This is important, since it bears on the continuity or discontinuity of the thin and thick disks.

Ojha: In case of the thin disk we did not derive the z component. We found that in the nearer distance bins, the Gaussian fit (using SEM) shows that the thin disk-like population was itself divided into two components. However, we derived the π and θ components for the thin disk which are found to be $\sigma_\pi = 40$ km s^{-1}, $\sigma_\theta = 30$ km s^{-1}.

The TMGS: a Powerful Tool for Sounding the Structure of the Milky Way

By FRANCISCO GARZÓN, P. L. HAMMERSLEY,
T. MAHONEY AND X. CALBET

Instituto de Astrofísica de Canarias, E–38200 La Laguna, Tenerife, Spain

We present some results from the Two Micron Galactic Survey (TMGS), a project which has been running for the past five years at the IAC, mapping extended areas of the galactic plane and bulge with a dedicated IR camera. With the analysis made so far we have found a giant star-formation region which strongly suggests the existence of a central galactic bar, whose receding end is found at $l = 15°$ to $28°$. This region contains some of the most luminous IR stars in the Galaxy. Also the inner distribution of the stars on the plane shows that the vast majority of these belong to the young and old disk population and to a well defined inner bulge, with almost no stars found at the galactic plane in the classical spherical distribution of the bulge.

1. Introduction

Following an idea of the late Mike Selby, in 1988 we started a project aimed at surveying large areas of the sky in the near IR, using his new 7 InSb channel camera, developed and built by his group at the Blackett Laboratory of the Imperial College in London. The observational work began in 1989, after several comissioning periods at the 1.5-metre Carlos Sánchez Telescope m) during 1988 and 1989, and have been carried out continously since then by our group, making use of about 25% of the total available time at the telescope.

Detailed description of the survey itself can be found in Garzón *et al.* (1993), where we also demonstrated that TMGS is able to penetrate the dense shell of obscuring material surrounding the central star distribution. Analysis of the huge amount of data gathered during the several observations campaigns started in 1993 and concentrates on the structural separation of the different star populations present in the equatorial plane and bulge of the Milky Way. For the time being we have tried to avoid the use of models of star distribution as much as possible, since the TMGS contains sufficient information to warrant being interpreted in its own right, without making further assumptions. Even so, some attempts have been made to investigate the star distribution as seen by TMGS with detailed models of star distribution, such as that of Wainscoat *et al.* (1994; see Cohen 1994), a cooperation which is now under way.

It is worth saying that some of the studies which will be mentioned in the subsequent sections have been carried out by making extensive use of the databases of the TMGS and integrated flux surveys, such as the DIRBE experiment aboard *COBE*. The combination of these two types of information has proven to be an excellent method for analysing large-scale structures, since it combines the detailed information which can be achieved with point-source surveys, such as histograms of magnitude distribution, with the wide sky coverage obtained with the surface brightness maps.

Finally, the TMGS database is currently being cross-correlated with other existing databases in optical and IR wavelengths, which will greatly increase its usefulness for the structural analysis of the distribution of star populations in the Milky Way.

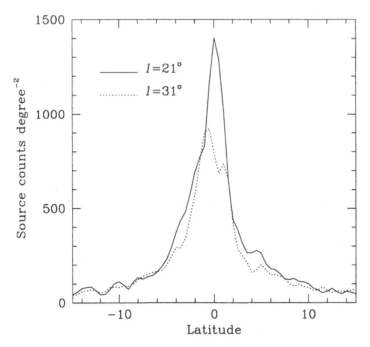

FIGURE 1. Cumulative star counts per square degree to $m_K = +8$.

2. The star formation region at $l = 15-28°$

When analysing the histograms of star counts in several regions along the galactic plane (Hammersely *et al.* 1994b), we found some striking features in the shape of the curves at specific positions spanning over longitudes from $l = 15°$ to $l = 28°$. Bassically, as can be seen from Figures 1 and 2, the large-scale distribution of stars at $l = 31°$ and $l = 21°$ coincides, except for a prominent spike which dominates the counts at $l = 21°$. The rest of the curve can be explained as an exponential disk and the contributions from the two spiral arms along the line of sight.

The spike at $l = 21°$ is big but much less than the one at $l = 27°$, as can be seen in Figure 3. In Hammersley *et al.* (1994b) we examine a number of possibilities that might account for that spike and conclude, for a number of reasons, that the most likely explanation for this feature is that it originates in a huge star-formation region containing some of the most luminous IR stars in the Galaxy. This region spans from $l = 28°$ to at least $l = 15°$, with a width of about 0.5° in b centred on the plane, and constitutes strong evidence for the existence of the galactic bar, since such large regions are present in most barred galaxies at the position where the bar interacts with the spiral arms. Also it allows us to calculate the position angle of the bar, which, from our data, turns out to be at $75° \pm 5°$, in fair agreement which that determined by Blitz & Spergel (1991), who place the near end of the bar in the first quadrant ($90° > l > 0°$).

3. The distribution of the stars on the plane

We have made extensive use of the TMGS database to investigate the large-scale distribution of stars in the plane of the Galaxy and have attempted to separate the different components, mainly the young and old disk populations. Some results are plotted in Figures 4 and 5, which show histograms of star counts at different galactic

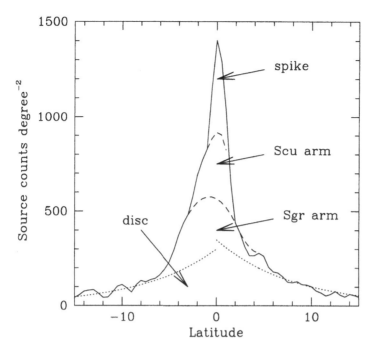

FIGURE 2. Cumulative star counts per square degree to $m_K=+8$ for $l=21°$. The various distributions are indicated, although the shapes are only approximate and extinction is not included.

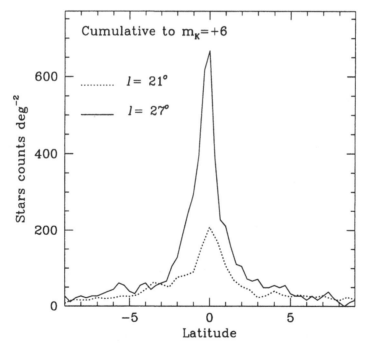

FIGURE 3. Cumulative star counts per square degree to $m_K=+6$ for $l=21°$ and $27°$. The $l=27°$ distribution is based on fewer scans than that at $l=21°$ and so appears to be noisier.

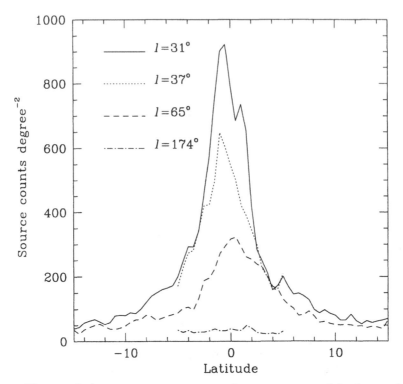

FIGURE 4. The cumulative star counts per square degree to $m_K = +8$ for four regions through the disk.

longitudes on the plane. In Figure 4, the wide wings of the exponential disk extending from about $b = 5°$ to at least $b = 15°$, in the $l = 31°$ and $65°$ histograms can be seen. In Hammersley *et al.* (1994a) we discuss the nature of this exponential feature and conclude that the stars are not local, that the scale *height* is about $7°$ to $8°$ in b, and that the same holds for both areas. We also show how close an exponential fit is to the measured counts and explain in Hammersley *et al.* (1995) that the observed differences in the fit for the positive and negative side can be attributed to the position of the Sun above the galactic plane and to the tilt of the local old disk.

In Figure 5 we show what we called in Hammersley *et al.* (1994a) the *inner* distribution, which is derived by substracting the calculated exponential fit to the measured counts at each region. The stars from the Sagittarius, Scutum and Perseus spiral arms are then prominent in the remaining distributions, as the local disk is removed. See Hammersley *et al.* (1994a) for details and a complete discussion.

The TCS is operated on the island of Tenerife by the Instituto de Astrofísica de Canarias in the Spanish Observatorio del Teide of the Instituto de Astrofísica de Canarias. The *COBE* datasets were developed by the NASA Goddard Space Flight Center under the guidance of the *COBE* Science Working Group and were provided by the NSSDC. We particularly wish to thank the late M.J. Selby, who died earlier in 1993, and without whom the TMGS would not have been possible.

This project has been supported by IAC Grant 28/86 and by the Science and Engineering Research Council. PLH has been supported by the Dirección General de Investigación Científica y Técnica, Spain.

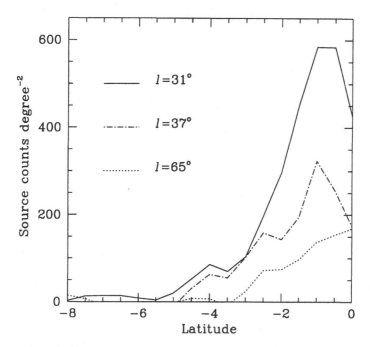

FIGURE 5. Cumulative star counts per square degree after subtracting the exponential function.

REFERENCES

BLITZ, L. & SPERGEL, D. N. 1991 *Astrophys. J.* **370**, 205.

COHEN, M. 1994 in *Science with Astronomical Near–Infrared Surveys* (ed. N. Epchetein, A. Omont, B. Burton & P. Persi), p. 181. Kluwer.

GARZÓN, F., HAMMERSLEY, P. L., MAHONEY, T., CALBET, X., SELBY, M. J. & HEPBURN, I. 1993 *Mon. Not. R. Astron. Soc.* **264**, 773.

HAMMERSLEY, P. L., CALBET, X., MAHONEY, T., GARZÓN, F. & SELBY, M. J. 1994a *Mon. Not. R. Astron. Soc.* submitted.

HAMMERSLEY, P. L., GARZÓN, F., MAHONEY, T. & CALBET, X. 1994b *Mon. Not. R. Astron. Soc.* **269**, 753.

HAMMERSLEY, P. L., GARZÓN, F., MAHONEY, T. & CALBET, X. 1995 *Mon. Not. R. Astron. Soc.* **273**, 206.

WAINSCOAT, R. J., COHEN, M., VOLK, K., WALKER, H. J. & SCHWARTZ, D. E. 1992 *Astrophys. J., Suppl. Ser.* **83**, 111.

The Centre of the Milky Way

By ANTONIO ALBERDI[1,2]

[1]Laboratorio de Astrofísica Espacial y Física Fundamental, Apdo. 50727, E–28080 Madrid, Spain

[2] Instituto de Astrofísica de Andalucía, Apdo. 3004, E–18080 Granada, Spain

The galactic centre has been shown to be one of the brightest objects in the sky at different wavelengths. The centre of the Milky Way in many ways resembles a low-luminosity active galactic nucleus. However, the nature of the concentration of mass and luminosity at the galactic centre is still not completely explained. Some astronomers consider that the energetic activity results from matter accreting onto a black hole at the very centre with a mass of several million times that of the Sun, while others consider that this activity is due to bursts of star formation. A strong argument for the existence of a black hole is the presence at the dynamical centre of the Galaxy of a unique compact radio source, SgrA*. The spectrum and morphology of SgrA* suggest the existence of a non-uniform distribution of synchrotron-emitting particles and magnetic fields. The time variability of SgrA* as well as its apparent size at centimetre and millimetre wavelengths can be interpreted according to its intrinsic activity and to an anisotropic turbulent medium at the galactic centre, acting as a screen which broadens the shape of SgrA*. On the other hand, there is observational evidence that the far-IR luminosity of the galactic centre could be generated by IRS16, a cluster of luminous stars. There is phenomenological evidence in favour of a strong interaction of the IRS16 wind with other sources in its neighbourhood, interaction which significantly influences the morphology and kinematics of gas and dust at the galactic centre.

---

1. Introduction

We describe the morphology of the galactic centre region from large to smaller scales, describing finally SgrA*, the compact radio source at the very centre of the Milky Way. We use scales similar to those previously described by Blitz *et al.* (1993) and Zylka *et al.* (1992).

Considering first those regions which are further than 300 pc from the centre of the Milky Way: atomic and molecular gas constitute more than 90% of the mass of the interstellar material, and are detected through their emission at radio wavelengths. Atomic hydrogen is observed through the 21-centimetre hyperfine line, whereas the molecular hydrogen H_2 is not directly observable, and is traced primarily using the 2.6-millimetre rotational transition of the CO and the CS molecule. The atomic hydrogen is confined within a disk of 1.8 Kpc. According to the orbits of the HI clouds, assuming an angle with respect to the line of sight of 16 degrees and assuming the existence of a stellar bar, we can establish a value of 2.4 Kpc for the corotation regime and a value of 200 pc for the inner Lindblach resonance (ILR) (Blitz *et al.* 1993, and references therein).

1.1. *The inner 300 pc*

The molecular gas has been traced with the help of the CO and CS emission. The central parts (the bulge) are denser and have been traced with the CS molecule. The CO disk has a size of 300 pc, and sits within the central 3 Kpc hole. In principle, the star formation might have been very efficient due to the concentration of high-density, highly pressured molecules; however, there is a relative paucity in the star-formation rate. One explanation for that could be the high-velocity dispersion of the molecular clouds, which would require these to be very massive clouds ($\sim 10^6$ solar masses) in order to be gravitationally bound (Spergel & Blitz (1992)). The most spectacular star-forming clouds such as SgrA and SgrB2 have masses of this order.

1.2. *The inner 30 pc*

Although this zone is far away from the ILR region, this is an active star-forming region. Different explanations have been given in order to explain this apparent controversy: (i) the existence of multiple individual supernova explosions; (ii) the existence of a smaller bar or another internal ILR, as has been suggested for other galaxies; (iii) the existence of a magnetic field in the disk which is coupling the disk with the coronal gas. There is further support for the existence of these strong fields coming from the radio continuum observations of filamentary structures (emitting synchrotron radiation) and a poloidal field which is connecting the gas disk with the hotter coronal gas (Sofue *et al.* 1989; Yusef-Zadeh *et al.* 1984).

1.3. *The central 10 pc*

The main features in the central 10 pc are SgrA and a stellar cluster, together with a molecular cloud of around 10^5 solar masses. SgrA has two main components: (i) SgrA East, with a size of $\sim$8 pc, which shows a non-thermal spectrum and is powered through supernova explosions; and (ii) SgrA West, with a size of 2 pc, whose emission is essentially thermal. SgrA West is placed inside a highly turbulent circumnuclear disk, which rotates with a typical velocity of $\sim$110 $Km \cdot s^{-1}$, and it is the reservoir of gas which feeds the nucleus. SgrA West has a spiral-like pattern, with three arms of ionised gas, and a compact radio source—SgrA*—at the very core of the galactic centre. The circumnuclear disk contains 60 solar masses of ionised gas and 600 solar masses of atomic gas (Yusef-Zadeh & Wardle 1993).

1.4. *The central parsec*

At this linear scale, there are significant collisions between the individual stars which produce tidal disruption in many of them. There is a central star cluster whose centre has been considered to be the infrared complex IRS16. IRS16 is placed 0.04 pc east of SgrA*, and is formed by at least 24 components at $2\mu m$ (Eckart *et al.* 1992). This complex appears to be the source of a strong wind with a velocity of 700 $Km \cdot s^{-1}$ and an inferred mass loss rate of 4×10^{-3} solar masses per year. These values are calculated from the broad HeI, Brα and Brγ lines. Moreover, a dozen HeI emission-line stars within 0.5pc of SgrA* have been found. It has been suggested that these stars are either blue supergiants or Wolf–Rayet-like stars with heavy mass loss ($\sim 10^{-4} - 10^{-5}$ solar masses per year) and outflow velocities of around 1000 $Km \cdot s^{-1}$.

1.5. *SgrA*

SgrA* is a unique, compact radio source at the very centre of the Milky Way. There is observational evidence that the galactic centre contains a point-mass black hole of $\sim 10^6$ solar masses coming from the observed velocity distribution of the stars and the measured velocity of the ionised streamers (Melia *et al.* 1992). Indeed, it is widely accepted that the radio emission from SgrA* is coming from a massive black hole accreting from the winds of the HeI stars.

2. The blue stellar cluster IRS16

The blue stellar cluster IRS16 near the galactic centre appears to be surrounded by the northern arm of SgrA West to the east, by the gravitational potential of a 10^6-solar-mass black-hole candidate (SgrA*), by another member of the hot cluster (IRS16NW) to the west and by the supergiant star (IRS7) to the north. All these sources are related through their physical interaction with the galactic centre outflow (Yesef-Zadeh 1993).

As a result of the interaction between the winds from IRS16 and the sources in the neighbourhood of the galactic centre, the following peculiar structures are produced:

(i) The supergiant star IRS7 shows a cometary tail due to the interaction between a global flow coming from the IRS16 complex and the wind coming from the cool atmosphere of the supergiant star (Yusef-Zadeh & Melia 1992).

(ii) The northern arm of the orbiting ionised gas inside the circumnuclear disk (CND) shows strong waviness, mainly due to Rayleigh–Taylor instabilities as a result of the ram pressure produced by the IRS16 wind. This ram pressure also produces strong polarisation in the vicinity of SgrA*, and it appears as if the CND and SgrA West were collimating the wind outflow from IRS16 (Yusef-Zadeh & Wardle 1993).

(iii) There is clear asymmetry in the distribution of ionised gas and blue stellar objects in the centre of the Milky Way. The predominant emission of IRS16 is distributed to the east of SgrA*, whereas the ionised gas emerges from the vicinity of SgrA* towards the western side. This ionised gas is distributed as a chain of blobs which collide with the orbiting gas forming a mini-cavity, i.e. a hole in the distribution of orbiting gas, which has been detected (Zhao *et al.* 1991a).

(iv) The compact radio source at the very centre of the Milky Way, the black-hole candidate SgrA*, which is apparently accreting from the winds coming from the IRS16 complex; there is experimental evidence (see below) to suggest that scattering processes are affecting SgrA*. An excellent candidate for the "scattering screen" is the ionised wind associated with IRS16, a highly turbulent medium.

3. The compact radio source SgrA*

3.1. *The structure of SgrA*

For the last 20 years, several groups have been trying to determine the structure and sky position of SgrA* using VLBI techniques at decimetre and centimetre wavelengths (e.g. Kellermann *et al.* 1977; Lo *et al.* 1981; Lo *et al.* 1985; Jauncey *et al.* 1989; Marcaide *et al.* 1992; Marcaide *et al.* 1993; Lo *et al.* 1993; Alberdi *et al.* 1993). The VLBI observations have shown SgrA* as a compact flat-inverted spectrum source, which also exhibits flux density variability, and with an angular size that scales with observing wavelength according to a λ^2-law. It is generally believed that the structure seen at centimetre wavelengths does not correspond to the intrinsic structure of SgrA*, but rather to the scattering disk reflecting the interaction of its radio emission with interstellar electrons in the inner region of the Galaxy. This λ^2 dependence has been confirmed by the observations of OH masers towards the galactic centre, confirming that the scattering must be taking place very close to it (van Langevelde *et al.* 1992).

Our recent VLBI observations at $\lambda1.35$cm showed that the observed surface brightness distribution of SgrA* is best modelled by an elliptically shaped Gaussian component with parameter values: flux density = 1.07 ± 0.15Jy, major axis of FWHM = 2.58 ± 0.08 milliarcseconds (mas) in position angle = $79° \pm 6°$, and axial ratio = 0.5 ± 0.2. The angular size of this model agrees well with the expected scattering size of SgrA* (Alberdi *et al.* 1993; Marcaide *et al.* 1993).

Melia *et al.* (1992), under the assumption that SgrA* is a black hole accreting material spherically from its surroundings, showed that the mass of the hypothetical black hole is correlated with both the radio spectral index and the critical wavelength below which the intrinsic source size predominates over the scattered source size due to the effects of the interstellar medium. They also state that the current observations are consistent with a mass for a black hole of 10^6 solar masses and a critical wavelength in the millimetre regime. This result is in agreement with other observational results:

- Gwinn *et al.* (1991) showed that the refractive scintillations of SgrA* were absent at λ=1.3 and 0.8 mm over timescales shorter than 24 hours. Since such scintillations were quenched for a source size greater than about 0.1AU, the observations of Gwinn *et al.* provided a bound on the intrinsic size of SgrA* and its description as a quiescent galactic nucleus with $T_b < 0.5 \times 10^{12}$K.

- Zhao *et al.* (1991b) claimed that at wavelengths shorter than 3.6 cm the flux density variability is mainly intrinsic and not associated with the refractive interstellar scintillation, since the variability amplitudes seem to increase with increasing frequency.

- Krichbaum *et al.* (1993) and Krichbaum *et al.* (1994) report VLBI observations of SgrA* at millimetre wavelengths. At 43 GHz, they find a size of $\theta_{43} = 0.75 \pm 0.08$ mas and a brightness temperature of $T_{B,43} = 1.7 \times 10^9 K$; at 86 GHz, they find a size of $\theta_{86} = 0.22 \pm 0.19$ mas and a brightness temperature of $T_{B,86} = 2.2 \times 10^9 K$. The size of SgrA* at both wavelengths is larger than the value expected according to the scattering law ($\theta_{43}^{scat} = 0.53 \pm 0.02$ mas and $\theta_{86}^{scat} = 0.13 \pm 0.01$ mas). These results suggest that the intrinsic structure of SgrA* was detected for the first time.

However, caution is advisable when considering the above results. As a matter of fact, Backer *et al.* (1993) at 7 mm and Rogers *et al.* (1994) at 3 mm—at different observing epochs than Krichbaum *et al.*—found angular sizes of $\theta_{43} = 0.67 \pm 0.03$ mas and $\theta_{86} = 0.16 \pm 0.01$ mas, consistent with those values expected from the λ-squared dependence of size observed at longer wavelengths. It is important to emphasise that the flux density of the source is extremely variable, and that the results of Rogers and Backer were obtained at different epochs than those of Krichbaum *et al.* For example, at 43 GHz, the flux density measured by Krichbaum *et al.* was 0.8 Jy less than during the measurements of Backer *et al.*

As we have mentioned above, variability has been reported for SgrA* at high frequencies. We have made three different VLBI observations of SgrA* at λ1.3 cm (see Figure 1). Our first observations were made on the epoch 1985.11 and were restricted to the baseline Haystack–Green Bank. We fitted the visibility amplitudes by a circular Gaussian model of angular size of 1.8 ± 0.1 mas and flux density 1.2 ± 0.4 Jy. Our second observations—with the participation of the VLBA, Y27 and Goldstone—were made on the epoch 1991.47, and the third epoch was made on the epoch 1992.90, with the same array, i.e. VLBA+Y27+Goldstone. The last two epochs provide similar results, and the brightness distribution of SgrA* is best modelled by an elliptically shaped Gaussian component with the following parameters: flux density=1.07 ± 0.15 Jy, major axis of FWHM=2.58 ± 0.08 milliarcseconds in position angle =$79° \pm 6°$, and axial ratio=0.5 ± 0.2. Comparing these results, we obtain an upper limit for the source expansion of about 0.01 ± 0.02 mas· yr^{-1}, supporting the idea that the structure of SgrA* has not changed on this timescale.

Eckart *et al.* (1992) detected—through speckle infrared observations—a faint source coincident with SgrA*, within the systematic relative positional uncertainty of ± 0.3 arcseconds. The source is detected on the K-band ($K = 13.7 \pm 0.6$) and on the H-band ($H = 13.9$). The authors claimed that, after correcting for extinction, this infrared

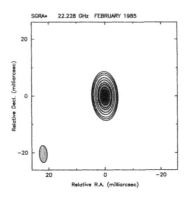

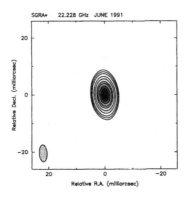

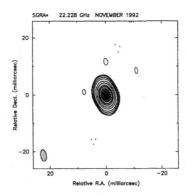

FIGURE 1. λ1.3 cm VLBI images of SgrA*. Contour levels are 0.5, 1, 2, 5, 10, 20, 30, 50, 70 and 90% of the peak brightness. The peak flux densities are 0.97 Jy/beam, 0.79 Jy/beam and 0.65 Jy/beam, respectively. The maps have been convolved with a beam of 6 × 2.8 milliarcseconds at P.A. 3° for the first two epochs, and a beam of 4.2 × 2.1 milliarcseconds at P.A. 8.3° for the third epoch.

counterpart has the Rayleigh–Jeans tail of a blackbody of temperature $T > 7000K$ and a luminosity of $4 \pm 2 \times 10^5$ solar luminosities.

3.2. *The spectrum of SgrA**

SgrA* has an inverted multidecade spectrum (spectral index 1/3, $S \propto \nu^{1/3}$) which is characteristic of the optically thick synchrotron radiation for a potential dependence of the electron distribution. This spectrum is typical of compact radio sources, with the usual core-jet morphology at milliarcsecond angular scales. In this case, the multirange spectrum requires the existence of a second component in the spectrum in the sub-millimetre regime and FIR, and a third component in the NIR.

However, Duschl & Lesch (1994) consider that SgrA* is driven by a mono-energetic distribution of electrons emitting synchrotron radiation. The mono-energetic frequency originates from the equilibrium between the energy gain (through magnetohydrodynamic waves or reconnection processes) and the energy loss (through synchrotron or inverse Compton radiation). In this case, the spectrum can be explained as optically thin synchrotron radiation from a mono-energetic distribution of electrons. Two other features in the galactic centre, the Bridge and the Arc, exhibit the same inverted spectrum ($S_\nu \propto \nu^{1/3}$, Lesch *et al.* 1988). In this case, the variability of the spectrum is interpreted as due to a variable energy input from the accretion disk.

Consideration of a monoenergetic electron distribution could globally explain the spectrum, but certain problems remain unresolved, such as the origin of the X-ray spectrum or the fact that the cooling energy is the same along the whole region, while this is very inhomogeneous.

3.3. *Model for SgrA**

The recent observations of SgrA*, both in the radio and NIR regime, favour the interpretation of the galactic centre as a low-luminosity active galactic nucleus. The kinematics of gas and stars suggests the existence of a black hole of around 10^6 solar masses and a galactocentric radius smaller than 0.1 pc. The luminosity of SgrA* is lower than 7×10^5 solar luminosities, far from the Eddington limit, and the low level of activity can be explained by a low central accretion rate, between 10^{-7} and 10^{-9} solar masses per year (Falcke *et al.* 1993). The last mm-VLBI results, which appear to show an elongated structure, also suggest the AGN scenario (black hole + accretion disk + jet) as the most plausible for SgrA*. In this scenario, the flat radio spectrum would be produced by the synchrotron emission coming from the jet, the increase in the sub-millimitre emission would come from the nozzle, the break in the infrared would be provided by the black hole, and the new component responsible for the NIR emission would originate from the accretion disk (Falcke *et al.* 1993).

There is a second model, developed by Melia *et al.*, which considers emission from heated gas accreting at a rate of 10^{-4} solar masses per year from the stellar winds emanating from the infrared complex IRS16. Recent mm-VLBI results argue against the Melia *et al.* hypothesis and are consistent with the accretion-disk and jet model of Falcke *et al.* (1993).

The AGN-like model and the mono-energetic electron distribution model of Duschl & Lesch can be counterchecked through variability measurements. If there is variability, according to the Duschl–Lesch model there will be a global shift of the spectrum, simultaneous at all frequencies, while, according to the AGN-like model, the outburst will be shifting progressively towards lower frequencies.

We would like to acknowledge the useful comments of Dr. Benjamín Montesinos and

Dr. Lucas Lara during the preparation of this manuscript. This work has been partially supported by the Spanish DGICYT Grant PB89–0009, by a research project from the Junta de Andalucía, and by an Acción Especial from the Consejo Superior de Investigaciones Científicas.

REFERENCES

ALBERDI, A., LARA, L., MARCAIDE, J. M., ELÓSEGUI, P., SHAPIRO, I. I., COTTON, W. D., DIAMOND, P. J., ROMNEY, J. D. & PRESTON, R. A. 1993 *Astron. Astrophys.* **277**, L1.

BACKER, D. C., ZENSUS, J. A., KELLERMANN, K. I., REID, M., MORAN, J. M. & LO, K. Y. 1993 *Science* **262**, 1414.

BLITZ, L., BINNEY, J., LO, K. Y., BALLY, J. & HO, P. T. P. 1993 *Nature* **361**, 417.

DUSCHL, W. J. & LESCH, H. 1994 *Astron. Astrophys.* **286**, 431.

ECKART, A., GENZEL, R., KRABBE, A., HOFFMAN, R., VAN DER WERF, P. P. & DRAPATZ, S. 1993 *Nature* **355**, 526.

FALCKE, H., MANNHEIM, K. & BIERMANN, P. L. 1993 *Astron. Astrophys.* **278**, L1.

GWINN, C. R., DANEN, R. M., MIDDLEDITCH, J., OZERNOY, L. M. & TRAN, T. K. 1991 *Astrophys. J., Lett.* **381**, L43.

JAUNCEY, D. L. *et al.* 1989 *Astron. J.* **98**, 44.

KELLERMANN, K. I., SHAFFER, D. B., CLARK, B. G. & GELDZAHLER, B. J. 1977 *Astrophys. J., Lett.* **214**, L61.

KRICHBAUM, T. P. *et al.* 1993 *Astron. Astrophys.* **274**, L37.

KRICHBAUM, T. P., SCHALINKSI, C. J., WITZEL, A., STANDKE, K. J., GRAHAM, D. A. & ZENSUS, J. A. 1994 in *Nuclei of Normal Galaxies: Lessons from the Galactic Centre.* NATO ASI Series. Kluwer.

LESCH, H., SCHLIKEISER, R. & CRUSIUS, A. 1988 *Astron. Astrophys.* **200**, L9.

LO, K. Y., BACKER, D. C., ECKERS, R. D., KELLERMANN, K. I., REID, M. & MORAN, J. M. 1985 *Nature* **315**, 124.

LO, K. Y., BACKER, D. C., KELLERMANN, K. I., REID, M., ZHAO, J. H., GOSS, W. M. & MORAN, J. M. 1993 *Nature* **362**, 38.

LO, K. Y., COHEN, M. H., READHEAD, A. C. S. & BACKER, D. C. 1981 *Astrophys. J.* **249**, 504.

MARCAIDE, J. M. *et al.* 1992 *Astron. Astrophys.* **258**, 295.

MARCAIDE, J. M., ALBERDI, A., LARA, L., ELÓSEGUI, P., SHAPIRO, I., COTTON, W., ROMNEY, J. & PRESTON, R. 1993 in *Sub-Arcsecond Radio Astronomy* (ed. R. J. Davis & R. S. Booth), p. 50. Cambridge University Press.

MELIA, F., JOKIPII, J. R. & NARAYANAN, A. 1992 *Astrophys. J., Lett.* **395**, L87.

ROGERS, A. E. E. *et al.* 1994 *Astrophys. J., Lett.* **434**, L59.

SOFUE, Y., REICH, W. & REICH, P. 1989 *Astrophys. J., Lett.* **341**, L47.

SPERGEL, D. N. & BLITZ, L. 1992 *Nature* **357**, 665.

VAN LANGEVELDE, H., FRAIL, D., CORDES, J. & DIAMOND, P. 1992 *Astrophys. J.* **396**, 686.

YUSEF-ZADEH, F. 1993 in *Nuclei of Normal Galaxies: Lessons from the Galactic Centre.* NATO ASI Series. Kluwer.

YUSEF-ZADEH, F. & MELIA, F. 1992 *Astrophys. J., Lett.* **385**, L41.

YUSEF-ZADEH, F., MORRIS, M. & CHANCE, D. 1984 *Nature* **310**, 557.

YUSEF-ZADEH, F. & WARDLE, M. 1993 *Astrophys. J.* **405**, 584.

ZHAO, J. H., GOSS, W. M., LO, K. Y. & EKERS, R. D. 1991a *Nature* **354**, 46.

ZHAO, J. H., GOSS, W. M., LO, K. Y. & EKERS, R. D. 1991b in *Relationships between AGN and Starburst Galaxies* (ed. A. V. Filippenko). ASP Conference Series, vol. 31, p. 295. ASP.

ZYLKA, R., MEZGER, P. G. & LESCH, H. 1992 *Astron. Astrophys.* **261**, 119.

SiO Maser Survey of IRAS Bulge Sources

By HIDEYUKI IZUMIURA[1]†, I. YAMAMURA[2],
S. DEGUCHI[3], N. UKITA[3], O. HASHIMOTO[4],
T. ONAKA[2] AND Y. NAKADA[5]

[1]SRON Groningen, Landleven 12, P.O. Box 800, 9700 AV Groningen, The Netherlands

[2]Department of Astronomy, University of Tokyo, Japan

[3]Nobeyama Radio Observatory, National Astronomical Observatory, Japan

[4]Department of Applied Physics, Seikei University, Japan

[5]Kiso Observatory, University of Tokyo, Japan

We surveyed 313 colour-selected *IRAS* bulge sources in the SiO maser emissions from the J=1-0, v=1 and 2 transitions with the Nobeyama 45-metre telescope. The survey resulted in 222 detections (Nakada *et al.* 1993; Izumiura *et al.* 1994,1995). The detection rate exceeded 80% for the brighter samples (Table 1). We discuss the observed SiO maser characteristics and the kinematics of the bulge in terms of the l–v diagram.

Line-of-sight velocity: The radial velocity of the SiO maser generally agrees well with the central velocity of the OH maser given by te Lintel Hekkert *et al.* (1991) within 3 $km\,sec^{-1}$ in 29 common sources, which is consistent with the results found by Jewell *et al.* (1991) for the solar neighborhood stars.

Maser intensity: We examined the integrated SiO intensity against the *IRAS* 12μm flux density. On average, our sample shows much stronger SiO intensities for a particular 12μm flux than found by Jewell *et al.* (1991) for stars in the solar neighbourhood. The highest observed value of the SiO intensity for a fixed 12μm flux is ten times as high as theirs. The cause of this large discrepancy should be clarified.

Completeness of our survey: We gave priority to the $4° \leq |b| < 5°$ and $7° \leq |b| < 8°$ regions. The sampling is almost complete in terms of the galactic longitude for the region of $l > -8°$ due to the declination limit at Nobeyama. In terms of the 12μm flux, our survey is almost complete down to $F_{12} \sim 1.5$ Jy in the $7°$–$8°$ strips. According to White-lock *et al.* (1991) our survey reached the far side of the bulge, as far as 4 kpc away from the galactic centre. Our sample, however, may be biased to the intrinsically brighter objects in the far-side. In the $4°$–$5°$ strips sampling is complete down to $F_{12} = 3$ Jy, so our survey is almost complete in the near side of the bulge (see Nakada *et al.* 1991). The far side has also been substantially sampled as far as 10 kpc from the Sun.

Implications of the l–v diagram : Izumiura *et al.* (1993) suggested a possible asymmetry in the l–v diagram. It has turned out to come mostly from sources in the $7°$–$8°$ strips in quadrant "I" of Figure 1a. The $4°$–$5°$ strips sources are distributed rather symmetrically around the origin in the diagram, in spite of the somewhat smaller distance coverage. The asymmetry is not due to the sampling effect but probably real. This suggests that either the $7°$–$8°$ strips have a velocity structure significantly different from the $4°$–$5°$ strips, or the $4°$–$5°$ strips will show a similar feature if the sampling becomes deeper. We might be seeing an extra kinematical component which becomes evident only away from the plane. Furthermore, there seems to be a hole centred at around $l = 4.5°$, V=75 $km\,sec^{-1}$ ("+" in Figure 1b). It is most evident in $3° < l < 6°$. If the distribution is random in the vertical direction on the figure for a range spanning 300 $km\,sec^{-1}$ in this lon-

† On leave from Department of Astronomy and Earth Sciences, Tokyo Gakugei University.

Colour Range	Flux Range (Jy)			
log(F$_{25}$/F$_{12}$)	1.0-2.0	2.0-4.0	4.0-8.0	8.0-16.0
−0.15-0.15	57 (24/42)	66 (44/67)	73 (74/101)	81 (43/53)

TABLE 1. Detection rates of the SiO masers (%) (N$_{detected}$/N$_{observed}$)

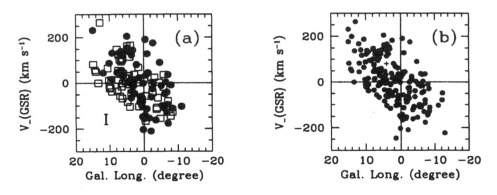

FIGURE 1. (a) l–v diagram for the 4°-5° (filled circles) and 7°-8° (open squares) regions (all sources with F$_{12}$ ¿ 10 Jy are plotted); (b) same as in "a" but for the entire bulge region.

gitude range, the probability of finding such a hole is below 0.2%. If it follows a gaussian distribution around the linear regression line (V< km sec^{-1}>= −20.7 + 8.5 ∗ l <degree>) fitted to the rest part of the diagram, the probability becomes far less than 0.1% assuming a velocity dispersion of 83 km sec^{-1}found for the rest. This hole may be related to the bar structure in the bulge (Blitz & Spargel 1991; Nakada *et al.* 1991) which would require considerable stellar streaming motion.

REFERENCES

BLITZ, L. & SPARGEL, D. 1991 *Astrophys. J.* **379**, 631.

IZUMIURA, H., DEGUCHI, S., HASHIMOTO, O., NAKADA, Y., ONAKA, T., ONO, T., UKITA, N. & YAMAMURA, I. 1994 *Astrophys. J.* **437**, 419.

IZUMIURA, H., ONO, T., YAMAMURA, I., OKUMURA, K., ONAKA, T., DEGUCHI, S., UKITA, N., HASHIMOTO, O. & NAKADA, Y. 1993 in *Galactic bulges* (ed. H. Dejonghe & H. J. Habing). IAU Symposium 153, p. 303. Kluwer.

IZUMIURA, H. *et al.* 1995 *Astrophys. J., Suppl. Ser.* in press.

JEWELL, P. R., SNYDER, L. E., WALMSLEY, C. M., WILSON, T. L. & GENSHEIMER, P. D. 1991 *Astron. Astrophys.* **242**, 211.

NAKADA, Y., DEGUCHI, S., HASHIMOTO, O., IZUMIURA, H., ONAKA, T. SEKIGUCHI, K. & YAMAMURA, I. 1991 *Nature* **353**, 140.

NAKADA, Y., ONAKA, T., YAMAMURA, I., DEGUCHI, S., UKITA, N. & IZUMIURA, H. 1993 *Publ. Astron. Soc. Jpn.* **45**, 179.

TE LINTEL HEKKERT, P., CASWELL, J. L., HABING, H. J., HAYNES, R. F. & NORRIS, R. P. 1991 *Astron. Astrophys., Suppl. Ser.* **90**, 327.

WHITELOCK, P., FEAST, M. & CATCHPOLE, R. 1991 *Mon. Not. R. Astron. Soc.* **248**, 276.

The Formation of Disk Galaxies

By GERHARD HENSLER AND MARKUS SAMLAND

Institut für Astronomie und Astrophysik, Universität Kiel, D-24098 Kiel, Germany

Simple evolutionary models of the Milky Way Galaxy are concerned either with the chemical evolution by the application of one-zone approximations or with the dynamical evolution of both the stellar component or a single-phase galactic gas. Although both cases can achieve excellent agreement for particular regions and with specific observational results, they fail if their results are combined for understanding the global evolution of the Galaxy because they lack self-consistency. Only if models take the multi-phase character of the interstellar medium and the gas–star interactions into account can they describe the energetical and dynamical evolution of galaxies in a physically reasonable manner. Vehement dynamical effects play a prominent role during the early phases of galactic evolution because the star-formation rate burst and the gravitational potential still allowed a large-scale outflow of hot supernova-ejected gas. These so-called chemo-dynamical models can account for the different observational issues, e.g. for the effective yields and for the stellar metallicity distributions in bulge and disk, self-consistently within one global evolutionary scenario.

1. Introduction

Massive disk galaxies consist at present of stars and a sufficient amount of gas that accounts for a present non-negligible star-formation rate (SFR). The stellar content can in principle be subdivided into at least three distinct populations: the halo, the bulge and the disk, which differ in their mean properties, briefly summarized as follows. While an extended stellar spheroid, the halo, is presumably pressure supported by means of its large anisotropic velocity dispersion and shows a large range in the metallicity, below solar, the disk stars tend to be more metal-rich and rotationally supported with large regular velocities but low velocity dispersions. The existence of cold, thin galactic disks visible by small velocity dispersions of newly formed stars below 10 km s^{-1} as well as by a strong concentration of cool molecular gas to the galactic equatorial plane invokes dissipation during the galaxy's collapse. The bulge as a third component forms the central almost spheroidal stellar part. With regard to the formation of these structures, three questions have to be addressed: (i) Are they formed separately? (ii) What are their ages and thus the sequence of formation? (iii) What are their evolutionary timescales?

The early phases of galactic evolution, thus also of disk galaxies, cannot yet be studied directly by resolved observations of distant systems. One possible approach to detect the evolutionary scenario is to look for evolutionary signatures in age, dynamics, and chemistry of the still existing stars, probably deducible from the stellar populations within our own Milky Way, a presumably "normal" disk galaxy. Such studies were initiated by Baade 1946 and have continued until now as the so-called population concept (e.g. Sandage 1986a). The results of different authors lead in principle from the kinematical and chemical reasons (e.g. Sandage & Fouts 1987) to a coherent collapse picture of a primordial gaseous protogalactic cloud which formed its first stars in a spheroidal component, the halo, followed by the formation of a rotationally supported component, the galactic disk. However, conclusions for the collapse and formation timescale differed immensely between the various authors. While Eggen *et al.* (1962) derived a very short timescale of the order of the free-fall time of an isothermal sphere, Yoshii & Saio (1979) concluded that the galactic collapse should have been lasted for a few billion years. Although a global trend of a radial metallicity gradient seems visible for globular clusters

233

(GCs) in the halo, their abundance scatter is large while their individual intrinsic homogeneity is striking. On the other hand, two clues to galactic evolution seem to be hidden by the fact that the halo field stars start at much lower metal abundances than the GCs, and that a younger population of GCs are attributed to the disk formation.

There is still a controversial debate over which type and structure the predecessors of the present galaxies had and to what extent the morphological differences between the Hubble-type galaxies have been structured by interactions of these protogalactic clouds (and thus by nurture, Silk 1986) in an early and therefore narrower universe. The contrary view attributes the differences to the initial conditions of the protogalaxies, i.e. to their nature. For a long time, galaxy interactions by close encounters or direct mergers as well as the cosmological evolution of large-scale density fluctuations could only be studied by means of dissipationless N-body simulations. Although detailed observations require the probably dominant action of dissipation at least during the early stage of galaxy formation, not until recently could more refined numerical schemes also take dissipational effects into account. These simulations were applied to galaxy formation in the cosmological context with special emphasis on the influence of a dissipationless dark-matter (DM) component. Recent simulations of density fluctuations up to scales of galaxy clusters (Katz & Gunn 1991; Katz 1992; Steinmetz & Müller 1994) confirmed the already widely accepted assumption that giant elliptical galaxies are formed on short timescales due to impacts and dissipational merging of massive gaseous protogalaxies.

Giant disk galaxies, on the other hand, although they are not unaffected by neighbouring galaxies, form as almost isolated complexes and are only slightly perturbed by infalling low-mass satellite galaxies of Magellanic type or less mass. Spiral galaxies can suffer only a mild infall of additional mass on a percentage level, because a larger mass fraction would alter the global disk structure (Toth & Ostriker 1992). Nevertheless, it has not yet been studied to what extent such dissipational effects would enhance the star formation as it is plausible for merging galaxies and is simulated for interactions (e.g. Scalo & Struck-Marcell 1986; Mihos *et al.* 1991). These less massive impacts are able to provide a heating of the galactic disk. Although global features such as bulge-to-disk ratios and rotation curves of disk galaxies can be deduced from these cosmological models and can be compared with observations, the spatial resolution of these global cosmologically dedicated simulations is still nowhere near sufficient to provide more detailed comparisons with the internal structures of galaxies like thick-to-thin disk evolution or star-formation timescales. Neither the spatial nor the internal resolution of components can allow for detailed heating processes and lead to a rapid cooling and strong dissipation, and thus to too-fast disk formation. This fact has not been obtained from more refined dissipative N-body simulations of protogalactic cloud systems (Theis & Hensler 1993).

Nonetheless, because spiral galaxies seem not to have undergone drastic collisions with massive protogalactic aggregates, one can proceed to further, more spatially resolved studies of the evolution of disk galaxies, under the assumption of almost isolated systems. Hydrodynamical simulations of the coherent collapse scenario by Larson (1975, 1976) and by Burkert and Hensler (1987a, 1988) yield very rapid collapses within only two free-fall times (τ_{ff}), the galactic disk formation after another τ_{ff} with too-rapid gas consumption by star formation, in contradiction to rates derived from observations (Sandage 1986b). In addition, several observationally determined structural details cannot be understood within the framework of the simple collapse picture.

Another kind of study neglects detailed dynamical effects to concentrates on particular regions, trying to follow the temporal evolution of mass fractions and metal abundances. If one assumes star formation to depend on the density of the cool gas phase through a Schmidt-like power law, one would expect the first few stars which are formed at low densities in a collapsing protogalactic gas cloud to be spread throughout the whole galaxy, i.e. as in the galactic halo and the solar vicinity. These stars, however, are characterised by large peculiar velocities. Not until dissipation allows the gas to settle down into the equatorial plane is a rotationally supported disk formed Burkert *et al.* 1992) where the velocity dispersions of subsequently born stars have dropped so that disk and halo stars can be distinguished from their kinematics. Gas consumption by star formation and the ejection of stellar nucleosynthesis products allows for the chemical evolution of a region. Its simple analytical description by a one-zone model of a closed box gives a linear relation between the time-dependent metallicity $Z(t)$ and the initial-to-temporal gas ratio $ln[M_{g,0}/M_g(t)]$. The slope is then determined by the yield Y, the metallicity release per stellar population. It is already at that stage plausible that this simple scenario cannot hold, if sufficient amounts of protogalactic gas survive the collapse and fall into the disk.

In the solar vicinity at least three severe problems arise when we consider the metallicity distribution (MD) of G dwarfs, stars that live long enough to trace the evolution of a galaxy. On the one hand, as was pointed out decades ago (van den Bergh 1962; Schmidt 1963) and descibed in more detail later (Pagel & Patchett 1975), the solar neighbourhood does not contain very metal-poor G dwarfs in the disk population. Out of 132 G dwarfs, no star has so far been observed below $[O/H] \approx -0.6$ (Pagel 1989), but instead the MD increases steeply below $[Fe/H] \approx -1.0$ (see Figure 4). Similarly, the age–metallicity relation of solar-neighbourhood stars (Twarog 1980; Carlberg *et al.* 1985; Nissen *et al.* 1985) shows the same effect, in the sense that the oldest stars with approximate ages of around 12 Gyr have metallicities of $[Fe/H] \geq -0.6$. On the other hand, a yield of $Y_0 \approx 2 Z_\odot$ for a normal stellar population does not fit the observed G-dwarf data with a slope of $Y = 0.4 Z_\odot$ only (Pagel 1987; or $Y = 0.5 Z_\odot$, Pagel 1989).

In order to solve this well-known G-dwarf problem various influences on the evolution of the solar neighbourhood have been invoked. Some models start from a finite metal enrichment by means of an initial starburst (Truran & Cameron 1971; Tinsley 1981), others required accretion of pristine halo gas (Lynden-Bell 1975; Chiosi 1980; Lacey & Fall 1983; Clayton 1986; Matteucci & Francois 1989). And even if an analytical infall rate is taken into account, the models are devoid of self-consistency. Simple closed-box models introduced a global reduction of the yield and fit the MD with a so-called effective yield Y_{eff} or as an open zone with an outflow of metal-enriched hot gas at later stages (Hensler & Burkert 1991; Meusinger 1992). Most recently Malinie *et al.* (1993) proposed a model of inhomogeneous chemical evolution (similar to that studied by Burkert & Hensler (1987b). However, they also require a metal pre-enrichment of 0.3 $Z_\odot$ and even then could not fit Y_{eff} by more than 0.58 $Z_\odot$. Multi-zone models (a more appropriate expression than "multi-phase"; Ferrini *et al.* 1992) couple different regions of the Galaxy (e.g. halo and disk) by mass exchange and try to approach the observational data by artificially varying the parameters. However, even their results deviate significantly because the dynamical action of the (really multi-phase) ISM as well as influences from other regions than those under consideration are neglected. Other approaches explain the low Y_{eff} by a steeper initial mass function (IMF) in the massive-star range at former times. However, all these *ad hoc* assumptions have not yet been properly elaborated and understood and could, furthermore, hold only locally since Y_{eff} conflicts with the larger one in the bulge of approximately 2 $Z_\odot$ and, on the other hand, exceeds by far the value of halo GCs (0.025 $Z_\odot$ on average, Pagel 1987). Only a bimodality of the IMF with different amplitudes in halo, bulge and disk could account for the discrepancies between their MDs and evolutionary history. Until now, however, no theoretical expertise has provided a physically reasonable handling of this.

From the determined parameters (e.g. Tendrup 1988; Rich 1988; Frogel 1988), the dominant stellar population in the bulge is the oldest one of metal-rich stars in the Galaxy. The differential MD of the galactic bulge stars (Pagel 1987; Pagel 1989) can be interpreted controversially although it agrees with the simple model of an effective yield of 1.8 $Z_\odot$. Moreover, from the limited insight into the bulge's structure the question has to be addressed as to the manner by which the bulge might fit into the global collapse scenario and, moreover, to what extent it interacts with the other components or whether it evolves isolated like a closed box. Although it is reasonable to expect a very early formation as the denser central part of the protogalactic cloud, this and its influence and interaction on the formation of the other components of disk galaxies are still in a state of speculation.

In summary, we would stress that the kind of models cited for reproducing the different MDs of the galactic components requires an artificial fine-tuning of different parameters or of prescribed interactions between the components, without taking any exact but only plausible approximations of local physical processes into account. An elegant manner, a much more sophisticated stage, and, moreover, the most reliable results would be achieved, if one could produce all the peculiarities of the different stellar populations by one single model that deals with the global galactic evolution dynamically and energetically and could thus account for the galactic components self-consistently, by avoiding any artificial ad hoc assumptions on local evolutionary differences.

With regard to the representation of hydrodynamical models, their reality has also to be considered with a degree of scepticism. Since only a single gas phase was treated in these above-mentioned simulations, effects of the ISM which are caused by phase transitions and absolutely require the simultaneous existence of different gas phases could not be considered. As we already know, the simple approximation of a constant heating rate leads to the two-phase description of the ISM (Field *et al.* 1965) and could explain the coexistence of cold interstellar clouds and a warm thermally stable phase. The star-forming sites are embedded in large molecular clouds where the massive stars deliver stellar winds and radiation and so influence the star-formation process in the sense of self-regulation (or positive feedback). Thus it is obviously totally independent of the parameterisation with the gas density (Cox 1983; Köppen *et al.* 1995). In addition, a hot gas phase as found one decade later is contributed by supernova ejecta (McKee & Ostriker 1977) and interacts by means of compression or evaporation of embedded clouds and by the formation of cooled condensed material.

Since gas phases cannot exist in isolation and will not be produced in equilibrium, energetic and materialistic processes have to act in order to achieve equilibrium states. In addition, the dynamical behaviour should differ between the gas phases according to their achievable momentum densities. While the hot intercloud medium (ICM) can normally not be pressure-confined to the explosion region and expands vehemently, the cool phase with low velocity dispersion has to be heated up or accelerated by momentum transport from hot streaming motions. This can be considered most convincingly in starbursts of both massive and dwarf galaxies, respectively. In comparison to those still observable starburst epochs, the early phases of Galaxy formation will also proceed in those high-energetical phenomena (Theis *et al.* 1992; Hensler *et al.* 1993). This means, however, that locally confined models are neglecting the dominant process of dynamical interactions between different regions within the same galaxy. Parameterisations of time-dependent gas infall and/or outflow for otherwise non-dynamical box models are, on the other hand, introducing artificial assumptions but *not self-consistent* dynamical issues.

2. The chemo-dynamical treatment

"Nature" in astrophysics is really complex since it not only consists of a variety of different components but also ones which interact energetically and materialistically by multiple processes. Since these processes are non-linearly coupled we should bear in mind that that we cannot extrapolate from the time-dependent behaviour of one single process to its evolution in the network of interactions. In order to approach reality, however, it is most important to take all the relevant processes into account and to analyse the temporal behaviour of counteracting ones, such as evaporation and condensation, star formation and stellar radiation, etc.

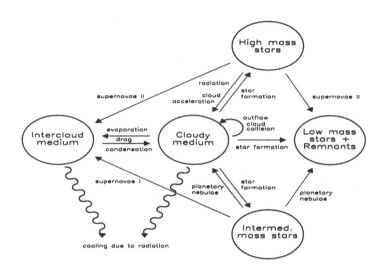

FIGURE 1. The different gaseous and stellar components are coupled via mass, momentum and energy exchanges. The most important interaction processes of the chemodynamical treatment are shown in this diagram.

For the purpose of modelling the global galactic evolution with emphasis on small-scale structures and a spatial resolution of about 200 pc we propose our so-called chemo-dynamical prescription (Hensler 1987) as the most sophisticated and thoroughly researched treatment of galactic evolution. It couples the dynamical treatment of a multiphase ISM and a multi-component stellar population with star-gas interactions by means of star formation, stellar radiation, and stellar mass loss, as well as phase transitions due to heating and cooling. First one-dimensional calculations have already been applied to the evolution of massive galaxies (Theis *et al.* 1992) and to dwarf galaxies in order to explain the structures of dwarf ellipticals (Hewnsler *et al.* 1993; Hensler *et al.* 1995). Two-dimensional models which are applied in order to understand the structures of disk galaxies such as our MWG have recently been completed (Samland 1994) and will be shortly be published in a comprehensive paper (Samland *et al.* 1995). A brief overview of the most important issues is given below in comparison to galactic properties.

3. The model

3.1. *The collapse phase and the halo*

We start from a protogalaxy with a purely gaseous, rotating, two-phase Kuzmin-Plummer sphere with a mass of $3.7 \cdot 10^{11} \, M_\odot$, a radius of 50 kpc, and an angular momentum of about $2 \cdot 10^7 \, M_\odot \, pc^2 \, Myr^{-1}$. An according spin parameter of $\lambda = 0.05$ can be derived. The contribution of a DM component has not been taken into account in the model presented because its gravitative contribution seems to be small up to distance scales of less than 20–30 kpc. Additional models with DM halos (Samland & Hensler 1995) show only a shortening of the first collapse phase but due to the above-mentioned self-regulation mechanisms almost the same subsequent evolution.

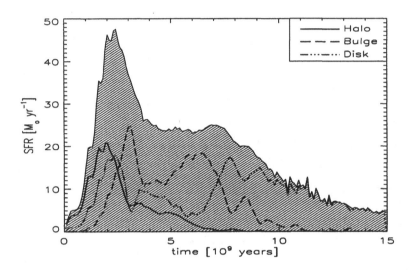

FIGURE 2. The total star-formation rates of the different galactic regions as a function of time.

In the following we wish to summarise only the global features of the most adequate model and refer the interested reader to more comprehensive papers or those concerning particular galaxy components. The system evolves differently from simple hydrodynamical models. The cool phase (CM), consisting of clouds with an initial mass fraction of 99% but a volume filling factor of only 20% and a temperature of 8000 K, starts to dissipate energy due to cloud collisions so that the initial equilibrium state is left and the collapse starts. While the purely hydrodynamical single gas phase models allows the galaxy to collapse rapidly within 1–2 free-fall times τ_{ff} ($\tau_{\mathrm{ff}} \approx 2.5 \cdot 10^8$ yr) the detailed consideration of heating and cooling of different gas phases in the chemo-dynamical treatment decelerates the violent collapse and extends the phase to approximately $2 \cdot 10^9$ yr. The ICM starts with 10^6 K, 1% mass fraction, and cools very slowly. While the SFR in the halo starts first but is small, collapsing gas concentrates towards the equatorial plane. Thus, the SFR within a flat cylinder of 2 kpc height representing a disk is rising drastically within the first 2.5 billion years and approaches 20 $M_\odot \, yr^{-1}$, the same amount as the total halo but with a much smaller volume and half a billion years later.

Figure 2 demonstrates that the SFR in the halo is only a short episode. Because of the collapse the gas streams through each specific volume in the halo so that the gas density at first increases but then the gas vanishes again. The SFR reacts accordingly and drops to almost zero.

3.2. *The bulge*

The condensation of the gas which collapses because of its low angular momentum radially towards the central region called "bulge" rises the density there by more than two orders of magnitude, thus enhancing the central SFR with a slight delay of 1 Gyr with respect to the halo and 500 Myr after the disk. Although the increase of the SFR in the bulge produces SNII explosions of the first stars and enlarges the ICM metallicity, Z_{CM} persists at $\approx 10^{-3}\, Z_\odot$ for the next $4\tau_{ff}$. Not until a high ICM density allows the material to condense onto the clouds (Begelman & McKee 1990) does the metal enrichment of the CM set in and increase simultaneously with the enhancement of the SFR due to further infalling clouds. This leads to the apparent similarity of the MD to closed-box models with $Y_{eff} \approx 2\, Z_\odot$ and has been interpreted accordingly (Matteucci & Brocato 1990). A principle spheroidal component could be formed.

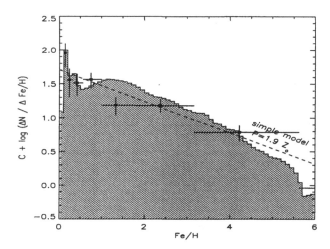

FIGURE 3. Differential distribution of [Fe/H] of bulge stars. The hatched area represents the model distribution, the crosses are data from Rich (1988), and the dashed line the effective yield for a closed-box model.

The sharp peak of the SFR is reached because the SNeII not only enlarge the velocity dispersion of the clouds, thereby heating up the CM, but also the hot supernova-expelled gas cannot be confined to the central region but expands from the centre. These processes diminish the further infall of CM, thus reducing the SFR. From this behaviour, we expect a metallicity distribution of the bulge stars with only an extremely small fraction of low-metallicity stars but, on the other hand, a steep increase of the star number at around solar metallicity in agreement with observations (Rich 1988). Figure 3 shows a comparison of the models with observations. A direct logarithmic comparison of our results with observational data (Rich 1988; McWilliam & Rich 1994), however, reveals clearly the absence of very low-metallicity stars in contradiction to strict closed-box MDs (see Samland & Hensler 1995a, 1995b).

3.3. *The disk*

The stars which are formed in the first star-formation peak will not remain in the disk plane because of their large peculiar velocity during the collapse phase. Therefore, the long-living stars will today not be sampled with the disk stars in particular. After this initial peak still infalling clouds settle towards the centre or due to their angular momentum vertically towards the equatorial plane, where hot metal-rich gas that stems from the preceding SFR-peak population pollutes the clouds with metals by means of condensation. The hot gas layer in the equatorial plane is supported by pressure rather than rotation. The initially low density of a cooler disk component provides only a low SFR and a small star count at low metallicities. The metal enrichment of the CM before the increase of the SFR after 7 Gyr leads to a steep increase of star counts at already higher metallicities than expected for a homogeneous evolution. The variations of the star counts with [Fe/H] as shown in Figure 4 can be attributed to variations of the SFR (Figure 2) in combination with different slopes of $[Fe/H]_{CM}(t)$ and can be directly compared with them. Therefore, only the stars born later than 6 Gyr after the onset of the galactic collapse will form the so-called disk population and are included in the G-dwarf sample (Pagel 1987, Pagel 1989).

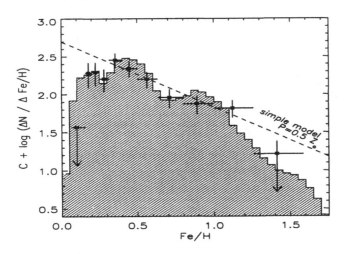

FIGURE 4. Differential distribution of [Fe/H] of solar-neighbourhood G dwarfs. The hatched area represents the model distribution, the crosses are data from Pagel (1989), and the dashed line the effective yield for a closed-box model.

4. Conclusions

Here we presented results of the chemo-dynamical models of galactic evolution which are dedicated to demonstrate the capability of this specific treatment. Problems arise if the stellar populations are considered in particular while their properties should be understood in a global evolutionary scenario. Simple models could achieve quite perfect agreement with observations under special (i.e. artificial) but not self-consistently prescribed physical parameters which have to be locally adapted, as mentioned in the Introduction, such as gas-infall rates, IMF variations, etc. In contrast, the chemo-dynamical parameterisations are fixed to the same most appropriate form everywhere in the whole galactic model. In addition, from the investigations of the effects of different processes with respect to their dependence on the parameterisation the main issue is their tendency for self-regulation. We can therefore state that the chemo-dynamical models provide the most reliable results to understand the evolution of galaxies, because they can reproduce the specific observational facts self-consistently within the global framework of the gas-dynamics plus mutual interactions between gas phase and between gas and stars.

Although the model presented is not intended to represent our Milky Way Galaxy perfectly, it can convincingly demonstrate the scenario of the formation of disk galaxies in general. One major issue of these studies has been the striking fact that the different galactic components, i.e. halo, bulge, and disk, do not evolve separately from each other. Since mass and energy exchanges between the different galactic regions vary temporarily during the galactic evolution, simple one-zone models fail to take these dynamical phases into account. From the results we would like to summarize that the MDs of bulge *and* disk stars are both well represented within one global model and even bearing the reasons of variations. We conclude that the temporal sequence of the formation of the galactic components is the following: the halo is formed after already 3 Gyr, followed by the bulge between 2–7 Gyr after the collapse commenced; and after 6 Gyr the formation is mainly and finally concentrated to the disk until now.

The authors are gratefully indebted to Andi Burkert, Joachim Köppen, Christian Theis and Jay Gallagher for stimulating and helpful discussions and large influences to our studies. M.S. gratefully acknowledges support by the DFG. The calculations were performed at the computer centers RZ in Kiel, HLZR in Jülich, and LRZ of the Bayerische Akademie der Wissenschaften in München.

REFERENCES

BAADE, W. 1946 *Publ. Astron. Soc. Pac.* **58**, 249

BEGELMAN, M. C. & MCKEE, C. F. 1990 *Astrophys. J.* **358**, 375

BURKERT, A. & HENSLER, G. 1987a *Mon. Not. R. Astron. Soc.* **225**, 21

BURKERT, A. & HENSLER, G. 1987b in *Nuclear Astrophysics* (ed. W. Hillebrand *et al.*), p. 159. Springer.

BURKERT, A. & HENSLER, G. 1988 *Astron. Astrophys.* **199**, 131

BURKERT, A., TRURAN, J. S. W. & HENSLER, G. 1992 *Astrophys. J.* **391**, 651

CARLBERG, R. G., DAWSON, P. C., HSU, T. & VANDEN BERG, D. A. 1985 *Astrophys. J.* **294**, 674

CHIOSI, C. 1980 *Astron. Astrophys.* **83**, 206

CLAYTON, D. 1986 *Publ. Astron. Soc. Pac.* **98**, 968

COWIE, L. L., MCKEE, C. F. & OSTRIKER, J. P. 1981 *Astrophys. J.* **247**, 908

COX, D. P. 1983 *Astrophys. J.* **265**, L61

EGGEN, O. J., LYNDEN-BELL, D. & SANDAGE, A. 1962 *Astrophys. J.* **136**, 748

FERRINI, F., MATTEUCCI, F., PARDI, C. & PENCO, U. 1992 *Astrophys. J.* **387**, 138

FIELD, G. B., GOLDSMITH, D. W. & HABING, H. J. 1965 *Astrophys. J., Lett.* **155**, L49

FROGEL, J. A. 1988 *Annu. Rev. Astron. Astrophys.* **26**, 51

HENSLER, G. 1987 *Mitt. Astron. Ges.* **70**, 141

HENSLER, G. & BURKERT, A. 1991 in *Chemical and Dynamical Evolution of Galaxies* (ed. F. Matteucci *et al.*), p. 168. Proc. Elba Workshop.

HENSLER, G., THEIS, C. & BURKERT, A. 1993 in *The Feedback of Chemical Evolution on the*

Stellar Content in Galaxies (ed. D. Alloin & G. Stasinska). Proceedings 3rd DAEC Meeting, p. 229.

HENSLER, G., THEIS, C. & GALLAGHER, J. S. 1995 *Astrophys. J.* , in preparation

KATZ, N. 1992 *Astrophys. J.* **391**, 502

KATZ, N. & GUNN, J. E. 1991 *Astrophys. J.* **377**, 365

KÖPPEN, J., THEIS, C. & HENSLER, G. 1995 *Astron. Astrophys.* , in press

LACEY, J. & FALL, S. M. 1983 *Mon. Not. R. Astron. Soc.* **204**, 791

LARSON, R. B. 1975 *Mon. Not. R. Astron. Soc.* **173**, 671

LARSON, R. B. 1976 *Mon. Not. R. Astron. Soc.* **176**, 31

LYNDEN-BELL, D. 1975 *Vistas Astron.* **19**, 229.

MALINIE, G., HARTMANN, D. H., CLAYTON, D. D. & MATHEWS, G. J. 1993 *Astrophys. J.* **413**, 633

MATTEUCCI, F. & BROCATO, E. 1990 *Astrophys. J.* **365**, 539

MATTEUCCI, F. & FRANCOIS, P. 1989 *Mon. Not. R. Astron. Soc.* **239**, 885

MCKEE, C. & OSTRIKER, J. P. 1977 *Astrophys. J.* **218**, 148

MCWILLIAM, A. & RICH, R. M. 1994 *Astrophys. J., Suppl. Ser.* **91**, 749

MEUSINGER, H. 1992 *Astron. Astrophys.* **266**, 190

MIHOS, J. C., RICHSTONE, D. O. & BOTHUN, G. D. 1991 *Astrophys. J.* **377**, 72

NISSEN, P. E., EDVARDSSON, B. & GUSTAFSSON, B. 1985 in *Production and Distribution of C,N,O Elements* (ed. I. J. Danziger *et al.*), p. 131. ESO Publ.

PAGEL, B. E. J. 1987 in *The Galaxy* (ed. G. Gilmore & B. Carswell), p. 341. Reidel.

PAGEL, B. E. J. 1989 in *Evolutionary Phenomena in Galaxies* (ed. J. E. Beckman & B. E. J. Pagel), p. 201. Cambridge University Press.

PAGEL, B. E. J. & PATCHETT, B. E. 1975 *Mon. Not. R. Astron. Soc.* **172**, 13

RICH, R. M. 1988 *Astron. J.* **95**, 828

SAMLAND, M. 1994 PhD thesis, University of Kiel.

SAMLAND, M. & HENSLER, G. 1995a in *Unsolved Problems of the Milky Way* (ed. L. Blitz). IAU Symposium 169, in press.

SAMLAND, M. & HENSLER, G. 1995b *Astron. Astrophys.* , submitted

SAMLAND, M., HENSLER, G. & THEIS, C. 1995 *Astrophys. J.* , submitted

SANDAGE, A. 1986a *Annu. Rev. Astron. Astrophys.* **24**, 421

SANDAGE, A. 1986b *Astron. Astrophys.* **161**, 89

SANDAGE, A. & FOUTS, G. 1987 *Astron. J.* **92**, 74

SCALO, J. M. & STRUCK-MARCELL, C. 1986 *Astrophys. J.* **301**, 77

SCHMIDT, M. 1963 *Astrophys. J.* **137**, 758

SILK, J. 1986 in *Spectral Evolution of Galaxies* (ed. C. Chiosi & A. Renzini), p. 3. Reidel.

STEINMETZ, M. & MÜLLER, E. 1994 *Mon. Not. R. Astron. Soc.* , submitted

TENDRUP, D. M. 1988 *Astron. J.* **96**, 884

THEIS, C., BURKERT, A. & HENSLER, G. 1992 *Astron. Astrophys.* **265**, 465

THEIS, C. & HENSLER, G. 1993 *Astron. Astrophys.* **280**, 85

TINSLEY, B. M. 1981 *Astrophys. J.* **250**, 758

TOTH, G. & OSTRIKER, J. P. 1992 *Astrophys. J.* **389**, 5

TRURAN, J. S. W. & CAMERON, A. G. 1971 *Astrophys. Space Sci.* **14**, 179

TWAROG, B. A. 1980 *Astrophys. J.* **242**, 242

VAN DEN BERGH, S. 1962 *Astron. J.* **67**, 486

YOSHII, Y. & SAIO, H. 1979 *Publ. Astron. Soc. Jpn.* **31**, 339

3D N-body Simulations of the Milky Way

By ROGER FUX, L. MARTINET AND D. PFENNIGER

Geneva Observatory, ch. des Maillettes 51, CH–1290 Sauverny, Switzerland

1. Introduction

While the existence of a central bar in our Galaxy now seems to be well established, its parameters—position angle, extension, axis ratio, angular speed, etc.—remain controversial. The large amount of photometrical and stellar kinematical data now becoming available within $\sim 30°$ of the galactic centre should provide potentially new constraints on these parameters. Unfortunately, a detailed barred model of the Milky Way, which would offer a powerful framework for interpreting such observations, does not yet exist. In a parallel paper (Fux *et al.* 1995, hereafter FMP) we have reported on our first attempt at constructing a 3D, dynamically self-consistent, so far dissipationless, barred model of the Galaxy by following the time evolution of a set of 400 000 particles, initially distributed according to a plausible axisymmetric mass model with isotropic Maxwellian velocities. This simulation, which we will refer to here as our "standard" run, spontaneously forms a bar with a semi-major axis of 5 kpc after about 1 Gyr (see Figure 1).

The reader should refer to FMP to complete this non-exhaustive presentation.

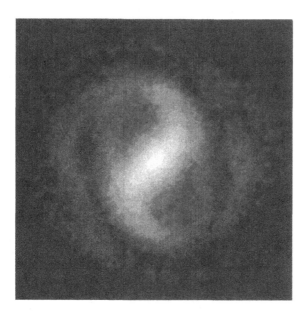

FIGURE 1. Face-on view of the visible particles in the standard run at $t = 1.37$ Gyr. The full size of the box is 24 kpc.

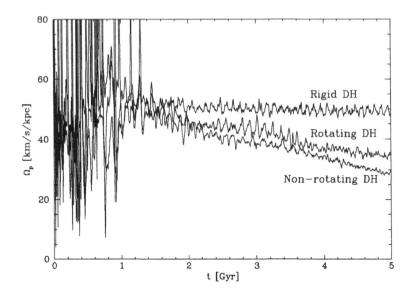

FIGURE 2. Time behaviour of the bar angular speed as a function of the dynamical properties of the dark halo (DH). The rotating DH curve refers to our standard run.

2. Pattern speed and resonances

The imposed isotropic velocity dispersion in the initial conditions of the standard run automatically leads to a *rotating* dark halo (DH), with a mean azimuthal velocity of about 100 km s^{-1} at the solar galactocentric distance R_o. We have also carried out two further runs with identical initial mass distributions but dynamically different DHs: the first with a live, but initially *non-rotating*, DH (the details of velocity attribution in this more complicated case are not given here), and the second with a *rigid* DH. All these runs develop quite similar bars, except for the time evolution of their pattern speed Ω_p, given in Figure 2. In the rigid DH case, besides fluctuations over short timescales, Ω_p remains almost constant during the whole run, betraying the absence of angular momentum transfer between the visible components sustaining the bar and the DH. In the live DH cases, the bar–DH interaction causes the bar to slow down at a rate $|d\Omega_p/dt| \approx 5$ km s^{-1}kpc^{-1}Gyr^{-1}. Furthermore, the torque exerted by the initially non-rotating DH on the bar is obviously somewhat stronger than for the rotating one.

In the standard run, the corotation before 2 Gyr lies between 4 and 5 kpc, the ILR slightly above 1 kpc and the OLR very near the Sun's position. Clearly, as the bar slows down, these resonances are moving outwards.

3. Disc kinematics on the solar circle

We have computed several kinematical quantities in the standard run all around the solar circle in order to find some steady correlation with the position angle α relative to the bar (see Figure 3a). After a first investigation, a fair sin-like behaviour has emerged for the mean radial velocity $\overline{U}(\alpha)$ of the disc particles, as shown in Figure 3b.

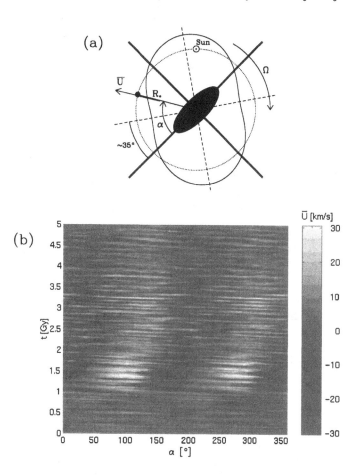

FIGURE 3. a) Definition of the angle α and mean radial velocity $\overline{U}(\alpha)$ of the disc particles on the solar circle, averaged over $t > 2$ Gyr. The dotted line represents the solar circle and the departure of the solid line from this line is proportional to $\overline{U}$, with the convention that $\overline{U} > 0$ toward the anti-centre. The dashed lines give the angular positions of maximum and minimum $\overline{U}$. A plausible location of the Sun is also indicated. b) Time behaviour of $\overline{U}(\alpha)$ for the whole standard run.

From 1 to 2 Gyr, corresponding to the period of strong spiral arms described in FMP, the phase of this sin moves progressively to higher values of α, and then settles into a more or less constant value, with the first minimum of $\overline{U}$ occurring at $\alpha \approx 35°$. The amplitude of the sin then approaches 10 km s^{-1} on the average, with fluctuations of the same order of magnitude over kinematical timescales. If we put the Sun in this final picture in a way such that the near end of the bar lies in the first galactic quadrant, as is now commonly accepted for the Milky Way bar, we would expect an outward streaming motion of the disc stars in the solar neighbourhood.

4. Other results

• The line-of-sight velocity dispersion that a fictive observer moving along the solar circle would measure in the direction of Baade's Window at the point closest to the galactic centre depends within 30% of its angular position relative to the bar, ranging from 110 km s^{-1} ($\alpha = 90°$) to 160 km s^{-1} ($\alpha = 0$) at $t = 1.2$ Gyr.

• The bar pumps disk stars into a peanut-shaped bar/bulge on a timescale of 1 Gyr. The resulting "isophotes" seen from the solar circle a few gigayears after the bar formation qualitatively resemble *COBE's* image whenever the near end of the bar lies at positive galactic longitude.

REFERENCES

FUX, R., MARTINET, L. & PFENNIGER, D. 1995 in *Unsolved Problems of the Milky Way* (ed. L. Blitz). IAU Symposium 169, in press. (FMP)

Elmegreen: Is the local expansion that you find the result of a spiral arm in our neighbourhood?

Fux: The expansion at $\alpha \sim 135°$ obtained when averaging over time is probably connected to the bar itself. However, the fluctuations of $\overline{U}(\alpha)$ over kinematical timescales may result from the transient and regenerative nature of the spiral arms, like their rapid phase evolution relative to the bar as they wound before disappearing.

Carney: Paul Schechter has been obtaining radial velocities for carbon stars in the outer two quadrants of the Galaxy, and I believe he does not see any signs of bar-induced "expansion". He may even have a negative gradient. Is there evidence supporting the expansion?

Fux: There is the paper by Blitz & Spergel (1991, *ApJ*, **370**, 205) where such an expansion seems to make sense. I agree that this apparent contradiction deserves some reflection.

Serrano: What is the difference between a bar and a bulge ?

Fux: The distinction between bars and bulges and the precise dynamical connection between them are not yet clear. Both are triaxial and rotate fast. Could triaxial bulges be weak bars? In other words, is the difference simply a question of axis ratio and of evolution? Moreover, quasi-prolate bars seen end-on can appear as quasi-spherical bulges. In numerical experiments, e.g. Combes *et al.* (1990, *A&A*, **233**, 82) and this paper, box- or peanut-shaped bulges result from resonance effects in the disk produced by a bar.

Galactic Bars and Stable Perfect Elliptic Disks

By STEPHEN LEVINE

Instituto de Astronomía, UNAM, Ensenada, B. C. México
P.O. Box 439027, San Diego, CA 92143, U.S.A.

To understand how a bar can influence the evolution of our Galaxy, we need to know what kinds of bars can survive for long periods of time. Particulate representations of perfect elliptic disks with maximum angular momentum are constructed for N–body stability analysis using a marching scheme to compute the distribution function weights for a selected set of orbits. A quiet start technique was used to populate each orbit. As seen previously, those elliptic disks with moderate ellipticity were stable, and the roundest ones developed spiral instabilities.

1. Introduction

Observations of other spiral galaxies show that a significant fraction of them are barred (de Vaucouleurs 1963), implying that bars are a long-lived phenomenon. In light of recent work showing that our Galaxy may have a central bar (Blitz & Spergel 1991; Blitz *et al.* 1993), and the fact that a bar can have a substantial effect upon the type and distribution of orbits present in the Galaxy (Teuben & Sanders 1985), we would like to know what kinds of bars are able to survive for long periods of time.

N-body simulations have already shown that barf-orming instabilities are quite common and typically strong (e.g. Ostriker & Peebles 1973). It is, however, difficult to say much about the overall properties of the orbits in such bars, whereas with analytic and semi-analytic bar models such as Freeman's bars (Freeman 1966), and the perfect elliptic disks, we can easily compute many of these properties of interest. In this contribution we set out to test the global stability of a non-rotating bar model, the perfect elliptic disks, which are the two-dimensional limiting case of the perfect ellipsoids (de Zeeuw 1985). The surface density of the perfect elliptic disks is given by

$$\Sigma(x,y) = \frac{\Sigma_0}{(1+m^2)^{3/2}}, \qquad m^2 \equiv \frac{x^2}{a^2} + \frac{y^2}{b^2},$$

which is non-singular in the centre, and shows that the density is stratified on concentric, similar ellipses. The perfect ellipsoids are, in fact, integrable potentials, where the equations of motion are separable, implying that we can find isolating integrals, or bounding surfaces, that uniquely specify each orbit (de Zeeuw 1985).

To test for global stability, a discrete, self-consistent model is constructed, and then loaded into an N-body integrator and allowed to evolve. This problem has already been tackled by Levine & Spark (1994, LS), using the method of simulated annealing and a quiet-start technique to create discrete representations of perfect elliptic disks. An alternative method for solving the self-consistent problem uses a marching scheme described by de Zeeuw *et al.* (1987) (ZHS), which we couple with a modification of the quiet-start technique developed by LS. The initial results from this are qualitatively the same as those shown in LS. There it was seen that the most elliptical perfect elliptic disks were subject to bending instabilities and the roundest to spiral instabilities, while those in between were stable. Each of these methods for constructing self-consistent models has different strengths, and will probably be used in different situations, so it is reassuring to see that they give consistent results.

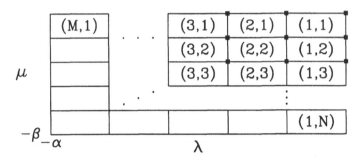

FIGURE 1. The ordering of the grid cells, and their respective orbits in the elliptical space in which the equations of motion separate. Nine of the outermost cells and orbits are shown explicitly, with the orbit turning points labelled by filled squares. The cells are labelled like a matrix to emphasise that the marching scheme can proceed in either direction.

2. Constructing a self-consistent model

When constructing a self-consistent model of a given potential, we are faced with trying to find a set of orbits and a distribution weighting function that together reproduce the overall density distribution. There are two general approaches to the problem: we can either select a set of orbits, and then compute the distribution function weights for each orbit; or we can choose the weights, and then try to find a compatible set of orbits.

In the first method, the problem is substantially better constrained if there are n orbits already chosen and we need only find n weights. In the second case, we have n weights, and are trying to find at least n orbit specifiers ($2n$ in the case of our two-dimensional, integrable potential). The first approach has been used (Schwarzschild 1979; Statler 1987; Teuben 1987; ZHS) to demonstrate that solutions to the self-consistent problem do exist. If we wish to construct a discrete representation of a potential for an N-body simulation, then the second method has the advantage of allowing us to insist that each orbit should contain an integral number of particles. LS showed that such problems could be attacked successfully using the second approach, and here we show that the first method can be adapted for this as well.

Because the density distribution of the perfect elliptic disks is not axisymmetric, the orbits in this potential divide into two families: "loop" orbits, which circulate about the centre, and "box" orbits, which oscillate in x and y and resemble Lissajous figures. In the maximum streaming case, the loop-orbit population is easily and uniquely determined (ZHS; LS), leaving only the problem of choosing the box orbits.

First, the area of the disk is divided up into a grid in the coordinates (λ, μ) in which the equations of motion separate. The upper right corner of each cell is designated the turning points of a box orbit (Figure 1). In each cell we compute the integrated surface density minus the sum of the integrated orbital surface densities of all the loop orbits. For the outermost box orbit and grid cell, the weighting function is then equal to the cell's integrated surface density divided by the non-normalised integrated orbital surface density of the box orbit paired with that cell. The now normalised contribution of this orbit is then subtracted from each of the inner cells, and the process repeated for the next outermost orbit and cell, working inwards until weights have been computed for all the orbits. ZHS have previously shown that positive definite solutions do exist.

The next step is to place particles upon each orbit in a quiet manner, so as to minimise random noise in the initial conditions, and make it easier to watch for the growth of instabilities. As is described in more detail in LS, the basic idea of this quiet-start technique is to place particles in a grid like manner on the torus in action–angle space corresponding to each orbit—thus distributing the particles evenly in the phase space of the orbit—and then construct a map from that space to position and velocity coordinates. In the work of LS, each orbit had the same weight, and hence the same number of particles.

Now that each orbit can have a different weight, it is necessary to modify the method slightly. For each orbit, the integrated normalised density determines how many particles should be placed on the orbit. This number is rounded to the nearest integer, and an attempt is made to factor this into two roughly equal divisors. The second nearest integer is also factored, in case the closest integer is prime, or has two very different sized divisors, and, if the number is greater than 10, the more nearly equal pair of divisors is taken to indicate the number of grid lines in each of the two dimensions. If the number is less than or equal to 10, then the factors of the nearest integer are used, even if the integer is itself prime. We have found that this method leads to an overall difference between the desired number of particles and the computed number of less than 0.08% for models with 50 000 particles, and a produces a good approximation of the potential.

3. Discussion

Using the marching scheme, we constructed models for the maximum streaming perfect elliptic disks with axis ratios of 0.910, 0.640 and 0.305 for direct comparison with the results in LS, since each demonstrated a different kind of behaviour in the earlier work. The models were constructed using a grid of 900 cells and box orbits, with grid lines spaced at equal square root intervals. The most circular model ($b/a = 0.910$) was violently unstable and quickly developed spiral arms. The model with axis ratio 0.305 was stable, and the model with $b/a = 0.640$ was marginally stable—an instability appears, but grows slowly. These results are shown in Figure 2 and mirror those found by LS.

The ZHS method for determining the weighting function for a given set of box orbits depends upon our being able to choose the set of loop orbits first and independently. It has already been shown (ZHS) that we can do this in the case of maximum streaming. The question then arises as to whether or not we can apply this method to non-maximum streaming cases, where the selection of loop orbits is not so clear. In fact, we can, since the linear density along the minor axis—outside of the foci of the elliptic coordinates in which the equations of motion separate—is comprised solely of loop orbits, so we have a constraint that depends only on, and restricts only the selection of loop orbits.

Thanks to L. S. Sparke, P. T. de Zeeuw and J. A. Sellwood.

REFERENCES

BLITZ, L., BINNEY, J., LO, K. Y., BALLY, J. & HO, P. 1993 *Nature* **361**, 6411.

BLITZ, L. & SPERGEL, D. 1991 *Astrophys. J.* **379**, 631.

DE VAUCOULEURS, G. 1963 *Astrophys. J., Suppl. Ser.* **8**, 31.

DE ZEEUW, P. T. 1985 *Mon. Not. R. Astron. Soc.* **216**, 273.

DE ZEEUW, P. T., HUNTER, C. & SCHWARZSCHILD, M. 1987 *Astrophys. J.* **317**, 607. (ZHS)

FREEMAN, K. C. 1966 *Mon. Not. R. Astron. Soc.* **133**, 47.

LEVINE, S. E. & SPARKE, L. S. 1994 *Astrophys. J.* **428**, 493. (LS)

OSTRIKER, J. P. & PEEBLES, P. J. E. 1973 *Astrophys. J.* **186**, 467.

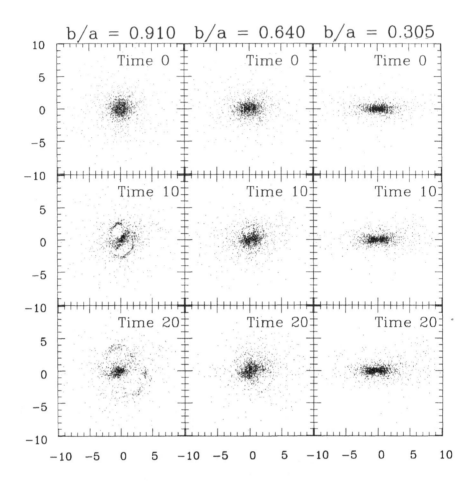

FIGURE 2. Positions of 2500 of the 50 000 particles are shown at times 0, 5, 10, 15 and 20 for axis ratios 0.910 (*left*), 0.640 (*centre*) and 0.305 (*right*)

SCHWARZSCHILD, M. 1979 *Astrophys. J.* **232**, 236.

SELLWOOD, J. A. 1981 *Astron. Astrophys.* **99**, 362.

STATLER, T. 1987 *Astrophys. J.* **321**, 113.

TEUBEN, P. J. 1987 *Mon. Not. R. Astron. Soc.* **227**, 815.

TEUBEN, P. J. & SANDERS, R. H. 1985 *Mon. Not. R. Astron. Soc.* **212**, 257.

Chaboyer: How long did your integrations last for the "stable" case?

Levine: The integrations were run for 3–6 half-mass crossing times, and in one of the "stable" cases extended to 15–20 crossing times.

Shore: What happens if you "kick" it—embed it in a halo?

Levine: Provided the "kick" is not too extreme, I'd expect the results to be qualitatively similar to what we have already seen. In general, embedding these disks in a halo would likely help stabilise them, although in that case we would no longer have the separable potential that we set out to test.

The Age of the Galactic Inner Halo and Bulge

By LAURA K. FULLTON AND BRUCE W. CARNEY

Department of Physics & Astronomy, University of North Carolina, Chapel Hill, NC
27599-3255, U.S.A.

New colour-magnitude diagrams (CMDs) of three globular clusters near the galactic centre are used to infer the age of the inner regions of the Galaxy relative to the more well-studied outer halo. The age of the galactic bulge relative to these clusters is determined in a *self-consistent* way using the *Hubble Space Telescope* CMD of Baade's Window (Hotzman *et al.* 1993) and the distance to the galactic centre found by Carney *et al.* (in preparation).

1. Introduction

One of the important outstanding questions concerning the formation of the Galaxy is the relative age and age dispersion of the stars and clusters in its central regions. Deriving ages from observations of field stars is inherently difficult due to the variations in metallicity, distance, reddening and perhaps even age itself, which are invariably present in any field sample. For these reasons, the most reliable ages can be derived for objects in which these spreads are minimised, namely star clusters. The galactic globular-cluster system contains some of the oldest identifiable objects in the Galaxy, and a synthesis of many careful studies of outer-halo globular-cluster colour-magnitude diagrams (CMDs) has led to the general conclusion that these clusters have a mean age of $\sim$ 15 Gyr with dispersion of $\sim$ 3–5 Gyr (cf. Sarajedini & Demarque 1990; van den Berg *et al.* 1990; Carney *et al.* 1992, CSJ). In part because they are more challenging observationally, the globular clusters near the galactic centre have been less well studied. Yet it is precisely these clusters which can help answer the important questions regarding the formation chronology of the galactic inner halo and disk: Are the inner regions of the Galaxy younger, older or the same age as the outer halo? What are the enrichment timescales of the galactic halo, disk, thick disk and bulge? To begin to answer these questions, we have undertaken a study of several globular clusters which lie within 4–5 kpc of the galactic centre. Here we present results for three of the clusters in our sample: NGC 6723, NGC 6352 and NGC 5927.

2. The sample

The clusters in our sample, NGC 6723, NGC 6352 and NGC 5927, represent a range of metallicities ([Fe/H] = -1.09, -0.60, -0.30, respectively, Zinn 1985) and distances from the galactic centre (R_{GC} = 2.4, 3.9, 4.7 kpc). We determined their ages based on the magnitude differences between their main-sequence turn-offs (MSTOs) and horizontal branches (HBs). The magnitude differences, ΔV_{HB}^{TO}, were determined from our CMDs for these clusters. Details concerning the photometry will be presented elsewhere (Fullton, *et al.*, in preparation; Fullton & Carney, in preparation).

3. Method of age determination

In deriving cluster ages, we followed the procedure of CSJ. We determined the magnitude of the zero-age horizontal branch (ZAHB) for our clusters from the CMDs. To

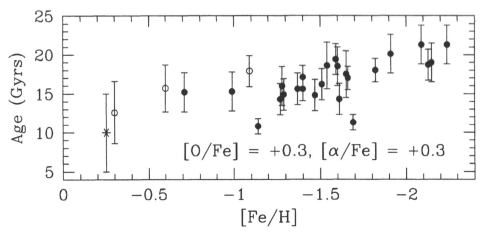

FIGURE 1. Age-metallicity relationship for globular clusters (circles) and the bulge (asterisk). Closed symbols are from CSJ; open symbols are from the present work.

determine the cluster distances, we used the relation

$$< M_V(RR) >= 0.15[Fe/H] + 1.01 \qquad (3.1)$$

from CSJ. Because this equation properly refers to the mean HB level and not the ZAHB, we applied the correction given by CSJ to convert the ZAHB magnitudes into mean HB magnitudes. With the cluster distances, we computed the absolute magnitude of the turn-off from ΔV_{HB}^{TO}. Following Carney (1983), we adopted a bolometric correction of -0.21. For isochrones, we used the Revised Yale Isochrones (Green, *et al.* 1987). We chose Y = 0.23 and computed effective heavy-element metallicities using the values $[\alpha/Fe] = +0.3$ and $[O/Fe] = +0.3$ to allow for the fact that the isochrones assume scaled solar mixtures, while we know (e.g. Brown *et al.* 1990; Sneden *et al.* 1994) that the globulars generally have enhanced $[O/Fe]$. Salaris *et al.* (1993) have shown that the use of such effective metallicities is a good approximation. Armed with the bolometric MSTO magnitudes, we computed the ages by interpolation in the grid of isochrones.

4. Cluster Ages

We find that 47 Tuc, NGC 6352 and NGC 6723 are coeval to within the errors, while NGC 5927 appears younger, although its age is more uncertain due to field-star contamination from the galactic bulge present in the ground-based cluster CMD. As part of a larger collaboration, we have recently obtained *HST* WFPC2 images of NGC 5927. Our preliminary CMD from these images is consistent with ΔV_{HB}^{TO} determined from the ground-based CMD, but final confirmation of our age for this cluster awaits completion of photometry of the new *HST* data.

Figure 1 is a reproduction of Figure 20 of CSJ with the three clusters in our sample added (open symbols). The figure shows that the age–metallicity relationship found by CSJ appears to continue into the inner 5 kpc of the Galaxy. This result does not contradict the result of Chaboyer *et al.* shown earlier at this meeting, since their sample did not include more than one cluster as close to the galactic centre as our three. The ages derived for the most metal-poor clusters ([Fe/H] $\sim$ -2.0) are probably 2–3 Gyr too old because the RR Lyraes upon which their distances are based may be evolved (Storm *et al.* 1994). The absolute-age scale is set by the zero point of Equation 3.1 which has been determined by statistical parallaxes of field RR Lyraes (Layden *et al.* 1994). This zero point is inconsistent with that determined by the absolute magnitude-*vs.*-period relationship for the galactic Cepheids. Adoption of the Cepheid zero point would revise the ages of *all* the clusters in the figure downwards by 4–5 Gyr (Walker 1992). However, the *relative* ages and thus the slope in the figure are independent of the zero point adopted. The relative ages depend solely on the slope of Equation 3.1 and the assumption that [O/Fe] and [α/Fe] = +0.3 independent of metallicity (CSJ).

5. Age of the bulge

Lee (1992) has claimed that the bulge is the Galaxy's oldest component based on the mean metallicity of the RR Lyraes in Baade's Window (BW) and his models of the horizontal branch. In this respect, NGC 6723 is an important test cluster for Lee's hypothesis. Its metallicity is comparable to the mean metallicity derived by Walker & Terndrup ([Fe/H] = -1.0; 1991) for the BW RR Lyraes, and it lies $\sim$ 2.4 kpc directly below the galactic centre in a region of low reddening. The large age we derived for NGC 6723 lends support to Lee's claim that the RR Lyraes in BW may be as old as some of the oldest globulars. How does this age compare with that of the bulk of the bulge population?

The magnitude of the bulge MSTO can be measured from the BW CMD of Holtzman *et al.* (1993), adopting their estimated extinction due to reddening for the *HST* F555W filter. To convert this into an absolute magnitude, we need to know the distance to the Baade's Window field. Based on K-band photometry of 58 RR Lyraes in BW, Carney *et al.* (in preparation) have determined the bulge distance using the M_K *vs.* log P relation for the RR Lyraes with the same zero point adopted in Equation 3.1. Using this distance, the helium abundance reported by Minniti elsewhere in this volume—Y = 0.28—and the mean metallicity of bulge K-giants found by McWilliam & Rich ([Fe/H] = $-$ 0.25 ([O/Fe] = 0.0, 1993), the mean bulge age was determined by interpolating in the same grid of isochrones used for the clusters. The derived age, 10 $\pm$ 5 Gyr, is plotted in Figure 1 as an asterisk. It should be stressed that as far as possible this age is on a scale which is fully consistent with that of the clusters.

6. Conclusions

• The age–metallicity relation found by CSJ appears to continue into the inner halo of the Galaxy. On average, the more metal-rich clusters appear to have formed later than the metal-poor ones.

• In Figure 1, the bulge lies on the continuation of the age-metallicity relationship defined by the globulars. This suggests that star formation and chemical evolution did not proceed especially rapidly at the galactic centre despite the higher density of gas there.

• The age we have derived for the bulge is some 8 Gyr younger than that derived for NGC 6723, our template cluster for the BW RR Lyraes. It should be kept in mind that although the error bars are large, they do not overlap. At the maximum age for the bulge and the minimum age for NG 6723 the ages differ by 1 Gyr. If NGC 6723 and the RR Lyraes represent the first stages of bulge formation, then taking the ages at face value implies a relatively long timescale for bulge formation, in agreement with our interpretation of Figure 1.

The work presented here is based on preliminary results from my dissertation at the University of North Carolina. I would like to thank my advisor, Bruce Carney, for his patience, advice and support, and for sharing his results on the Baade's Window RR Lyraes in advance of publication.

REFERENCES

BROWN, J. A., WALLERSTEIN, G. & OKE, J. B. 1990 *Astron. J.* **100**, 1561.

BUONANNO, R., CORSI, C. E. & FUSI PECCI, F. 1989 *Astron. Astrophys.* **216**, 80.

CARNEY, B. W. 1983 *Astron. J.* **88**, 623.

CARNEY, B. W., STORM, J. & JONES, R. V. 1992 *Astrophys. J.* **386**, 663. (CSJ)

GREEN, E. M., DEMARQUE, P. & KING, C. R. 1987 *The Revised Yale Isochrones*. Yale University Observatory.

HOLTZMAN, J. A., *et al.* 1993 *Astron. J.* **106**, 1826.

LAYDEN, A. C., HANSON, R. B. & HAWLEY, S. L. 1994 *Bull. Am. Astron. Soc.* **26**, 911.

LEE, Y.-W. 1992 *Astron. J.* **104**, 1780.

McWILLIAM, A. & RICH, M. R. 1994 *Astrophys. J., Suppl. Ser.* **91**, 749.

SALARIS, M., CHIEFFI, A. & STRANIERO, O. 1993 *Astrophys. J.* **414**, 580.

SARAJEDINI, A. & DEMARQUE, P. 1990 *Astrophys. J.* **365**, 219.

SNEDEN, C., KRAFT, R. P., LANGER, G. E., PROSSER, C. F. & SHETRONE, M. D. 1994 *Astron. J.* **107**, 1773.

STORM, J., CARNEY, B. W. & LATHAM, D. W. 1994 *Astron. Astrophys.* **290**, 443.

VANDENBERG, D. A., BOLTE, M. & STETSON, P. B. 1990 *Astron. J.* **100**, 445.

WALKER, A. R. 1992 *Astrophys. J., Lett.* **390**, L81.

WALKER, A. R. & TERNDRUP, D. M. 1991 *Astrophys. J.* **378**, 119.

ZINN, R. 1985 *Astrophys. J.* **293**, 424.

Shore: How did you choose the [O/Fe] value?

Fullton: I chose that value based on cluster measurements in the literature. Many clusters do not have measured oxygen abundances and many of those that do show a range of [O/Fe] ratios between stars *within* the cluster (Sneden *et al.* 1994). If this spread is interpreted as being due to mixing, and we expect the measured [O/Fe] ratios to go down, not up, as a result, then the upper limit seen for the clusters of [O/Fe] $\sim +0.3$ dex seems reasonable for the primordial ratio for all the clusters. I didn't have time to say that I have obtained high-resolution echelle spectra of three stars in NGC 6723 and one in NGC 5927 from which I am in the process of determining individual elemental abundances, including oxygen.

Shore: It's actually an observational question, because in computing isochrones the CNO opacities are very important in this temperature range, at the TO point.

Buonanno: Did you check the ratio R $=$ N(HB)/N(RGB) for the clusters you presented here?

Fullton: Not yet, but I intend to if possible. The upper portions of some of our CMDs are not well populated, especially in the case of NGC 5927, so we may not get good statistics. I should mention that I assumed $Y = 0.23$ because it is consistent with big-bang production scenarios and with observational evidence. If the helium fraction of the metal-rich clusters I have studied is larger than for the other clusters, then that would reduce their ages slightly, steepening the gradient shown in Figure 1.

The Formation of the Galactic Bulge: Clues from Metal-Rich Globular Clusters

By DANTE MINNITI

European Southern Obs., Karl-Schwarzschild-Str. 2, D–85748 Garching bei München, Germany

We compare the kinematics, spatial distribution, metallicities and ages of the globular clusters with the different components of the Milky Way: disk, bulge and halo. It is concluded that the metal-rich globular clusters with $R \leq 3$ kpc are associated with the galactic bulge rather than the thick disk. We discuss the implications for the formation of the inner Milky Way.

1. Introduction

1.1. *The galactic globular cluster system*

There are only ~150 globulars in the galaxy. This limits the statistics in studies of the globular cluster system.

Frenk & White (1980, 1982) analysed a sample of about 60 clusters with high quality data at the time to study the kinematics and dynamics of the globular cluster system. They found the metal-rich clusters to be very concentrated towards the galactic centre: all clusters with [Fe/H] ≥ -0.8 lie within the solar circle. Indeed, looking at the spatial distribution, it is tempting to conclude that the metal-rich globulars are associated with the bulge.

Zinn (1985) found a bimodal metallicity distribution in a homogeneous sample of about 110 clusters, with peaks at [Fe/H] ~ -1.5 and -0.5. From the spatial distribution and kinematics, he concluded that the metal-rich clusters ([Fe/H] ≥ -0.8) form a disk system, with the rest belonging to a halo system. This division between halo and disk was supported by Armandroff (1989), who associated the disk clusters with the thick disk. No clusters were associated with the bulge at that time, because the bulge was thought to be very metal-rich (mean [Fe/H] $= +0.3$, after Rich 1988), and have a very large velocity dispersion ($\sigma = 113$ km s^{-1}, after Mould 1983). Also, very few clusters within 2 kpc of the galactic centre had been studied.

1.2. *The galactic bulge*

In the last few years, our ideas about the bulge structure, kinematics and metallicity have changed. In particular, McWilliam & Rich (1994) revised the mean metallicity of the bulge downwards by 0.55 dex. The mean metallicity of Baade's window (at $R = 0.5$ kpc) is now [Fe/H] $= -0.25$. Minniti (1994) obtained spectra for about 700 bulge giants in three different windows, deriving accurate radial velocities and metallicities. The lower abundance scale is confirmed by these data: we find a mean [Fe/H] $= -0.5$ at $R = 1.5$ kpc. Also, Minniti (1994) finds that the bulge velocity dispersion drops from $\sigma \approx 110$ km s^{-1} at $R = 0.5$ kpc to $\sigma \approx 60$ km s^{-1} at $R = 2.0$ kpc, and that the bulge is consistent with an oblate isotropic rotator with rotation velocity $V \approx 110$ km s^{-1}. In the light of these new data, we can re-examine the comparison in kinematics, spatial distribution, ages and metallicity of the bulge and disk with that of the metal-rich globulars.

255

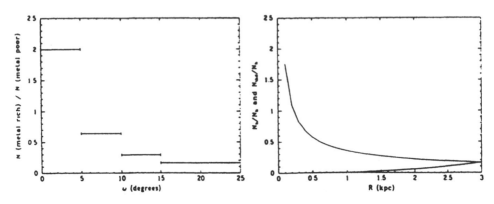

FIGURE 1. (*a*) Observed relative numbers of metal-rich to metal-poor clusters *vs.* galactocentric distance; *b*) Expected relative numbers predicted by our simple galactic model. The upper and lower curves represent N_{bulge}/N_{halo}, and $N_{thick\ disk}/N_{halo}$, respectively.

2. Evidence for a bulge population of metal-rich globulars

2.1. *Spatial distribution*

When looking at the spatial distribution of metal-rich globulars and the bulge, a direct comparison of the scale-lengths and scale-heights is not constructive, since clusters were preferentially destroyed in the inner galaxy (Aguilar et al. 1988). Instead, what one can do is to compare the relative numbers of metal-rich to metal-poor clusters as function of distance. This is plotted in Figure 1a, using data of Zinn (1990). The ratio $N_{metal\ rich}/N_{metal\ poor}$ increases steeply towards the galactic centre, as expected if the metal-rich clusters follow the bulge rather than the thick disk. Let us check this by looking at a very simple galactic model, with the following density laws:

• Halo: $\rho_h \sim r^{-n}$, with $n = 3.0 - 3.5$, from halo globulars, RR Lyraes and field BHB stars (Zinn 1985; Saha 1985; Preston *et al.* 1991).
• Bulge: $\rho_b \sim r^{-m}$, with $m = 3.65 - 4.2$ from K and M giants, *IRAS* sources, and the integrated K light (Terndrup 1988; Frogel 1988).
• Thick disk: $\rho \sim e^{-h\ /r}e^{-h\ /Z}$, with $h_r = 3.5$ kpc, $h_Z = 1.0$ kpc, from star counts (Ojha *et al.* 1994).

The results of the expected N_{bulge}/N_{halo} and $N_{thick\ disk}/N_{halo}$ are shown in Figure 1b, normalised to $N_i/N_{halo} = 0.16$ at $R = 3$ kpc. Since the relative increase of the halo power law is always faster than an exponential law towards $R = 0$, the thick disk underpredicts the number of globulars in the inner 3 kpc, and overpredicts the number of outer globulars. On the other hand, the functional form of the ratio agrees with the hypothesis of the clusters belonging to the bulge.

2.2. *Kinematics*

Armandroff (1989) measured the rotation and velocity dispersion for metal-rich globulars to be $V_{rot} = 177 \pm 25$ and $\sigma = 58 \pm 11$, which are different than the bulge values (Section 1). The metal-rich globulars seem to rotate faster than the bulge, but this rotation signal is dominated by the clusters *outside* of 3 kpc. Furthermore, Armandroff (1989) followed the constant rotation solution method developed by Frenk & White (1980), which may not apply to the bulge: the available data are more consistent with the model of Kent (1992).

Fortunately, the line-of-sight velocity dispersion of the disk globulars is not affected by

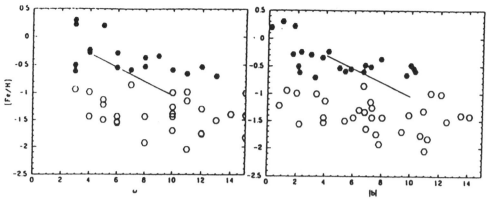

FIGURE 2. (a) Globular cluster metallicity *vs* angular distance from the galactic centre; (b) Globular cluster metallicity *vs* galactic latitude. Filled circles indicate clusters with [Fe/H] ≥ −0.8. The mean bulge metallicity is plotted as a solid line from the data of Terndrup (1988), rescaled according to McWilliam & Rich (1994).

the rotation solution. It is then legitimate to compare directly the velocity dispersions of the different systems. For the metal rich systems with $\omega \leq 15$ degrees (R ≤ 2.1 kpc):

  * Metal-rich globulars: $\sigma = 77 \pm 14$ km s^{-1}
  * Metal-rich bulge giants: $\sigma = 72 \pm 4$ km s^{-1} at 1.5 kpc

These velocity dispersions are indistinguishable. Also, the inner metal poor systems give:

  * Metal-poor globulars: $\sigma = 126 \pm 20$ km s^{-1},
  * Metal-poor halo giants: $\sigma = 109 \pm 13$ km s^{-1} at 1.5 kpc

These values suggest that the metal-poor stars and clusters in the inner few kiloparsecs are just the natural inner extension of the halo. Note that the velocity dispersion of the old disk increases towards the galactic centre (Lewis & Freeman 1990). Then, the inner disk velocity dispersion is expected to be much larger than the observed numbers:

  * Old disk within 3 kpc: $\sigma \sim 100$ km s^{-1}

which would seem to rule out the disk as host of the metal-rich globular cluster system.

2.3. *Metal abundances*

The mean metal abundance of the inner (R ≤ 3 kpc) metal-rich globular clusters is [Fe/H] = −0.49±0.05 (Armandroff 1989). With the new metallicity scale for the bulge, based on the abundances derived by McWilliam & Rich (1994) and by us, there is no appreciable difference between the mean abundances of the bulge and metal-rich globulars. This is illustrated in Figure 2, which shows the bulge and cluster metallicities as function of angular distance from the galactic centre and as function of galactic latitude.

2.4. *Ages*

Star clusters are excellent probes of the ages of different stellar populations, since they have very well defined sequences in the colour-magnitude diagrams. This is because all member stars are coeval and have similar chemical composition. However, it is very difficult to measure ages of the inner metal-rich globulars, due to heavy reddening, and to foreground contamination by the disk and bulge. Nevertheless, several studies indicate that these clusters are younger than the halo globulars, with ages $\sim 10 - 13$ Gyr (e.g. Hodder *et al.* 1993; Demarque & Lee 1992; Fullton 1995, this volume).

It is even more difficult to obtain the mean age of the galactic bulge, because of the same problems, added to the finite depth along the line of sight and to the wide metallicity range. However, most recent studies seem to converge towards a mean age of ~ 10 Gyr for the bulge (Terndrup 1988; Holtzmann *et al.* 1993), although it is not clear if there is an age range. Even though we cannot be sure that the bulge and the metal-rich clusters have the same age, it is clear that the current determinations are consistent.

3. Discussion

It seems clear that the metal-rich globular clusters with $R \leq 3$ kpc are associated with the galactic bulge rather than with the thick disk.

An important implication is that the bulge must be younger than the halo. This is in direct contradiction with the picture of Galaxy formation presented by Lee (1992), who proposed that the bulge is 1.5 Gyr older than the halo, based on RR Lyraes in Baade's window. We argue that the RR Lyraes, like the metal-poor globulars, represent the inner extension of the halo and do not tell us anything about the mean age of the bulge.

3.1. *How did the bulge form?*

An exhaustive discussion of different scenarios of bulge formation can be found in Wyse & Gilmore (1992). We can now speculate about the origin of the bulge, using the argument that the metal-rich globulars belong to this component, trying to complement the discussion of Wyse & Gilmore (1992):

• Large merging event or strong interaction?

As argued by Ashman & Zepf (1992) among others, multiple peaks in the metallicity distribution of globular clusters may be the result of a late merging event. Such violent episodes can lead to the formation of globulars in the inner regions of a galaxy. Since the metallicity distribution of globulars is bimodal in the Milky Way, it is worth exploring this possibility further.

• Strong bar instability heating up the inner disk?

Pfenniger & Norman (1990) suggested that a strong bar potential in the inner disk could create a concentration of stars in the centre, which then are heated out of the galactic plane by resonances. It would be worth checking whether such a scenario will produce metallicity gradients such as those observed in the bulge, and if it would affect in the same way the orbits of the globulars associated with the bulge.

• From the gas clouds ejected by a violent starburst?

Sofue & Habe (1992) suggested such a scenario to explain the formation of bulges and the origin of Hubble morphological types. It would be interesting to check whether such a process can form metal-rich globular clusters.

• Dissipational collapse, using leftover gas from the formation of the halo?

Carney *et al.* (1990) pointed out that most of the gas left over from halo formation would have sunk towards the bulge, due to its low angular momentum. Thus, the picture of Eggen *et al.* (1962), where the protogalactic cloud contracts at the same time as it is forming stars, seems very atractive. Such a scenario will reproduce the observed metallicity gradient naturally. The metal-rich clusters will also be more concentrated than the metal-poor clusters, as observed. The strong dependence of kinematics on metallicity (Minniti 1994) also seems to support this scenario.

I would like to thank Tim Bedding for his comments and help with this paper.

REFERENCES

AGUILAR, L., HUT, P. & OSTRIKER, J. P. 1988 *Astrophys. J.* **335**, 720.

ARMANDROFF, T. 1989 *Astron. J.* **97**, 375.

ASHMAN, S. E. & ZEPF, K. M. 1992 *Astron. J.* **384**, 50.

CARNEY, B. W., LATHAM, D. W. & LAIRD, J. B. 1990 *Astron. J.* **99**, 752.

DEMARQUE, P. & LEE, Y. W. 1992 *Astron. Astrophys.* **265**, 40.

EGGEN, O. J., LYNDEN-BELL, D. & SANDAGE, A. 1962 *Astrophys. J.* **136**, 748.

FRENK, C. S. & WHITE, S. D. M. 1980 *Mon. Not. R. Astron. Soc.* **193**, 295.

FRENK, C. & WHITE, S. D. M. 1982 *Mon. Not. R. Astron. Soc.* **198**, 173.

FROGEL, J. A., TERNDRUP, D., BLANCO, V. & WHITFORD, A. 1990 *Astrophys. J.* **353**, 494.

HODDER, P. J. C., NEMEC, J. M., RICHER, H. B. & FAHLMAN, G. G. 1992 *Astron. J.* **103**, 460.

HOLTZMAN, J., ET AL. 1993 *Astron. J.* **106**, 1826.

KENT, S. M. 1992 *Astrophys. J.* **387**, 181.

LEE, Y. W. 1992 *Astron. J.* **104**, 1780.

LEWIS, J. R. & FREEMAN, K. C. 1990 *Astron. J.* **97**, 139.

McWILLIAM, A. & RICH, R. M. 1994 *Astrophys. J., Suppl. Ser.* **91**, 749.

MINNITI, D. 1994 *Astrophys. J.* submitted.

MINNITI, D., RIEKE, M. & OLSZEWSKI, E. 1994 in *IR Astronomy with Arrays* (ed. I. S. McLean), p. 107. Kluwer.

MOULD, J. 1983 *Astrophys. J.* **266**, 255.

OJHA, D. K., BIENAYMÉ, O., ROBIN, A. C. & MOHAN, V. 1994 *Astron. Astrophys.* **290**, 771.

PFENNIGER, D. & NORMAN, C. 1990 *Astrophys. J.* **363**, 391.

PRESTON, G. W., SCHECTMAN, S. A. & BEERS, T. C. 1991 *Astrophys. J.* **375**, 121.

RICH, R. M. 1988 *Astron. J.* **95**, 828.

SAHA, A. 1985 *Astrophys. J.* **289**, 310.

SOFUE, Y. & HABE, A. 1992 *Publ. Astron. Soc. Jpn.* **44**, 325.

TERNDRUP, D. M. 1988 *Astron. J.* **96**, 884.

WYSE, R. F. G. & GILMORE, G. 1992 *Astron. J.* **104**, 144.

ZINN, R. 1985 *Astrophys. J.* **293**, 424.

ZINN, R. 1990 *J. R. Astron. Soc. Can.* **84**, 89.

Carney: One of the key questions is whether the formation and evolution of the bulge is related to the disk, the halo or is independent. Rosie Wyse and Gerry Gilmore proposed a nice test: the mass *vs.* angular momentum distributions. Your data are an excellent contribution to that problem. Do you have any results yet?

Minniti: I do not have any definitive results yet on this test. However, the idea of a causal relationship between the formation of the halo and the bulge looks very atractive, since the gas lost from the halo formation would sink to the bulge due to its low angular momentum (Carney *et al.* 1990).

Kaluzny: Could you comment on the age of giants forming your sample?

Minniti: I have not determined ages directly for the giants of my sample. Other recent results seem to give a mean age for the bulge of $t \approx 10$ Gyr.

The Helium Abundance of the Galactic Bulge

By DANTE MINNITI

European Southern Obs., Karl-Schwarzschild-Str. 2, D–85748 Garching bei München,
Germany

We measure the helium abundance in several fields towards the galactic bulge. We find a mean
He abundance of $Y = 0.28 \pm 0.02$, similar to that of the metal-rich globular clusters.

The absolute He abundance of the inner regions of the Galaxy is a key ingredient for models of galactic chemical evolution. However, previous He-abundance determinations in the galactic bulge give contradictory results. Terndrup (1988) and Renzini (1994) obtain $Y = 0.30 - 0.35$, while the counts of Davidge (1991) imply $Y \leq 0.22$.

This paper presents new measurements of the He abundance of the galactic bulge based on IR photometry of five bulge fields. The fields were chosen from among some well-known windows towards the galactic bulge. These windows are Sgr1 (at l, $b = 0°, -3°$), Ter2 ($-4°, 2°$), Baade's window ($1°, -4°$), F588 ($8°, 7°$) and F589 ($11°, 3°$). They cover a wide range of distances (from 0.3 kpc to 1.6 kpc), to study possible effects of disk contamination and the He gradient in the bulge.

The He abundances are derived following the R' method (Buzzoni et al. 1983), which is a variant of the R method proposed by Iben (1968). First we construct luminosity functions (LFs) in K' for all the fields, down to $\sim$2 magnitudes below the horizontal branch (HB). Next, we fit a power law to the LFs to obtain the total number of stars, following the procedure outlined by Davidge (1991). Once we have the relative numbers of HB and RGB+AGB stars, we apply the calibrations given by Buzzoni et al. (1983) and Caputo et al. (1987) to obtain the final helium fractions. Renzini (1994) found that this method is independent of ages for stellar populations older than $T \sim 10^8$ yr.

The findings are: (i) the mean He abundance of the galactic bulge is $Y = 0.28 \pm 0.02$, and (ii) there is no significant He gradient in the Galactic bulge. This value of Y is similar that of the metal-rich globular clusters such as 47 Tuc, and somewhat higher than that of the metal-poor halo globulars. Therefore, the bulge is more chemically evolved than the halo, as expected from its higher metallicity, $[Fe/H] = -0.25$ (McWilliam & Rich 1994). These conclusions do not seem to be dependent on reddening, disk contamination or different depths along the line of sight.

A full account of this work will appear in *Astronomy & Astrophysics*.

REFERENCES

BUZZONI, A., FUSSI PECCI, F., BUONANNO, R. & CORSI, C. 1983 *Astron. Astrophys.* **128**, 94.

CAPUTO, F., MARTÍNEZ ROGER, C. & PÁEZ, E. 1987 *Astron. Astrophys.* **183**, 228.

DAVIDGE, T. 1991 *Astrophys. J.* **380**, 116.

IBEN, I. 1968 *Nature* **220**, 143.

MCWILLIAM, A. & RICH, R. M. 1994 *Astrophys. J., Suppl. Ser.* **91**, 749.

RENZINI, A. 1994 *Astron. Astrophys.* **285**, L5.

TERNDRUP, D. M. 1988 *Astron. J.* **96**, 884.

Kinematics and Metallicity of the Galactic Globular Cluster System

By EMILIO J. ALFARO[1], J. CABRERA-CAÑO[1,2],
A. J. DELGADO[1] AND K. A. JANES[3]

[1]Instituto de Astrofísica de Andalucía (CSIC), Apdo. 3004, E–18080 Granada, Spain

[2]Universidad de Sevilla, Apdo. 1065, E–41080 Sevilla, Spain

[3] Boston University, 725 Commonwealth Ave., Boston, MA 02215, U.S.A.

An analysis of the spatial, kinematical and chemical structure of galactic globular clusters (GGCs) is presented, following the methodology proposed by Morrison et al. (1990). The results of this analysis lead us to suggest that GGCs lend themselves to being represented as two populations, whose main differences are their rotation speeds and concentration towards the galactic plane, whilst their chemical abundances and ages overlap.

1. Introduction

Over the last decade a series of observational results have been obtained which have revolutionised the paradigm of our Galaxy, and substantially altered our ideas about the formation and early evolutionary states of the Milky Way.

In particular, most of this research refers to the existence of a stellar population with characteristics which are intermediate between those of the halo and the disk. Following Majewski, we refer to this stellar population as intermediate Population II.

Norris et al. (1985) reviewed the role played by selection effects on the kinematically selected sample of Eggen et al. (1962) and concluded that a population of low-metallicity ([Fe/H]< −1.0) stars, with orbital eccentricity (e) of less than 4 does exist and seems to be representative of a third stellar population. Since then, numerous studies have been made to determine the main observational characteristics of this third stellar population and how it might fit in with the formation of the Galaxy (e.g. Gilmore & Wise 1985, 1986; Sandage & Fouts 1987; Carney et al. 1989; Morrison et al. 1990, Majewski 1993; Morrison 1993; Zinn 1993; Beers & Sommer-Larsen 1995; and Majewski, Carney, Agostinho et al. and Ojha et al. in this volume). These authors have attempted to identify the main physical properties of this third stellar population and establish observational patterns which might allow it to be characterised as an independent galactic component. The results, when not contradictory, seem to show a complex fauna of subsystems which appear to share certain physical properties but not others. The key question therefore hinges on defining the range of the physical variables associated with the intermediate Population II, such as as age, kinematics and metallicity.

It will be useful to mention here some results that might help us find our way. Norris et al. (1985) and Morrison et al. (1990) found a group of stars with metallicities as low as [Fe/H]= −1.6, with disk kinematics. Morrison (1993) goes even further in this direction by claiming that the metal-weak thick disk is kinematically different from the metal-rich thick disk. Although the calibration in the metallicity of the samples of Morrison et al. has been questioned (Twarog & Anthony-Twarog 1994), this result has been revalidated by Beers & Sommer-Larsen (1995), who analysed a compilation of various star samples from the literature and identified stars with disk kinematics and abundances as low as [Fe/H]= −2.0 and perhaps even lower. These authors report that the distribution of stars within 1 kpc of the galactic plane can be modelled by two components with different rotation velocities (halo + disk) for the whole range of metallicities in the sample. The proportion of stars of each component varies with the metallicity until [Fe/H]= −1.5, beyond which the proportion is constant. On the other hand, several papers have reported the existence of a very metal-poor population with a flattened distribution, giving rise to the concept of a "dual halo." The need to knock these results into some sort of shape has been stressed by Majewski (this volume), who proposes that the different "flattened subpopulations" are no more than different representations of one and the same population.

In the present contribution we make a further attempt to tackle analysis of the spatial, kinematic and chemical structure of the galactic globular cluster (GGC) system in light of these new hypotheses, following a similar methodology to that used by Morrison (1990) and Beers & Sommer-Larsen (1995) to analyse the kinematic behaviour of field stars. GGCs provide a sample of objects whose usefulness and appropriateness for the study of the halo structure has been called into question. Their detractors base their arguments on two basic objections: (i) the sample is composed of few objects (approximately 100 with reliable measurements of distance, metallicity and radial velocity), and (ii) it is still an open question whether the GGC system is truly representative of the stellar halo population. The first question is easy to deal with. 100 objects qualify in statistical terms as a "small sample," but the estimated parent population is only of order 160–200 (Harris 1976), so the current sample is fairly close to the total population. The statistically significant properties derived from this sample should be considered as representative of the GGC system, at least for galactic regions where the bias caused by extinction is not strong. The key question, however, lies in considering whether the properties derived from GGCs can be extrapolated to the entire halo. There is no clear answer to this question, which also depends on whether we are able to clearly delimit the subpopulations found in the galactic halo. An "optimistic" viewpoint should lead us to analyse the GGC system and compare our results with those derived from other samples of halo field stars. Such a comparison would help us to establish the similarities and differences and analyse their possible origin.

2. Kinematics *vs.* metallicity in the GGC system

For the present study we have used the list of 93 GGCs with reliable determinations of apparent magnitudes of horizontal-branch stars, metallicities, reddening and radial velocities inside a galactocentric radius of 20 kpc compiled by Armandroff (1989, 1992). The absolute magnitude calibration of Liu & Janes (1990) has been used to redetermine values for the heliocentric and galactocentric distances, adopting here a value of Av/E(B–V)= 3.2 and $R_\odot$ = 8.5 kpc. Following the method of Frenk & White (1980), we have determined the rotation velocity for different metallicity groups in ten-object bins. The results are shown in Figure 1.

GGCs with metallicity of less than −1.8 show a mean rotation velocity comparable to that of metal-rich ([Fe/H]> −1) clusters, although dispersion along the line of sight was greater, and clusters with metallicity between −1.0 and −1.8 show mean rotation velocities close to zero with a relative minimum at [Fe/H]= −1.5. The 10 objects grouped in this metallicity bin show a retrograde rotation velocity, although it is not significantly different from zero given the estimated error bar. Figure 2 shows the result of the same analysis when the sample of clusters is grouped into bins of 18 objects. This plot shows, in general terms, similar features to those mentioned above. The most representative feature of this analysis is the existence of a group of very metal-poor clusters ([Fe/H]< −1.8) with high mean rotational velocity.

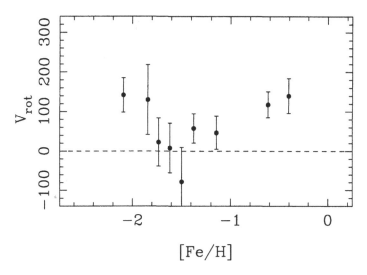

FIGURE 1. V_{rot} *vs.* [Fe/H] for 93 globular clusters inside a galactocentric radius of 20 kpc. Data have been grouped in bins of about ten objects

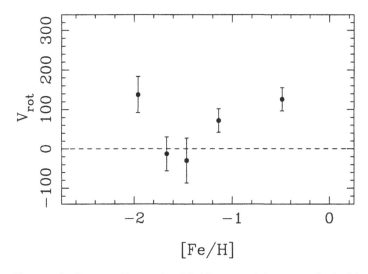

FIGURE 2. Same as Figure 1, with bins containing around 18 objects each

In order to analyse this result in more detail, we carried out an individual estimation of V_{rot}—following the prescription of Morrison *et al.*—for the GGCs in a galactic region defined by $|\ cos\ \lambda\ | > 0.5$, where λ is the angle between the heliocentric cluster position and l= 90, b= 0. This restricts the sample to those clusters which are located in directions such that the observed radial velocity has a component due to presumed rotation in excess of 110 km s^{-1}. Figure 3 shows a V_{rot}-[Fe/H] diagram for the 25 clusters which verify this condition, once the data are trimmed following Morrison (1993). In this figure we have marked with a circle those objects situated at $|\ Z\ |< 3$ kpc. For metallicities greater than -1.1 all the objects in this subsample have rotation velocities in excess of 90 km s^{-1}, but we also notice how clusters with [Fe/H] values as low as -2.0 have estimated rotation velocities as high as 120 km s^{-1}.

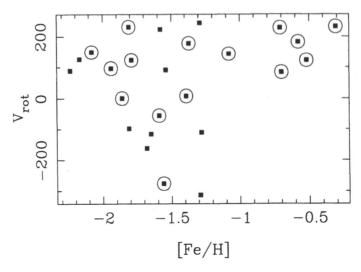

FIGURE 3. V_{rot} *vs.* [Fe/H] for 25 globular clusters located in those galactic regions where $| \cos \lambda | > 0.5$ ($\cos\lambda=\cos b.\sin l$). Open circles denote the objects with $| Z | \leq 3$ kpc

Cluster	V_{rot}	[Fe/H]	$\cos\lambda$	$\cos\phi$	Z
NGC 2298	230	-1.81	-0.88	-0.45	-3.0
NGC 2808	178	-1.37	-0.96	-0.73	-1.7
NGC 4372	149	-2.08	-0.84	-0.98	-2.5
NGC 6362	143	-1.08	-0.54	-0.95	-2.0
NGC 5286	122	-1.79	-0.74	-0.88	1.6

TABLE 1.

In particular, five clusters in our sample with metallicities of [Fe/H]< -1 and $| Z |< 3$ kpc display a rotation velocity in excess of 120 km s^{-1}(Table 1). These clusters, together with the six cited by Majewski (1994) for which spatial velocities have been determined, form a group of 11 objects which are representative of a metal-poor rotating disk.

The metallicity and rotation-velocity histograms for this subsample of 25 clusters are shown in Figure 4, where we have also represented those objects with $| Z |< 3$ kpc (filled area). These plots would seem to suggest that the GGC sample lends itself to being separated into two populations: (i) a flattened halo with a rotation velocity in excess of 90 km s^{-1}, covering the complete metallicity range, and (ii) a non-rotating halo with chemical abundances lower than -1.0, centred around [Fe/H]$= -1.5$.

3. Discussion

The studies of the GGC system by Zinn (1993) and van den Bergh (1993) point towards the existence of two subsystems of clusters with metallicities lower than [Fe/H]$= -1$, which display different spatial, kinematic and chemical properties and may be representative of two age groups, if we accept the hypothesis that this variable is the second parameter controlling horizontal-branch morphology. Recent analyses of dwarf field stars (Majewski, this volume; Beers & Sommer-Larsen 1995) also suggest that there are two kinematically separate populations in which the component with rotating-disk kinemat-

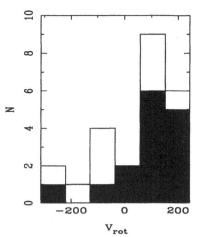
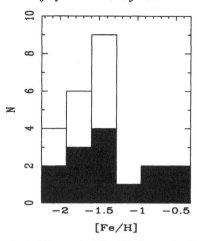

FIGURE 4. Histograms in V_{rot} and [Fe/H] for the sample in Figure 3. The filled area includes the clusters with $|Z| \leq 3$ kpc

ics shows a broad metallicity distribution and appears more concentrated towards the galactic plane. Zinn (1993) based his analysis on the division of the cluster system into two groups defined by the [Fe/H]–HB morphology diagram. However, the idea that this diagram allows reliable age-group differences to be established has been questioned by Catelan & Freitas-Pacheco (1993), based on an analysis of horizontal-branch evolution models. Buonnano (1993) also provided evidence to suggest that age might not be the only "second parameter" controlling horizontal-branch morphology.

In view of this situation, we have undertaken our analysis of the GGC system following the methodology described by Morrison *et al.* (1990) and later developed by other authors (Beers & Sommer-Larsen 1995). By applying this method to the GGC sample we thus limit the study to galactic regions where the observed radial velocity is representative of the rotational component of the spatial velocity of the cluster. The results show that the GGC system could be composed of two superimposed galactic components: the first being an oblate system with a high rotation velocity and a wide range of metallicities, while the second is more spherical, apparently non-rotating, and limited to a narrower range of chemical abundances. Fullton & Carney (this volume) have determined the age of certain clusters in the internal region of the halo and report that they are younger (10–13 Gyr) than the most metal-poor clusters located in regions outside the solar circle. Given that these clusters are associated with the oblate component of our analysis it seems evident that this population covers a wide age range. However, Minnitti (1995, this volume) has presented evidence suggesting that some of these young clusters in the internal regions of the halo could belong to the galactic bulge.

We conclude that the properties of the GGC system are not in contradiction with a view of the galactic halo composed of two populations, as proposed by Majewski (this volume). The oblate, disk-like rotating group of globular clusters displays similar properties as the intermediate Population II reviewed by Majewski, and could be considered as being its cluster counterpart.

REFERENCES

ARMANDROFF, T. E. 1989 *Astron. J.* **97**, 375.

ARMANDROFF, T. E. 1992 *Astron. J.* **104**, 164.

BEERS, T. C. & SOMMER-LARSEN, J. 1995 *Astrophys. J., Suppl. Ser.* **96**, 175.

BUONANNO, R. 1993 in *The Globular Cluster-Galaxy Connection* (ed. G. Smith & J. Brodie). ASP Conference Series, vol. 48, p. 131. ASP.

CARNEY, B. W., LATHAM, D. W. & LAIRD, J. B. 1989 *Astron. J.* **97**, 423.

CATELAN, M. & FREITAS PACHECO, J. A. 1993 *Astron. J.* **106**, 1858.

EGGEN, O. J., LYNDEN-BELL, D. & SANDAGE, A. R. 1962 *Astrophys. J.* **136**, 748.

FRENK, C. S. & WHITE, S. D. M. 1990 *Mon. Not. R. Astron. Soc.* **193**, 295.

GILMORE, G. & WYSE, R. F. G. 1985 *Astron. J.* **90**, 2015.

GILMORE, G. & WYSE, R. F. G. 1986 *Nature* **322**, 806.

HARRIS, W. E. 1976 *Astron. J.* **81**, 1085.

LIU, T. & JANES, K. E. 1990 *Astrophys. J.* **354**, 273.

MAJEWSKI, S. R. 1993 *Annu. Rev. Astron. Astrophys.* **31**, 575.

MAJEWSKI, S. R. 1994 in *Astronomy from Wide-Field Imaging* (ed. H. T. MacGuillwray *et al.*). IAU Simposyum 161, p. 425. Kluwer.

MORRISON, H. L. 1993 *Astron. J.* **105**, 539.

MORRISON, H. L., FLYNN, C. & FREEMAN, K. C. 1990 *Astron. J.* **100**, 1191.

NORRIS, J., BESSEL, M. S. & PICKLES A. J. 1985 *Astrophys. J., Suppl. Ser.* **58**, 463.

SANDAGE, A. R. & FOUTS, G. 1987 *Astron. J.* **93**, 74.

TWAROG, B. A. & ANTHONY-TWAROG, B. J. 1994 *Astron. J.* **107**, 1371.

VAN DEN BERGH, S. 1993 *Astrophys. J.* **411**, 178.

ZINN, R. 1993 in *The Globular Cluster-Galaxy Connection* (ed. G. Smith & J. Brodie). ASP Conference Series, vol. 48, p. 38. ASP.

Majewski: How much overlap is there between the group of Rodgers & Paltoglou retrograde clusters and the sample of retrograde clusters you have described (i.e. what fraction do they have in common)?

Alfaro: Only two clusters in our sample (NGC 1851 and NGC 3201) are also included among the Rodgers & Paltoglou list of clusters with high retrograde rotation. This low proportion of common objects can be explained by the fact that we have related our sample to high | *cos λ* | values and both samples present slightly different abundance values for some clusters. NGC 362 and NGC 5139, in our sample, also show retrograde rotation for the same interval of chemical abundances.

Bellazzini: Concerning inner-halo GCs as a hot tail of the classical metal-rich thick-disk population, in one of our posters we presented a result pointing towards the complete opposite. I will not illustrate it now because the result is too complex to explain in just a few words, but a simple KS test comparing disk clusters ([Fe/H]> −0.8) and halo clusters lying inside the solar circle (so the two samples share the same radial range), shows that the two samples are extracted from two different populations in the R_{GC}/Z_{GP} parameter at a confidence level of > 99%, the disk sample being the flatter. So, in my view, the inner-halo clusters are perhaps a somewhat flattened system, but a clearly different set with respect to metal-rich disk clusters.

Alfaro: Our analysis suggests that the inner halo could be populated by two populations, mainly separated by their kinematic behaviour. While the metal-rich clusters all belong to the disk population, some globular clusters with [Fe/H]< −1.0 also display kinematic and spatial properties which are compatible with those of a disk population. The fact that the most metal-rich clusters tend to be located in the very inner halo and more concentrated towards the galactic plane could indicate either the existence of a weak metallicity gradient or that they belong to the bulge, as has been suggested by Minniti in this meeting.

The Milky Way
and the Local Group:
Playing with Great Circles

By FLAVIO FUSI PECCI[1], M. BELLAZZINI[2]
AND F. R. FERRARO[1]

[1] Osservatorio Astronomico di Bologna, V. Zamboni 33, I-40126, Bologna, Italy

[2] Dipartimento di Astronomia, V. Zamboni 33, I-40126, Bologna, Italy

The small group of recently discovered galactic globular clusters (Pal 12, Ter 7, Rup 106, Arp 2) significantly younger than the average cluster population of the Galaxy are shown to lie near great circles passing in the proximity of most satellite galaxies of the Milky Way. Assuming that these great circles are in some way preferential planes of interaction between the Galaxy and its companions, we identified along one of them another candidate "young" globular cluster, IC4499. Within this observational framework, the possibility that the sample of young globulars found in the halo of the Galaxy could have been captured from a satellite galaxy or formed during a close interaction between the Milky Way and one of its companions is briefly discussed.

1. The origin of galactic young globulars

The idea that many galactic satellites appear to be situated along a few great circles (planes or streams) and that this evidence may be somehow related to their origin has been an unsettled but persistent benchmark in the last twenty years or so for many studies of the stellar systems in the Local Group (e.g. Kunkel & Demers 1976; Kunkel 1979; Lynden-Bell 1976, 1982, who found remarkable great circles connecting most of the dwarf satellites of the Milky Way). This topic has recently received a strong new impulse (Majewski 1993), in particular since the latest discovery of the Sagittarius dwarf spheroidal galaxy (Sgr–dSph) currently being disrupted and absorbed by the Milky Way (Ibata *et al.* 1994).

Our specific interest in this subject originates from the early detections made of a small set of globular clusters which appear to be significantly younger (3–5 Gyr) than the bulk of the cluster population studied so far in the Galaxy, on the basis of their turn-off locations, i.e. Pal 12 , Ruprecht 106, Arp 2 and Terzan 7 (see Buonanno *et al.* 1995 and references therein). Here we present some preliminary results of a study aimed to test the hypothesis that these clusters could have been captured by a satellite galaxy of the Milky Way or formed during a close interaction between the Galaxy and one or more of its satellites, following an early suggestion by Lin & Richer (1992). The work has been carried out by analysing spatial complanarities and kinematical and chemical compatibilities between groups of objects, in order to find patterns which can be relics of tidally generated tails or bridges, preserving memory of their common origin. Following this approach we have apparently picked up another "young" candidate, i.e. IC 4499, (Ferraro *et al.* 1994). Hence it may be worth reporting some results very briefly as a possible guide for further observations. The basic data sources for the galactic globular clusters are the lists presented by Thomas (1989), Zinn (1985) and Armandroff (1989) while the data on Local Group galaxies are drawn from van den Bergh (1993b) and Zaritsky (1994).

2. Preliminary results

The main suggestions emerging from this still qualitative analysis are:

• The known young globular clusters of the Galaxy are found to lie on two great circles. The first, here called MP-1, contains Ter 7, Arp 2 and some other galactic globular clusters, the Sgr–dSph Sag, UMi Dra, and Carina, and the Magellanic Clouds. The objects belonging to this group are shown to be kinematically compatible with a "common" origin. The second, MP-2, passing through the MCs and the Dra UMi dSph's, contains also Rup 106, Pal 12 and grossly fits the extension of the Magellanic Stream. Along this great circle we picked up another young globular cluster candidate, IC4499. It is also worth noting that all the five young globulars (with perhaps the exception of Pal 12) are found to be on *plunging* orbits, and two of them (Rup 106 and IC4499) are found to be prograde, according to the method adopted by Kinman (1959) and Van den Bergh (1993a).

• Within this framework, two main hypothesis can be considered: (i) young clusters formed because of the interaction/merging between the Galaxy and one or more of its satellites which induced star formation (Mihos *et al.* 1992); or (ii) they were captured from satellite galaxies which are known to contain young globulars, such as the Magellanic Clouds, or from the debris of a disrupted galaxy, like perhaps Sgr–dSph.

• Clues are also found that the location on preferential planes may be a general characteristic of the Local Group members are also found. An extensive presentation of this and previous results, together with a deeper discussion of the topic, will be included in a forthcoming paper (Fusi Pecci *et al.* 1995).

We warmly thank Roberto Buonanno, Carlo Corsi, Ivan Ferraro, Harvey Richer and Greg Fahlman, members of the group involved in the whole project, for the lively collaboration and discussions.

REFERENCES

ARMANDROFF, T. E. 1989 *Astron. J.* **97**, 375.

BUONANNO, R., CORSI, C. E., FUSI PECCI, F., RICHER, H. B. & FAHLMAN, G. G. 1995 *Astron. J.* **109**, 650.

FERRARO, I., FERRARO, F. R., FUSI PECCI, F., BUONANNO, R., CORSI, C. E., RICHER, H. B. & FAHLMAN, G. G. 1994 *Astron. J.* submitted.

FUSI PECCI, F., BELLAZZINI, M., FERRARO, F. R. & CACCIARI, C. 1995 *Astron. J.* submitted.

IBATA, R. A., IRWIN, M. J. & GILMORE, G. 1994 *Nature* **370**, 194.

KINMAN, T. D. 1959 *Mon. Not. R. Astron. Soc.* **119**, 559.

KUNKEL, W. E. 1979 *Astrophys. J.* **228**, 718.

KUNKEL, W. E. & DEMERS, S. 1977 *Astrophys. J.* **214**, 21.

LIN, D. N. C. & RICHER, H. B. 1992 *Astrophys. J., Lett.* **388**, L57.

LYNDEN-BELL, D. 1976 *Mon. Not. R. Astron. Soc.* **174**, 695.

LYNDEN-BELL, D. 1982 *Observatory* **102**, 202.

MAJEWSKI, S. R. 1993b in *Galaxy Evolution: The Milky Way Perspective* (ed. S. R. Majewski). ASP Conference Series, vol. 49, p. 5. ASP.

MIHOS, J. C., RICHSTONE, D. O. & BOTHUN, G. D. 1992 *Astrophys. J.* **400**, 153.

THOMAS, P. 1989 *Mon. Not. R. Astron. Soc.* **238**, 1319.

VAN DEN BERGH, S. 1993a *Astrophys. J.* **411**, 178.

VAN DEN BERGH, S. 1993b , preprint.

ZARITSKY, D. 1994, preprint.

ZINN, R. 1985 *Astrophys. J.* **293**, 424.

Dynamical Families in the Galactic Globular Cluster System

By MICHELE BELLAZZINI[1], F. FUSI PECCI[2]
AND F. R. FERRARO[2]

[1] Dipartimento di Astronomia, Via Zamboni 33, I–40126, Bologna, Italy

[2] Osservatorio Astronomico di Bologna, Via Zamboni 33, I–40126 Bologna, Italy.

The inner-halo galactic globular clusters (IHGGCs) lying inside the solar circle ($R_{GC} < 8Kpc$, [Fe/H] < -0.8) do not display any record of the *natural correlation* between cluster absolute integrated magnitude and central density which clearly exists for the other GGCs (i.e. both the outer-halo GGCs, with $R_{GC} > 8Kpc$ and [Fe/H] < -0.8, and the disk GGCs, with [Fe/H] > -0.8). Assuming that this correlation could be primordial or originated by cluster internal evolution, it is argued that the IHGGCs, as a subsystem, experienced a chaotic dynamical evolution not shared, for instance, by the disk GGCs, although the radial distributions of the two groups are nearly the same.

1. Introduction

The present dynamical status of any GGC must store (at least in principle) important information about the effects due to any interaction with the environment. The fact that all of the ~ 30 clusters which are thought to have undergone core collapse are within 10 kpc of the galactic centre (Chernoff & Djorgovski, 1989) does indeed prove that environmental effects accelerate dynamical evolution of the globulars. Djorgovski & Meylan (1993) have found two remarkable global properties of the galactic GC system: (i) cluster intrinsic structural parameters ($Logr_c$, $Log\rho_0$, c, etc., here respectively: core radius, central luminosity density and concentration parameter as defined in Djorgovski 1993) are correlated with each other and with cluster integrated absolute magnitude, M_V, i.e. more luminous clusters are more centrally concentrated; and (ii) clusters located at large galactocentric distances (R_{GC}) are looser. M_V and R_{GC}, are found to be uncorrelated with each other. Assuming that the correlation between cluster concentration and cluster integrated magnitude is controlled by cluster internal evolution or by primordial conditions, while any correlation between cluster structural parameters and, for example, R_{GC} must be somehow related to environmental effects, the properties of different cluster families in the Galaxy are analysed, based on the above framework. Data were drawn from Djorgovski (1993). The whole sample (113 GGCs) was then divided into four groups:

• Disk GCGs: having [Fe/H] ≥ -0.8 (Zinn 1985).

• Inner-halo GGCs: having [Fe/H] < -0.8 and lying inside the solar circle, adopted to be at $R_{GC} = 8Kpc$.

• Outer-halo GGCs: having [Fe/H] < -0.8 and presently lying outside the solar circle but inside 40 Kpc. The few clusters with $R_{GC} > 40Kpc$ are not discussed here. Post core collapse clusters were excluded from the analysis.

2. Results

The main result of our analysis is shown in Figure 1: both "outer" and "disk" clusters show a noticeable correlation in the plane $|M_V|$ *vs.* $Log\rho_0$ while no sign of correlation

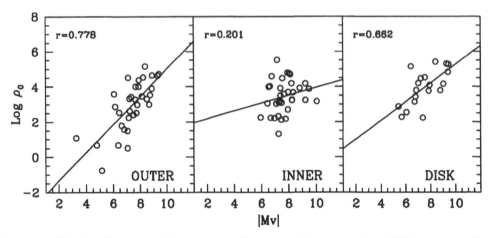

FIGURE 1. The distributions of the three considered samples in the plane $|M_V|$ - $Log\rho_0$. Continuous lines are linear regression lines through the points, r is the linear correlation cofficient. For the sake of simplicity the absolute value of M_V is used.

is found for the "inner" sample in the same plane. Note that the same result is found also in the $|M_V|$ - $Logr_c$ and $|M_V|$ -c planes; it has robust statistical significance and is confirmed by bivariate analysis. Since "disk" and "inner" clusters mostly live in the same dense environment, it is important to understand why they keep such a different memory of their dynamical history. In our view, a possible explanation is related to the fact that they have very different orbits. In particular, "disk" clusters are probably orbiting circularly near the plane of the Galaxy and they are rotating with approximately the same velocity as their stellar neighbours (Zinn 1985; Armandroff 1989), so minimising the shocking effects of the galactic disk (Aguilar 1993). However, "inner" clusters are known to have a small net rotation compared to that of the galactic disk (Armandroff 1989; Zinn 1993) and they have a more spread-out distribution of heights above the galactic plane, suggestive of orbits having large inclinations with respect to the plane. Consequently, they had probably undergone a great number of passages across the densest parts of the galactic disk, suffering several episodes of fast-induced dynamical evolution which caused a loss of memory of the original ranking settled by their internal evolution.

An extensive presentation of the results obtained and a deeper discussion of the topic will be included in a forthcoming paper (Bellazzini *et al.* 1995).

REFERENCES

AGUILAR, L. A. 1993 in *Galaxy Formation: The Milky Way Perspective* (ed. S. R. Majewski). ASP Conference Series, vol. 49, p. 155. ASP.

ARMANDROFF, T. E. 1989 *Astron. J.* **97**, 375.

BELLAZZINI, M., FUSI PECCI, F. & FERRARO, F. R. 1995 in preparation.

CHERNOFF, D. & DJORGOVSKI, S. G. 1989 *Astrophys. J.* **339**, 904.

DJORGOVSKI, S. G. 1993 in *Structure and Dynamics of Globular Clusters* (ed. S. G. Djorgovski & G. Meylan). ASP Conference Series, vol. 50, p. 373. ASP.

DJORGOVSKI, S. G. & MEYLAN G. 1993 *Bull. Am. Astron. Soc.* **25**, 885.

Blue Straggler Stars in Galactic Globular Clusters as Tracers of Dynamical Effects on Stellar Evolution

By FRANCESCO R. FERRARO

Osservatorio Astronomico di Bologna, V. Zamboni 33, I–40126, Bologna, Italy

A summary is presented of the most recent results we obtained from the study of blue straggler stars in galactic globular clusters. The peculiar radial distribution of the blue stragglers found in M3 is also discussed.

1. Blue straggler stars in galactic globular clusters

As is well known, galactic globular clusters (GGCs) are the oldest simple stellar population in the Universe, so they could be used to set constraints on the origin and the age of the Universe. Moreover, they are considered the natural laboratory where theoretical models of stellar evolution can be tested and the ideal stellar aggregates where one can investigate environmental effects on stellar evolution.

Within this framework, over the last few years our group has dedicated particular attention to the study of a possible link between the photometric properties of special groups of stars in the CMD and the structural parameters of clusters, finding, for example, quite convincing evidence for the dependence of the horizontal-branch tail extension on the central density of the cluster (see Fusi Pecci et al. 1993).

I shall here present some results we obtained from the systematic study of the blue straggler stars (BSS) in GGCs. The BSS were first detected by Sandage (1953) in M3. Since then, BSS have been found in other GGCs, open clusters, dwarf spheroidal and in the field, suggesting that they are probably a normal population of any stellar aggregate. Many possible explanations have been proposed to model BSS (see Livio 1993). In particular, there is a growing belief that binaries can be related via various mechanisms to the origin of BSS in GGCs. In this scenario the detection of BSS in both high- and low-density GGCs indicates the possibility that different mechanisms could be at work in generating BSS in different environments. In order to investigate any possible correlations between the detection of BSS and the structural parameters of the clusters, we compiled a catalogue to include all the BSS detected in GGCs. In the updated version of our catalogue (Fusi Pecci et al. 1992, 1993) we have listed ~ 630 BSS candidates in 26 GGCs. From the analysis of this catalogue we obtained the following results:

- the BSS Luminosity Function (LF) in dense GGCs ($Log \rho_0 > 3$) is different from that in low density GGCs ($Log \rho_0 < 3$) at 3σ level of confidence;
- the possible detection of candidate BSS-descendants located at the red HB edge in various clusters with blue HB population;
- in a sample of six well-studied GGCs, the BSS mean ridge line seems to be shifted with respect to the Zero Age Main Sequence by increasing amounts with increasing metallicity.

Moreover we found (Ferraro et al. 1994) a quite interesting behaviour plotting the number of the detected BSS (N_{BSS}) as a function of sampled light (L_S) expressed in unit of $10^4 L_\odot$ (see Figure 1). It is quite interesting to see that if we divide the cluster set into

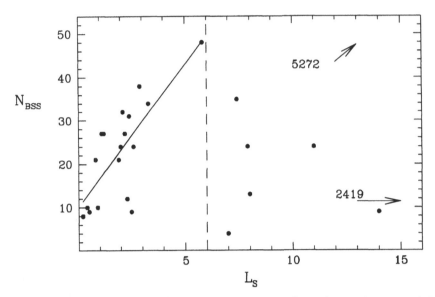

FIGURE 1. The observed number of blue straggler stars (N_{BSS}) as a function of the sampled light (L_S) expressed in units of $10^4 L_\odot$. The positions of two clusters (NGC 5272 and NGC 2419) are indicated by arrows, since their coordinates fall outside the axis range. The vertical dashed line divides the two samples showing different behaviour with increasing L_S.

two subgroups: group a for $L_S < 6 \times 10^4 L_\odot$, and group b for $L_S > 6 \times 10^4 L_\odot$, then in group a a strong correlation is evident between N_{BSS} and L_S ($r = 0.76$), while no correlation can be found for clusters in group b.

An obvious explanation for this result could be the existence of a bias in the BSS search, which would imply an increasing incompleteness as the sampled light increases. This may well have occurred, but *a priori* there is no obvious reason to explain why BSS should be lost as light sampling increases, and the detected difference could be due to an "intrinsic" difference in the BSS origin or survival.

In particular, it is quite interesting to note that the cluster concentration is different for the two subgroups: the mean central density is $< Log\rho_0 >= 1.5 \pm 0.3$ and $< Log\rho_0 >= 3.3 \pm 0.5$ for group a and b, respectively.

Hence, though the difference is not very strong on statistical grounds, one could conclude that while in loose clusters the number of the detected BSS is proportional to the sampled light, in the intermediate- to high-density globulars N_{BSS} drops with respect to this relation, and the actual number of BSS depends on various other factors.

This evidence is compatible with an interpretative scenario where in loose clusters the BSS form from primordial binaries and their number is roughly proportional to the sampled luminosity. As we increase the cluster concentration and the number of cluster members, primordial binaries become less and less efficient as BSS progenitors while collisional binaries take place, and the BSS number actually depends on a variety of other parameters (star interactions, binary system formation and destruction, etc.).

Admittedly, these results are all inconclusive, since GGCs listed in our catalogue have been observed by different authors in a quite inhomogeneous way and the surveys may still be very incomplete.

Up to now, M3 is the only GGCs which has been properly surveyed in the inner and

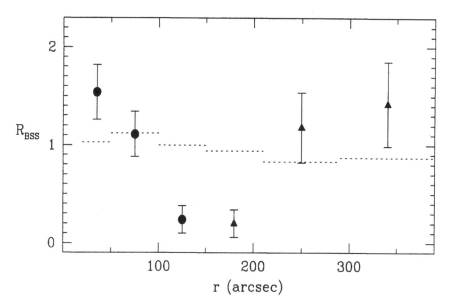

FIGURE 2. The relative frequency of BSS in M3 is plotted as a function of the radial distance from the cluster centre. The data points at $r < 150''$ (dots) are CCD data and those at $r > 150''$ (triangles) are from photographic data. The horizontal lines show the relative frequency of the SGB stars used as a comparison population.

outer regions (see below) and which has a sufficiently complete and homogeneous sample of BSS.

2. M3: a very special case

M3 is historically important in BSS studies, since it was the first GGC in which BSS were detected. Moreover, it has been the target of intensive study by many groups. In particular we obtained:

• a PHOTO-sample covering $2.5' < r < 7'$ with $B < 18.6$ (see Buonanno *et al.* 1994 for more details)

• a CCD-sample based on a series of frames taken in very good observation conditions ($FWHM < 0.8''$) at the CFHT covering the inner regions of the cluster ($r < 3.5'$) (see Ferraro *et al.* 1993).

In order to make the two samples sufficiently homogeneous, ensuring the maximum radial extension ($20'' < r < 7'$), we limit our analysis to the bright BSS region ($B < 18.6$).

To study the BSS radial distribution, we divided the surveyed area into annuli; then the number of BSS and normal SGB stars were counted and normalised to the total light sampled in that annulus. By so doing, we can define the relative frequencies of BSS and SGB stars, R_{BSS} and R_{SGB}, respectively. They were obtained by computing the ratio of (the number of stars per annulus)/(total number of stars) to the fractional sampled luminosity, i.e. $R_{BSS} = \frac{N}{N_{(\ )}}/\frac{L}{L}$ and $R_{SGB} = \frac{N}{N_{(\ )}}/\frac{L}{L}$.

Figure 2 shows the relative frequencies as a function of the radius. As can clearly be seen, R_{SGB} is essentially uniform over the area surveyed, while R_{BSS} is clearly bimodal, showing a clear-cut dip in the range $100'' < r < 210''$ ($4r_c < r < 8r_c$). The significance of this dip can be assessed by noting that only five BSS are observed in a radial interval where 20 are "expected" in order to fill up the dip; hence the observed number of stars represents a 3.4σ fluctuation with respect to the expected number.

The observed dip strongly suggests a peculiar absence of bright stars from the BSS LF, probably due to a differential survival mechanism guided by the cluster dynamics. In the picture of the BSS as products of binaries evolution, this result is fully compatible with the conclusions of a recent theoretical study by Hut *et al.* (1992) on the radial distribution of binaries in a GGC with a certain amount of primordial binaries. A possible scenario emerged from this study: in the core a population of recycled (by collisions) primordial binaries is expected; the intermediate zone from a few to several dozen core radii will be depleted in binaries by mass segregation effects (causing the observed dip). Some binaries escape into the halo where they are placed on "parking orbits" from which they slowly drift back into central regions.

The observed radial behaviour of the BBS in M3 has also been reproduced by a theoretical simulation specifically computed for M3 by Sigurdsson *et al.* (1994).

REFERENCES

BUONANNO, R., CORSI, C.E., BUZZONI, A., CACCIARI C., FERRARO, F. R. & FUSI PECCI, F. 1994 *Astron. Astrophys.* **209**, 69.

FERRARO, F. R., FUSI PECCI, F. & BELLAZZINI, M. 1995 *Astron. Astrophys.* **294**, 80.

FERRARO, F. R., FUSI PECCI, F., CACCIARI, C., CORSI, C. E., BUONANNO, R., FAHLMAN, G. G. & RICHER, H. B. 1993 *Astron. J.* **106**, 2324.

FUSI PECCI, F., FERRARO, F. R., BELLAZZINI, M. DJORGOVSKI, S. G., PIOTTO, G. & BUONANNO, R. 1993 *Astron. J.* **105**, 1145.

FUSI PECCI, F., FERRARO, F. R. & CACCIARI, C. 1993, in *Blue Stragglers* (ed. R. A. Saffer). ASP Conference Series, vol. 53, p. 97. ASP.

FUSI PECCI, F., FERRARO, F. R., CORSI, C. E., CACCIARI, C. & BUONANNO, R. 1992 *Astron. J.* **104**, 1831.

HUT, P., MCMILLAN, S. & ROMANI, R. W. 1992 *Astrophys. J.* **389**, 527.

LIVIO, M. 1993, in *Blue Stragglers* (ed. R. A. Saffer). ASP Conference Series, vol. 53, p 3. ASP.

SANDAGE, A. 1953 *Astron. J.* **58**, 61.

SIGURDSSON, S., DAVIES, M. B. & BOLTE, M. 1994 *Astrophys. J., Lett.* **431**, L115.

Kaluzny: Your sample of BSS is very inhomogeneous. For some clusters only relatively small fields located in outside regions were observed, while for others the photometry for the whole cluster is available. Could this have any influence on your conclusion?

Ferraro: I agree with you that the BSS sample is quite inhomogeneous, and I noted earlier that the results I have presented are not conclusive. However, they should be considered to be a first step indicating that BSS in different environments could have different origins. Of course, in order to reach firm conclusions we need more complete and extensive surveys, like in M3.

Carney: Your M3 data are impressive. Have you computed $(B-V)/(B+V+R)$ for the region where BSS are less common ($100''-200''$) it vs. where they are more common (outside $200''$)? It would be an interesting test of the sensitivity of the horizontal-branch morphology to BSS and the environment in general.

Ferraro: That is a good point. Bright RGB and HB stars are saturated in the CCD frames that we used to study the BSS population. We are now reducing short exposures to sample the HB properly, so that $(B-V)/(B+V+R)$ can be computed.

Effects of Chaos in the Galactic Halo

By WILLIAM J. SCHUSTER AND CHRISTINE ALLEN

Instituto de Astronomía, Universidad Nacional Autónoma de México, Mexico

To study the way in which the principal periodic orbits in a galactic potential determine orbital structure, horizontal and vertical surfaces of section, i.e. (dR/dt, R) and (dz/dt,z), are being used to explore the potential of Allen & Santillán (1991) and to investigate possible vertical structure in the galactic halo. The chaotic "scattering" process due to the nearly spherical mass distribution close to the Galactic centre in conjunction with the confinement of the chaotic orbits produces a vertical segregation of both chaotic and non-chaotic orbits in the halo. Certain z_{max}, z_{min} are preferred by the chaotic orbits over others as a result of the conservation of the total orbital energy and of the interaction and confinement of the chaotic orbits by the principal families of periodic orbits (Figure 1). Some of these periodic orbits have been identified. Correlations between the structure found in the observed W distribution and that of the numerically determined z_{max}, z_{min} histograms are shown for our sample of 280 halo stars (Schuster et al. 1993). W is the star's velocity perpendicular to the galactic plane and z_{max}, z_{min} the maximum distances above or below the galactic plane, respectively, reached by the star in the course of its orbit. This vertical structure may explain certain puzzling observations of the galactic halo, such as conflicting c/a values for the shape of the halo, and unusual velocity dispersions and/or distributions near the galactic poles. These results are in good agreement with Hartwick's (1987) two-component model for the halo.

Our main conclusions are as follows: (i) chaotic scattering of galactic halo stars can serve to hide or destroy correlations involving kinematic parameters, such as a possible chemical gradient in the halo; (ii) due to these chaotic processes, as well as to the coupling of vertical and horizontal motions, the W velocity component of halo stars does not give a good indication of the maximum height above or below the galactic plane reached by a halo star in the course of its orbit; (iii) vertical Poincare sections (Figure 2) provide a good means of studying the effects of chaos as well as the velocity dispersions produced at the galactic poles by different galactic potentials; (iv) the observed W histogram for our 280 halo stars, the calculated z_{max} histogram, and various vertical Poincare sections all show structure, evidence for the interaction and confinement of chaotic orbits by families of periodic orbits; (v) such interactions and confinements may help to explain the non-Gaussian "clumping in velocity space" that has been observed for halo stars near the galactic poles (Norris 1986), the possible existence of "moving groups" within the halo, as well as the differing shapes and velocity dispersions obtained for the galactic halo from different stellar samples.

REFERENCES

ALLEN, C. & SANTILLÁN, A. 1991 *Rev. Mex. Astron. Astrofis.* **22**, 255.

HARTWICK, F. D. A. 1987 in *The Galaxy* (ed. G. Gilmore & B. Carswell), p. 281. Reidel.

NORRIS, J. 1986 *Astrophys. J., Suppl. Ser.* **61**, 667.

SCHUSTER, W. J., PARRAO, L. & CONTRERAS-MARTÍNEZ, M. E. 1993 *Astron. Astrophys., Suppl. Ser.* **97**, 951.

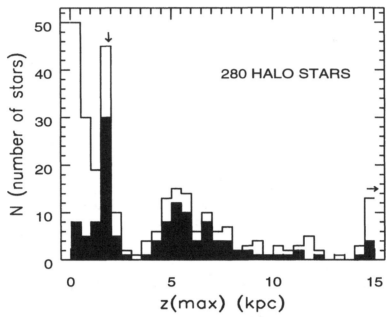

FIGURE 1. The calculated z_{max} histogram for the 280 halo stars obtained using the galactic potential of Allen & Santillán (1991) with each orbit integrated for at least 15 radial or vertical oscillations. A minimum in the histogram is observed over the interval 2.5–3.0 kpc and a secondary maximum over 1.5–2.0 kpc. This secondary peak is produced mainly by stars with chaotic orbits (shaded area).

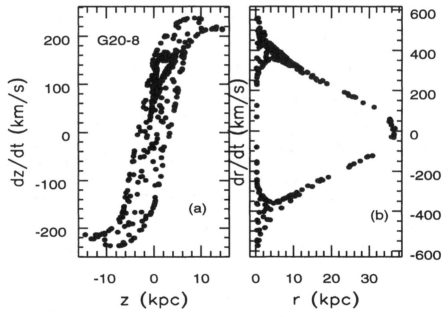

FIGURE 2. Vertical and horizontal Poincare sections for the halo star G20-8, which has a chaotic orbit. The vertical Poincare section (a) plots the W=dz/dt velocity against the vertical height each time the star crosses the Sun's radial distance travelling outward, while the horizontal Poincare section (b) the galactocentric radial velocity against the radial position each time the star crosses the galactic plane travelling upwards.

Disk-Shocking on Globular Clusters Simulated by the Perturbation Particles Method

By STEPHANE LEON[1], F. COMBES[2] AND F. LEEUWIN[3]

[1] IRAM, Núcleo Central, E–18012 Granada, Spain

[2] DEMIRM, Observatoire de Paris, 61 Av. de l'Observatoire, F–75014, Paris, France

[3] Scuola Normale Superiore, 7 Plazza dei Cavalieri, I–56126 Pisa, Italy

Globular clusters (GCs), the oldest stellar systems known, are an ideal tool for testing theoretical considerations on the N–Body problem with long-range gravity forces. In a first approximation, these systems can be considered as collisionless. External gravitational interactions can significantly alter the evolution of GCs, and prevent or accelerate the core collapse. External perturbation can occur from the tidal interaction of giant molecular clouds (transverse to the GC orbit) or from the crossing of the galactic thin disk. This latter has been found to largely dominate the heating of GC (Chernoff et al. 1986).

Most approaches to this problem have used a certain number of approximations, the main one being the impulsive approximation (Ostriker et al. 1972; Chernoff et al. 1986). This approximation assumes that the GC stars have no time to move inside the GC during the travel time of the cluster through the disk (Ostriker et al. 1972). However, this assumption is particularly poor for GCs with low eccentricities or for stars with short crossing time in the core.

Here we use a numerical technique, the so-called "perturbation particles" method (PP, for details, see Leeuwin et al. 1993), to analyse the weak perturbation of a single disk-crossing. Comparing the short duration of the crossing with the relaxation time we assume a collisionless system. Such systems are driven by the collisionless Boltzmann equation (CBE). The PP technique allows the CBE to be solved numerically; the underlying idea is simple: considering a system whose unperturbed distribution function f_0 is analytically known, we decompose the distribution function of the perturbed system by $f = f_0 + f_1$. To compute the perturbed part of the distribution, f_1, we use particles whose mass m_i will represent the local perturbation. Their motion is computed by a normal N-body code, but with a number of particles much reduced, the main advantage of PP being a substantial reduction of the numerical noise. The choice of f_s, the sampling density, is crucial in order to avoid undersampling of phase space during the evolution of the system. Because there is no rule to deal with, f_s will be of the same form as f_0, but more extended in space. For the unperturbed distribution function f_0 we use a Plummer sphere, with potential Φ_P given by

$$\Phi_P(r) = -\frac{GM_{cl}}{\sqrt{r^2 + b^2}}$$

The parameter b fixes concentration of the GC, and is linked to the core radius r_c by $r_c \sim 0.56b$. The vertical gravitational force K_z of the disk at height z and radius R takes the form (Van der Kruit 1987): $K_z = -4G\sigma \arctan(\sinh(z/z_e))$, where σ is the surface density at the radius R and z_e is the characteristic scale height of the plane. We adopt the value of 325 pc given by Van der Kruit (1987).

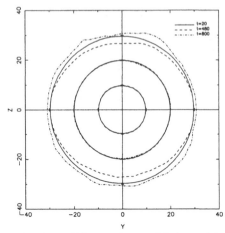

FIGURE 1. Isophotes for three different times with intensities of three initial radii
$(r = 9, 19, 29$ pc$)$.

We performed one run at $R = 6$ kpc with a test globular cluster ($M_{GC} = 0.5 \times 10^5 M_\odot$ and $r_t = 30$ pc). We checked that by taking a parameter $b_s = 3b$ for the sampling distribution we minimise sampling error, since perturbations are maximum in the halo. We start from $z = 3000$ pc above the plane down to -4000 pc at a velocity $V_z = 170 km.s^{-1}$. At the end of the simulation the GC reaches a new equilibrium state. During the period of travel some transient features appear. The cluster exhibits radial oscillations, particularly in the core where the period fits the theory quite well (Sobouty 1984). The GC presents transient flattening during the crossing (see Figure 1), which can reach the ellipticity $\epsilon = 0.1$ ($\epsilon \equiv 1 - b/a$). This compression has already been noted by Chernoff *et al.* (1986). Projected halo density can fall by two magnitudes, corresponding to the first oscillation where the outer parts are flattened, and the cluster compressed. As reported in the literature (Spitzer & Chevalier 1973; Chernoff *et al.* 1986), after crossing, the halo undergoes an expansion—typically of 3% for the radius including 90% of the mass. The total energy increase is about 0.1% of the total potential energy of a Plummer. The impulsive approximation tends to underestimate the energy increase. The choice of z_e can affect the energy increase by a factor of 2-3 times the energy increase; taking a lower value, 200 pc, increases strongly the strength of the shock.

Djorgovski & Meylan (1994) have recently shown a strong dependence of cluster parameters (e.g. c, r_c) on the distance to the galactic centre R_G and to the galactic plane Z_g. This could be the result of gravitational shocks, whose efficiency could have been overlooked (cf. Weinberg 1994). PP provides a useful and precise direct tool to describe the weak effects of a single disk-crossing which is neither adiabatic nor impulsive.

REFERENCES

CHERNOFF, D. F., KOCHANEK, C. S. & SHAPIRO, S. L. 1986 *Astrophys. J.* **309**, 183.

DJORGOVSKI, S. & MEYLAN, G. 1994 *Astron. J.* **108**, 1292.

LEEUWIN, F., COMBES, F. & BINNEY, J. 1993 *Mon. Not. R. Astron. Soc.* **262**, 1013.

OSTRIKER, J. P., SPITZER, L. & CHEVALIER, R. A. 1972 *Astrophys. J., Lett.* **176**, L51.

SOBOUTI, Y. 1984 *Astron. Astrophys.* **140**, 82.

VAN DER KRUIT, P. C. 1987 *Astron. Astrophys.* **192**, 117.

WEINBERG, M. D. 1994 *Astron. J.* **108**, 1403.

The Horizontal-Branch Morphology

By ROBERTO BUONANNO
AND GIACINTO IANNICOLA

Osservatorio Astronomico di Roma, I–00040 Monte Porzio Catone (Roma), Italy

We examine the horizontal-branch (HB) morphology presented by galactic globular clusters, making extensive use of the parameter B2–R/B+V+R introduced by Buonanno (1993) to describe the colour distribution of the stars along the HB. It is found that both the age and the central density of the clusters have an important influence on the HB morphology.

A new, improved estimate of the age of galactic globular clusters is derived from the HB morphology.

1. Introduction

Understanding the horizontal-branch (HB) morphology of galactic globular clusters, i.e. the distribution in colour of the stars presently populating the HB, is a tantalising astrophysical problem.

Seventy-five years ago, Shapley (1919) realised that the HB morphology differs from cluster to cluster, and later Arp *et al.* (1952) clearly established that the location in colour of the HB stars is correlated with the position of the red-giant branch (RGB) in the CM diagram and therefore with the metal abundance (the *first* parameter). They noticed, in fact, that in M92 only the blue side of the RR Lyrae instability strip is populated, whereas the HB of M3 is populated on both the red and the blue side of the interval.

In the 1960s, Sandage & Wallerstein (1960), Van den Bergh (1967), Faulkner (1966) and Sandage & Wildey (1967) noticed that some intermediate metal-poor clusters show very different HB types. Sandage & Wallerstein (1960), in particular, noticed that clusters like M13 and M22 present HBs which are too blue for their metallicity and concluded that "in addition to chemical composition, the age may be affecting the correlation." This is the origin of the problem of the second parameter, $2^{nd}P$, required for the description of the HB morphologies which are actually observed.

In the late 1970s, Searle & Zinn (1978) proposed a picture for the formation of the Galaxy and its early evolution based in part on the evidence that the strength of the second parameter shows a systematic variation with R_G, the galactocentric distance. This outline was supported by the progressive addition of new data which seem to indicate that the tightly bound clusters of the inner halo are largely coeval, while the outer halo formed over a period of several billion years by the merging of protogalactic fragments. The loosely bound clusters are the relics of such fragments. (Zinn 1980; Lee *et al.*1988; Lee *et al.* 1994, hereafter LDZ).

In conclusion, the above-mentioned picture identifies the age as the $2^{nd}P$, which controls the distribution of the stars along the HB. However, although this scenario is consistent with most of the existing data, it is admittely based on only a handful of clusters for which reliable ages have been determined. In addition, Fusi Pecci *et al.* (1993, hereafter FPAL) and Buonanno (1993, hereafter B1) have shown that the presence of blue tails in the HB is often correlated with the cluster central density and it has been suggested that the age is only *one* $2^{nd}P$ and that the observed HB morphology is the result of the global combination of different phenomena.

Recently, the role played by some candidate $2^{nd}Ps$ has been weakened by analysis of the periods of RR Lyrae variables made by Lee (1992) and Lee *et al.* (1994). These papers have shown that if the blue HB morphologies of the clusters in the inner halo are due to a larger helium abundance in comparison with those in the outer halo, one would expect to find *increased* periods for the RR Lyrae, in contrast to the observations (similar considerations exclude a larger core rotation for clusters in the inner halo).

It seems, in conclusion, that age and central density (or another equivalent cluster structural parameter) remain the most attractive candidates to be identified with the $2^{nd}P$. In order to check how the two above-mentioned candidates influence the colour distribution of the stars along the HB, we compiled a catalogue of 56 galactic globular clusters largely coincident with the compilation of LDZ. By duly taking into account the effects of density, we will show that the HB morphology is an important age indicator.

2. HB morphology parameters

With a view to providing a quantitative description of the star distribution along the HB, Rosino (1965) defined the parameter $B/B + R$, where B and R are the numbers of blue and red HB stars, respectively. An improved parameter was proposed by Zinn (1986) and by Lee (1989), who defined the quantity $B - R/B + V + R$, where V is the number of variables on the HB. This index ranges from -1 for clusters with only red HB stars to $+1$ for clusters with only blue HB. Although this index is highly informative, it presents the disadvantage of being not particularly sensitive for HBs with most stars in the blue region.

In addition to the Lee & Zinn parameter, FPAL proposed several new parameters characterising the HB morphology. Two of them are particularly informative about the mass which a star loses to pass froms M_i to M_{ZAHB}: L_t, i.e. the total length of the HB and $(B - V)_{peak}$, i.e. the de-reddened colour of the peak of the HB stars distribution. Along the same lines, in a study aimed at detecting colour gradients in the field blue HB stars, Preston *et al.* (1991) introduced the parameter B_W, i.e. the number of HB stars in the colour interval $-0.02 < (B - V)_0 < 0.18$. This parameter, obviously informative of the presence of HB blue tails, becomes less and less useful the more the HB shifts to lower temperatures. To overcome the main drawbacks of the quoted parameters, B1 introduced a new index, $B2 - R/B + V + R$, to characterise the HB morphology, having defined $B2$ as the complement of B_W to B, i.e. $B2 = B - B_W$.

The behaviour of the index $B2 - R/B + V + R$ *vs.* $B - R/B + V + R$ is illustrated in Figure 1 of B1. This figure shows that clusters with blue HB tails appear compressed into the interval $0.9 < B - R/B + V + R < 1.0$ (and are therefore practically indistinguishable from one another), while the index $B2 - R/B + V + R$ spans a much larger interval. For instance, the old index classifies clusters like Arp 2, NGC 5897, NGC 6341 and NGC 6218 in the same HB-morphology class ($B - R/B + V + R \sim 0.88$) whereas $B2 - R/B + V + R$ ranks them according to the values $-0.22, 0.09, 0.45$ and 0.77.

3. New insights into the HB morphology

3.1. *The catalogue*

For the present work we have used the catalogue compiled by FPAL with some updating and revisions. The data on the abundances were taken by Zinn (1985), while for the structural parameters we adopted the compilation of Webbink (1985). L_t and $(B-V)_{peak}$ are taken from FPAL (we have measured a few new entries). To calculate $B2 - R/B + V + R$, and $B - R/B + V + R$ we have counted the values B2, B, V, R directly from the published CM diagrams, in general finding an excellent correspondence with the independent counts of LDZ.

3.2. $(B2 - R)/(B + V + R) vs. (B - R)/(B + V + R)$

One of the most interesting results obtained by LDZ is that, for any given [Fe/H], the mean HB type tends to become redder and the scatter in HB type becomes larger with increasing R_G, as in Searle & Zinn (1978).

In order to interpret these findings, LDZ computed theoretical isochrones produced by their code for synthetic HBs under the assumption of fixed mass-loss along the RGB. They concluded that the observed systematic variation in HB morphology can be explained if the stellar population in the inner halo ($R < 8 Kpc$) is relatively old and coeval, whereas the mean age of the clusters in the outer halo is lower and largely non-uniform.

However, having already considered that the index $B2 - R/B + V + R$ offers a better resolution for clusters with HB blue tails, and considered that some 50% of clusters in the inner halo have assigned $B - R/B + V + R > 0.5$, it is useful to have a better insight into the problem. This can be done following B1, who replotted the same data points of LDZ using the new HB-morphology index. The plot $B2 - R/B + V + R$ vs. [Fe/H] for clusters with $R_G < 8 Kpc$ reveals immediately that clusters in the inner halo which form a narrow sequence in the plane $B - R/B + V + R$ vs. [Fe/H], show a large scatter when the index $B2 - R/B + V + R$ is used to classify the HB morphology. If one accepts that the inner halo clusters are coeval, then the dispersion above mentioned indicates that *another parameter, which is not the age, is necessary to explain the different blue HB morphologies*. In the following we will try to identify this additional second parameter.

3.3. $(B - V)_{peak}$ vs. $(B2 - R)/(B + V + R)$ and $log \rho_0$

Figure 1 shows the correlation between $(B - V)_{peak}$ and $B2 - R/B + V + R$. The linear relationship of this figure is very interesting, because it indicates that the HB morphology as measured by the new index depends mostly on the peak of the colour distribution of the HB stars, and that the dispersion in colour plays only a marginal role. LDZ came to the same conclusion on the basis of their synthetic HB calculations.

A look at Figure 1, however, shows at once that the data points present a large scatter around the mean line: should one find a correlation of the residuals with some physical parameter, then the new candidate $2^{nd}P$ would soon be identified.

Therefore, having computed a linear best fit to the data in Figure 1, we computed the difference $(B2 - R/B + V + R) - [B2 - R/B + V + R]_{fit}$ and plotted in Figure 2 these residuals vs. $log \rho_0$, the central mass density. Again, a clear correlation emerges, showing that the higher the central density, the bluer the HB morphology becomes. Note that this effect is not due to a group of clusters possibly peculiar under some aspects. In fact, there seems to be a physical mechanism which tends to alter the distribution of the stars along the HB. This trend is even clearer when we select those clusters with $[Fe/H] = -1.6 \pm 0.3$, i.e. clusters in the range of metallicity where the maximum sensitivity to any variation of mass loss is expected. This plot confirms the correlation between high central density and HB morphology. Let it be clear that we are not implying that the age is not a $2^{nd}P$, but that structural parameters (i.e. central mass density) *and* age are the most important among the possible $2^{nd}Ps$.

The conclusions of this section, which is based entirely on empirical arguments, are the following:

(i) The HB morphology depends mostly on the location of the peak of the colour distribution.

(ii) For any given $(B - V)_{peak}$ the range of $B2 - R/B + V + R$ is correlated with

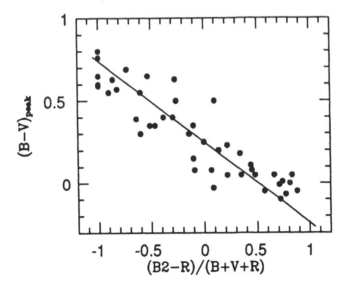

FIGURE 1. The peak of the de-reddened colour HB distribution, $(B-V)_{peak}$, plotted *vs.* the HB morphology parameter. Note that measurement errors alone cannot account for the dispersion along the linear relationship.

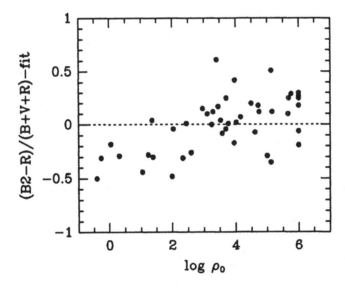

FIGURE 2. The residuals along the linear relationship of Figure 1, plotted *vs.* the central mass density of the cluster. The trend with $log\rho_0$, clearly visible in the figure, is strengthened plotting only clusters with $-1.9 < [Fe/H] < -1.3$.

$log\rho_0$, the central mass-density. This correlation is particularly strong for clusters of intermediate metallicity (i.e. $[Fe/H] = -1.6 \pm 0.3$)

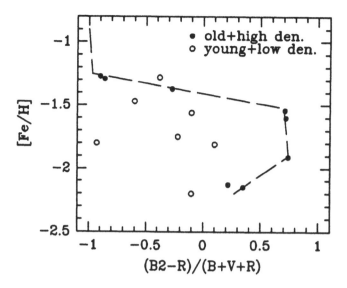

FIGURE 3. The HB morphology index, plotted *vs.* the metallicity for clusters of different characteristics (circles: clusters which are relatively young and with a low central density; dots: clusters which are relatively old and with a high central density).

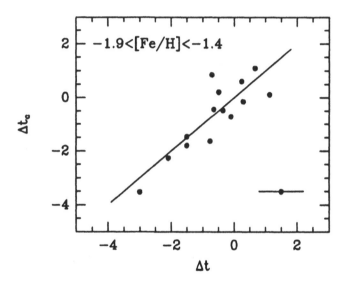

FIGURE 4. The cluster relative ages obtained through the analysis of the turn-off regions (abscissa), plotted *vs.* the cluster ages derived via Equation 2. The *rms* with respect to the 45° line gives $\sigma \sim 1$ Gyr

4. Horizontal-branch morphology as an age indicator

The effects of the central density in designing the HB morphology (bluer HBs corresponding to higher $log\rho_0$), may be interpreted as an effect of dense cluster environments on the rate of stellar interactions and, then, on the amount of mass lost by HB progenitors (Buonanno *et al.* 1985). As we mentioned before, this does not exclude age from

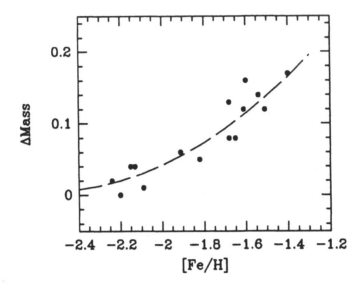

FIGURE 5. The difference between the stellar initial mass and the ZAHB mass, plotted *vs.* metallicity for nearly coeval clusters. The best fit to the data points reported in the figure as a dashed line gives an empirical estimate of the quoted mass loss.

also playing an important role, because the mean mass of the HB stars decreases with the age in correspondence with the decrease of mass along the RGB.

A complex picture is therefore emerging where, in addition to [Fe/H] (the *first parameter*), both the age and the central concentration have important effects on the HB morphology. In order to use this to date a given globular cluster, the effects of the age and of $log\rho_0$ must be disentangled.

The first step, then, is to select a sample of clusters with reliable ages. Buonanno & Iannicola (1994) reviewed the techniques used to determine relative cluster ages, and reported that the method proposed by Buonanno *et al.* (1994, hereafter BCPF), is the one which avoids the main drawbacks of the most commonly used techniques, namely ΔV_{HB}^{TO} (the magnitude difference between the horizontal branch and the turn-off at the colour of the latter) and $\Delta(B-V)_{TO-RGB}$ (the colour difference between the turn-off and the base of the red-giant branch).

In short, BCPF defined the parameter $\Delta V_{0.05}$ as the distance in magnitude between the HB and $V_{0.05}$, the point on the main sequence defined by van den Bergh *et al.* (1990), and then used $\Delta V_{0.05}$ to determine the ages of clusters with well-populated HBs. These ages, in turn, were used to determine the empirical relationship

$$\Delta(B-V)_{TO-RGB} = f([Fe/H], \Delta t)$$

where Δt is the relative age with respect to an arbitrary cluster.

BCPF preliminarily obtained

$$\Delta(B-V)_{TO-RGB}(ref) = 0.325(\pm0.008) + 0.049(\pm0.005)[Fe/H]$$

with a linear best fit to the data available for 12 clusters used as reference for the age, and

$$\delta/\Delta t = -0.006 \pm 0.001 mag/Gyr$$

where $\delta = \Delta(B-V)_{TO-RGB} - \Delta(B-V)_{TO-RGB}(ref)$. This relationship is strictly

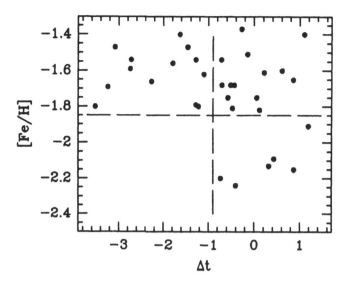

FIGURE 6. The (relative) age of globular clusters as found in this paper, plotted *vs.* their metallicity. Dashed lines visualise the indication that the Galaxy produced (or captured) GCs of any metallicity for a couple of gigayears. Subsequently, only clusters more metal-rich that $[Fe/H] \sim -1.8$ were created.

valid only for clusters with $[Fe/H] < -1$ because the helium abundance seems not yet to be well-established for very metal-rich clusters.

It is possible, at this point, to group clusters using two parameters: age and central mass density. In particular we defined four groups with nearly the same number of objects, namely: A—clusters with low density and old age; B—clusters with high density and old age; C—clusters with high density and young age; and D—clusters with low density and young age.

It is interesting to note that clusters of different regions show distinct HB morphologies. In particular, clusters in groups B and D show for the same value of [Fe/H] a dramatic change of the index $B2 - R/B + V + R$, as illustrated in Figure 3. We therefore concluded that, in order to understand fully the HB morphology, the three main parameters which concur in determining such morphology, i.e. metallicity (the first parameter), central mass-density and age (the second parameters), must be disentangled.

A linear best fit to the data points in the interval of metallicity $-1.9 < [Fe/H] < -1.4$ led to the following equation

$$(1) \qquad \frac{B2 - R}{B + V + R} = .17(\pm 0.015)\log\rho_0 + .29(\pm.025)\Delta t - .10(\pm.057)$$

Isolating the age, Equation 1 gives

$$(2) \qquad \Delta t = 3.45\frac{B2 - R}{B + V + R} - .58\log\rho_0 + .34$$

(valid for $-1.9 < [Fe/H] < -1.4$; Δt in Gyr).

Written in this form, the equation discloses the importance of the term in $\log\rho_0$. In fact, one immediately sees that a difference $\Delta\log\rho_0 \sim 3.5$ between two clusters is equivalent, in terms of HB morphology, to a difference $\Delta t \sim 2$ Gyr.

Finally, in order to show the accuracy of Equation 2, we plotted in Figure 4 the ages in input *vs.* the ages obtained via Equation 2. The rms computed with respect to the

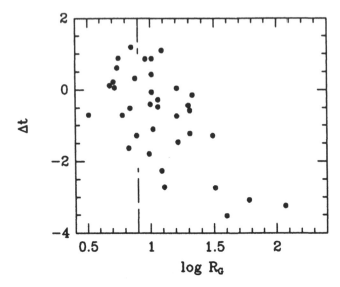

FIGURE 7. The age of the same clusters of Figure 6, plotted *vs.* R_G, the galactocentric distance. Note that only clusters with reliable age estimate (i.e. clusters more metal-poor than $[Fe/H] \sim -1.4$ are reported.

45° line plotted in the figure gives $\sigma \sim 1$ Gyr. Considering that this is of the order of the intrinsic error of the ages in input, we conclude that errors of the calibrators are the dominant source of uncertainty.

An additional step is worthy of mention. Having selected a sample of coeval clusters, one can use the models of Buser & Kurucz (1978, 1992) to transform $(B - V)_{peak}$ into $(log T_e)_{peak}$ for these clusters and then use Figure 1 of FPAL to compute ΔM, i.e. the average mass loss for cluster stars before reaching the ZAHB.

Figure 5 shows ΔM *vs.* [Fe/H] for the sample of coeval clusters.

5. Conclusions

The main result presented in this paper is that the HB morphology of intermediate-metallicity clusters is an excellent indicator of the age, provided that we take into consideration the effects of the cluster environments. Actually, one can easily infer this age by simply counting the number of HB stars in different colour intervals and by applying Equation 2.

Using this procedure, we derived the age of about 30 globulars in the interval $-1.9 < [Fe/H] < -1.4$ and plotted these ages *vs.* [Fe/H] and $log R_G$, the galactocentric distance, respectively in Figures 6 and 7 (the ages of six metal-poor clusters, $[Fe/H] < -1.9$, appearing in these figures were taken directly by BCPF).

The major implications of Figures 6 and 7 are that the Galaxy apparently produced globular clusters of all metallicities for a couple of gigayears, and that after such early periods, only clusters more metal-rich than $[Fe/H] < -1.8$ were produced by (or coalesced to) the Galaxy. In addition, a gradient "age *vs.* R_G" seems to be well defined, with almost constant dispersion in age. This supports the Searle & Zinn (1978) scenario of a protogalaxy which was formed from inside out over several gigayears.

REFERENCES

ARP, H. C., BAUM, W. A. & SANDAGE, A. R. 1952 *Astron. J.* **75**, 4.

BUONANNO, R. 1993 in *The Globular Cluster-Galaxy Connection* (ed. G. Smith & J. Brodie). ASP Conference Series, vol. 48, p. 131. (B1)

BUONANNO, R., CORSI, C. E. & FUSI PECCI, F. 1985 *Astron. Astrophys.* **145**, 97.

BUONANNO, R., CORSI, C. E., FUSI PECCI, F., RICHER, H. B. & FAHLMANN, G. G. 1993 *Astron. J.* **105**, 184.

BUONANNO, R., CORSI, C. E. FUSI PECCI, F., RICHER, H. B. & FAHLMANN, G. G. 1994 *Astrophys. J., Lett.* **430**, L121.

BUONANNO, R., CORSI, C. E., PULONE, L. & FUSI PECCI, F. 1994, in preparation. (BCPF)

BUONANNO, R. & IANNICOLA, G. 1994 in *Nuclei in the Cosmos*. Third International Symposium on Nuclear Astrophysics, in press.

BUSER, R. & KURUCZ, R. 1978 *Astron. Astrophys.* **70**, 555.

BUSER, R. & KURUCZ, R. 1992 *Astron. Astrophys.* **264**, 557.

CHIEFFI, A. & STRANIERO, O. 1989 *Astrophys. J., Suppl. Ser.* **71**, 47.

FAULKNEN, J. 1966 *Astrophys. J.* **144**, 978.

FUSI PECCI, F., FERRARO, F., BELLAZZINI, M., DJORGOVSKY, S., PIOTTO, G. & BUONANNO, R. 1993 *Astron. J.* **105**, 1145. (FPAL)

HURLEY, D. J. C., RICHER, H. B. & FAHLMANN, G. G. 1989 *Astron. J.* **98**, 2124.

LEE, Y. W. 1989 PhD thesis, Yale University.

LEE, Y. W. 1992 *Astron. J.* **104**, 1780.

LEE, Y. W., DEMARQUE, P. & ZINN, R. 1988, in *Calibration of Stellar Ages* (ed. A. G. D. Philip), p. 14. L. Davis Press, Schenectady.

LEE, Y. W., DEMARQUE, P. & ZINN, R. 1994 *Astrophys. J.* **423**, 248. (LDZ)

PRESTON, G. W., SHECTMAN, S. A. & BEERS, T. C. 1991 *Astrophys. J.* **375**, 121.

ROSINO, L. 1965 *Kl. Veroff. Bamberg Sternw.* 4, 9, No 40.

SANDAGE, A. R. & WALLERSTEIN, G. 1960 *Astrophys. J.* **131**, 598.

SANDAGE, A. R. & WILDEY, R. 1967 *Astrophys. J.* **150**, 469.

SEARLE, L. & ZINN, R. 1978 *Astrophys. J.* **225**, 357.

SHAPLEY, H. 1919 *Astrophys. J.* **49**, 96.

VANDENBERG, D. A., BOLTE, M. J. & STETSON, P. B. 1990 *Astron. J.* **100**, 445.

VAN DEN BERGH, S. 1967 *Astron. J.* **72**, 70.

WEBBINK, R. F. 1985 in *Dynamics of Star Cluster* (ed. J. Goodman & P. Hut). IAU Symposium 113, p. 541. Reidel.

ZINN, R. 1980 *Astrophys. J., Suppl. Ser.* **42**, 19.

ZINN, R. 1985 *Astrophys. J.* **293**, 424.

ZINN, R. 1986 in *Stellar Populations* (ed. C. A. Norman, A. Renzini & M. Tosi), p. 732. Cambridge University Press.

Shore: Did you find any correlation between the presence of "HB blue tails" and that of millisecond pulsars (MSPs)?

Buonanno: Observations have shown that GCs contain a large pulsar population, even though only a small fraction of them are currently observable in the radio surveys. Considering that several GCs in which MSPs have been discovered fall in a range of metallicity where the HB is not particularly sensitive to the the enviromental parameters (47 Tuc, Terzan 5, M4, NGC 6760, NGC 6539, etc.), we have not yet looked for any correlation. However, as the number of MSPs is expected to scale with ρ_c^α with $\alpha \sim 0.5$, one can probably anticipate that such correlation exists.

Carney: Your mass loss *vs.* metallicity results could be compared to A. Cox's results for RR Lyrae masses (from double-mode systems) with those of model turn-off masses. Have you had a chance to check that yet?

Buonanno: The stellar pulsation theory, as applied to double-mode pulsators by Cox *et al.*, leads to typical mass $0.55\ M_\odot$ and $0.65\ M_\odot$ for the Oosterhoff I and Oosterhoff II clusters, respectively. Adopting a metallicity of $[Fe/H] = -2$ for the OoII globulars and $[Fe/H] = -1.4$ for the OoI and using the models of Chieffi & Straniero (1989) for the turn-off masses, one obtains a mass-loss difference $\Delta M \sim 0.1 M_\odot$ for these two metallicities. The relation shown in Figure 6 gives $\Delta M \sim 0.13$ for the same two values of metallicity.

Chaboyer: When you fit $\Delta(B - V)\,vs.[Fe/H]$ for coeval GCs, what is the dispersion about the fit — i.e. what is your error estimate in the derived ages?

Buonanno: The relation you quote is based primarily on the selection of a sample of coeval clusters. This selection is based on the "vertical" parameter $\Delta V_{0.05}$ which gives intrinsic errors ranging from $\Delta t = \pm 0.5$ Gyr (for well-studied clusters) to $\Delta t = \pm 2$ Gyr.

Martínez: Do you think there is some influence of density in the production of RR Lyrae in GCs?

Buonanno: Yes. My idea is that the environment, affecting the mass-loss rate, may shift red HBs to the blue, and then favour the production of RR Lyrae.

Surdin: Why do you use in your plots the value of $log \rho_0$? It seems to me that from a dynamical point of view it is more reasonable to use ρ_h (half-mass density) because it is constant during a long period of the cluster evolution.

Buonanno: Preliminary tests seem to indicate that most of the correlations are lost passing from $log \rho_0$ to $log \rho_h$.

Ages of Galactic Globular Clusters from the New Yale Isochrones

By BRIAN CHABOYER[1], P. DEMARQUE[2],
D. B. GUENTHER[3], M. H. PINSONNEAULT[4]
AND L. L. PINSONNEAULT[5]

[1]Canadian Institute for Theoretical Astrophysics, Toronto, Ontario, M5S 1A7, Canada

[2]Department of Astronomy, Yale U. P.O. Box 208101, New Haven, CT 06520–8101, U.S.A.

[3]Department of Astronomy and Physics, Saint Mary's U. Halifax, NS, B3H 3C3, Canada

[4]Deparment of Astronomy, Ohio State U. 174 W. 18th Ave. Columbus, OH 43210–1106, U.S.A

[5]Houston, Texas, U.S.A.

A new grid of theoretical isochrones based on the Yale stellar evolution code using the OPAL and Kurucz opacities has been constructed. The grid of isochrones spans a wide range of metallicities, helium abundances and masses. The construction of the isochrones is described and the isochrones are compared to galactic globular cluster observations. A solar calibrated mixing length ($\alpha = 1.7$) yields a good fit to globular cluster colour-magnitude diagrams. Ages for 40 globular clusters are determined using the $\Delta V(TO - HB)$ method and the formation of the halo is discussed.

1. Introduction

There have been important recent advances in the calculation of stellar opacities (Iglesias & Rogers 1991; Kurucz 1991, and private communication). For this reason, we have embarked on a project to construct a new grid a theoretical isochrones which use the latest avaliable physics, and cover a wide range of physical parameters. The scientific goals of this project are threefold: (i) to test the theory of stellar structure and evolution by using state of the art physics to construct isochrones and compare to observations; (ii) to provide a uniform grid of isochrones which can be used to determine stellar ages; and (iii) to upgrade the building blocks for population synthesis to study distant galaxies and galaxy formation. In my talk, I will only discuss the first two points.

2. Isochrone construction

New stellar evolution models have been calculated which extend from the zero-age main sequence to the tip of the giant branch (the construction of core ^{4}He burning stars is still in progress). These models use the OPAL high temperature opacities (Iglesias & Rogers 1991) and the low temperature opacities are from Kurucz (1991). The nuclear reaction rates are from Bahcall & Pinsonneault (1992). Stellar models with masses ranging from 0.5 $M_\odot$ to 7.0 $M_\odot$ were calculated. The heavy-element composition of the models varied from $Z = 2 \times 10^{-6}$ to $Z = 0.10$. For each metallicity, two to four different ^{4}He abundances were used. One of the great uncertainties in present stellar evolution calculations is the treatment of convection. For this reason, we have varied two parameters which deal with convection: (i) the mixing length, and (ii) the amount of over-mixing beyond the convective core.

The conversion from stellar models to isochrones was performed using the method of equal evolutionary points (Prather 1976). The construction of the isochrones themselves is independent of the choice of colour calibration. In this paper, the semi-empirical colour

Z	[Fe/H]	$[\alpha/\text{Fe}]$	Y	Mixing Length	over-shoot (H_p)	Age (Gyr)
2×10^{-6}	-4.3	0.4	0.20, 0.23, 0.26	1.5, 1.7, 2.0	0.0	$4 - 22$
2×10^{-5}	-3.3	0.4	0.20, 0.23, 0.26	1.5, 1.7, 2.0	0.0	$4 - 22$
6×10^{-5}	-2.8	0.4	0.20, 0.23, 0.26	1.5, 1.7, 2.0	0.0	$4 - 22$
2×10^{-4}	-2.3	0.4	0.20, 0.23, 0.26	1.5, 1.7, 2.0	0.0	$2 - 22$
6×10^{-4}	-1.8	0.4	0.20, 0.23, 0.26	1.5, 1.7, 2.0	0.0	$2 - 22$
2×10^{-4}	-1.3	0.4	0.20, 0.23, 0.26	1.5, 1.7, 2.0	0.0	$2 - 22$
4×10^{-4}	-0.9	0.3	0.20, 0.23, 0.26	1.5, 1.7, 2.0	0.0	$0.5 - 22$
7×10^{-4}	-0.6	0.2	0.20, 0.23, 0.26	1.5, 1.7, 2.0	0.0	$0.5 - 22$
0.01	-0.4	0.2	0.25, 0.27, 0.30, 0.35	1.7	0.0, 0.1	$0.05 - 22$
0.02	$+0.0$	0.0	0.25, 0.27, 0.30, 0.35	1.5, 1.7, 2.0	0.0, 0.1	$0.05 - 22$
0.04	$+0.3$	0.0	0.32, 0.34	1.7	0.0	$0.05 - 22$
0.06	$+0.5$	0.0	0.36, 0.40	1.7	0.0	$0.05 - 22$
0.10	$+0.7$	0.0	0.44, 0.52	1.7	0.0	$0.05 - 22$

TABLE 1. Parameters for our grid of isochrones

calibration of Green *et al.* (1987) was used to transform from the theoretical luminosities and temperatures to observational magnitudes and colours. The uncertainties are largest at the lowest temperatures. The stellar models use a scaled solar heavy-element mixture. The effects of α-element enhancement are taken into account by adjusting the relationship between [Fe/H] and total Z (Salaris *et al.* 1993). The full isochrone grid is tabulated in Table 1. These isochrones will be available via anonymous ftp (release date around January, 1995); contact demarque@astro.yale.edu for further details.

3. Isochrone fits

The isochrones have been compared to observed colour-magnitude diagrams (CMDs) for a variety of globular clusters. A typical example is shown in Figure 1, which shows the fit to M68 ([Fe/H] = -2.1) for two different mixing lengths: $\alpha = 1.7$ and $\alpha = 2.0$. The 16 Gyr $\alpha = 1.7$ isochrone provides a superb fit from the lower main sequence, through the main sequence turn-off and subgiant branch, all the way to the tip of the red-giant branch. The $\alpha = 2.0$ (and $\alpha = 1.5$) isochrones do not provide a good match to the observations for any reasonable values of the reddening and distance modulus. Similar conclusions are reached for globular clusters with higher metallicities, though our fit to shape of the 47 Tuc ([Fe/H] = -0.7) CMD is not as good. Our solar calibrated models require a mixing length of $\alpha = 1.7$. Thus, one value of the mixing length provides a good match to the Sun and metal-poor stars.

4. ΔV(TO − HB) ages

A commonly used age indicator which is insensitive to reddening and uncertainties in our treatment of convection and stellar atmospheres is the difference in magnitude between the turn-off and the horizontal branch, ΔV(TO − HB) (e.g. Iben & Renzini 1984). The ages derived using ΔV(TO − HB) are well understood theoretically as they do not depend on the model colours. As we have not yet calculated horizontal-branch models, the turn-off magnitudes from our isochrones have been combined with the observed RR Lyr magnitudes ($M_v = 0.20$[Fe/H] $+ 1.06$, Carney *et al.* 1992; Skillen *et al.* 1993) to calculate ages for 40 halo globular clusters which have observed ΔV(TO − HB) values. An

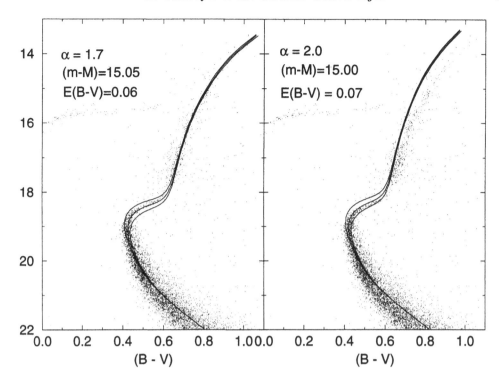

FIGURE 1. The observations of M68 (dots) by Walker (1994) are compared to the best fitting theoretical isochrones with ages of 14, 16 and 18 Gyr (solid lines). Isochrones with a mixing length of 1.7 are shown in the left panel, while those with $\alpha = 2.0$ are shown in the right panel. The reddening and distance modulus used to shift the isochrones are shown on the graph.

age spread of ~ 4 Gyr is clearly present. There is no significant correlation between age and galactocentric distance, but an age-metallicity relationship does exist in the outer halo, and is shown in Figure 2. From this, we can conclude that the halo of our Galaxy formed over an extended period of time, with an gradual enrichment in the heavy-element content in the outer halo.

BC is supported by NSERC of Canada. Research supported in part by NASA grants NAG5–1486, NAGW–2136, NAGW–2469 and NAGW–2531 to Yale University.

REFERENCES

BAHCALL, J. N. & PINSONNEAULT, M. H. 1992 *Rev. Mod. Phys.* **64**, 885.

CARNEY, B. W., STORM, J. & JONES, R. V. 1992 *Astrophys. J.* **386**, 663.

GREEN, E.M., DEMARQUE, P. & KING, C. R. 1987 *The Revised Yale Isochrones and Luminosity Functions* Yale Univ. Obs., New Haven.

IBEN, I., JR. & RENZINI, A. 1984 *Phys. Rept.* **105**, 329.

IGLESIAS, C. A. & ROGERS, F. J. 1991 *Astrophys. J.* **371**, 408.

KURUCZ, R. L. 1991 in *Stellar Atmospheres: Beyond Classical Models* (ed. L. Crivellari, I. Hubeny & D. G. Hummer), p. 440. Kluwer.

PRATHER, M. 1976 PhD thesis, Yale University.

SALARIS, M., CHIEFFI, A. & STRANIERO, O. 1993 *Astrophys. J.* **414**, 580.

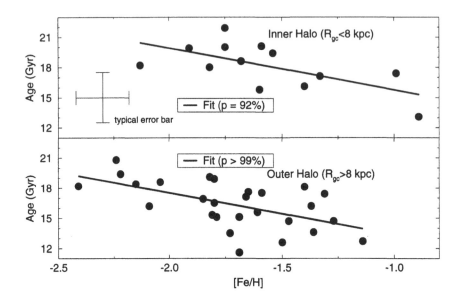

FIGURE 2. Ages of globular clusters as a function of metallicity for the inner halo (upper panel) and outer halo (lower panel). The solid line shows a least squares fit to the data. The correlation is significant at greater than the 99% confidence level for the outer halo clusters. In contrast, the significance of the correlation for the inner halo clusters is only 92%.

SKILLEN, I., FERNLEY, J. A., STOBIE, R. S. & JAMESON, R. F. 1993 *Mon. Not. R. Astron. Soc.* **265**, 301.

Aparicio: The observed luminosity function is a powerful tool to test the goodness of stellar evolutionary models. How good is the agreement with your models?

Chaboyer: Our models do a good job of predicting the relative numbers of stars on the red-giant branch. At present, there are few observed luminosity functions which extend from the main sequence up the giant branch. The only one I am aware of was recently published by M. Bolte. There appears to be a discrepancy between the theory and observations. More observations are needed to access the significance of this problem.

Aparicio: What is the critical mass for the helium shell flash in your models?

Chaboyer: I have not examined this issue.

Carraro: A point about the Green *et al.* (1987) colour transformation. Basically they were built up to study the metal-poor globular clusters population. At what level of confidence can they be used for supersolar, younger clusters (e.g. NGC 6791)?

Chaboyer: I have not looked into this question too closely, as I have only been working on the metal poor part of this project.

Kaluzny: It seems to me that your isochrones do not provide a good fit for the lower main-sequence. Some systematic deviations are visible on the M_V vs. $(B - V)$ and M_V vs. $(V - I)$ CMDs.

Chaboyer: The CMDs I have compared my isochrones to were picked because they included extensive giant branches. These CMDs are not that deep, so that there is a fair bit of scatter in the observations on the lower main sequence. I don't see any large systematic deviations between the theory and observations on the lower main sequence (see Figure 1). It would be best to compare the isochrones to some deep CMDs in order to fully address this question.

AGB and RGB Stars as Tracers of the Early and Intermediate Star-Formation History

By ANTONIO APARICIO AND CARME GALLART

Instituto de Astrofísica de Canarias, E-38200, La Laguna, Tenerife, Canary Islands, Spain

1. Introduction

The Milky Way and Andromeda galaxies are the largest members of the Local Group, and their evolution is affected by the evolution of their host as a whole. At the same time, they themselves play an important role in the evolution of the Local Group. Considerable information can be obtained for the Local Group, but little is known about the distances and the full star-formation history of its galaxies.

RGB and AGB stars are the keys to trace the full star-formation history of nearby galaxies. These stars are usually the most prominent population of dwarf spheroidal galaxies, but it has been shown (Gallart *et al.* 1994; Aparicio & Gallart 1994) that they are also observable in dwarf irregular galaxies. This will open the door to the study of the earliest star-formation processes taking place in these galaxies. The star-formation history of the Local Group galaxies is a crucial piece of information for answering basic questions about the evolutionary history of the group.

2. Some basic questions about the Local Group

• Can the star formation history of dwarf spheroidals (dSphs) be explained as a function of their distances to the major spirals?

• Could the fact that dwarf irregulars (dIrs) are not found in the vicinity of the major spirals be due to these galaxies inhibiting star formation in their neighbours after some hundred million years?

• If the star formation in dIr is produced by bursts, are these bursts driven by the passage of the dIr close to a spiral?

• Is the Local Group a bound system? How many of the neighbouring galaxies are actually bound to it? Moreover, are Andromeda and the Milky Way a bound pair?

3. How to get answers

To study the star-formation history of the dwarf galaxies and to determine their distances and radial velocities, both are necessary pre-requisites for attaining a full understanding of the dynamical and evolutionary history of the Local Group and its members, including star-formation and chemical enrichment processes.

The full star-formation history of the dwarf galaxies, including the oldest populations, must be inferred if global information is to be obtained. For dIrs, only the youngest populations have usually been analysed. But, as we have pointed out, it is also possible to study their oldest stars.

4. Ways to do it

• The colour-magnitude (CM) diagram with red colours (V, R and I) reveals the old

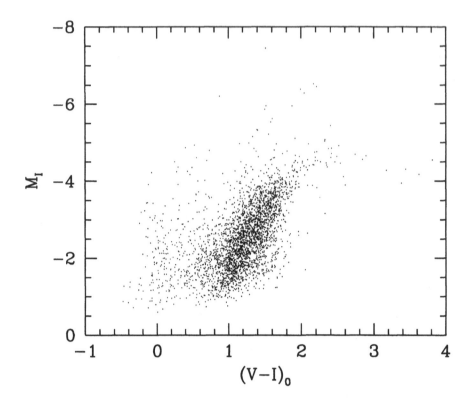

FIGURE 1. M_I vs. $(V - I)_0$ CM diagram of the Pegasus dwarf irregular galaxy, taken from Aparicio & Gallart (1994). Stars formed from the first generations up to the present are revealed.

and intermediate-age stars: the AGBs and the RGBs. Examples of this are the CM diagrams obtained for NGC 6822 and the Pegasus D IG by Gallart *et al.* (1994) and Aparicio (1994). Figure 1 shows the CM diagram of Pegasus, taken from Aparicio & Gallart (1994).

• Synthetic CM diagrams built with varying SFR and chemical enrichment law, to analyse the star-formation history in detail, by comparing them with the observational CM diagrams.

REFERENCES

APARICIO, A. 1994 *Astrophys. J., Lett.* **437**, L27.

APARICIO, A. & GALLART, C. 1995 in *The Local Group: Comparative and global properties.* III ESO/CTIO workshop, in press.

GALLART, C., APARICIO, A., CHIOSI, C., BERTELLI, G. & VÍLCHEZ, J. M. 1994 *Astrophys. J., Lett.* **425**, L9.

Radial Velocities of Nearby RR Lyrae Stars

By DAVID MARTÍNEZ[1], E. J. ALFARO[1]
AND A. HERRERO[2]

[1]Instituto de Astrofísica de Andalucía (CSIC), P. Box 3004, 18080 Granada, Spain

[2]Instituto de Astrofísica de Canarias, La Laguna, Tenerife, Spain

A survey of chemical abundances and radial velocities for a spatially selected sample of RR Lyrae stars has been started. Assuming an absolute magnitude of −.6 for these objects, this leads to a near spherical volume of 2 kpc radius to be explored. Special emphasis is devoted to those RR Lyrae stars contained in the Input Catalogue of the *Hipparcos* mission. Radial velocities, proper motions and abundances for these objects should provide us with a useful database for studies of galactic structure.

1. Introduction

The RR Lyrae stars have been used extensively for studying the halo and thick disk structure. Of the roughly 6000 catalogued field objects (Kholopov 1985), the widest range of types is located in a region of about 2.5 kpc around the Sun, where a surplus of metal-rich variables is found, that is not present in the globular cluster or Baade's Window (Walker & Terndrup 1991). Nevertheless, the actual sample is very heterogeneous and reveals important gaps in their observational data. This introduces an important bias when RR Lyraes are used in order to obtain information about the kinematical and chemical structure of the Milky Way. Recently, Layden (1994) determined chemical abundances and radial velocities for an *in situ* sample of RR Lyrae stars, thus improving the size and quality of the current database in the solar vicinity. We are continuing this work by determining radial velocity estimates for those stars in the Layden catalogue for which measurements are poor or even lacking, also including RRc-type variables which were not studied by Layden. Special emphasis will made on those RR Lyrae stars included in the Input Catalogue of the *Hipparcos* mission. The astrometric data provided by *Hipparcos*, together with reliable chemical abundances and radial velocities from ground based telescopes, should produce a high quality set of local RR Lyraes, especially suitable for galactic studies.

2. The sample

A sample of about 350 nearby RR Lyrae stars has been selected from the General Catalogue of Variable Stars (Kholopov 1985), including all catalogued stars up to a cutoff limiting magnitude of $\bar{V}$ =13. RR Lyraes of type c are also included. Stars with absolute galactic latitudes lower than 10 were removed from the input set due to the high extinction in this galactic region.

Assuming an absolute magnitude, independent of metallicity, of −0.6, the selected sample is distributed in a near spherical volume, 2 kpc in radius, centred on the Sun. 181 RR Lyraes out of this sample are contained in the *Hipparcos* Input Catalogue, for which accurate proper motions and parallaxes will be published in the near future. A preliminary catalogue containing all the available kinematical and chemical data on these stars has been tailored with a view to selecting the highest priority targets.

Name	φ	Vr	σ	N
CH AQL	0.85	28	11.5	2
ST BOO	0.18	-9.5	27.6	2
AE BOO	0.85	94	5.6	3
TV CRB	0.11	-170	2.6	3
UY CYG	0.86	48	13.6	3
XZ CYG	0.82	103	2.8	2
SU DRA	0.52	145	2.1	2
XZ DRA	0.56	-5	–	1

TABLE 1. Preliminary results of radial velocity of RR Lyrae stars

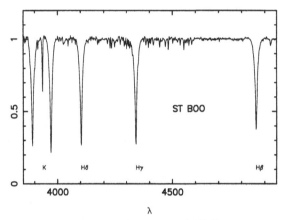

FIGURE 1. Spectrum of ST Boo

3. Observational plan

The main objective of this project is to obtain chemical abundances and radial velocities for a large sample of RR Lyrae stars. The ΔS index (Suntzeff *et al.*, 1994) will be used to estimate the chemical abundance. This will allow us to obtain reliable metallicity estimates (error around .15 dex) as well as to calculate radial velocities from spectra of moderate integration times. A wavelength range between 3500–5000 Åwith a resolution of about 3 Ådefines the spectral characteristics of our data. Intermediate aperture telescopes—such as the Isaac Newton Telescope (INT) at La Palma and the 2.2-meter telescope at Calar Alto—and CCD receptors provide high quality spectra (signal-to-noise ratio > 40) with integration times of less than 20 minutes. One observation run was already performed last May with the 2.2-metre telescope at Calar Alto. IRAF package is used both for reducing the frames as well as for obtaining radial velocities and equivalent widths from the calibrated spectra. The preliminary results of the first observation run are shown in Table 1.

REFERENCES

KHOLOPOV, P. N. 1985 *General Catalogue of Variable Stars, 4th Ed.*. Nauka, Moskow.

LAYDEN, A. 1994 *Astron. J.* **108**, 1016.

SUNTZEFF N. B., KRAFT, R. P. & KINMAN, T. D. 1994 *Astrophys. J., Suppl. Ser.* **367**, 528.

WALKER, A. R. & TERNDRUP, D. M. 1991 *Astrophys. J.* **378**, 119.

Dust in Globular Clusters:
10 μm Imaging of ω Centauri

By LIVIA ORIGLIA[1], F. R. FERRARO[2]
AND F. FUSI PECCI[2]

[1]Osservatorio Astronomico di Torino, Strada Osservatorio 20, I–10025 Pino Torinese, Italy

[2]Osservatorio Astronomico di Bologna, Via Zamboni 33, I–40126 Bologna, Italy

Interstellar gas and dust should be present in the central region of the most massive and concentrated clusters due to mass-loss processes occurring in the post-MS stages of stellar evolution. For the first time, we secured 10 μm imaging of a 35" × 35" field in the central region of ω Cen. We detected concentrated thermal emission, in good positional coincidence with an *IRAS* point source, which can be attributed to the presence of dust in the circumstellar medium of evolved red stars.

1. Introduction

Interstellar matter should be present in galactic globular clusters (GGCs) due to mass loss processes of Population II stars, which is required to explain, for example, the observed morphology of the HB in the GGC colour-magnitude diagrams, the pulsational properties of the RR Lyrae stars, the absence of AGB stars brighter than the RGB tip (e.g. Fusi Pecci & Renzini 1976; Renzini 1977; Fusi Pecci *et al.* 1993, and references therein) and the presence of emission features in the spectra of RGB and AGB stars (reviewed by Dupree 1986). With the assumption that gas is swept out only when the cluster passes through the disk every 0.1 Gyr, one might expect a few 10–100 M$_\odot$ of intracluster matter, particularly in the central regions of the most massive clusters.

Recently, some evidence of dust in the intracluster medium has been found. Some clusters have *IRAS* emission in their core (Lynch & Rossano 1990); IR excesses around luminous giants and long period variables in 47 Tuc (Frogel & Elias 1988; Gillet *et al.* 1988) and multicolour polarisation of some clusters (Forte & Mendez 1989; Minniti *et al.* 1992) have also been detected.

2. Results

2.1. *IR Imaging*

Using the new ESO's mid-IR camera TIMMI (Käufl *et al.* 1992) mounted at the ESO 3.6-metre telescope, on June 2, 1994, we obtained a 10 μm detection in the N filter of the central region of ω Cen, and precisely in a 35" × 35" field about 70" south and 10" east from the cluster centre. A bright blob slightly southeast of the field centre with a peak emission at 24 mJy arcsec^{-2} over the local background has been detected. The integrated flux over a circle of 5" diameter is ~230 mJy. Other diffuse emission at about 5 mJy arcsec^{-2} brighter than the local background is also present all over the frame.

In order to check the presence of luminous red stars in the selected field and their contribution at the 10 μm emission, on June 24, 1994, we also secured J and K images using ESO's near-IR camera IRAC2 (Moorwood *et al.* 1992) mounted at the ESO/MPI 2.2-metre telescope. Within 10" of the 10 μm blob we found a red luminous star with J-K ~1.1 and K-N~0.1 which implies a F$_{10\mu m}$ <60 mJy. The estimated 10 μm flux of the bright blob cannot be simply explained in terms of photospheric emission by the red star which should contribute for $\leq 20\%$, and thus the excess could be due to the presence of a dusty circumstellar envelope.

2.2. *IRAS data*

Inside the ω Centauri core radius (4 pc) there is the *IRAS* point source 13237-4713 with 12 μm flux of $\sim$360 mJy. There is also evidence of extended emission at 12 μm within the central 4' with a total flux of $\sim$9 Jy. The other *IRAS* fluxes at 25, 60 and 100 μm are only upper limits or of low quality, those at 60 and 100 μm being also contaminated by cirrus emission. The *IRAS* point source is located at only 15" W and 23" N from the bright blob detected at 10 μm. Taking into account that the ellipse of uncertainty of the *IRAS* coordinates is 50" $\times$ 11" and the ground telescope pointing accuracy is $\geq$ 10", the probability is high of there being a true coincidence between the two detected emissions.

3. Discussion and conclusion

Assuming coincidence between 10 μm and IRAS emissions and using a $\sim \nu B_\nu$ dust emissivity, we obtained a colour temperature of $T_{10-12\mu m} \sim 190$ K, which is typical of a warm dust component observed in the circumstellar envelope of red, evolved stars, at 10–100 stellar radii (see, e.g., van der Veen & Habing 1988).

We also estimated the amount of dust at this temperature: M_{dust} $(T = 190K)$ $\sim 3 \times 10^{-7} M_\odot$, which is relatively small compared to the expected $\sim 10^{-3}$ $M_\odot$ if the stars lost 0.1–0.2 $M_\odot$ and assuming a gas to dust ratio of 100. Most of the dust could be indeed cooler than 190 K and more diffuse within the intracluster medium, that is at a larger distance than a few tens of stellar radii within the circumstellar envelope.

Warm dust detected at 10 μm seems indeed to trace the presence of an occuring mass loss event, while cooler and more diffuse dust in the intracluster medium might be interpreted as remnants of circumstellar envelopes at the end of their expansion process.

REFERENCES

DUPREE, A. K. 1986 *Annu. Rev. Astron. Astrophys.* **24**, 377.

FORTE, J. C. & MÉNDEZ, M. 1989 *Astrophys. J.* **354**, 222.

FROGEL, J. A. & ELIAS, J. H. 1988 *Astrophys. J.* **324**, 823.

FUSI PECCI, F., FERRARO, F. R., BELLAZZINI, M., DJORGOVSKI, G. S., PIOTTO, G. & BUONANNO, R. 1993 *Astron. J.* **106**, 1145.

FUSI PECCI, F. & RENZINI, A. 1976 *Astron. Astrophys.* **46**, 447.

GILLET, F. C., DEJONG, T., NEUGEBAUER, G., RICE, W. L. & EMERSON, J. P. 1988 *Astron. J.* **96**, 116.

KÄUFL, H. U., JOUAN, R., LAGAGE, P. O., MASSE, P., MESTREAU, P. & TARRIUS, A. 1992 *The Messenger* **70**, 67.

LYNCH, D. K. & ROSSANO, G. S. 1990 *Astron. J.* **100**, 719.

MINNITI, D., COYNE, G. V. & CLARIÁ, J. J. 1992 *Astron. J.* **103**, 871.

MOORWOOD A., FINGER G., BIEREICHEL P., DELABRE B., VAN DIJSSELDONK, A., HUSTER G., LIZON J-L., MEYER, M., GEMPERLEIN H. & MONETI A. 1992 *The Messenger* **69**, 61.

RENZINI, A. 1977, in *Advanced Stellar Evolution* (ed. P. Buovier & A. Maeder), p. 149. Saas-Fee, Geneva Observatory.

VAN DER VEEN, W. E. C. J. & HABING, H. J. 1988 *Astron. Astrophys.* **194**, 125.

Optical and Near-IR Photometry of the Bulge Globular Cluster NGC 6553

By MARIA D. GUARNIERI[1]†, P. MONTEGRIFFO[2],
S. ORTOLANI[3], A. MONETI[4],
A. BARBUY[5] AND E. BICA[6]

[1]Universitá degli Studi di Torino, Italy

[2]Dipartimento di Astronomia di Bologna, Bologna, Italy

[3]Osservatorio Astrofisico di Padova, Padova, Italy

[4]European Southern Observatory, La Silla, Chile

[5]Universidade de Sao Paulo, Brazil

[6]Universidade Federal do Rio Grande do Sul, Brazil

The bulge region of our Galaxy contains a number of globular clusters, the study of which can provide important information about protogalactic conditions after the initial collapse of the spheroid. The age, and dynamical and chemical conditions of the bulge can be also investigated. Unfortunately, they are hardly observable in the optical wavelength due to the heavy extinction in the direction of the galactic plane. For these reasons we have started a systematic study of a sample of metal-rich clusters in the infrared using the new NICMOS3 camera, 256×256 pixels, mounted at the MPI/ESO telescope at La Silla (Chile). Here, we present the first results of combined visual and near-infrared photometry in one of the most metal-rich globular clusters, NGC 6553 ($[Fe/H] \sim -0.2$). The optical counterparts come from Ortolani *et al.* (1990).

1. The colour-magnitude diagrams

In the left panel of figure 1 we show the (K, V–K) CMD for all the stars of the sample. This diagram confirms the peculiar nature of NGC 6553 with a very red, compact horizontal branch also in the IR diagrams. The giant branch is well defined in the whole extension up to K $\sim$ 6; a parallel redder and lower sequence is also visible in all the diagrams. Further investigations will confirm if this feature is due to a differential reddening or to the background field. The steep giant branch in the IR diagrams shows that the use of IR colours greatly reduces the blanketing effects of the cool stars in metal-rich clusters. The horizontal branch in Figure 1 can be localised at about K = 12.25 mag and V–K = 4.35. This measurement is much more accurate than the luminosity level obtained using optical CMDs; where it appears strongly tilted. From these values $V_{HB} = 16.60$ is deduced in a good agreement with Ortolani *et al.* (1990). Assuming $M_V^{HB} = 0.8$ for $[Fe/H] = -0.2$ (Fusi Pecci, F., private communication) we get a distance modulus of 13.35 corresponding to d = 4.8 Kpc from the Sun.

Another approach, not dependent on the data, makes use of the new Padua isochrones (Bertelli *et al.* 1994). For an age of $\sim$14 Gyr, the mean level of the horizontal branch is at $M_K = -1.45$, i.e. a dereddened distance modulus of 13.4 or about 4.8 Kpc, in excellent agreement with the above determination.

The right panel in Figure 1 shows the dereddened $(J - K)_0$ *vs.* $(V - K-)_0$ colour-colour diagram. Superimposed (filled circles) are the giant stars in 47 Tuc observed by Frogel *et al.* (1981). The sequence for the field giants (solid line) is from Frogel *et al.*

---

† Present address: Osservatorio Astronomico di Torino, Strada Osservatorio 20, 10025 Pino Torinese, Torino, Italy

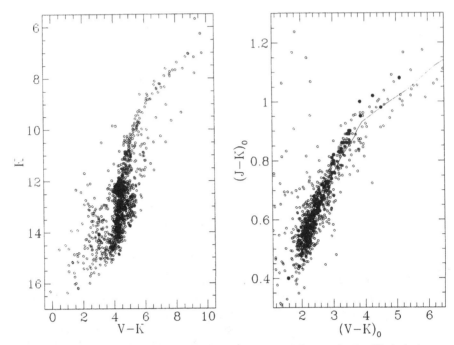

FIGURE 1. IR photometric diagrams of NGC 6553. In the right panel, the filled circles represent the giants in 47 Tuc observed by Frogel *et al.* (1981), and the line is the sequence for field giants given by Frogel *et al.* (1978)

(1978). There is a clear shift to higher $(V-K)_0$ values of the NGC 6553 points compared with 47 Tuc, while the locus of the field stars fits very well. If the shift is only an effect of metallicity (Ferraro *et al.* 1994; Bertelli *et al.* 1994) this indicates a nearly solar abundance for NGC 6553.

The details concerning data acquisition and reduction, and the full astrophysical discussion will be published in a separate paper (Guarnieri *et al.* 1994).

REFERENCES

BERTELLI, G., BRESSAN, A., CHIOSI, C., FAGOTTO, F. & NASI, E. 1994 *Astron. Astrophys., Suppl. Ser.* **106**, 275.

FERRARO, F. R., FUSI PECCI, F., GUARNIERI, M. D., MONETI, A., ORIGLIA, L. & TESTA, V. 1994 *Mon. Not. R. Astron. Soc.* **266**, 829.

FROGEL, J. A., PERSSON, S. E., AARONSON, M. & MATTHEWS, K. 1978 *Astrophys. J.* **220**, 75.

FROGEL, J. A., PERSSON, S. E. & COHEN, J. G. 1981 *Astrophys. J.* **246**, 842.

GUARNIERI, M. D., MONTEGRIFFO, P., ORTOLANI, S., MONETI, A., BARBUY, A. & BICA, E. 1994 in preparation.

ORTOLANI, S., BARBUY, B. & BICA, E. 1990 *Astron. Astrophys.* **236**, 362.

New VI CCD Photometry of NGC 1851

By IVO SAVIANE[1], G. PIOTTO[1], F. FAGOTTO[3],
S. ZAGGIA[1], M. CAPACCIOLI[1,2] AND A. APARICIO[3]

[1] Dipartimento di Astronomia, Università di Padova, Italy

[2] Osservatorio Astronomico di Capodimonte, Napoli, Italy

[3] Instituto de Astrofísica de Canarias, Spain

We present VI CCD photometry for nine fields towards the centre of NGC 1851 covering an area of 22×22 arcmin^2 , out to 1.2 tidal radii. These were observed during two nights in February 1993 at the ESO 3.5-metre NTT telescope + EMMI. Our sample consists of $\sim 20\,500$ stars reaching V = 22 or I = 21.5 at the 50% completeness level. These data have been collected within a project of mapping the stellar distribution of a sample of galactic GCs, in order to detect the effects of dynamical evolution and to reconstruct the stellar initial mass function (IMF). The mass function (MF) slope of NGC 1851 is $x = 1.0 \pm 0.5$, in agreement with the general scenario of an MF depending on the position in the galactic gravitational potential and metal content of the cluster. We also plan to check the influence of environmental parameters such as concentration on the evolution of their stellar populations. As a by-product of this research we can exploit our database to investigate the evolved stages of stellar evolution of low-mass stars.

NGC 1851 is an intermediate-metallicity ($[Fe/H] = -1.3$) globular cluster (GC) located at a fairly high galactic latitude ($l = 244^0$, $b = -35^0$), thus suffering relatively low extinction ($E_{B-V} = -0.02$); it is located at a galactocentric distance of 17.2 kpc and at a distance from the galactic disk of 6.9 kpc. NGC 1851 has a high concentration ($c = 2.15$) and extended tidal radius ($r_t = 9.4\,arcmin$).

Figure 1 shows the composite (all nine fields) colour-magnitude diagram (CMD) where also superimposed is a synthetic CMD to be discussed below. The most interesting feature in the CMD of Figure 1 is the bimodal appearance of the HB, as already shown by Stetson (1981; see also discussion in Walker 1992). Despite its metallicity, the HB of NGC 1851 shows both a red clump typical of very metal-rich clusters and an extended blue HB typical of very metal-poor clusters. But NGC 1851 is not a classical second parameter cluster, as the parameters suggested by the years—such as He, [CNO/Fe], rotation, magnetic field strengths, age or some other yet unknown factor that affects mass loss—act just in one direction, producing either a redder or a bluer HB. Over the past few years, many authors have suggested that the morphology of the GGC HBs can be explained in terms of age acting as a second parameter. However we point out that age differences cannot explain the HB of NGC 1851 as age is a "canonical" second parameter. In order to better clarify this point, we have run a set of simulations.

The observed CMD has been compared with synthetic ones, generated by Monte Carlo simulations based on the following assumptions: (i) evolutionary isochrones are taken from Bertelli *et al.* (1994), which are based on the models with chemical composition $[Z, Y] = [0.001, 0.230]$; the conversion from the theoretical to the observational plane is made convolving the library of stellar spectra by Kurucz (1992, private communication); (ii) the initial mass function is the Salpeter with $x = 1.35$; (iii) star formation is assumed to take place in an almost instantaneous initial burst. In these preliminary simulations we did not include the observational errors and no attempt has been made to reproduce the number of stars in a given interval of magnitude or evolutionary phase. For the purpose of looking at the bimodal distribution of HB stars these approximations do not

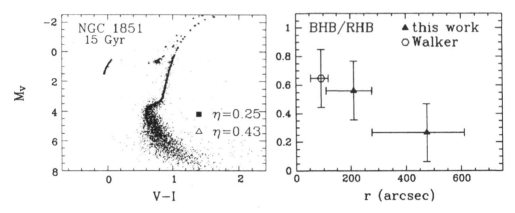

FIGURE 1. *Left*: synthetic CMD for the case of variable η superimposed to the observed CMD for NGC 1851. The difference in luminosity between the turn-off and the HB is reproduced for an age of 15 Gyr. Assuming this age it turns out that the red HB is reproduced by a mass loss efficiency of $\eta = 0.25$ while the blue HB by $\eta = 0.43$. *Right*: The ratio BHB/RHB *vs.* the radial distance from the cluster centre (also plotted is the value from Walker's frame). Note the radial trend: BHB stars are more numerous in the central part of the cluster.

affect the evidence that *a single age is not able to fit the red and blue part of the HB at the same time*. Instead, we need a bimodal mass loss to reproduce the HB morphology, as shown by the synthetic CMD of Figure 1. Why a bimodal mass loss mechanism in the same cluster? Rood *et al.* (1993) propose that stellar rotation could be the cause, at least for NGC 2808. We cannot enter here into details; however, we note that not only is the HB of NGC 1851 bimodal, but there might also be a gradient in the ratio of the BHB and the RHB stars, in the sense that BHB stars are more abundant in the centre of the cluster; unfortunately, the small number of stars makes it difficult to prove that the gradient is statistically significant. If the gradient is real, it could be a further indication that the stellar environment in high concentration clusters could in some way affect the mass loss during the RGB phase, as discussed in Djorgovski & Piotto (1993), who have found similar colour and population gradients in all the PCC clusters in their sample, and in Fusi Pecci *et al.* (1993), who have shown that the HB morphology is related to the cluster concentration, with higher concentration clusters having bluer HBs.

REFERENCES

BERTELLI, G., BRESSAN, A., CHIOSI, C., FAGOTTO, F. & NASI, E. 1994 *Astron. Astrophys., Suppl. Ser.* **106**, 275.

DJORGOVSKI, S. & PIOTTO, G. 1993 in *Structure and Dynamics of Globular Clusters* (ed. S. G. Djorgovski & G. Meylan). ASP Conference Series, vol. 50, p. 203. ASP.

FUSI PECCI F., FERRARO, F. R., BELLAZZINI, M., DJORGOVSKI, S., PIOTTO, G. & BUONANNO, R. 1993 *Astron. J.* **105**, 1145.

ROOD, R. T., CROCKER, D. A., FUSI PECCI, F., FERRARO, R. F., CLEMENTINI, G. & BUONANNO, R. 1993 in *The Globular Cluster Galaxy connection* (ed. G. H. Smith & J. P. Brodie). ASP Conference Series, vol. 48, p. 218. ASP.

STETSON, P. B. 1981 *Astron. J.* **86**, 687.

WALKER, A. R. 1992 *Publ. Astron. Soc. Pac.* **104**, 1063.

An Atlas of Synthetic Spectra for Metallicity Calibration of SMR Stars

By MARIA L. MALAGNINI[1] , M. CHAVEZ[2,3]
AND C. MOROSSI[4]

[1]Dipartimento di Astronomia, Università degli Studi di Trieste, Via Tiepolo, 11, I-34131 Trieste, Italy

[2]International School for Advanced Studies, Strada Costiera 11, I-34014 Trieste, Italy

[3]Instituto Nacional de Astrofísica, Optica y Electrónica, Apdos. Postales 51 y 216, CP 72000, Puebla, México

[4]Osservatorio Astronomico di Trieste, Via Tiepolo, 11, I-34131 Trieste, Italy

We present the atlas of synthetic spectra, covering the wavelength range 4850–5400 Å , computed starting from the 1993 release of atmosphere models and codes by Kurucz (CD-ROM 13 and CD-ROM 18). In cool stars, the selected wavelength region is characterised mostly by magnesium and iron absorptions. The spectra were synthesised for nine values of effective temperature in the range 4000–6000 K, surface gravities 1.5, 3.0, and 4.5 dex, at metallicities −0.50, 0.5, and 1.0 dex. Solar metal-abundance spectra are available in the same temperature range for seven gravity values, from 1.5 to 4.5 with a step of 0.5 dex.

The high resolution used in the computations (resolving power $\lambda/\Delta\lambda$ = 250,000) makes the atlas suitable for chemical-abundance analyses and for metallicity calibration of SMR stars for which high-quality high- and low-resolution spectroscopic data are now available.

In Figures 1 and 2 we give some examples of synthetic spectra, computed for the atmospheric parameters typical of G dwarfs and giants, zoomed in the regions dominated by iron (Fe I) absorptions.

In each plot, the high-resolution spectrum from the atlas is given, together with two examples of the same spectrum degraded to the resolution of observations now available for abundance analyses of SMR stars (McWilliam & Rich, 1994; McQuitty et al. 1994).

By deriving suitable spectral indices from the synthetic spectra, the role of magnesium and iron spectral features in modelling the chemical evolution of galactic globular clusters can be investigated (Chávez et al. 1995).

REFERENCES

Chavez, M., Malagnini, M. L. & Morossi, C. 1995 *Astrophys. J.* **440**, 210.

Kurucz, R. L. 1993 CD–Rom 13 and 18.

McQuitty, R. J., Jaffe, T. R., Friel, E. D. & Dalle Ore, C. M. 1994 *Astron. J.* **107**, 359.

McWilliam, A. & Rich, R.M. 1994 *Astrophys. J., Suppl. Ser.* **91**, 749.

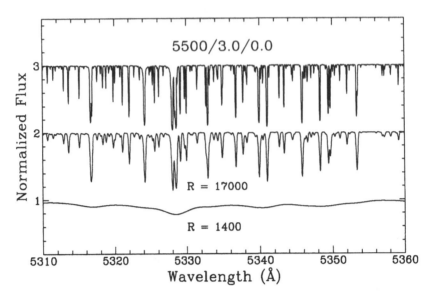

FIGURE 1. Synthetic spectrum computed for T_{eff} = 5500 K, log g = 3.0 dex, and [M/H] = 0.0 dex in the iron region (the spectra computed at $\lambda/\Delta\lambda$ = 250 000 and 17 000 are vertically shifted by 2.0 and 1.0, respectively)

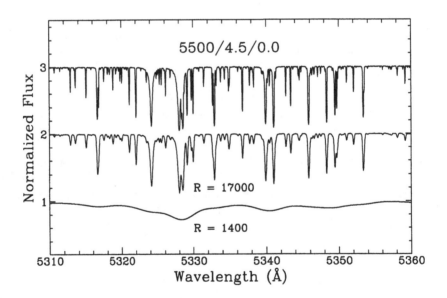

FIGURE 2. Synthetic spectrum computed for T_{eff} = 5500 K, log g = 4.5 dex, and [M/H] = 0.0 dex in the iron region (the spectra computed at $\lambda/\Delta\lambda$ = 250 000 and 17 000 are vertically shifted by 2.0 and 1.0, respectively)

Halo Field Stars and Globular Clusters: Clues to the Origin of the Milky Way

By BRUCE W. CARNEY

Department of Physics & Astronomy, University of North Carolina, Chapel Hill, NC
27599-3255, U.S.A.

Recent studies of the metallicities and kinematics of field stars and of globular clusters have revealed signs of population substructure. In addition to the "thick-disk" population, the metal-poor stars and clusters also show signs of an "accreted" component and a "disk-like" component.

1. Introduction

The formation of our Milky Way Galaxy is a topic akin to archaeology's study of the rise of civilisation. Out of the debris left behind from the prior epochs we attempt to discern the early history, but the remnants of the earliest eras are few, and hard to place into context. Until M31 collides with the Milky Way, both will remain disk systems. But has the Milky Way always had a disk shape? The spherical distribution of the ancient, metal-poor globular clusters suggests (but certainly does not prove!) that the protogalaxy was much rounder than it is now. If it first reached the self-gravitating stage as a rounder and larger cloud of gas, how rapid was its early evolution? How quickly did it assume a disk shape? And when did it burst into brilliance as star formation began in earnest?

No modern attempt to answer these questions can neglect the fundamental impact of Eggen $et\ al.$ (1962), who envisioned the protogalaxy as a nearly spherical gas cloud undergoing free-fall collapse. The rapid timescale was estimated qualitatively by the lack of evidence for an age spread amongst the globular clusters that then good main sequence photometry. It was also estimated more quantitatively by the apparent correlation between planar galactic orbital eccentricity (nominally an adiabatic invariant) and metallicity (nominally a time indicator). The failure of the adiabatic assumption is most easily understood in a rapidly changing potential, as in free-fall collapse. The implication that the eccentricity $vs.$ metallicity correlation involves such a short timescale was disputed by Isobe (1974) and by Saio & Yoshii (1979), and there is now known to be a several-billion-year age spread amongst the globular clusters (Sarajedini & King 1989; VandenBerg $et\ al.$ 1990; Carney $et\ al.$ 1992). Nonetheless, the basic model seems sound enough. If not collapse, then some sort of contraction probably occurred. Can we see signs of this?

To answer this, we seek to identify stellar populations from those earliest phases of the Galaxy's formation. However, we should keep clearly in mind what is meant by a stellar population. Certainly mean metallicity and mean velocities and velocity dispersions are relevant. But there is more. If a gas cloud with an initial metallicity, $[m/H]_0$, undergoes star formation, then if the gas and stars do not become separated, the simple "closed box model" of Searle & Sargent (1972) predicts a metallicity distribution function like that in Figure 1 (modified to show the distribution in $[m/H]$, or log Z, instead of —as is usually done— in Z, following Laird $et\ al.$ 1988). The "population" shows a mean metallicity and a metallicity spread. The mean metallicity in the simple model is set by the "yield", the ratio of heavy elements trapped in long-lived stars to the total mass. The yield is set by stellar physics, including the initial mass function, and so probably

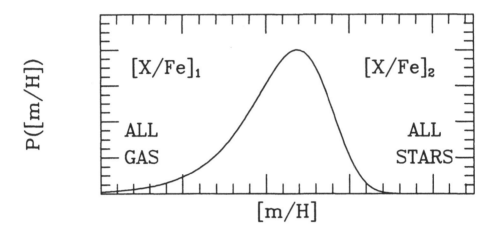

FIGURE 1. The prediction in log Z of a simple closed box model for the metallicity distribution function.

does not vary much, *and* by the possible separation of the gas from the stars, which can vary to a great degree. The Galaxy's halo population is probably metal-poor not because it began with low metallicity, but because it lost its gas, snuffing out star formation and chemical enrichment at an early stage. Furthermore, the element-to-iron ratios are set by the nucleosynthesis sources. If the star formation occurs rapidly enough, only massive star supernovae are thought to contribute to the enrichment, and $[X/Fe]_1$ should show signs of enhanced oxygen and other "α" elements relative to iron. If the entire phase of star formation was rapid, $[X/Fe]_2$ will equal $[X/Fe]_1$, but if it occurred slowly enough so that other nucleosynthesis sources could enrich the gas, such as from AGB stars and Type Ia supernovae, then $[X/Fe]_2$ will differ from $[X/Fe]_1$.

One of the best illustrations of the importance of Figure 1 is Zinn's (1985) work on the two populations of globular clusters. The metallicity distribution shows two distinct peaks, one at $[Fe/H] = -1.5$, and one at -0.5, indicative of two independent ensembles evolving from gas into stars. If the two populations were related, a smooth metallicity distribution would be more likely. The metal-rich globulars show a disk-like spatial distribution and strong net rotation, and thus are probably not related in any direct way to the "classical" metal-poor spherically-distributed globular clusters. Majewski (1993) has thoroughly reviewed the many ideas regarding the thick disk population's place in the Galaxy's chronology, and it will not be discussed further here.

2. Accretion *vs.* Contraction

The evidence in favour of a purely coherent, monolithic evolution of the Galaxy's classical halo population has been weakened over the past 16 years, beginning with the study of the galactic globular cluster system by Searle & Zinn (1978). Briefly, they noted that the horizontal branch (HB) morphology of the globular clusters seems to change as a function of galactocenteric distance, with the redder horizontal branches being more prevalent at large distances. Metallicity is thought to be the primary variable governing the HB morphology, but Searle & Zinn noted that the globular clusters show no sign of a radial metallicity gradient (at least beyond the solar circle). If age is the "second

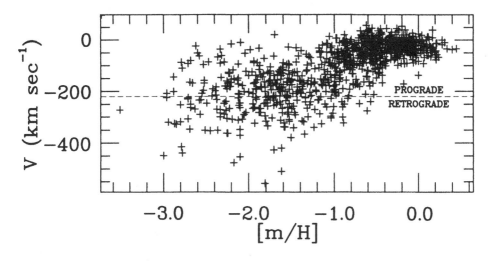

FIGURE 2. The angular momentum (V velocity) *vs.* metallicity relation from the data of
Carney *et al.* (1994a)

parameter", the clusters with redder HBs would be younger by a few gigayears. Such
an age spread is supported by main sequence turn-offs. However, such an age spread in
a contracting mixture of gas and stars ought to produce a metallicity gradient, which
is not observed. Searle & Zinn suggested that the independent formation and chemical
evolution of "protogalactic fragments", and their subsequent accretion by the Galaxy
would explain the apparently contradictory observations.

Studies of the much more numerous field halo stars have also strengthened the importance of accretion. The work of Suntzeff *et al.* (1991) on the field RR Lyraes and of
Preston *et al.* (1991) on the field blue HB stars strengthens the Searle & Zinn (1985)
arguments that there is a change in HB morphology without a significant metallicity
gradient beyond the solar circle.

Studies of kinematically unbiased samples of metal-poor field stars (Norris *et al.* 1985;
Norris 1986), of proper motion-selected samples (Carney *et al.* 1990b; Ryan & Norris
1991), and of complete *in situ* sample stars with measured proper motions (Majewski
1992) also suggest that accretion of metal-poor stars played a major role in the early
evolution of the Galaxy. The primary evidence takes two forms. First, in a coher-
ent/monolithic model with a large age spread, we should see clear signs of a "spin-up":
the net rotation about the galactic centre should correlate with age (i.e. with metallicity). Such signs are not seen, as discussed by Carney *et al.* (1990b). Figure 2 shows the
latest version of their work, incorporating the expanded sample reported by Carney *et
al.* (1994a). The V velocity is zero for circular motion about the galactic centre, and
the dashed line separates prograde from retrograde rotation. There is no real sign of
evolution in V from the metal-poor to the metal-rich domains; only a blending of two apparently distinct populations is seen. One is the disk-like sample, which here is primarily
thick disk. The other is the near-zero angular momentum metal-poor halo population.

Perhaps an even more fundamental objection to the "spin-up" is the result of Majewski
(1992). For stars lying more than 5 kpc from the plane, he found $<V> = -275 \pm 16$
km s^{-1}, a clear sign of retrograde rotation and impossible to reconcile with a prograde
disk. (This assumes a local circular velocity, Θ_0, of 220 km s^{-1}.) Majewski's results

have been criticised since they are vulnerable to the adopted distance scale, but, as will be discussed below, our new results support his claims.

3. Two metal-poor components?

It is hard to believe that *all* of the metal-poor stars now found in the solar neighbourhood had their origins in accretion events. It seems more plausible that at least some local metal-poor stars may have formed in the protodisk, and if we can find them we can learn more about the protodisk's shape, chemical enrichment, and early evolution generally.

There are signs that the metal-poor stars and globular clusters may actually belong to two, and perhaps more, distinguishable populations. Hartwick (1987) noticed that he could not model the metal-poor ([Fe/H] < -1) RR Lyraes in the solar neighbourhood with a single spherical-distributed component. He required a flattened (c/a ≈ 0.6) component as well, which was dominant near the Sun. Sommer-Larsen & Zhen (1990) found a similar result in their study of 118 metal-poor ([Fe/H] < -1.5) kinematically unbiased field stars. They claimed that most of the total mass in metal-poor stars, and about 60% at the solar position, was in a spherical component (c/a $= 0.85 \pm 0.12$), while the remainder was in a flattened distribution, with $|Z| < 3$ kpc, and that its rotation was too low to explain the flattening.

The metal-poor globular clusters also show some signs of being divisible into two components. Rodgers & Paltoglou (1984) noted that for [Fe/H] ≈ -1.5, the <V> velocity of the globular clusters (deduced solely from radial velocity data) is -292 ± 41 km s^{-1}. Could this signify a distinct population? More recent work by van den Bergh (1993a,b) showed that the retrograde globular clusters, not just those with [Fe/H] ≈ -1.5, are predominantly Oosterhoff I clusters, with <P>$_{ab} \approx 0.55$ day, and that this might suggest a unique merger remnant. Zinn (1993) returned to the HB morphology data of clusters to divide his sample for study. Basically, using theoretical HB models to guide him, he identified a sample of clusters whose HB morphologies were those expected for a constant (and large) age. Clusters whose HB morphologies were redder than expected for their [Fe/H] values were classified as "young halo", under the assumption that age is the second parameter. While these clusters do seem to be younger than the "old halo" clusters, the "young" clusters do not appear to be coeval, 1–3 Gyr younger than the "old halo" clusters. Zinn then studied two important properties that test the origins of these two samples: the angular momentum content and metallicity gradient, the presence of which would support the coherent/monolithic formation process. The 19 "young halo" clusters do not show a metallicity gradient, and have <V> $= -284 \pm 74$ km s^{-1}, which is a retrograde signature at slightly less than 1σ significance. The "old halo" clusters do show a weak metallicity gradient and a distinct (3σ) prograde rotation, with <V> $= -150 \pm 22$ km s^{-1}. The simplest interpretation of these results is that some of the Galaxy's globular clusters (the "old halo" ones) participated in its formation directly, while the remainder formed in lower density environments further from the protogalaxy over a longer period of time, and were later accreted.

The data from a large sample of proper motion stars appear to support this hypothesis. Carney *et al.* (1994a) have measured mean metallicities ([m/H]), kinematics, and galactic orbits for 1048 stars. Since we are unable to assign relative ages to all of these field stars, we seek an alternative criterion. If we are testing a protodisk *vs.* an accretion origin, distance from the plane should serve us well, and in particular we use $< |Z_{max}| >$, the average maximum distance from the plane that a star reaches during about 15 galactic orbits. Our criterion is thus very similar to that employed by Majewski (1992), and, as he did, we find a strong retrograde signature, $<V> = -270 \pm 21$ km s^{-1}, based on a sample of 33 stars $< |Z_{max}| > \geq 5$ kpc. While we are also vulnerable to the criticism of the effects of the distance scale choice, we note that, consistent with the discussion in Laird *et al.* (1988a), our distances are likely to be slight underestimates, hence our retrograde signature may also be underestimated. It is also worth comparing this $<V>$ to that for the high-velocity stars closer to the plane. If we take $< |Z_{max}| > \leq 2$ kpc then for three different criteria to define a halo sample, we find prograde rotation: $<V> = -191 \pm 7$ km s^{-1} (defined by 154 stars with $[m/H] \leq -1.5$); -203 ± 5 km s^{-1} (defined by 145 stars with e ≥ 0.8, where the eccentricity e $= (R_{apo} - R_{peri})/(R_{apo} + R_{peri})$, and R_{apo} and R_{peri} are the average apogalacticon and perigalacticon distances); and -172 ± 12 km s^{-1} (for 53 stars with $[U^2 + W^2]^{1/2} \geq 200$ km s^{-1}). The separation into retrograde and prograde samples is real, and is found using the same distance scale applied to both samples. The sample with $< |Z_{max}| > \geq 5$ kpc also does not show a radial metallicity gradient, -0.005 ± 0.006 dex/kpc, while the three "low-halo" samples defined above all show weak signs of such a gradient: -0.015 ± 0.013; -0.015 ± 0.013; and -0.028 ± 0.016 dex/kpc, respectively. Further, there is weak evidence that the "high-halo" sample may be younger than the "low-halo" sample, which could tie our sample and those of Majewski (1992) and Zinn (1993) together. Schuster & Nissen (1989) used $uvby\beta$ photometry to estimate ages for individual metal-poor stars. Accepting those stars in their sample that also are in ours, and only those where the metallicities estimated by the two groups agree well, we find that for the five stars with $< |Z_{max}| > \geq 4$ kpc, $<[m/H]> = -1.64$ and $< t_9 > = 13.2 \pm 0.8$ Gyr, where $< t_9 >$ is the mean age in gigayears. For the 29 stars with $< |Z_{max}| > \leq 2$ kpc, $<[m/H]> = -1.71$ and $< t_9 > = 15.7 \pm 0.3$ Gyr. The number statistics are admittedly poor, but at least consistent with the retrograding, no metallicity gradient, "high-halo" population being younger.

More recently, Norris (1994) has expanded on his Monte Carlo simulations of our original proper motion sample. Norris & Ryan (1991) utilised the Monte Carlo method to try to determine whether a discrete thick-disk component or an "extended" disk component better match the run of $<V>$, $\sigma(U)$, $\sigma(V)$, and $\sigma(W)$ *vs.* $[m/H]$ in the data of Carney *et al.* (1990a). The "extended" component has the four kinematic variables being functions of $[m/H]$, whereas the "discrete" thick disk had no such metallicity dependence. Both models represented the data fairly well, although the "extended" model did somewhat better. To test these models more thoroughly, one must explore the most metal-poor domains.

Very metal-poor *disk* stars do exist. Morrison *et al.* (1990) first identified possible "metal-weak disk stars" from *DDO* photometry of a kinematically unbiased sample of G and K giants about 4 kpc from the Sun. Down to $[Fe/H] \approx -1.6$, their data showed signs of a disk component with a scale height of 1.4 ± 0.7 kpc, and $<V> = -70 \pm 15$ km s^{-1}. Ryan & Lambert (1994) have obtained high-resolution spectra of many of the Morrison *et al.* (1990) giants, and found that those claimed to be halo stars do indeed have low metallicities but the metal-poor giants with disk kinematics appear to have higher metallicities, $[Fe/H] > -1$, closer to the thick disk mean. This has weakened the case for very metal-poor stars with disk kinematics, but there is further evidence. First, the absolute proper motion and radial velocity of the globular cluster M28 obtained by Rees & Cudworth (1991) indicate clear disk-like kinematics: $\Pi = -36 \pm 5$, $\Theta = -36 \pm 20$, and $Z = -35 \pm 15$ km s^{-1}. The cluster's metallicity is low, $[Fe/H] \approx -1.5$. Beers & Sommer-Larsen (1994) have been studying a very large sample of local kinematically unbiased stars, and their modelling of the galactic rotational velocities ($v_{rot} = V + \Theta_0$) indicates clearly the need for two components for metallicities $[Fe/H] \leq -0.6$. By restricting the sample to stars with $|Z| \leq 1$ kpc, and stars whose v_{rot} value derived solely

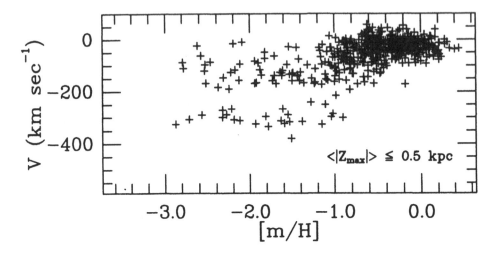

FIGURE 3. The distribution of V velocity (i.e., galactic rotational velocity, v_{rot}, or angular momentum) as a function of metallicity in the Carney *et al.* (1994a) sample for stars with $< |Z_{max}| > \le 0.5$ kpc

from the projection of the radial velocity exceeds 100 km s^{-1}, they find that a metal-pure halo model cannot describe the distribution of v_{rot} values. A disk component added to the halo component is required, and in fact appears to be required for metallicities below -2.0. Finally, even in our kinematically-biased study of proper motion stars, we find low-metallicity stars with disk-like kinematics. Figure 3 shows our sample when restriction is made to those stars with $< |Z_{max}| > \le 0.5$ kpc, consistent with disk stars. Some of the stars clearly do not belong to the classical disk since they are in retrograde rotation. (The gap near $v_{rot} = 0$; V $= -220$ illustrates the kinematic biases. If a star has a small $< |Z_{max}| >$ value, it has a small W velocity, generally. If the star has $v_{rot} \approx 0$ km s^{-1}, then all of its energy of motion must come from the U velocity: it must have nearly a purely radial orbit. The distribution of orbital shapes of halo stars puts very few stars into purely radial, planar orbits, and so these stars are rare, causing the gap in the diagram.) The fact that we see stars approximately confined to the plane, and with V velocities not so different from the Sun in our kinematically biased sample means metal-poor disk kinematics stars exist, and in fairly substantial numbers. Stars with V ≈ 0 are much less likely to be included in a proper motion sample than stars with V $<< 0$. We are working on a method to account for the bias and assign proper weights to the stars in our program, and Figure 4 illustrates some preliminary results. Based on a methodology developed by Luis Aguilar (Aguilar et al. 1994), Figure 4 shows the distribution of relative weights for stars in our sample north of $-2°$ declination and with $\mu \ge 0.23''$/yr. The clump of six stars near log(weight) $= 4.5 - 5.5$ and [m/H] < -2 have high weights because they are disk stars. Their high weights relative to the other metal-poor stars suggests, in fact, that very metal-poor disk stars may be as common in the solar neighbourhood as very metal-poor halo (i.e. high-velocity) stars. We have begun the detailed modelling of the weighted sample but, suffice to say, the unweighted mean kinematics of these stars clearly suggest disk kinematics: $< |U| > = 44 \pm 12$ km s^{-1}, $<V> = -68 \pm 12$ km s^{-1}, and $< |W| > = 46 \pm 17$ km s^{-1}, where the errors are those of the mean for the six stars.

On the other hand, Norris (1994) has suggested that the most metal-poor "protodisk"

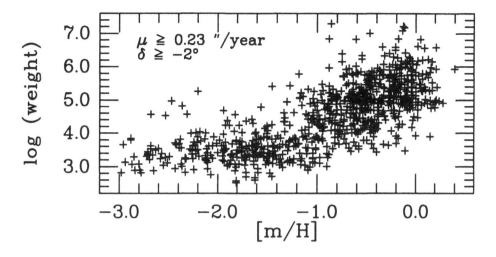

FIGURE 4. Calculated relative weights for stars in the proper motion-based sample obtained by Carney *et al.* (1994a). High weights indicate low velocities relative to the Sun.

stars may be much hotter than the above samples suggest. His argument is based upon the work of Norris & Ryan (1991), and that neither the discrete nor the extended thick disk model explained the upturn in $\sigma(W)$ in the lowest metallicity bin in our previous proper motion sample (Carney *et al.* 1990a) and, in part, this provoked the recent contribution by Norris (1994). Here he has included the effect of infall on the metallicity distribution function, which basically adds a higher, flatter tail to the metal-poor side of Figure 1. The normalisation of the degree of infall is arbitrary, as is the variation of the kinematics with metallicity, but Norris finds an excellent match between the models and the data if there is a "discrete", or, as he calls it, an "accreted" halo, and a "contracted halo", or, as I prefer, a "protodisk" component. To explain the upturn in $\sigma(W)$, however, Norris is required to make the oldest, presumably most metal-poor extension of the protodisk very hot dynamically, with $<V> = -200$ km s^{-1}, and $\sigma(U) = \sigma(V) = \sigma(W) = 135$ km s^{-1}. This protodisk component is rounder than the accreted halo, to which he assigns $<V> = -240$ km s^{-1}, $\sigma(U) = 180$, $\sigma(V) = 80$, and $\sigma(W) = 80$ km s^{-1}! While this does involve two populations of metal-poor stars, for two reasons we are reluctant to accept this distribution of kinematics among the accreted and protodisk components. The first reason in simply all the evidence discussed above: the very metal-poor disk stars detected, even in the kinematically biased proper motion sample, are not that hot dynamically. Second, we believe that the issue in dispute, the upturn in $\sigma(W)$ at the lowest metallicities, is caused by a change in the "accreted" halo component, and not the protodisk component. In Figure 5 we show the metallicity histograms for stars in our sample with $< |Z_{max}| >$ in two ranges: $2 - 4$ kpc, and > 4 kpc. The peak in the lower $< |Z_{max}| >$ metallicity distribution at [m/H] ≈ -1.5 is typical of all halo samples (see Laird *et al.* 1988b), including those in our survey defined by, for example, V ≤ -150 km s^{-1}, e ≥ 0.8, or $[U^2 + W^2]^{1/2} \geq 200$ km s^{-1}. But for the stars with $< |Z_{max}| > > 4$ kpc, the metallicity seems to shift. It is *these* stars that are responsible for the upturn in $\sigma(W)$ in the lowest metallicity bin on our old data (and in our new survey as well). These 46 stars are also in retrograde rotation, with $<V> = -255 \pm 17$ km s^{-1} (-270 ± 21 km s^{-1} if we use the 33 stars with $< |Z_{max}| > \geq 5$ kpc). If we further restrict the samples to the most metal-poor stars ([m/H] ≤ -2.0), the $<V>$ values become -293 ± 26 km

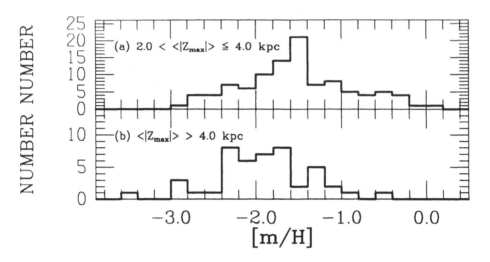

FIGURE 5. The metallicity histograms for stars in two ranges of $< |Z_{max}| >$, taken from the Carney *et al.* (1994a) sample

s^{-1} (20 stars; $< |Z_{max}| > \geq 4$ kpc) and -321 ± 30 km s^{-1} (15 stars; $< |Z_{max}| > \geq 5$ kpc). This retrograde rotation is certainly not the signature of a protodisk population.

4. Yet another metal-poor population?

Preston *et al.* (1994) have presented strong evidence for the existence of a sample of blue, metal-poor dwarfs. The stars are bluer than globular cluster turn-offs, and hence are either younger or are the field equivalents to blue stragglers found in many globular clusters. However, their estimated densities relative to metal-poor field horizontal branch stars far exceeds ratios encountered within globular clusters, so these blue metal-poor field dwarfs apparently belong to a young, metal-poor population. The kinematics of the sample are also unusual, being intermediate between the traditional halo and disk or even thick disk populations. Preston *et al.* (1994) estimate $<V> \approx -92$ km s^{-1}, and an isotropic velocity ellipsoid: $\sigma(U) = \sigma(V) = \sigma(W) \approx 90$ km s^{-1}. Their interpretation of the kinematics, metallicities, and young ages is that the stars represent the accretion by the Milky Way of a dwarf galaxy, a good prototype being the Carina dwarf, in which about 80% of the stars are as young as the stars identified by Preston *et al.* (1994). They also point out that the local density of this population is low, amounting to perhaps a few percent of the local metal-poor star density. The low density, coupled with the less extreme kinematics, indicates that such stars would be rare even in large surveys of proper motion stars such as our own. This does, of course, assume that the victim was dominated by a young population, like Carina. If a much larger old population had underlain the young metal-poor population, however, such stars, with low metallicities and intermediate kinematics, might appear in our survey of proper motion stars. A quick examination of the distribution in V velocities for metal-poor stars (Figure 6) does not, however, reveal a prominent sign of stars with $V \approx -90$ km s^{-1}. This does not contradict the interpretation by Preston *et al.* (1994) for a recent merger event, but says only that the victim was probably not dominated by an underlying old population.

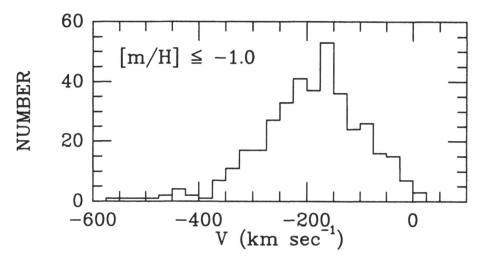

FIGURE 6. The metallicity histogram for stars with $[m/H] \leq -1.0$ in the Carney *et al.* (1994a) sample. No strong sign of a population with $<V> = \approx -90$ km s^{-1} is seen.

5. Summary

The globular clusters and the field stars appear to contain at least three populations, probably quite distinct in terms of their origins. One is probably due to a variety of minor accretion events, as suggested by Searle & Zinn (1978) to explain the age spread among the globular clusters and the absence of a metallicity gradient. Zinn (1993) and Carney *et al.* (1994b) claim that this population has a modest retrograde net rotation, consistent with a separate history from the formation of the Galaxy's disk. On the other hand, very metal-poor clusters and field stars whose kinematics are more consistent with a protodisk have been found. These stars seem to be more closely confined to the plane, show a radial metallicity gradient, and have prograde net rotation. The associated globular clusters seem to be older, on average, than the sample associated with the accretion events. Finally, Preston *et al.* (1994) have found signs of the accretion of a metal-poor object in which star formation had occurred up to recently (a few gigayears). Hence the accretion also occurred recently. Such events should not really surprise us. The Magellanic Clouds are doomed to such a fate within 5–10 Gyr, and depending upon its dark matter content (which may be considerable; cf. Suntzeff 1994), the newly-discovered Sagittarius dwarf (Ibata *et al.* 1994) may be in the process of merging with the Galaxy's disk. Accretion by the Milky Way of small satellite galaxies clearly has played a role in the formation of stellar populations, and may have played a vital role in the formation of the Milky Way Galaxy.

I gratefully acknowledge the support of the US NSF grant AST 92–21237, without which this work could not have been done. It is also always a pleasure to thank my colleagues: Dr. David W. Latham, Dr. John B. Laird, Dr. Tsevi Mazeh, and Dr. Luis Aguilar.

REFERENCES

AGUILAR, L. A., CARNEY, B. W., LATHAM, D. W. & LAIRD, J. B. 1994, in preparation.

BEERS, T. C. & SOMMER-LARSEN, J. 1994 *Astrophys. J.* in press.

CARNEY, B. W., AGUILAR, L., LATHAM, D. W. & LAIRD, J. B. 1990a *Astron. J.* **99**, 201.

CARNEY, B. W., LAIRD, J. B., LATHAM, D. W. & AGUILAR, L. A. 1994a in preparation.

CARNEY, B. W., LATHAM, D. W. & LAIRD, J. B. 1990b *Astron. J.* **99**, 572.

CARNEY, B. W., LATHAM D. W., LAIRD, J. B. & AGUILAR, L. A. 1994b *Astron. J.* **107**, 2240.

CARNEY, B. W., STORM, J. & JONES, R. V. 1992 *Astrophys. J.* **386**, 663.

EGGEN, O. J., LYNDEN-BELL, D. & SANDAGE, A. R. 1962 *Astrophys. J.* **136**, 748.

HARTWICK, F. D. A. 1987 in *The Galaxy* (ed. G. Gilmore & B. Carswell), p. 281. Reidel.

IBATA, R. A., GILMORE, G. & IRWIN, M. 1994 *Nature* **370**, 194.

ISOBE, S. 1974 *Astron. Astrophys.* **36**, 333.

LAIRD, J. B., RUPEN, M. P., CARNEY, B. W. & LATHAM, D. W. 1988 *Astron. J.* **96**, 1908.

MAJEWSKI, S. R. 1992 *Astrophys. J., Suppl. Ser.* **78**, 87.

MAJEWSKI, S. R. 1993 *Annu. Rev. Astron. Astrophys.* **31**, 575.

MORRISON, H. L., FLYNN, C. & FREEMAN, K. C. 1990 *Astron. J.* **100**, 1191.

NORRIS, J. 1986 *Astrophys. J., Suppl. Ser.* **61**, 667.

NORRIS, J. E. 1994 *Astrophys. J.* **431**, 645.

NORRIS, J., BESSELL, M. S. & PICKLES, A. J. 1985 *Astrophys. J., Suppl. Ser.* **58**, 463.

NORRIS, J. E. & RYAN, S. G. 1991 *Astrophys. J.* **380**, 403.

PRESTON, G. W., BEERS, T. C. & SHECTMAN, S. A. 1994 *Astron. J.* **108**, 538.

PRESTON, G. W., SHECTMAN, S. A. & BEERS, T. C. 1991 *Astrophys. J.* **375**, 121.

REES, R. F. & CUDWORTH, K. M. 1991 *Astron. J.* **102**, 152.

RODGERS, A. W. & PALTOGLOU, G. 1984 *Astrophys. J., Lett.* **283**, L5.

RYAN, S. G. & LAMBERT, D. L. 1994 *Astron. J.* submitted.

RYAN, S. G. & NORRIS, J. E. 1991 *Astron. J.* **101**, 1835.

SAIO, H. & YOSHII, Y. 1979 *Publ. Astron. Soc. Jpn.* **91**, 553.

SARAJEDINI, A. & KING, C. R. 1989 *Astron. J.* **98**, 1624.

SCHUSTER, W. J. & NISSEN, P. E. 1989 *Astron. Astrophys.* **222**, 69.

SEARLE, L. & SARGENT, W. L. W. 1972 *Astrophys. J.* **173**, 25.

SEARLE, L. & ZINN, R. 1978 *Astrophys. J.* **225**, 357.

SOMMER-LARSEN, J. & ZHEN, C. 1990 *Mon. Not. R. Astron. Soc.* **242**, 10.

SUNTZEFF, N. B. 1994. Private communication.

SUNTZEFF, N. B., KINMAN, T. D. & KRAFT, R. P. 1991 *Astrophys. J.* **367**, 528.

VAN DEN BERGH, S. 1993a *Mon. Not. R. Astron. Soc.* **262**, 588.

VAN DEN BERGH, S. 1993b *Astron. J.* **105**, 971.

VANDENBERG, D. A., BOLTE, M. & STETSON, P. B. 1990 *Astron. J.* **100**, 445.

ZINN, R. 1993. In *The Globular Cluster-Galaxy Connection* (ed. G. H. Smith & J. P. Brodie). ASP Conference Series, vol. 48, p. 38. ASP.

Janes: One could simplify the terminology a bit by defining the galactic components by *process*. In that respect, one might lump the Preston *et al.* merger even in with the earlier accretion part of the halo as just a continuation of that process.

Carney: Yes, that's true, and certainly it helps in distinguishing the galaxy-formation ideas. On the other hand, Preston *et al.* do seem to have detected a specific event with characteristic metallicities and kinematics, parameters we must be aware of as we try to get at the evolution of the "real" protodisk.

Shore: How much of a dynamical smearing might have resulted from one of these collision events? Using the lunar analogy, do the infall events wipe out the signal from other earlier events?

Carney: Quinn *et al.* (1993: *Astrophys. J.* , **403**, 74) have discussed some of these issues, referring specifically to how much bombardment the disk can sustain before it is destroyed. It takes a pretty big hit to destroy the disk. Smaller fragments will settle into the disk and spread out in some fashion—primarily along the capture trajectory. The various mergers shouldn't perturb each other much, since the overall gravitational potential should not change significantly.

Shore: When people like Quinn or Dupraz do collisional models of blips with larger galaxies, they always get shells and jets and isolated portions of phase space as signatures. Is there any such observation for the halo stars?

Carney: Good question. This is more than just a search for moving groups, a subject about which much has been written and little proven. However, Majewski *et al.* (1994: *Astrophys. J., Lett.* , **427**, L37) have found evidence of an unusual "clump" in phase space, and we seem to see a similar trace in our data.

Majewski: Do you see any reasons for distinguishing between the lower, flattened halo and a more ubiquitous thick disk which includes a very metal-poor tail? That is, is there any physical reason for two distinct populations?

Carney: No, I do not see any reason to separate the thick disk from the disk-like very metal-poor stars. On the other hand, neither do I see any evidence that means they must be linked. I still am uncertain as to what the thick disk really is and its origin.

Majewski: I note that you seem to be using terminology alluding to an ELS-type, contracted/dissipative origin for the lower halo or "protodisk", whereas my impression was that your earlier work leaned toward an origin for the thick disk through dynamical heating of the thin disk via satellite mergers. Do you have a preference toward either model based on your new data?

Carney: Actually, the implication of our previous work was that the stellar thick disk we see is mostly the stars of the dwarf galaxy victim, not stars formed as a result of the merger. We based that conclusion on the apparent constancy of the peak of the metallicity distribution function in our data as a function of $< |Z_{max}| >$ or V velocity. In other words, there was no obvious metallicity gradient. Add to that the same metallicity peak for the thick-disk globular clusters at small galactocentric distances, and the argument seemed compelling, to us at least. However, the lower halo does seem to show a radial metallicity gradient. The situation now appears to be more complicated, and in my own mind the question is still open.

Alfaro: What do you think about the possibility of seeing the very metal-poor pro-grade rotation population in the globular cluster system?

Carney: I think that's what Bob Zinn (1993) has claimed he sees, but it turns out to be very hard for him to distinguish between his "young" and "old" globular clusters at the lowest metallicity levels, say [Fe/H] < -2, because the HB morphology begins to lose its sensitivity to age. Perhaps alternative HB-morphology classification schemes would be useful to explore that question.

Chernin: My question is about the spin angular momentum of the "retrograde" halo component as a whole. Are there any reasons from observational data to assume that the total spin vector of this component is essentially anti-parallel to the spin vector of the main halo component? So how "retrograde" exactly is this component?

Carney: That's a very good question. I can't answer that for you in detail here except to say that the means of the two other velocity components, U and W, are quite close to zero.

Wide Binaries Among Halo Subdwarfs

By EDUARDO L. MARTÍN[1], R. RÉBOLO[2]
AND M. R. ZAPATERO OSORIO[2]

[1]Astronomical Institute "Anton Pannekoek", Kruislaan 403, 1098 SJ, Amsterdam, The Netherlands

[2]Instituto de Astrofísica de Canarias, Vía Láctea s/n, E–38200 La Laguna, Tenerife, Spain

We present results of a search for wide companions to halo subdwarfs of known metallicity. We have taken CCD I-band frames in order to detect faint stars at angular separations of 1"–30" from the subdwarfs. The target sample includes over 1000 stars, and we have already observed about 50%. For a small subset of this sample (about 50 stars), astrometry of CCD images taken three years apart has allowed us to identify a handful of common proper motion pairs. Spectroscopic observations of three of them have been made and show that one companion is a late K-type star and two are M-type. The radial velocity and overall characteristics confirm that in all likelihood these are very low mass secondaries of halo stars. The coolest companions found in this survey could be near the faint end of the halo main sequence.

1. Introduction

One of the major probes of the history and structure of the Galaxy is the study of Population II (halo) stars. They provide us with perspective on chemical and dynamical evolution, and they can be found in significant numbers within 200 pc of the Sun. However, the census of nearby halo stars is far from being complete, in particular regarding cool stars (late K- and M-type). There are two main observational problems in identifying cool halo stars. (i) They are intrinsically faint, although their expected magnitudes are well within the detection limits of modern CCDs in 1-metre class telescopes. (ii) Metallicity indices, which have been commonly used for classifying metal deficient stars, are not well calibrated for spectral types later than about K2. This is due to the difficulty of modelling cool stellar atmospheres where molecules become prevalent, and the lack of standard stars. Population I late K and M dwarfs can be understood with the aid of reference M stars in open clusters of known age, distance and metallicity. However, for Population II there are no nearby clusters to compare with.

Wide binary stars are like mini-clusters because they offer us the possibility of studying two stars of different mass, but with the same age, distance and chemical composition. Hence, by considering a large enough sample we may start addressing general questions like the statistical mass function or the binary frequency. However, a necessary first step is to carefully identify real physical binaries. Since we are interested in wide binaries (angular separations larger than 1") with faint secondaries, the probability of having chance alignments is non-negligible. In Section 2 we show that proper motion measurements are an effective means of separating real companions from background stars.

Our focus on wide binaries comes from the fact that the components can be spatially resolved with conventional CCDs in earth-bound telescopes. Hence, we can obtain photometry and spectroscopy for each star. Another positive property of these binaries is that they are sufficiently apart for each component to have evolved independently. The only disadvantage may be that we do not expect to find many of them, although this can be compensated by taking a sufficiently large sample of targets to look at.

In this paper our main aim is to present the first positive results of our search for faint companions. We have identified four *bona fide* halo wide binaries out of a subset of 51 targets, and we have obtained spectroscopy of three of the companions. An early presentation of this project can be found in Martín & Rébolo (1992).

2. Search technique and early results

Our sample of targets includes all the stars than we have been able to find in the literature which meet the following requisites: (i) G and K spectral types; (ii) subdwarf main sequence with metallicity measured spectroscopically or via narrow band colours, [m/H] must be ≤ -1 (but see Morrison *et al.* 1992); (iii) m_V in the range 8–15, and declination higher than -30°. The prime sources of targets were the papers by Laird *et al.* (1988) and Schuster & Nissen (1989), which contain over 1000 suitable stars.

The search method for detecting a faint companion to each target star can be divided in three steps:

(i) Two or more I-band CCD images are taken with either the 1-metre Jacobus Kapteyn telescope (JKT) at the La Palma observatory, or the 0.8-metre IAC telescope (IAC80) at the Tenerife observatory. Average seeing is 1" at the JKT and 1.4" at the IAC80. The dynamical range of the CCD detectors allow us to detect companions up to eight magnitudes fainter in m_I than the target subdwarf. So far we have taken I CCD frames of about 500 stars.

(ii) For those targets with any faint companion closer than 30" we plan to take second-epoch CCD frames two or three years after the first observations. Since all the subdwarfs are high-velocity stars, any truly bound companion must share the same velocity, and hence the distance between the stars must be conserved (orbital motions are negligible in timescales of a few years). Unless the proper motion vector is perpendicular to the pair axis, we can easily see if the stars have common proper motions (CPM) just from their relative distance, but we also check with the other stars in the field. Up to now we have taken second-epoch CCD frames of the 11 candidate binaries which were observed for the first time in 1991 (Martín & Rébolo 1992, Table 1). The results have been that only four pairs have common proper motions. Out of six companions at angular separations $\leq$10" we find that four are CPM, and out of six at larger separations we find none with CPM. The binary frequency in this small subset of our sample is 8%.

iii) The last step we plan is to make follow up observations of the CPM companions. These include optical spectroscopy, optical and IR photometry, and trigonometric parallaxes if necessary. We have already obtained VRI CCD colours and low- and high-spectral-resolution spectroscopy for three CPM companions. All of them have radial velocities consistent with their primaries, and they show molecular absorption bands indicating that they are cool.

3. Final remarks

The reddest of the four CPM companions found so far has $(V-I)_{Cousins}$=2.2, and it is 3.9 magnitudes fainter in the I-band than its primary, which is the star G176–46. This is an outstanding multiple system; Latham *et al.* (1992) reported that G176–46 is a triple spectroscopic system composed of a very close binary (P_{orb}=10 days) with masses $\sim$0.6 and 0.4 $M_\odot$, and a more distant component with mass $\sim$0.7 $M_\odot$ ($P_{orb} \sim$300 yr). Therefore, the faint companion found by us is the fourth (!) component, and the less massive. It is intriguing to think how this fascinating multiple system formed, as it involves a relatively large mass for the parental original cloud and a complicated redistribution of angular momentum.

The second reddest CPM companion has $(V-I)_{Cousins}$=2.0, and it is 2.3 magnitudes fainter in the I-band than the primary (G116-009). Probably the most remarkable thing about this object is its large separation; 10.2", which at the distance of 257 pc estimated by Laird *et al.* (1988) implies a projected separation of 2621 AU. Ryan (1992) noted that very wide binaries are unlikely to survive in the dense environment of a globular cluster, and hence it would be unlikely that the wide binary G116–009 was formed in a globular cluster.

We stress the fact that we have not yet found any companion anywhere near to our contrast threshold of eight magnitudes difference in I. The colour of the faint companion to G176–46 is similar to those of stars at the bottom of the observed main sequence in globular clusters of similar metallicity. Note that the metallicity of G176–46 is [m/H]=-1.7±0.4 (Laird *et al.* 1992), similar to the globular ωCen. Assuming a distance of 130 pc to G176–46 (Laird *et al.* 1988), we find M_V=12.2, which is similar to some stars studied by Monet *et al.* (1992), e.g. LHS 192, LHS 3259, LHS 3628, etc. These stars define a subluminous main sequence about two magnitudes fainter in M_V than Population I dwarfs. Figure 2 of Monet *et al.* 2 indicates that the end of that MS may be at $M_V \sim$14.5 and (V-I)$\sim$2.8, and our results imply that these stars are moderately metal-deficient ([m/H]$\sim$-1.5).

Our search technique can readily detect companions down to the bottom of the halo MS because we can detect companions four I magnitudes fainter with respect to their primaries than the G176–46 system. This survey is complementary to the investigations of globular clusters as we can obtain mutual benefits. We can learn from the statistics in the globulars, and they can use our companions as reference stars for photometric and spectroscopic properties. The spectral class of our two faintest companions is clearly M because of the presence of molecular bands, and we tentatively assign subclasses M1 and M2 to the secondaries of G116–009 and G176–46, respectively. These types are only an indication of what the spectra looks like in comparison with the classification scheme for Population-I M dwarfs. However, a new M classification may be necessary for metal-deficient stars, and our stars are arguably good standards for it. In a forthcoming paper we will present the spectra of these objects and the results of model atmosphere fitting with different effective temperatures.

E.M. thanks E. Oblak for useful discussions and acknowledges support from the European Human Capital and Mobility grant CHRX–CT93–0329.

REFERENCES

LAIRD, J. B., CARNEY, B. W. & LATHAM, D. W. 1988 *Astron. J.* **95**, 1843.

LATHAM, D. W., MAZEH, T., STEFANIK, R. P., DAVIS, R. J., CARNEY, B. W., LAIRD, J. B., TORRES, G. & MORSE, J. A. 1992 *Astron. J.* **104**, 774.

MARTÍN, E. L. & RÉBOLO, R. 1992 in *Complementary Approaches to Double and Multiple Star Research* (ed. H. A. McAlister & W. I. Hartkopf). IAU Colloquium 135, ASP Conference Series, vol. 32, p. 336. ASP.

MONET, D. G., DAHN, C. C., VRBA, F. J., HARRIS, H. C., PIER, J. R., LUGINBUHL, C. B. & ABLES, H. D. 1992 *Astron. J.* **103**, 638.

MORRISON, H. L., FLYNN, C. & FREEMAN, K. C. 1990 *Astron. J.* **100**, 1191.

RYAN, S. G. 1992 *Astron. J.* **104**, 1144.

SCHUSTER, W. J., NISSEN, P. E. 1989 *Astron. Astrophys.* **222**, 69.

Oxygen Abundances of Metal-Rich Halo Stars

By JOHN B. LAIRD AND ZHEN XIAO

Department of Physics & Astronomy, Bowling Green State University, Bowling Green, OH–43403, U.S.A.

We present oxygen abundances for a sample of nine stars with $-0.9 <$ [Fe/H] < -0.5, most of which have kinematics typical of metal-poor halo stars. The observed [O/Fe] abundances show a scatter which is larger than the expected observational errors. These data suggest that the halo was inhomogeneous and that there is not a unique break point in the [O/Fe]–[Fe/H] relation.

1. Introduction

The Carney–Latham survey of high-proper-motion stars (see Carney *et al.* 1994) has revealed a number of stars that have halo kinematics but metal abundances [m/H] > -1, the traditional halo–disk boundary. We have obtained high-resolution echelle spectra of nine stars from that survey with $-0.9 <$ [Fe/H] < -0.5, to measure their element-to-element abundances and thus probe their chemical history. All but one of these stars have large orbital eccentricities and UVW velocities, indicating they are very probably members of the halo population rather than the disk.

2. Results

The derived [Fe/H] abundances agree well with the low-S/N [m/H] measurements of Carney *et al.* Thus we confirm their result that the halo includes stars with metal abundances significantly higher than the traditional limit, reaching at least [Fe/H] $= -0.5$.

Oxygen abundances have been derived using the O I 7771-5 triplet. Figure 1 shows a plot of [O/Fe] *vs.* [Fe/H], with the present values shown as filled circles. Results from the literature were chosen using only field dwarf stars with [O/Fe] measured using the OI triplet. The recent work of Edvardsson *et al.* (1993) and Tomkin *et al.* (1992) provides a large and fairly uniform data set, and their values were used in preference to other values for the same stars. Additional stars were included from Barbuy & Erdelyi-Mendes (1989), Bessell *et al.* (1991), Clegg *et al.* (1981), García López *et al.* (1993) and Sneden *et al.* (1979). (For consistency, the results of Edvardsson *et al.* (1993) were restored to the original OI triplet scale instead of the corrected [OI] scale.)

The stars in the present sample show a large range of [O/Fe], larger than the estimated observational scatter, with some values characteristic of halo stars, and others similar to the disk stars of the same [Fe/H]. It is not possible to assign individual stars unambiguously to one population or another, and, in addition, the present sample was subject to a strong kinematic bias, since it was chosen from a proper motion survey. Nevertheless, the eight high-velocity stars observed here have kinematics typical of metal-poor halo stars, including the other metal-poor stars in Figure 1. The stars with lower [O/Fe] are not distinct from nor less extreme in their kinematics than other halo stars.

The range of [O/Fe] in the present data suggests that the halo was not homogeneous when these metal-rich stars formed. Furthermore, these data greatly weaken the case for a sharp break in the [O/Fe]–[Fe/H] relation at [Fe/H] ~ -1, as is commonly adopted. The O/Fe ratio is usually regarded as an indicator of the chemical evolution timescale, since most of the O is produced more quickly in type II supernovae and the Fe primarily in

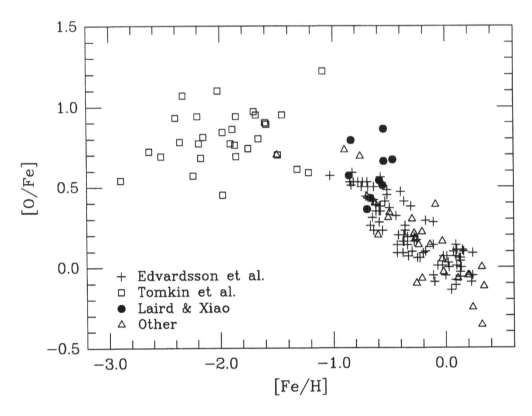

FIGURE 1. [O/Fe] *vs.* [Fe/H]

longer-lived type Ia supernovae, with lifetimes typically 1 Gyr or longer (Wyse & Gilmore 1988). If so, these data suggest that at least some parts of the halo reached relatively high metal abundance very quickly, while other parts of the halo formed metal-rich stars significantly later in time.

We gratefully acknowledge the support of the National Science Foundation under grant No. AST 92–21711.

REFERENCES

BARBUY, B. & ERDELYI-MENDES, M. 1989 *Astron. Astrophys.* **214**, 239.

BESSEL, M. S., SUTHERLAND, R. S. & RUAN, K. 1991 *Astrophys. J., Lett.* **383**, L71.

CARNEY, B. W., LATHAM, D. W., LAIRD, J. B. & AGUILAR, L. 1994 *Astron. J.* **107**, 2240.

CLEGG, R. E. S., LAMBERT, D. L. & TOMKIN, J. 1981 *Astrophys. J.* **250**, 262.

EDVARDSSON, B., ANDERSEN, J., GUSTAFSSON, B., LAMBERT, D. L., NISSEN, P. E. & TOMKIN, J. 1993 *Astron. Astrophys.* **275**, 1.

GARCÍA LÓPEZ, R. J., RÉBOLO, R., HERRERO, A. & BECKMAN, J. E. 1993 *Astrophys. J.* **412**, 173.

SNEDEN, C., LAMBERT, D. L. & WHITAKER, R. W. 1979 *Astrophys. J.* **234**, 964.

TOMKIN, J., LEMKE, M., LAMBERT, D. L. & SNEDEN, C. 1992 *Astron. J.* **104**, 1568.

WYSE, R. F. G. & GILMORE, G. 1988 *Astron. J.* **95**, 1404.

The Globular Cluster System and the Formation of the Galactic Halo

By VLADIMIR SURDIN

Sternberg Astronomical Institute, 13, Universitetskii Prospect, 119899, Moscow, Russia.

The reconstruction of the primordial globular cluster system gives us an evaluation of its initial mass of about $10^8 M_\odot$. It is not enough to explain an origin of the stellar component of the galactic halo even if we take into account the low efficiency of gravitationally bound systems formation. For the recent value of cluster formation efficiency (CFE $\approx$ 10%) extrapolated into the halo formation epoch, we can expect to obtain a total stellar mass connected with the survived globular cluster population of $\sim 10^9 M_\odot$. This is less than the mass of the stellar component of the galactic halo ($\sim 10^{10} M_\odot$). To solve this discrepancy we assume the existence of another population of low-mass star clusters which was completely destroyed by stellar evaporation to the present. These clusters may have formed during the turbulent collapse of the Galaxy. From a self-consistent scenario of the formation and early evolution of these clusters we obtain a good prediction of the total mass of the galactic halo.

1. Initial total mass of globular cluster system

To understand the globular-cluster formation process and its connection with the origin of the galactic halo, we must reconstruct the globular cluster mass spectrum. This spectrum as observed today must be a result of both the original mass spectrum of cluster formation and the subsequent dynamical evolution of the clusters due to processes such as early mass loss, tidal shocking by the galactic bulge and disk, dynamical friction and stellar evaporation coupled to the surrounding tidal field (Surdin 1978, 1979; Aguilar *et al.* 1988; Chernoff & Weinberg 1990; Surdin 1993). In principle, these dynamical effects are rather strong: some observable correlations between globular cluster parameters and their galactocentric distances (R_g) appear as a result of the dynamical evolution of the cluster system (Surdin 1994). However, both tidal shocking and dynamical friction (most effective for massive clusters) are relatively ineffective outside of about 5 kpc. Then we may use massive ($\log(M/M_\odot) > 5$) clusters of the galactic halo ($R_g \geq 5$ kpc) for the reconstruction of the cluster mass spectrum.

In the Milky Way, as well as in M31, the number of clusters per unit mass fits a simple power-law distribution: $N \propto M^{-2}$ (Surdin 1979; Racine 1980). Considering this spectrum as the primordial one we can obtain an initial total mass for the globular cluster system: $M_{tot}(\text{GC}) \sim 10^8 M_\odot$ for the mass range of $6.5 \geq \log(M/M_\odot) \geq 2.5$. This is of the order of 1% of the stellar component of the galactic halo. Is there any opportunity to construct a unified scenario of common origin of the star populations of both the galactic halo and the globular cluster system?

2. Some parallels between globular- and open-cluster formation

Currently in the Milky Way, both open clusters and star associations are born in giant molecular clouds with very low total star-formation efficiency (SFE $\sim$ 1%). This is why after gas ejection from a star-formation region about 90% of new born stars form a gravitationally unbound system–association, and only 10% of stars form a long-life system–open cluster. The recent formation rates of both open clusters and are the same: $\nu_{OC} \approx \nu_A \approx (2\text{–}4)\cdot 10^{-4}\text{yr}^{-1}$ (Elmegreen 1983). But the characteristic mass of open clusters, $M_{OC} \sim 10^3 M_\odot$ (Danilov & Seleznev 1994), is less than that of an association ($\sim 10^4 M_\odot$). Then the mean star-formation rate through the open cluster channel is SFR(OC)$\simeq 0.3 M_\odot \mathrm{yr}^{-1}$ and through the association channel it is SFR(A)$\simeq 3 M_\odot \mathrm{yr}^{-1}$. The relative value of the formation rate of stars collected into bound systems I propose to call "efficiency of gravitationally bound systems formation" or simply "cluster-formation efficiency" (CFE):

$$\mathrm{CFE} = \frac{\mathrm{SFR(C)}}{\mathrm{SFR(C)} + \mathrm{SFR(A)} + \mathrm{SFR(I)}} \qquad (2.1)$$

where SFR(C), SFR(A) and SFR(I) are the star-formation rate in clusters, in associations and the formation rate of isolated stars. In the present-day Galaxy the value of SFR(I) is small (Zinnecker *et al.* 1993). So CFE $\simeq$ SFR(C)/SFR(A) $\simeq 10\%$.

Up to now, we has been no theoretical explanation for the particular value of CFE. This value depends on the star-to-gas distribution as well as on the distribution of stellar velocities at the moment of the gas ejection away from a star-formation region. So we have no any ideas about the value of CFE at the globular cluster formation epoch. There is only one way to take this key factor into account: by proposing its current value for globulars formation as well. Thus we suppose that each episode of star formation in the young Milky Way gives not only a globular cluster but a short life association too. For the present galactic value of the cluster-formation rate (CFE = 10%) we can evaluate the total amount of stars genetically connected with globulars: $M_{tot}(\mathrm{GC+A})$ = $M_{tot}(\mathrm{GC})/\mathrm{CFE} \simeq 10^9 M_\odot$. Nor is it enough to construct a stellar component of the galactic halo whose mass is $\sim 10^{10} M_\odot$ (Edmunds & Phillips 1984; Haud *et al.* 1985).

To solve this discrepancy we can assume the existence of another population of star clusters which once populated the primordial galactic halo, but were completely destroyed due to stellar evaporation to the present. From many available theories of cluster formation during the galaxy collapse process (*e.g.* Fall & Rees 1985; Murray & Lin 1992) I will consider the idea of Sabano & Tosa (1983) on cloud formation during the turbulent collapse of the Protogalaxy. Developing this idea I will give a self-consistent scenario of the formation and early evolution of low-mass clusters in the galactic halo, to obtain a good prediction of the total mass of the Population II stars.

3. Cluster formation in the turbulent protogalaxy

3.1. *Too big a jump*

There are certain observational arguments for a slow, turbulent-pressure-supported contraction of the Galaxy (Yoshii & Saio 1979). To understand a star-formation process in such conditions, Sabano & Tosa (1983) considered a slowly contracting, turbulent protogalaxy, whose gravitational energy disappears in shock waves. They have shown that the cooling time of dense after-shock gas is rather short. This creates the conditions necessary for the development of gravitational instability. Sabano & Tosa calculated the Jeans mass and radius ($M_J \approx 3 \cdot 10^5 M_\odot$ and $R_J \approx 50$ pc) and found it to be very close to the real masses and radii of present-day globular clusters. They conclude that this is a new globular-cluster formation scenario.

One important issue can be addressed. This kind of transition from theoretical values of Jeans parameters to observational values of globular cluster parameters is very common for theorists. But it makes too big a jump over several very important evolutional episodes: gas loss from the protocluster, the escape of fast-moving stars (cluster–association dichotomy), tidal truncation during the first flyby near the galactic centre, etc. In the present paper I would like to consider at least one of these.

Perhaps the most important episode of early cluster evolution is gas loss from the star-formation regions. In the present-day Galaxy, this process decreases the mass of these regions to 10^2 times (SFE $\sim 1\%$). We really must take this process into account for our globular cluster formation theory. In the galactic disk, cluster–association dichotomy decreases the mass of bound clusters by a factor of ten (CFE $\approx 10\%$). As a result, the transition takes place from GMC masses, $(10^5 - 10^6)M_\odot$, to open cluster masses, $(10^2 - 10^3)M_\odot$. But we do not know the exact value of CFE at the Galaxy formation epoch and see no way of theoretically evaluating the CFE. This problem is far from simple. So we will have to restrict our consideration to allow for the SFE value only. Fortunately, this is enough to be able to calculate the total mass of Population II stars.

3.2. *From gaseous cloud to star cluster*

Let us consider a gravitationally bound cloud of mass M_C and radius R_C. The gravitational energy of the cloud is $|U_C| = GM_C^2/R_C$. How many stars may be created in the cloud before it is destroyed by the stars themselves? We can express the star energy input to the cloud as

$$E = \mu \epsilon M_S, \tag{3.2}$$

where ϵ is the energy output of massive stars per unit of young stellar aggregate mass (M_S) and μ is the fraction of this energy conserved in the bubble of hot gas inside the cloud. Star formation breaks off when the cloud is disintegrated by stars under the natural condition:

$$|U_C| = E. \tag{3.3}$$

From this equation for $\epsilon = 10^{49} erg/M_\odot$ and $\mu = 0.1$ (Larson 1974; Tenorio-Tagle *et al.* 1990) we obtain:

$$M_S = 10^4 M_\odot \left(\frac{M_C}{10^6 M_\odot}\right)^2 \left(\frac{R_C}{10\text{pc}}\right)^{-1}. \tag{3.4}$$

For the current star-formation process inside giant molecular clouds $(M_{GMC} \approx 10^6 M_\odot; R_{GMC} \cdot 20$ pc). This equation gives a good enough prediction of the cluster/association mass: $M_S \approx 5 \cdot 10^3 M_\odot$. We may expect this formula to be equally valid for the ancient star-formation process too.

3.3. *The extinct star-cluster population of the galactic halo*

From Equation (3.4) we can obtain the total stellar aggregate mass which forms in each Sabano–Tosa cloud. If the equilibrium radius of the cloud is $R_C \approx 0.5R_J$, then $M_S \approx 400M_\odot$. Even if all these stars were bound in the cluster (CFE = 100%), it had no chance to survive during the cosmological time. The minimum mass of a Population II star cluster which did not completely evaporate in the Galaxy is (Surdin 1978):

$$M_{min} = 2 \cdot 10^4 \left(\frac{R_p}{1\text{kpc}}\right)^{-1} M_\odot, \tag{3.5}$$

where R_p is the perigalactic distance of the cluster orbit. So a cluster of mass $M_S = 400M_\odot$ can survive only outside 50 kpc from the galactic center only. Could some very distant low-mass globulars be the debris of that ancient population?

4. Formation of the galactic halo

4.1. *Field stars*

To evaluate the total number of Sabano–Tosa clouds (N) we can assume that the gravitational energy of the Milky Way disappeared in shock waves of the turbulent protogalaxy:

$$N = \left(\frac{M_G}{M_J}\right) \left(\frac{V_\infty}{V_t}\right)^2, \tag{4.6}$$

where M_G is the galactic stellar population mass, V_∞ is the mean galactic escape velocity (taking into account the dark corona) and V_t is the turbulent velocity. For $M_G = 10^{11} M_\odot$ (Haud *et al.* 1985), $V_\infty = 650 \pm 50 \,\mathrm{km\,s^{-1}}$, $M_J = 3 \cdot 10^5 M_\odot$ and $V_t = 100 \,\mathrm{km\,s^{-1}}$ (Sabano & Tosa 1983) we obtain $N \approx 2 \cdot 10^7$. All these clusters must be completely evaporated at the present to give a total mass of the halo field stars $M_H = M_S N = 8 \cdot 10^9 M_\odot$. This is in good agreement with observations (Edmunds & Phillips 1984; Haud *et al.* 1985).

4.2. *Globular clusters*

According to Equation (3.4), globular clusters must have massive gaseous progenitors: $M_C \sim (10^7 - 10^8) M_\odot$. Clouds as massive as this may have formed in pregalactic pancakes (e.g. Jones *et al.* 1981) at an earlier cosmological epoch (Peebles 1984).

My participation in the Workshop was made possible by financial support from the LOC.

REFERENCES

AGUILAR, L., HUT, P. & OSTRIKER, J. P. 1988 *Astrophys. J.* **335**, 720.

CHERNOFF, D. F. & WEINBERG, M. D. 1990 *Astrophys. J.* **351**, 121.

DANILOV, V. M. & SELEZNEV, A. F. 1994 *Astron. Astrophys. Trans.* **6**, No 2, 85.

EDMUNDS, M. G. & PHILLIPS, S. 1984 *Astron. Astrophys.* **131**, 169.

ELMEGREEN, B. G. 1983 in *The Nearby Stars in the Stellar Luminosity Function.* (ed. A. G. Davis Philip & A. R. Upgren). IAU Colloquium 76, p. 235. L. Davis Press.

FALL, S. M. & REES, M. J. 1985 *Astrophys. J.* **298**, 18.

HAUD, U., JOEVEER, M. & EINASTO, J. 1985 in *The Milky Way Galaxy.* (ed. H. van Woerden, R. J. Allen & W. B. Burton). IAU Symposium 106, p. 85. Reidel.

JONES, B. J. T., PALMER, P. L. & WYSE, R. F. G. 1985 *Mon. Not. R. Astron. Soc.* **197**, 967.

LARSON, R. B. 1974 *Mon. Not. R. Astron. Soc.* **169**, 229.

MURRAY, S. D. & LIN, D. N. C. 1992 *Astrophys. J.* **400**, 265.

PEEBLES, P. J. E. 1984 *Astrophys. J.* **277**, 470.

RACINE, R. 1980 in *Star Clusters* (ed. J. E. Hesser). IAU Symposium 85, p. 369. Reidel.

SABANO, Y. & TOSA, M. 1983 in *Theoretical aspects on structure, activity and evolution of galaxies*, p. 15. University of Tokyo.

SURDIN, V. G. 1978 *Sov. Astron.* **22**, 401.

SURDIN, V. G. 1979 *Sov. Astron.* **23**, 648.

SURDIN, V. G. 1989 *Astron. Nachr.* **310**, 381.

SURDIN, V. G. 1993 in *The Globular Cluster – Galaxy Connection.* (ed. G. H. Smith & J. P. Brodie). ASP Conference Series, vol. 48, p. 342. ASP.

SURDIN, V. G. 1994 *Astron. Lett.* **20**, 15.

TENORIO-TAGLE, G., ROZYCZKA, M. & BODENHEIMER, P. 1990 *Astron. Astrophys.* **237**, 207.

ZINNECKER, H., MCCAUGHREAN, M. J. & WILKING, B. A. 1993 in *Protostars and Planets. III.* (ed. E. H. Levy & J. I. Lunine), p. 429. The University of Arizona Press.

On the Origin of Halo Globular Clusters and Spheroid Stars

By ENRICO PESCE[1] AND MARIO VIETRI[2]

[1]Istituto Astronomico, Università di Roma, Via Lancisi 29, I–00100 Roma, Italy

[2]Osservatorio Astronomico di Roma, I–00040 Monte Porzio Catone (Roma), Italy

A scenario is presented for the simultaneous formation of halo globular clusters and spheroid stars, in which shocks occuring in the gaseous, primordial ($Z= 0$) Galaxy play the central role. Denser-than-average subcondensations are shown to cool quickly and implode. Small condensations ($M < M_l = 5 \times 10^4 M_\odot$) are returned to the hot surrounding phase by the reflected shock; large ones ($M > M_u = 2 \times 10^7 M_\odot$) fragment into stars in the dense shell behind the implosion front, thus generating the first spheroid stars. Intermediate clouds, suddenly cooling to $100 K$ because of collisional H_2 excitation, are compressed and left in approximate pressure equilibrium with the hot medium: they give rise to halo GCs. The dependence of the proposed formation mechanism upon overall galactic metallicity and galaxy rotation velocity is weak, so that a universal upper mass limit is deduced. A mass range for halo GCs ($2 \times 10^3 - 6 \times 10^5 M_\odot$) and a velocity dispersion for spheroid stars in agreement with observations are predicted.

1. Introduction

The formation of globular clusters is a difficult problem: their masses vary only by two orders of magnitude in contrast to the five orders of magnitude spanned by the masses of their parent galaxies and their luminosity functions are quite similar in all well-known galactic systems (Magellanic Clouds, M31, M87, M33, etc.), so that a universal physical formation mechanism is required.

Previous attempts concentrated on locating a particular time in the history of proto-galaxies at which the Jeans mass of the gaseous material equal the typical GC's mass. Fall & Rees (1985) proposed that gas of primordial composition would undergo a thermal instability leading to a two phase medium, the cold phase of which, unable to cool below the Lyman barrier ($T \simeq 9000 K$), would be imprinted by its Jeans mass ($\simeq 10^6 M_\odot$) and fragment into GCs. It was pointed out, however (Shapiro et $al.$ 1987), that H_2 formation via H^-, in conditions of non-equilibrium ionisation, would lead to runaway cooling to temperatures below $10^3 K$, and a reduction of the Jeans mass by four orders of magnitude.

In order to rescue this model, Kang et $al.$ (1990) postulate an intense UV-field from a pre-existing AGN or previous populations of stars to photodissociate H_2 and prevent the rapid cooling of the gas and the decrease of the Jeans mass to stellar values, but such a modified scenario is not free from objections and is at most marginally consistent with observations of metallicity. We propose a different solution (Vietri & Pesce 1995).

2. The model

We follow Fall & Rees (1985) in the description of the early Galaxy. The gravitational potential is due to a dark halo, with density distribution $\rho(R) = V_c^2/4\pi G R^2$, and $V_c = 220 \, km \, s^{-1}$, which is adequate for $3 \, kpc < R < 30 \, kpc$. Gas falls freely into the potential well, in time

$$\tau_{ff} = \left(\frac{\pi}{2}\right)^{1/2} \left(\frac{R}{V_c}\right) = 5.6 \times 10^6 yr \frac{R}{1 \, kpc}$$

and is heated to the virial temperature of the halo $T_h = \mu_h V_c^2/2k_B \simeq 1.7 \times 10^6 K$, and its density is determined by requiring that its free-fall time equals the cooling time:

$$\rho_h = 1.7 \times 10^{-24} \left(\frac{R}{1\,kpc}\right)^{-1} gm\ cm^{-3}.$$

3. Evolution of shocks and mass limits

Overdense regions in the protogalaxy cool more effectively and are compressed to higher densities by the hot surrounding gas. Let us consider one such region. The self-gravitational free-fall timescale $\tau_{grav} = (3\pi/32G\rho)^{1/2} = 5.4 \times 10^7 yr(R/1\,kpc)^{1/2}$ is very long at such low densities, compared to all other relevant timescales, so the cloud will initially evolve isochorically. We numerically followed the temperature history of one such subregion for $n = 1cm^{-3}$ and demonstrated that H_2-formation via H^- is efficient even at such low densities: it manages to cool the clouds in a time $\tau_{ic} \ll \tau_{ff}, \tau_{grav}$ to very low temperatures $T < 10^3\,°K$, so the hot surrounding medium drives shocks into any such cloud.

Three different types of clouds can be distinguished according to the cloud crossing time by the shock R_c/V_S, where $V_S \simeq 100 km s^{-1}$ in the conditions under investigation. If $R_c/V_S < \tau_{cool}$, where τ_{cool} is the cooling time of the post-shock layer, the cloud material does not have time to cool before the shock reaches the cloud centre and bounces back, and is then returned to the hot phase. In larger clouds, the shocked material has time to cool and form a cold, dense shell moving with the shock which in time τ_g (Vishniac 1983) will collapse because of self-gravity. If $R_c/V_S > \tau_g$, fragments of roughly stellar size will form, which collapse to make stars in time τ_g. These stars, moving at the shock speed, cannot be decelerated by the bounced shock, nor can they be bound by the cloud self-gravity, so they disperse as stars into the galaxy. The derived velocity dispersion for such fragments is close to the observed value $\sigma \simeq 100\,km s^{-1}$ for the spheroid stars. In the intermediate range, $\tau_{cool} < R_c/V_S < \tau_g$, the return shock decelerates the collapsing shell and leaves a much smaller, colder, denser cloud in approximate pressure equilibrium with the hot medium. These intermediate clouds are the progenitors of GCs. For a reasonable star-formation efficiency $\eta = 0.03$, the predicted GC masses, M_{GC}, are bracketed by:

$$\eta M_l = 1.5 \times 10^3 M_\odot \left(\frac{R}{1\,kpc}\right)^2 < M_{GC} < \eta M_u = 6 \times 10^5 M_\odot \left(\frac{R}{1\,kpc}\right)^{1/2};$$

the details of this calculation are in Vietri & Pesce (1995). Figure 1 shows these curves, together with masses of known galactic globular clusters: solid circles are halo GCs for which individual M/L_V have been determined (Pryor & Meylan, 1993), open triangles those for which a fiducial $M/L_V = 3$ was assumed, for lack of exact values. Most triangles lie outside $R = 20\,kpc$, where our model loses predictive power.

4. The dependence on galaxy size and metallicity

To the best of our knowledge, this is the first time an upper limit to the GC mass has been found and it can also be shown that $\eta M_u \propto V_c^{1/2}$, so that the largest allowed mass is quite a universal limit, almost independent of the host galaxy. The argument goes as follows. The density of the hot phase is determined under the assumption that its cooling time will be about equal to its collapse timescale τ_{ff}; since τ_{cool} is mostly determined by bremsstrahlung, for which radiative losses scale as $\propto \rho_h^2 T_h^{1/2}$, we have

$$\tau_{cool} \approx \frac{\rho_h T_h}{\rho_h^2 T_h^{1/2}} \simeq \tau_{ff} \approx \frac{R}{V_c} \Rightarrow \rho_h \approx V_c T_h^{1/2} \approx V_c^2$$

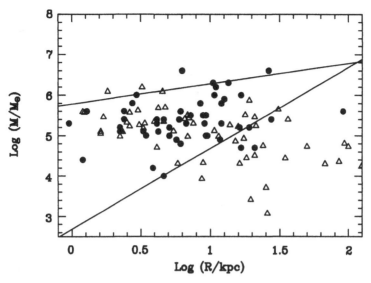

FIGURE 1. Limits obtained for GC masses as function of the galactocentric radius.

because the virial temperature $T_h \propto V_c^2$. Furthermore, since obviously $V_S \propto V_c$, we obtain

$$\tau_g \approx \left(\frac{1}{\rho_h V_S}\right)^{1/2} \approx V_c^{-3/2}$$

for the self-gravitational fragmentation timescale of the cold post-shock shell. Finally we have

$$M_u \approx \rho_h (V_S \tau_g)^3 \approx V_c^{1/2} .$$

Stated otherwise, using the Faber–Jackson relation $L \propto V_c^4$, we have that $M_u \propto L^{1/8}$, so that a variation of L over four orders of magnitude (i.e. ten astronomical magnitudes, more than from Cen A to the LMC), corresponds to a range of M_u of just a factor of three. On the contrary, in the theory of Fall & Rees, the mass limit decreases as $M_J \propto V_c^{-2}$, a much stronger and somewhat counterintuitive dependence.

In the scenario of Fall & Rees, the dependence of the value of the imprinted M_J on the cooling function makes the formation of GCs and stars a sort of threshold process: for $Z \leq 0.01 Z_\odot$ only GCs can be formed, whereas for higher Z only stars are expected, because gas can then cool below the Lyman barrier in a collapse timescale. This is in contrast with determinations of metallicity distributions for globulars and halo stars and with observations of galaxies still forming GCs. On the contrary, in our model, the presence of more metals in the initial gaseous medium simply implies a lower temperature ($T \simeq 10 K$) of the post-shock material, which then fragments into stars more easily. The upper mass is thus reduced to:

$$M_{GC} < \eta M_u \left(\frac{T_{ps}}{100\ K}\right)^{3/4} \simeq 10^5 M_\odot,$$

so that later GCs can form and are only a bit smaller than earlier ones.

REFERENCES

FALL, S. M. & REES, M. J. 1985 *Astrophys. J.* **298**, 18.

KANG, H., SHAPIRO, P. R., FALL, S. M. & REES, M. J. 1990 *Astrophys. J.* **363**, 488.

PRYOR, C. & MEYLAN, G. 1993 in *Structure and Dynamics of Globular Clusters* (ed. S. G. Djorgovski & G. Meyland). ASP Conference Series, vol. 50, p. 357. ASP.

SHAPIRO, P. R. & KANG, H. 1987 *Astrophys. J.* **318**, 32.

SILK, J. & WYSE, R. G. F. 1989 in *The Epoch of Galaxy Formation* (ed. C. S. Frenk, R. S. Ellis, T. Shanks, A. F. Heavens & J. A. Peacock). NATO ASI Series C, vol. 264, p. 285. Kluwer.

VIETRI, M. & PESCE, E. 1995 *Astrophys. J.* in press. (20 March 1995 issue).

VISHNIAC, E. T. 1987 *Astrophys. J.* **274**, 152.

Elmegreen: Can the H_2 shield itself against UV radiation?

Vietri: Easily. Actually, the real problem is exactly the opposite: enormous UV fluxes (Kang *et al.* 1990) are necessary to suppress the formation of H_2, so large as to conflict with constraints on stellar metallicity distributions if the UV field is of stellar origin, or to require the formation of a $10^9 M_\odot$ black hole before the formation of any star if non-thermal.

Lesch: Would you expect much more substructure when the shock moves into the cloud? What about Rayleigh-Taylor and other instabilities?

Vietri: The implosion shock is very stable. Potentially, it is subject to the R–T and Kelvin–Helmoltz instabilities; however, since the initial evolution is isochoric, mantaining the density of the cloud equal to that of the hot phase, and since the shock velocity, in the isothermal case, can be shown to tend to a constant, the R–T instability is suppressed in our scenario. We checked that the growth timescale of the K–H instability were longer than the shock crossing time of the cloud and that the implosion front were stable also against Vishniac's dynamic instability.

Surdin: Did you assume some particular value of the star- formation efficiency or did you compute it self-consistently?

Vietri: We cannot obtain it self-consistently and so we have adopted a reasonable value of the SFE parameter (Silk & Wyse 1989). It should be said that such a parameter is poorly known even in present age ISM conditions and it is subject to an indetermination of about one order of magnitude.

The Formation of Galaxies in Clusters

By CHRISTINE JONES

Harvard-Smithsonian Center for Astrophysics, Cambridge, MA 02138, U.S.A

Observations of clusters can provide information about the environmental influences on galaxies including the factors which govern a galaxy's morphological type and the formation efficiency of galaxies, which may provide clues as to how the Milky Way formed.

1. Introduction

Although our detailed knowledge of the Milky Way is in many ways much more extensive than that of any other galaxy, observations of other galaxies have been fundamental to understanding our own. Probably the most important and most widely known example is M31. Much of our understanding of galactic rotation, stellar populations and evolution and distance scales are based on observations of M31. As I discuss is this paper, observations of galaxies in clusters also can provide clues about the formation of our galaxy. After a brief review of the properties of galaxy groups and clusters, I will discuss three specific topics:

• The well-recognized segregation of galaxy morphological types in clusters and the probable reasons for that segregation.

• The formation efficiency of galaxies in different environments from groups to rich clusters.

• Related to the discussions of structure and metallicity in the Milky Way, I discuss, for elliptical galaxies, the halo mass distribution and the heavy element abundances and their distribution in the hot coronae around these galaxies.

2. A Brief Introduction to the Properties of Groups and Clusters

Clusters of galaxies are the largest gravitationally bound systems in the Universe and the largest objects which appear to be, at least approximately, in equilibrium. The luminous material in clusters falls primarily into two forms – the visible galaxies and the hot, x-ray emitting intracluster medium. The richest, densest clusters contain predominantly elliptical and lenticular (S0) galaxies, while in less dense clusters, up to half the galaxies are spirals. While the visible galaxies and the x-ray emitting hot gas are important components, most of the mass in rich clusters is "dark matter." Although this material has not been directly observed and its nature remains unknown, X-ray and visible light observations can determine the amount and distribution of this dark matter.

Rich (Abell-like) clusters have x-ray luminosities ranging from as low as those of individual bright galaxies to 1000 times brighter: $10^{42} - 10^{45}$ ergs sec^{-1} (Abramopoulos & Ku 1983; Jones & Forman 1984). Gas temperatures range from a few 10^7 to 10^8 K (Mushotzky et al. 1978; Edge 1989) and are comparable to the equivalent temperatures as measured by the velocity dispersions for the galaxies in the cluster. The gas densities in the cores of rich clusters lie in the range $10^{-2} - 10^{-3}$ cm^{-3} and the inferred cooling times of the gas can be as short as 10^9 years at the cluster center (Fabian et al. 1984, 1991). The gas mass is typically a few 10^{14} M$_\odot$ within the central few Mpc of rich clusters or about 25% of the total cluster mass. Compact (dense) groups of galaxies have predominantly early-type galaxy populations and are also bright in x-rays (Schwartz et al. 1980; Kriss et al. 1983). These groups have gas masses of $\approx 10^{12} - 10^{13}$ M$_\odot$ (within ≈ 0.5 Mpc)

329

and gas temperatures up to a few 10^7 K. Luminosities range up to 10^{44} ergs sec^{-1}, well into the regime populated by rich clusters. In general, the x-ray emission from compact Morgan groups is azimuthally symmetric and is centered on the bright D galaxy which dominates the group.

Present epoch clusters display a wide variety of properties which can be understood in a framework of cluster dynamical evolution. Gunn & Gott (1972) first noted that, while the dynamical timescale of Coma (a very rich cluster) was less than a Hubble time, most less dense clusters would have dynamical timescales greater than a Hubble time.

One of the indicators of the dynamical state of a cluster is the degree of substructure as demonstrated by either the galaxy or the gas distribution. The X-ray observations have been found to be particularly suited to studies of the structure of the cluster potential. Even the Coma cluster, long considered the prototype of a large core radius, relaxed system, has been found to be a multiple system undergoing mergers based on optical (Fitchett & Webster 1987) and X-ray observations (Briel *et al.* 1992).

From both optical and X-ray samples, the percentage of surveyed clusters with substructure is about 40% (see Jones & Forman 1992 for additional discussion). Although the X-ray and spectroscopic surveys show considerable substructure for all cluster luminosity classes at all redshifts, as West & Bothun (1990) emphasized, it is important to separate clusters whose substructure is a true vestige of smaller subsystems that have recently merged from others in which a small group lies outside or is falling into a virialized cluster. In the *Einstein* cluster sample, about 20% of clusters appear to be undergoing mergers of subsystems, while some of the X-ray luminous, hot clusters, such as A2256 or A85, contain a dominant, virialized cluster and a smaller, infalling group.

3. The Effect of Cluster Environment on Galaxy Morphology

Understanding why elliptical galaxies are the dominant inhabitants of cluster cores, while spirals dominate the galaxy population in the cluster outskirts and in the field, may lead to understanding how different types of galaxies form. Melnick & Sargent (1977) demonstrated the existence of population gradients with cluster radius. Dressler (1980) suggested that the morphological type depended fundamentally on the local galaxy density, with ellipticals formed in the highest density regions. However, in a reexamination of Dressler's sample of ≈ 6000 galaxies in 55 clusters, Whitmore *et al.* (1993) showed that the distance of the galaxy from the cluster center was more important in determining the galaxy morphology than was the local galaxy density. In particular, when the morphology-local galaxy density relation was examined using only those galaxies within 0.25 characteristic radii of the cluster center, no correlation of galaxy type with density was found. The apparent correlation of morphology with local density probably results from the good correlation of local galaxy density with cluster radius.

As Figure 1 shows, the fraction of elliptical galaxies is nearly constant in the outer regions of clusters comprising 10 – 15 percent of galaxies, and increases dramatically to 60 – 70 percent only within 0.5 Mpc of the cluster center. The fraction of S0 galaxies rises slowly as one approaches the cluster core, but falls within the 0.2 Mpc central region. Unlike that for the early type galaxies, the spiral fractions falls, at first slowly, then rapidly as one approaches the cluster center. Whitmore *et al.* (1993) proposed that the increase in the elliptical fraction in the cluster cores results from the destruction of large, late-forming disk and proto-disk galaxies. Thus it is the absence of spirals, and not the preferential formation of ellipticals, that leads to the larger fraction of elliptical galaxies at cluster centers. Since most elliptical galaxies form at very early epochs, before cluster

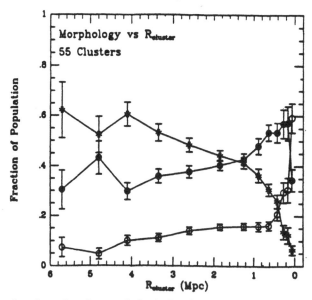

FIGURE 1. The fraction of each morphological galaxy type as a function of distance from the cluster center for the 55 clusters in the Dressler (1980) sample. Open circles denote ellipticals, filled circles S0's and asterisks are spirals and irregulars. (from Whitmore *et al.* 1993).

collapse, they could survive, even in cluster cores, during the extended epoch of cluster collapse.

One possible mechanism for the destruction of large disk galaxies would be tidal shear produced by the cluster gravitational field during and after the collapse of the cluster core. Whitmore (1994) suggested that the sharply peaked central gravitational potential about a cD galaxy also would increase the tidal shear. As Figure 2 shows, the material from the destroyed protogalaxies forms a significant fraction of the total intracluster gas observed within 2.0 Mpc of the cluster center.

4. The Formation Efficiencies of Galaxies in Clusters

The discovery of hot gas in clusters of galaxies completely changed our view of these systems. Rather than clusters consisting of galaxies moving through an empty void, the X-ray observations showed the presence of a diffuse, hot gas filling the intracluster volume. More importantly, it was realized that in the richer clusters, the hot gas was the dominant observable baryonic component, exceeding the luminous mass in galaxies by roughly a factor of five. In this section we discuss the relative baryonic contributions of the galaxies and the gas and what we can learn about galaxy formation in clusters.

4.1. *Correlation of Gas Mass and Stellar Mass*

One can use the X-ray surface brightness to determine physical cluster parameters. For a spherically symmetric cluster or subcluster, the radial x-ray surface brightness distribution is accurately parameterized by the expression

$$S(r) = S(0)(1 + (r/a_x)^2)^{-3\beta+1/2} \qquad (4.1)$$

where S(0) is the central surface brightness, a_x is the cluster core radius and, in the hydrostatic-isothermal β model (Cavaliere & Fusco-Femiano 1976), β is the ratio of the energy per unit mass in the galaxies to that in the gas. With the exception of the central

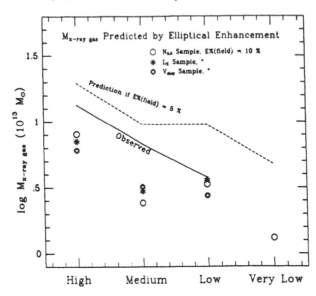

FIGURE 2. The predicted mass in the ICM from the destruction of disk-type galaxies (circles, asterisks, and stars), assuming the initial elliptical fraction is 10 percent. If the initial elliptical fraction is as low as 5 percent, the predicted mass contributed to the ICM is shown by the dashed lines. The solid line shows the observed mass of the ICM within a 2 Mpc radius for subsamples defined by their X-ray luminosity. (from Whitmore *et al.* 1993).

cusp in cooling flow clusters, this form for the surface brightness distribution has been found to be an adequate description to the extent that gas has been traced in clusters, out to 8-10 core radii (see Jones & Forman 1984).

This expression for the cluster surface brightness can readily be inverted to give the gas density distribution

$$n(r) = n(0)(1 + (r/a_x)^2)^{-3\beta/2} \qquad (4.2)$$

Formally, this inversion requires the assumption that the gas be isothermal. However, the count rate from a fixed mass (and volume) of gas in the *Einstein* or *ROSAT* energy bands changes by less than 10% over the full range of cluster temperatures from 2 to 15 keV. Hence, the cluster surface brightness as measured by either *Einstein* or *ROSAT* is nearly a direct measure of the gas density profile.

The gas mass in clusters is a few $\times 10^{14}$ solar masses within the central few Mpc. For clusters with measured gas temperatures and measured surface brightness profiles, Figure 3 shows the fraction of the virial mass which is hot gas. For the lowest luminosity clusters, the gas mass is a small fraction of the virial mass. However, in general as the virial mass of the cluster increases, so does the gas mass, such that, the fraction of the virial mass which is hot gas increases (e.g. White *et al.* 1993, David *et al.* 1995). The intracluster medium is typically 10% to 30% of the cluster virial mass. At large radii, the gas mass fraction in all clusters surveyed appears to be about 30% of the total mass. In an $\Omega = 1$ universe, for standard Big Bang nucleosynthesis models, the baryonic fraction of the total mass must be less than about 6% (Walker *et al.* 1991). More exactly,

$$0.04 < \Omega_b h_{50}^2 < 0.06 \qquad (4.3)$$

where Ω_b is $f_b\Omega$ and f_b is the baryon mass fraction. A baryon fraction of 30% $(= 30h_{50}^{-3/2})$

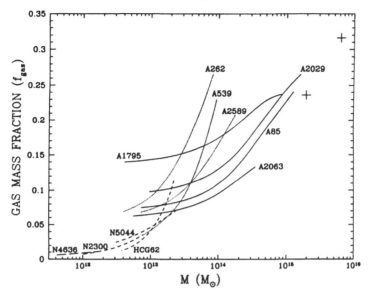

FIGURE 3. The fraction of the virial mass which is hot gas. The cluster virial mass is derived from the measured gas temperature assuming that the gas is isothermal. Central cooling regions are omitted. The two crosses mark the gas mass fractions in A2163.

places limits on Ω of

$$0.13h_{50}^{-1/2} < \Omega < 0.2h_{50}^{-1/2} \qquad (4.4)$$

We also rule out very small values of H_0 since clusters cannot be more than 100% baryons ($f_b = 0.30h_{50}^{-3/2} < 1$.) Thus we limit Ω to fall between 0.1 and 0.3.

To address the question of the efficiency of galaxy formation in different environments, the mass in intracluster gas can be compared with the cluster galaxy mass. As shown in Figure 4, for a small sample of clusters, the ratio of the mass in gas to that in galaxies varies systematically from poor to rich clusters (David *et al.* 1990). In particular, in groups and poor clusters the gas mass (as measured within five core radii) is comparable to the galaxy mass, while in the richest, hottest clusters, the gas mass is up to six times the mass in the galaxies. For a larger sample of clusters, Arnaud *et al.* (1992) compared the stellar mass and gas mass as measured within 3 Mpc radii. The gas mass increases as $M_{gas} \propto M_{stellar}^{1.9}$ (Arnaud *et al.* 1992). It is well known that the optical mass-to-light ratio increases with the size of the system (e.g., Blumenthal *et al.* 1984) and from poor to rich clusters, the ratio of x-ray emitting gas to virial mass increases (Figure 3). Thus, one can understand qualitatively why the ratio of gas mass to stellar mass should increase from poor to rich clusters.

These correlations of gas to galaxy mass can be used to compute the efficiency of galaxy formation and can provide important constraints on formation scenarios. The galaxy formation efficiency—the conversion of baryons from gas to stars in galaxies— can be written as $\epsilon = M_{stellar}/(M_{stellar} + M_{gas})$ or equivalently as $\epsilon = (1 + M_{gas}/M_{stellar})^{-1}$ if we make three assumptions: 1) that clusters are closed systems (no gas is gained or lost), 2) that the gas mass expelled from galaxies is small compared to that presently seen in clusters, and 3) that the two luminous components – galaxies and gas – comprise the total baryonic complement of clusters. Thus, by measuring $M_{gas}/M_{stellar}$, we determine that the efficiency of star (and galaxy) formation ranges from 50% for groups ($M_{gas}/M_{stellar} \approx 1$) to about 15% for the richest clusters ($M_{gas}/M_{stellar} \approx 6$). Alternatively, we can

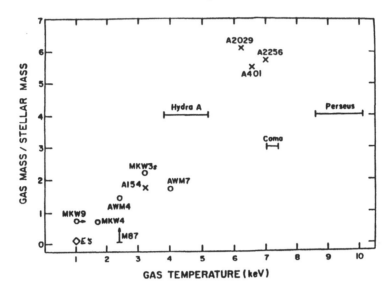

FIGURE 4. The ratio of the gas mass to the stellar mass plotted against the temperature of the ICM. The gas and stellar masses are evaluated within five core radii (from David *et al.* 1990)

assume that considerable mass is ejected by galaxies. If we estimate this ejected mass by making the extreme assumption that all the gas mass observed in poor clusters was originally expelled from galaxies, then the galaxy formation efficiency in groups is 100%, while that in the richest clusters remains essentially unchanged. Thus, the efficiency of galaxy formation decreases as one moves to richer systems. In other words, although the richest systems obviously produced more galaxies, they produced them less efficiently. Alternatively as Whitmore *et al.* (1993) has suggested, the richest clusters have more effectively destroyed their constituent galaxies, particularly the spirals.

David and Blumenthal (1992) calculated galaxy formation efficiencies within the context of cold dark matter scenarios with biasing (Blumenthal *et al.* 1984; Bardeen *et al.* 1986) In such scenarios, the initial fluctuations which will become groups are larger in amplitude than those which become clusters, since there is more power in the initial fluctuations on smaller linear scales. Biasing requires that galaxies form only in perturbations whose density exceeds a critical density threshold. Since the form of the power spectrum of density perturbations implies that smaller mass perturbations are generally of larger amplitude, galaxy perturbations in groups are more likely to exceed the critical threshold than are galaxy perturbations in rich clusters.

5. Dark Halos and Heavy Element Abundances in Early-Type Galaxies

The discovery of X-ray luminous hot coronae around early type galaxies (Forman *et al.* 1979; Nulsen *et al.* 1984; Forman, *et al.* 1985) dramatically changed our understanding of these galaxies. Prior to the X-ray observations, it was believed that the gas shed by evolving stars was expelled by galactic winds (e.g. Faber & Gallagher 1976). However, the X-ray observations showed extended hot coronae with luminosities up to 10^{42} ergs sec^{-1}, temperatures ≈ 1 keV (10^7K), and gas masses up to 10^{10} M$_\odot$. The presence of heavy halos was required to suppress galactic winds to permit the buildup of gas to the values observed (Forman *et al.* 1985; Fabian *et al.* 1986; Mathews & Loewenstein 1986; Sarazin & White 1988; David *et al.* 1991).

Spatially resolved X-ray spectra provide a measurement of the heavy element abundance in the hot gas. The hot corona, as the repository of the stellar mass loss (that not expelled from the galaxy or recycled into stars), contains a fossil record of past supernova activity in the galaxy. Since the radial flow time of the gas is long, the hot gas contains a history of the metals injected into it by evolving stars. As Loewenstein and Mathews (1991) emphasized, measuring radial abundance gradients determines both the history of the past supernova rate, as well as its present value. Abundance measurements also distinguish among different models for the gas dynamics in the corona.

We analyzed the *ROSAT* PSPC images of NGC 4472 and NGC 1399 (see Forman *et al.* 1993 or Jones *et al.* 1994 for details). Both galaxies exhibit extensive coronae in X-rays with emission visible to beyond $10'$. To derive the parameters needed for the mass determination, we generate radial profiles to determine the X-ray surface brightness distribution and extract spectra in annuli. For each annulus, the observed source spectrum was compared to models of an optically thin thermal spectrum. There is cooler gas in the inner regions, but in the outer regions the gas temperature is essentially constant.

Since the gas is essentially isothermal, we can directly determination the gas density distribution from the observed surface brightness distribution. For NGC 4472, we find

$$M(r) = 6.5 \pm 0.5 \times 10^{12} T_7 r_{100\mathrm{kpc}} M_\odot$$

Taking the optical luminosity for NGC 4472 as $L_B = 7.4 \times 10^{10} L_\odot$ for a distance of 16 Mpc, we constrain the mass-to-light ratio (in solar units) to lie in the range 78–93 for $r = 100$ kpc and in the range 54–65 for $r = 70$ kpc. Lauer (1985) gives a core mass-to-light ratio of 14. We find similar values for NCG1399, a mass-to-light ratio of 47–92 for $r = 110$ kpc. Hence, the *ROSAT* PSPC observations definitively demonstrate the existence of a massive dark halos surrounding these galaxies.

5.1. *The Distribution of Iron in the Hot Coronae of Bright Elliptical Galaxies*

As mentioned above, abundance measurements of the hot coronal gas in early-type galaxies can be used to derive the evolution of supernova activity since the supernova ejecta remain in the hot coronae for billions of years. Also, different evolutionary models of the hot coronae predict different abundances. For example, predicted abundances in long-lived wind models (D'Ercole *et al.* 1989) are several times the solar value and can readily be distinguished from those with nearly solar abundances expected in quasi-hydrostatic or inflow models (Loewenstein & Mathews 1987; David *et al.* 1991).

The results of our measurements of the iron abundance are shown in Figure 5 for NGC 4472 and NGC 1399.

5.1.1. *Implications of the Measured Abundance Gradients*

The present epoch supernova rate in elliptical galaxies is quite poorly constrained. Still less is known regarding the time dependence of this rate. Ciotti *et al.* (1991) argued that a rapid decline in the supernova rate with time is warranted based on either of two mechanisms for the origin of type Ia supernovae – Roche lobe overflow of an evolving star onto a white dwarf or the merger of two white dwarfs driven by gravitational radiation. Other models of the evolution of hot coronae (Loewenstein & Mathews 1987; David *et al.* 1990) derived a considerably weaker dependence of the supernova rate on time.

The nearly solar abundance of heavy elements over a wide range in radii observed in NGC 4472 and NGC 1399 eliminates models which predict heavy element abundances much in excess of the solar value such as the long-lived wind models.

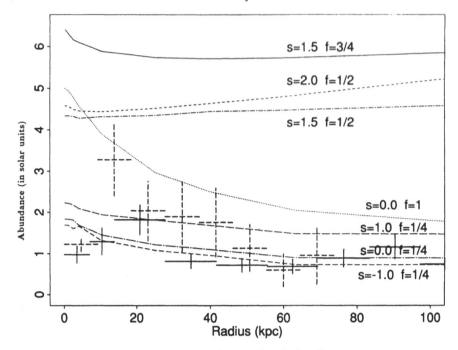

FIGURE 5. The distribution of the heavy element (iron) abundance measurements as a function of radial distance from the center of NGC 1399 (solid crosses) to that of NGC 4472 (Forman *et al.* 1993) and for model calculations (Loewenstein & Mathews 1991). The near-solar value of the iron abundance and its constancy throughout the corona for both NGC 1399 and NGC 4472 suggest a relatively constant supernova rate over time with a present epoch value at the low end of the range derived by van den Bergh & Tammann (1991). Beyond 2′, there is an apparent trend of decreasing abundance with increasing radial distance from NGC 4472. NGC 1399 shows a similar, but less significant fall in abundance with radius. Mushotzky *et al.* (1994) recently has reported a fall in iron abundance with radius for the bright elliptical NGC 4636.

A general investigation of the dependence of the heavy element abundances in hot coronae on the supernova rate was carried out by Loewenstein & Mathews (1991). In Figure 5, we have superposed several of their model calculations which show the derived radial behavior of the heavy element abundance for a family of galaxy models for a $10^{11} L_\odot$ galaxy with a heavy halo. It is apparent from Figure 6 that models with much higher supernova rates at earlier epochs conflict with the observations. In particular, models in which the supernova rate is characterized by $s \geq 1.5$ predict heavy element abundances in excess of that observed. The models which most closely describe the observations are those with constant or slowly varying supernova rates ($|s| < 1$) and even these require relatively modest values of the present epoch supernova rate. Comparing the curves for the two models with $s = 0$ in Figure 6 shows that present epoch values equal to van den Bergh & Tammann's (1991) rate ($f \approx 1$) produce iron abundances in excess of those observed, while predicted iron abundances for $f = 1/4$ (again for $s = 0$) are in good agreement with the observations.

In summary, the rejection of models with rapidly decreasing supernova rates also has implications for the origin of type Ia supernovae. The rejection of models with high iron abundances implies the rejection of models in which the type Ia supernova rate decreases substantially in time.

I would like especially to thank Emilio Alfaro and Antonio Delgado, all members of the organizing committee, and other conference participants for providing such a stimulating scientific conference in a delightful setting. I am very grateful to Brad Whitmore, Larry David and Bill Forman for their contributions to this paper. This work is supported by the Smithsonian Astrophysical Observatory and NASA contracts and grants NAS8–39073 and NAG8–1881.

REFERENCES

ABRAMOPOULOS, F. & KU, W. 1983 *Astrophys. J.* **271**, 446.

ARNAUD, M., ROTHENFLUG, R., BOULADE, O., VIGROUX, L. & VANGIONI-FLAN, E. 1992 *Astron. Astrophys.* **254**, 49.

BRIEL, U., EBELING, H., EDGE, A., HARTNER, G., HENRY, J., SCHWARZ, R. & VOGES, W. 1991 *Astron. Astrophys.* **100**, 999.

CAVALIERE, A. & FUSCO-FEMIANO, R. 1976 *Astron. Astrophys.* **49**, 137.

DAVID, L., ARNAUD, K., FORMAN, W. & JONES, C. 1990 *Astrophys. J.* **356**, 32.

DAVID, L. & BLUMENTHAL, G. 1992 *Astrophys. J.* **389** 510.

DAVID, L., FORMAN, W. & JONES, C. 1991 *Astrophys. J.* **369**, 121.

DAVID, L., JONES, C. & FORMAN, W. 1995 *Astrophys. J.* in press.

DRESSLER, A. 1980 *Astrophys. J.* **236**, 351.

EDGE, A. 1989 PhD thesis, Leicester University.

FABER, S. & GALLAGHER, J. 1976 *Astrophys. J.* **204**, 365.

FABIAN, A. C., NULSEN, P. E. J. & CANIZARES, C. 1984 *Nature* **310**, 733.

FABIAN, A. C., NULSEN, P. E. J. & CANIZARES, C. 1991 *Astron. Astrophys. Rev.* **2**, 191.

FABIAN, A., THOMAS, P., FALL, M. & WHITE, R. 1986 *Mon. Not. R. Astron. Soc.* **221**, 1049.

FITCHETT, M. & WEBSTER, R. 1987 *Astrophys. J.* **317** 653.

FORMAN, W., JONES, C. & TUCKER, W. 1985 *Astrophys. J.* **293**, 102.

FORMAN, W., SCHWARZ, J., JONES, C., LILLER, W. & FABIAN, A. 1979 *Astrophys. J., Lett.* **23**, L27.

JONES, C. & FORMAN, W. 1984 *Astrophys. J.* **276**, 38.

JONES, C. & FORMAN, W. 1992 in *Clusters and Superclusters* (ed. A. C. Fabian), p. 49. Cambridge University Press.

KRISS, G., CIOFFI, G. & CANIZARES, C. 1983 *Astrophys. J.* **272**, 439.

LOEWENSTEIN, M. & MATHEWS, W. 1987 *Astrophys. J.* **319**, 614.

MATHEWS, W. & LOEWENSTEIN, M. 1986 *Astrophys. J., Lett.* **306**, L7.

MELNICK, J. & SARGENT, W. 1977 *Astrophys. J.* **215**, 401.

MUSHOTZKY, R., SERLEMITSOS, P., SMITH, B., BOLDT, E. & HOLT, S. 1978 *Astrophys. J.* **225**, 21.

NULSEN, P. E. J., STEWART, G. & FABIAN, A. C. 1984 *Mon. Not. R. Astron. Soc.* **208**, 185.

SARAZIN, C. & WHITE, R. E. 1988 *Astrophys. J.* **331**, 102.

SCHWARTZ, D., SCHWARZ, J. & TUCKER, W. 1980 *Astrophys. J., Lett.* **238**, L59.

VAN DEN BERGH, S. & TAMMANN, G. 1991 *Annu. Rev. Astron. Astrophys.* **29**, 363.

WALKER, T. P., STEIGMAN, G., SCHRAMM, D. N., OLIVE, K. A. & KANG, H. 1991 *Astrophys. J.* **378**, 186.

WEST, M. & BOTHUN, G. 1990 *Astrophys. J.* **350**, 36.

WHITMORE, B. 1994 *Clusters of Galaxies.* XIVth Moriand Astrophysics Meeting.

WHITMORE, B., GILMORE, D. & JONES, C. 1993 *Astrophys. J.* **407**, 489.

Lesch: What does substructure in the X-ray images of clusters mean for the value of Ω? Did you use non-virialized clusters?

Jones: Richstone *et al.* (1992: *Ap.J.* **393**, 477) suggested that the large fraction of clusters observed with substructure at the present epoch required that Ω be large. In particular, in a low-density universe, clusters should evolve quickly so that little or no substructure should be observed at the present epoch. However surveys find that the fractions of clusters with substructure is about 40%. Evrard *et al.* (1994 *Ap.J. Letters* found from numerical simulations of cluster formation with Ω = 0.2, that fewer clusters with structure are found than the fraction observed. These simulations have generally assumed that the baryon fraction in clusters is 10%. Since we observe this fraction to be about 30%, it would be useful to have simulations with a high cluster baryon fraction and Ω of 0.3.

We attempt to select virialized clusters for our analysis. The clusters we choose when we compute the gas mass, galaxy mass, and total mass, do not appear from their X-ray images to have substantial substructure. Most show a single, symmetric cluster, while some have an associated smaller group. In the latter, we measure the larger component.

Franco: The intracluster gas pressure is of the same order of magnitude as in the solar neighborhood. This may imply that the pressure inside individual galaxies should be larger. Can you comment on this?

Jones: The gas density at the cluster center is about .01 to .001 cm^{-3} and falls with radius. The gas temperature of the ICM is 10^7 to 10^8K. If galaxies were stationary in the cores of clusters, there would be considerable outer pressure. However, since galaxies in clusters are moving at velocities of 300 to 1000 km/sec, there is substantial ram pressure, which as galaxies approach the core, strips them of their hot coronae.

Elmegreen: Can you rule out the model where ellipticals form by the mergers of spirals? Do elliptical-rich clusters have more gas left over from an inability to form spirals?

Jones: The Whitmore *et al.* results that I showed do not rule out the generation of ellipticals from spirals. Figure 2 shows that the rich, elliptical dominated clusters also have more gas in their intercluster medium which is consistent with being the gas that would otherwise have formed spirals. While it seems to us more likely that ellipticals formed early on as ellipticals in cluster cores and the more fragile, disked spirals either never formed or were quickly destroyed, one also could postulate that first spirals formed and then these either merged to form ellipticals or were destroyed.

Magnier: At the March ASCA meeting, people commented that the abundances were coming out systematically low compared to optical measurements. Along these lines, have you compared ASCA and *ROSAT* measurements.

Jones: We have compared *ROSAT* and ASCA measurements for five early-type galaxies and find good agreement between the gas temperatures and the abundances. Early on, it appeared that *ROSAT* was measuring higher abundances, but this was due to the use of difference values for the solar/cosmic iron abundance in the spectral fitting software.

von Linden: How do you know which elliptical galaxies are formed at the beginning of cluster formation and which ones are built through collisions of spiral galaxies.

Jones: In the Whitmore *et al.* interpretation, ellipticals form first. With regard to how one might distinguish different formation processes for ellipticals, Monique Arnaud has suggested that low mass ellipticals, which appear to have little iron in their hot X-ray coronae, may have formed from the mergers of spirals, while the high mass ellipticals with iron abundances approaching solar values would not be the results of mergers.

Halo Model of the Spiral Galaxy NGC 3198

By MASATOSHI TAGA[1] AND MASANORI IYE[2]

[1]Department of Astronomy, School of Sciences, Univ. of Tokyo, Bunkyo-ku, Tokyo 113, Japan

[2]Japan National Large Telescope Division, National Astronomical Observatory, Mitaka, Tokyo 181, Japan

We investigated the amount and the distribution of the mass of the dark halo of spiral galaxy NGC 3198 using a new approach: "galactic seismology". Assuming that the spiral pattern of the galaxy is the manifestation of the gravitational instability on the galactic disk, we constrain the mass of the halo so as to give consistency between the observed spiral pattern and the calculated spatial pattern of gravitationally unstable mode of the galactic disk. The result is that the amount of the mass of the halo should be small and that it must be very extended.

1. Introduction

The mass distribution of the dark halo of galaxies has been a major issue in the field of galactic astronomy and is still the subject of considerable controversy. The existence of dark halos around galaxies is inferred mainly from their rotation velocity, which is usually flat until the last observed points. It is, however, not straightforward to separate the contributions from the disks and the halos to the rotation velocity, because we do not know their mass-to-luminosity ratios accurately.

Here we present a new method to construct a galaxy model, which is free from assumptions about the mass-to-luminosity ratio or Toomre's Q value. This method is what we call the "galactic Seismology", a term analogous to "helioseismology" or "geoseismology". A galactic disk becomes gravitationally unstable under certain circumstances. We consider the spiral patterns of galaxies as the manifestations of gravitationally unstable modes, and put restrictions on models of galaxies so that the calculated spiral patterns are consistent with the observed spiral pattern.

Ueda et al. (1985) analysed the stability of an S0 galaxy NGC 3115 by global modal analysis, and concluded that to make the galactic disk stable so as not to develop spiral patterns on its disk, more than 30% of gravitational potential should be ascribed to the mass in the halo. Athanassoula et al.(1987) made use of Toomre's "swing amplification theory" (Toomre 1981), to show that about half the mass must be in the immobile halo. We here follow the approach of Ueda et al.(1985) and apply a global modal analysis to an Sc galaxy NGC 3198 to find the distribution of the halo mass of this galaxy.

2. Models of NGC 3198

For the purpose mentioned above, we chose NGC 3198—one of the nearest spiral galaxies. In order to analyse the gravitational instability on the galactic disk, we need to know the kinematic properties of the galaxy in detail. NGC 3198 has been observed intensively (e.g. Wevers et al. 1986; Bottema 1988; Begeman 1989), and the rotation velocity and the velocity dispersion curves are well determined (Figure 1). With these data, we construct mass models of this galaxy and calculate the gravitationally unstable modes.

The gravitational potential of spiral galaxies can be generally expressed as a superposition of those of its bulge, disk, and halo. In NGC 3198, there is no prominent bulge. One can therefore assume that the contribution of the bulge to the gravitational potential is

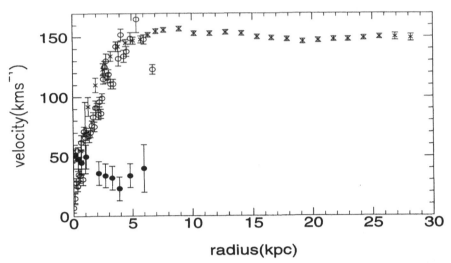

FIGURE 1. The rotation velocity and the velocity dispersion of NGC 3198. Open and filled circles are the rotation velocity and the velocity dispersion determined from optical lines by Bottema (1988). Crosses show the rotation velocity measured in HI 21-centimetre line by Begeman (1989).

negligible. In fact, the observed rotation velocity of NGC 3198 rises gradually outward from the centre. We therefore construct equilibrium models of NGC 3198 with the disk and halo only.

From these assumptions, we construct galaxy models with a compact mass component and an extended mass component. We express the potential of each component with Toomre's first model (Toomre 1962). Furthermore, we assume that the fraction f_1 ($0 \leq f_1 \leq 1$) of the mass of the compact component and the fraction f_2 ($0 \leq f_2 \leq 1$) of that of the extended component is in the flat, thin disk, and the rest $(1 - f_1, 1 - f_2)$ is in the three-dimensional halo. The gravitational potential in the galactic plane is, therefore, expressed as follows:

$$\psi_{compact}(r) = -\frac{GM_1}{a_1}\left(1 + \frac{r^2}{a_1^2}\right)^{-\frac{1}{2}} \tag{2.1}$$

$$\psi_{extended}(r) = -\frac{GM_2}{a_2}\left(1 + \frac{r^2}{a_2^2}\right)^{-\frac{1}{2}} \tag{2.2}$$

$$\psi_{disk}(r) = f_1\psi_{compact} + f_2\psi_{extended} \tag{2.3}$$

$$\psi_{halo}(r) = (1 - f_1)\psi_{compact} + (1 - f_2)\psi_{extended} \tag{2.4}$$

$$\psi_{total}(r) = \psi_{disk} + \psi_{halo} \tag{2.5}$$

$$= \psi_{compact} + \psi_{extended} \tag{2.6}$$

$$= -\sum_{i=1}^{2} \frac{GM_i}{a_i}\left(1 + \frac{r^2}{a_i^2}\right)^{-\frac{1}{2}} \tag{2.7}$$

were r is the radius from the centre, G is the gravitational constant, and M_i, a_i ($i = 1, 2$) are the characteristic mass and radius of each component. Note that the fraction of the disk in the extended component f_2 is expected to be small ($f_2 \simeq 0$) because the disk does not extend out so far.

In the present models, free parameters are f_1 and f_2. When f_i is given, M_i and a_i are determined by fitting the resultant rotation and velocity dispersion curves to the

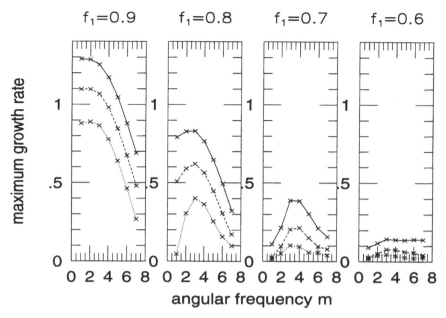

FIGURE 2. Growth rate of the most unstable mode *vs.* the angular frequency m for models with f_1=0.6, 0.7, 0.8, 0.9. In each panel, we show the growth rate of models with f_2=1.0 (solid line), 0.5 (dashed line) and 0.0 (dotted line). The growth rate is normalised by the dynamical timescale of the compact component, $(GM_1/a_1^3)^{\frac{1}{2}}$.

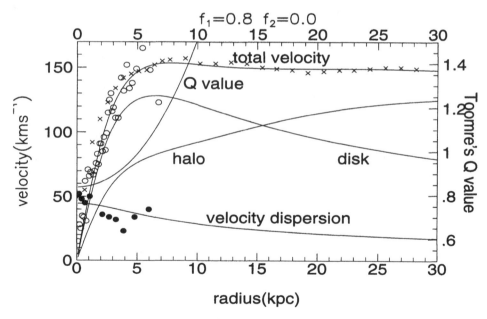

FIGURE 3. Rotation curve, velocity dispersion curve and Toomre's Q value for the equilibrium model with $f_1 = 0.8$, $f_2 = 0.0$. Observed data are also plotted.

observed data using the least square method via the relations,

$$\frac{v_0(r)^2}{r} = \frac{d\psi_{total}(r)}{dr} + \frac{dP_0(r)}{dr}, \tag{2.8}$$

$$P_0(r) = K\mu_0(r)^\Gamma = \mu_0(r)\sigma_0(r)^2, \tag{2.9}$$

where v_0 is the rotation velocity, P_0 is the pressure, σ_0 is the velocity dispersion, and μ_0 is the surface mass density of the galactic disk. In these relations, we assumed a polytropic relation for the equation of state.

Since we assume that the galactic disk is infinitesimally thin, surface mass density of the disk of these models that gives the potential (2.3) is expressed uniquely as

$$\mu_{disk}(r) = \sum_{i=1}^{2} \frac{f_i M_i}{2\pi a_i^2} \left(1 + \frac{r^2}{a_i^2}\right)^{-\frac{3}{2}}. \tag{2.10}$$

On the other hand, three-dimensional mass distribution of the halo cannot be determined solely from the potential (2.4). If we make the additional assumption that the halo is spherically symmetric, the solution takes the form of the potential of Plummer's sphere.

3. Results and conclusions

The observed spiral pattern of this galaxy that is to be compared with the calculated pattern of instability is analysed by Considère & Athanassoula (1988). They showed that the two-armed spiral pattern is dominant in this galaxy, so we seek models that develop the $m = 2$ spiral pattern as the most unstable mode.

Figure 2 shows the maximum growth rate for each model and for each angular wavenumber m. This figure shows that to make the two-armed spiral pattern dominant, one must have $f_1 > 0.8$, i.e. almost all the gravitational force must come from the disk itself. Otherwise the three- or four-armed unstable mode will be the most conspicuous.

This result suggests that the disk of NGC 3198 is a strongly self-gravitating system and that the disk–halo mass ratio inside the last point observed by HI 21-centimetre line at about 30 kpc from the centre should be smaller than three. This value, of course, becomes even smaller if we take the ratio inside the radius where we can observe in optical lines. Figure 3 shows the rotation velocity and velocity dispersion curves and Toomre's Q value of the model with $f_1 = 0.8, f_2 = 0.0$, which has the most massive halo among the models that develop a strong two-armed spiral pattern. This shows that the Q value becomes smaller than unity in the central 5 kpc, which in turn suggests that the disk may be dynamically unstable in this region.

REFERENCES

ATHANASSOULA, E., BOSMA, A. & PAPAIOANNOU, S. 1987 *Astron. Astrophys.* **179**, 23.

BEGEMAN, K. G. 1989 *Astron. Astrophys.* **223**, 47.

BOTTEMA, R. 1988 *Astron. Astrophys.* **197**, 105.

CONSIDÈRE, S. & ATHANASSOULA, E. 1988 *Astron. Astrophys., Suppl. Ser.* **76**, 365.

TOOMRE, A. 1962 *Astrophys. J.* **138**, 385.

TOOMRE, A. 1981 in *The Structure and Evolution of Normal Galaxies* (ed. S. M. Fall & D. Lynden-Bell), p 111. Cambridge University Press.

UEDA, T., IYE, M. & AOKI, S. 1985 *Astrophys. J.* **288**, 196.

WEVERS, B. M. H. R., VAN DER KRUIT, P. C. & ALLEN, R. J. 1986 *Astron. Astrophys., Suppl. Ser.* **66**, 505.

The Formation of cD Galaxies

By CARLOS GARCÍA-GÓMEZ[1] E. ATHANASSOULA[2]
AND A. GARIJO[1]

[1]Univ. Rovira i Virgili, E.T.S.E., Dep. Enginyeria Informàtica, 43006 Carretera de Salou s/n, Tarragona, Spain

[2]Observatoire de Marseille, 2, Place Leverrier, F–13248 Marseille Cedex 4, France

1. Introduction

We present the first results of a study of the formation of cD galaxies in the centre of rich clusters by means of N-body simulations. These galaxies are the brightest objects formed by stars in the Universe and resemble elliptical galaxies. Normally, they can be found in the centre of rich clusters. However, cD galaxies are brighter than a normal elliptical (Malumuth & Kirshner 1985) and their surface brightness profiles deviate in the external parts from the $r^{1/4}$ law (de Vaucouleurs law), showing a luminosity excess (Oemler 1976). Moreover, their velocity dispersion profiles are flatter than the profiles of normal galaxies (Carter *et al.* 1985).

Several theories have been proposed to explain the formation of these particular galaxies. All of them find certain difficulties when they try to explain the properties of cD galaxies, and it is possible that these theories are just complementary explanations. Some authors propose that cD galaxies can be formed in the centre of the potential well by accretion of material that has been stripped from the rest of the galaxies by tidal forces. These can be produced by the cluster potential or by encounters between galaxies (Richstone 1975). Another posibility could be the accumulation of galaxies in the bottom of the potential well, as they are slowed down by dynamical friction. This process is known as "galactic cannibalism" (Ostriker & Hausman 1977).

2. Simulations

We have performed two N-body simulations. In each of them, 50 galaxies are distributed randomly within a homogeneous sphere. Each galaxy is formed by 900 particles distributed according a Plummer. This makes a total of 45 000 particles. We use standard units where $R_g = 1 = G = M_g$. R_g and M_g are, respectively, the radius and the mass of a galaxy. The simulations are followed up to a Hubble time and we make the assumption that initially all the mass is concentrated around the galaxies. We use two different initial conditions. In Run A we study a collapsing system, while Run B is a virialised system.

In our simulations, a central giant galaxy is quickly formed starting from the galaxies which are initially near the centre. Later this galaxy grows up, accreting some of the cluster members, as in the galactic cannibalism picture (Hausman & Ostriker 1978). Moreover, tidal forces strip off some material from all the galaxies. This mass will be deposited in the centre of the potential well which is occupied by the central giant galaxy.

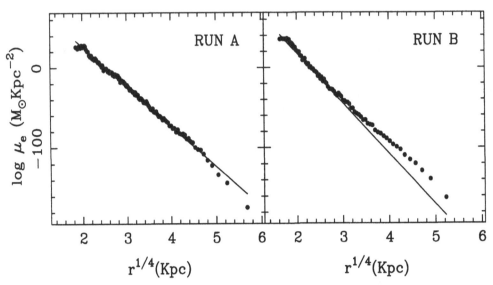

FIGURE 1. Surface density profiles for the central galaxies in Run A and Run B. Note the luminosity excess typical of cD galaxies for the central galaxy of simulation B.

3. Discussion

In both cases a central giant galaxy is formed, which could, at first sight, be identified with the cD galaxies in the centre of rich clusters. However, the properties of the resulting galaxies are strongly dependent on the initial conditions. Only in the case of initially virialised systems will it be possible to associate this galaxy with a proper cD.

This can be shown in Figure 1, where we represent the surface density profile of both giant galaxies. We can see that the central member of Run A follows a $r^{1/4}$ law all along its radius, but this is not the case for the central member of Run B. The orbital structure of the two galaxies is also different. The central galaxy in Run A is not an isotropic system because the radial and tangential velocity dispersion dominate over the azimuthal velocity dispersion. This is the result of the radial systematic infall of galaxies in this simulation. On the other hand, the galaxy resulting from initial virialised conditions is an isotropic system.

It can be concluded from these simulations that the formation of a cD galaxy will be possible in clusters with nearly virialised initial conditions. In these cases, tidal forces have more time to operate, stripping off the less bounded material from the galaxies. This mass is then accreted over a central galaxy formed by galactic cannibalism. This is the material which will later form the halo of the cD galaxy.

REFERENCES

CARTER, D., INGLIS, I., ELLIS, P.S., EFSTATHIOU, G. & GODWIN, J.G. *Mon. Not. R. Astron. Soc.* **212**, 471.

HAUSMAN, M.A. & OSTRIKER, J.P. 1978 *Astrophys. J.* **224**, 320.

MALUMUTH, E.M. & KIRSHNER, R.P. 1985 *Astrophys. J.* **291**, 8.

OEMLER, A. 1976 *Astrophys. J.* **209**, 693.

OSTRIKER, J.P. & HAUSMAN, M.A. 1977 *Astrophys. J., Lett.* **217**, L125.

RICHSTONE, D.O. 1975 *Astrophys. J.* **200**, 535.

Dynamical Collapse of Isothermal and Adiabatic Triaxial Protogalaxies

By PETER BERCZIK AND IGOR G. KOLESNIK

Main Astronomical Observatory of Ukrainian National Academy of Sciences, 252650, Golosiiv, Kiev–022, Ukraine

This article is first in the series of papers about the three-dimensional hydrodynamic models of triaxial galaxy formation. We use the most common triaxial model for initial conditions of protogalaxies. The body of protogalaxies is modelled by triaxial ellipsoids with semiaxes A, B and C. For initial velocity fields we propose the lineal distibution which is defined by initial constant angular velocity Ω_0. The total angular momentum does not coincide with the semiaxes. The velocity field can therefore be described with the formule: $\mathbf{V}_0(x, y, z) = [\Omega_0 \times \mathbf{r}]$. In this paper we discuss the series of models of isothermal and adiabatic collapse of pure gas triaxial protogalaxies.

The amount of data indicating that the majority of galaxies are triaxial is steadily increasing. First of all, elliptical galaxies are triaxial (Illingworth 1977; Binney 1985). In many cases, this conclusion is reached following analysis of the shape and brightness distribution in images of elliptical galaxies. The still more convincing data on the triaxial shape of ellipticals give their kinematical properties (Frank *et al.* 1989; Kormendy & Djorgovski 1989; de Zeeuw & Franx 1991). Minor axis dust lane galaxies (Bertola 1987) and polar-ring galaxies (Whitmore *et al.* 1990) must also be triaxial, because they can be stable only when their gravitational potential is triaxial. From the database about the bar-galaxies we conclude that in a considerable part of disk systems the gravitational potential is also essentially triaxial also (Binney 1992; Bertola *et al.* 1991).

Therefore, the triaxial shape is the important feature inherent to the galaxies of the different Hubble types. So the problem arises of what the origin of these triaxial galaxy shapes might be. The merger of galaxies is usually considered as a process that leads to a triaxial galaxy formation (Toomre & Toomre 1972). Certainly, these processes play some role in the phenomenon (Noguchi 1987; Gerin *et al.* 1990); Elmegreen *et al.* 1991), but we doubt that all the variety of known triaxial galaxies can be explained by merging alone. We start from the idea that a triaxial configuration already existed at a protogalactic stage of evolution. It is highly plausible that this configuration has an angular momentum vector, which does not coincide with any main axis of an initial figure (Chernin 1993; Hiotelis & Voglis 1989).

We can therefore expect that as a result of dynamical collapse a triaxial galaxy with complicated internal kinematics will be formed. There are cosmological arguments in favour of the triaxiality of protogalaxies, because perturbations in the length scales of galaxies are non-spherical (Berdeen *et al.* 1986). In this paper we consider the case of a protogalaxy being in an initial state a homogeneous triaxial ellipsoid, with an angular momentum directed arbitrarily in space. Simulations were carried out by the SPH method (Berczik & Kolesnik 1993; Monaghan 1992). The properties of gas galaxies, which can be formed as a result of isothermal or adiabatic collapse, are considered. The initial body of the protogalaxy is modelled by triaxial HI ellipsoids with semiaxes A, B and C. The mass of the protogalaxy is taken as $M_{HI} = 10^{12} \cdot M_\odot$. The initial temperature of gas is taken as T_0. At the beginning the orientation of ellipsoid semiaxes A, B and C coincides with the coordinate system axis (x, y, z). The initial velocity fields are defined through the angular velocity $\mathbf{V}_0(x, y, z) = [\Omega_0(x, y, z) \times \mathbf{r}]$. The initial angular

A	B	C	Ω_0/Ω_{cir}	$T_0 = 10^4$	10^5	$5 \cdot 10^5$	$10^4\,(adiab)$
100	90	50	(0.8, 0.0, 0.8)	-	-	B1	-
100	60	50	(0.5, 0.0, 0.5)	-	-	T1	-
100	75	50	(0.8, 0.0, 0.8)	-	TT1	-	-
100	75	50	(0.5, 0.0, 0.5)	-	-	-	TT1A
100	75	50	(0.8, 0.0, 0.4)	TT2	-	-	-
100	75	50	(0.4, 0.0, 0.8)	TT4	-	-	-
100	75	50	(1.0, 1.0, 1.0)	-	!32	-	!44

TABLE 1. The list of parameters

velocity is taken as being constant in space. Thereafter $\Omega_0(x, y, z) \equiv \Omega_0 = Const$. The orientation of initial angular velocity is defined by the components: $(\Omega_{0x}, \Omega_{0y}, \Omega_{0z})$. We investigate many different orientations of initial angular velocity. In Table we list the basic parameters of models. $\Omega_{cir} = V_{cir}/A$ and $V_{cir} = (G \cdot M_{HI}/A)^{1/2}$. The A, B and C, in the table are shown in kpc.

On the basis of our models, we may draw several general conclusions about the properties of forming galaxies:

• Isothermal collapse in the triaxial model leads to warping of the disk structure and internal bar.

• Evolution of the triaxial model generates a global spiral structure. In the isothermal model the spiral pattern is more sophisticated and there is a longer more duration.

• The adiabatic phase is important for bulge formation and its parameters.

• With adiabatic models the kinematical properties of the core and peripherial parts of protogalactic clouds are very different.

REFERENCES

BERCZIK, P. P. & KOLESNIK, I. G. 1993 *Kinematika i Fizika Nebesnykh. Tel.* **9**, No. 2, 3.

BERDEEN, J. M., BOND, J. R., KAISER, N. & SZALAY, A. S. 1986 *Astrophys. J.* **304**, 15.

BERTOLA, F. 1987 in *Structure and Dynamics of Elliptical Galaxies* (ed. T. de Zeeuw). IAU Symposium 127, p. 135. Reidel.

BERTOLA, F., VIETORI, M. & ZEILINGER, W. W. 1991 *Astrophys. J., Lett.* **374**, L13.

BINNEY, J. 1985 *Mon. Not. R. Astron. Soc.* **212**, 767.

BINNEY, J. 1992 *Annu. Rev. Astron. Astrophys.* **30**, 51.

CHERNIN, A. D. 1993 *Astron. Astrophys.* **267**, 315.

DE ZEEUW, T. & FRANX, M. 1991 *Annu. Rev. Astron. Astrophys.* **29**, 239.

ELMEGREEN, D. M., SUNDIN, M., ELMEGREEN, B. & SUNDELIUS, B. 1991 *Astron. Astrophys.* **244**, 52.

FRANX, M., ILLINGWORTH, G. & HECKMAN, T. 1989 *Astrophys. J.* **344**, 613.

GERIN, F., COMBES, F. & ATHANASSOULA, L. 1990 *Astron. Astrophys.* **230**, 37.

HIOTELIS, N. & VOGLIS, N. 1989 *Astron. Astrophys.* **218**, 1.

ILLINGWORTH, G. 1977 *Astrophys. J., Lett.* **218**, L43.

KORMENDY, J. & DJORGOVSKI, S. 1989 *Annu. Rev. Astron. Astrophys.* **27**, 235.

MONAGHAN, J. J. 1992 *Annu. Rev. Astron. Astrophys.* **30**, 543.

NOGUCHI, M. 1987 *Mon. Not. R. Astron. Soc.* **228**, 635.

TOOMRE, A. & TOOMRE, J. 1972 *Astrophys. J.* **178**, 623.

WHITMORE, B. C., LUCAS, R. A., MC ELROY, D. B., STEIMAN-CAMERON, T. Y., SACKETT, P. D. & OLLING, R. P. 1990 *Astron. J.* **100**, 1489.

The Formation of the Milky Way

By BRUCE W. CARNEY

Department of Physics & Astronomy, University of North Carolina, Chapel Hill, NC
27599–3255, U.S.A.

Introduction

The five days of this meeting have sped by too quickly, and we are near the meeting's end. I think I speak for all of us when I say that I have learned a great deal from the many talks, from the posters, and from the three previous discussion sessions. Given all the discussion, I feel it would be presumptious, and, more important, too time-consuming to try to summarise the many highpoints of the meeting. I prefer to spend our little remaining time discussing some of the key questions that remain, and perhaps try to focus more on what we should be doing once the meeting is over.

There is a a children's game, popular in America and probably elsewhere, called "20 Questions". Someone thinks of some object, person, place, or whatever, and the idea is, by means of "yes" and "no" answers, to guess the object. The "Formation of the Milky Way Galaxy" game has been played in somewhat this vein for quite some time. "Does it have a disk?" "Does it have a halo?" From these easy questions we have rapidly moved into the more difficult questions, and while I cannot produce a list of 20 specific questions whose answers would reveal all we wish to know about our home galaxy, I have been keeping track of some that seem to be bothering most of us here. I will summarise them here, and perhaps we can then begin to argue about possible answers or what we're going to have to do to obtain the answers.

Dark matter

1. What are its effects on the dynamical timescales we measure or seek to measure to establish the Galaxy's chronology?

2. Have we ruled out baryonic matter that might have (had) effects on the Galaxy's chemical evolution?

Angular momentum

3. Who ordered that? (With apologies to I. I. Rabi and Steve Shore.)

The halo

Mergers seem to be or have been important.

4. Are they or were they?

5. Have there been only a few major episodes of merging or many small ones?

6. How much of the halo stellar population emerged from gas clouds near, but not originally "part of" the Galaxy?

7. What fraction of the metal-poor stars were part of the original "protodisk"?

347

The thick disk

8. How should we, or in fact can we, fit the very metal-poor disk stars with the thick disk?

9. Is the thick disk part of the "protodisk" or is it a discrete merger artifact?

The thin disk

10. Did it form "inside-out"?

11. Why isn't it more metal-rich?

12. How can we reconcile the new shallow gradient in [O/H] with the older stars' steeper [O/H] and [Fe/H] gradients?

13. How can we explain the negligible slope and large scatter in the local disk's age–metallicity relation?

14. How much gas has the galactic disk accreted since its formation?

15. Is the "G-dwarf problem" really a problem?

The bulge

16. Is there a bulge or only a bar?

17. Can we really discuss it as a separate/separable population?

18. What do the models suggest about its origin?

19. What do the observations suggest about its origin?

History

20. What is it going to require to put high-precision *absolute*, or even *relative*, ages on the oldest and bulk of the four nominal stellar populations?

Discussion

Well, there are our current 20 questions, or perhaps I should say, here are 20 of our current questions. I await the comments, agreements, and disagreements from everyone here.

Chernin: The question of dark matter is not so unimportant because 90% of the mass of the Milky Way is probably dark. It is a big problem, anyway; but I'd like to say only one thing—about the effect of dark matter on the dynamic timescales. It was very nice to hear just this morning that the model by Eggen, Lynden-Bell, & Sandage is still alive. Basically, in the framework of this model, an interesting combination of universal physical constants was found that gives the size of the protogalaxy: $R = 2 \times 10^{23}$ cm (Antonov & Chernin 1975; Gott & Thuan 1976; Silk 1977; Rees & Ostriker 1977). It follows from comparison of two characteristic timescales, that of free fall and that of cooling of the protogalactic cloud, which is supposed to be in quasi-static equilibrium, as in the model of Eggen, Lynden-Bell, & Sandage. Dark matter may contribute some new features to this model, and help the model to survive new challenges in galactic cosmogony. When we add dark matter to the same volume as the gas in that old model we obtain another value for the timescales and for this characteristic length, and it will depend now upon the ratio of baryonic and dark matter densities: $R_{DM} = R(1 + \rho_{DM}/\rho_B)^{-1}$. Perhaps this ratio of dark matter density to baryonic matter density in the Universe is also a kind of universal constant, in fact no less fundamental than the proton-to-electron rest mass ratio. The two characteristic sizes, R and R_{DM}, agree within a factor of ten, and whereas the first one refers to the protogalaxy, the smaller new one refers to the Galaxy's final state. The question is: What is the physics that leads to this final universal characteristic spatial scale for galaxies like the Milky Way?

Shore: I'm tempted to quote Wittgenstein again: "That about which we can say nothing we should pass over in silence." But he said that at the end of the book, so it's appropriate. There's a question I'd like to ask the people here who do stellar dynamics. In many of the models that we use for clusters of galaxies, you throw in these Plummer potentials or some other static potential, you call that a galaxy, then throw dark matter into the system, and you put these little potentials inside this background and from that you calculate clustering. Now what happens when, in the process of going from the very extended distribution of mass to some centrally condensed thing that we think we're living in, you have a change in the smooth component of the background potential on a dynamical timescale? I remember from when I was knee-high to a cockroach that the thing this gave rise to was violent dynamical mixing and not just violent relaxation, but it was one of the conditions that you modelled with strange attractors. Now is this something that can just be neglected in the building of the Galaxy or is it important because all of these questions about dynamical formation of the halo might also be taking place on the same timescale as changes in the distribution of the dark matter—whatever the hell that means—and, consequently, the change in the broader component of the potential?

Elmegreen: I'd like to move onto another issue relating to the dark matter. At the IAU, and also brewing in the literature, is this idea that there's a lot of dark matter, maybe all of it, in the form of cold molecular material in the disk at large radii. This idea was started by Pfenniger, Combes, and Martinet, and observations by Lequeux and Allen. It's highly controversial—there are contrary observations and there are theoretical problems— but the arguments for it have to do with some observations that starburst galaxies have more gas per unit star than other galaxies, as if the interaction brought in this reservoir of dark gas. Also, if there is gas at large radii, we could just not know about it. It could be at 3 °K, and shielded inside the ionised layer that we know is out there. Also the ratio dark matter to bright matter seems to be less in early type galaxies than in late type galaxies, as if early types are more evolved and more of this dark matter gas has turned into stars. They're skirting the issue with several observations that all point in this direction, but it is all very uncertain. This would have enormous implications on everything, if it's true.

Carney: How does one reconcile the material out at the edge of our Galaxy with other matter we know about, such as the Magellanic Stream? Wouldn't you see signs of colliding streams in that case? Isn't there a problem with having a thin Magellanic Stream and such a massive outer gaseous disk?

Elmegreen: I don't see why that's a problem right now. In interaction models, usually the stream that follows the companion is from the original big galaxy, so that stream could have been the Milky Way's outer disk, and not so much the ram-pressure stripping of the LMC, but a tidal accretion of the Milky Way disk.

Carney: I'm sorry, but I must have phrased my question poorly. I'm concerned about the existence of the Magellanic Stream and the fact that it would have to pass through that putative disk. Wouldn't we see signs of that interaction?

Elmegreen: Yes, but the disk could go out to only 40 or 50 kpc, and then it wouldn't.

Franco: If there are large amounts of molecular gas in the outskirts of galaxies, this should appear in galaxies that have bridges, because that should be the first gas that is transferred from one galaxy to the other. I don't think that has ever been observed.

Shore: Isn't it fairly easy to hide that gas? Because you'd have to be able to see it in emission, which means that it would depend on the local emission measure and excitation conditions. So you've got the problem that, even if there is molecular material there, would you even be able to see it?

Franco: Well, if you excite it in these bridges, perhaps yes.

Tosi: From radio observations of irregular galaxies in the Local Group and elsewhere, the general idea is that H_2 does indeed surround all of them. There is a sort of "hole" in H_2 coinciding with the optical image of the galaxy. And then there is a large amount of molecular hydrogen around. The problem is that you can see it only in very special conditions. It's just an observational problem, and I think that is why Lequeux suggests that it is dark matter. He has seen so many H_2 halos.

Shore: Can you tell us more about this?

Elmegreen: Lequeux had a paper in 1993, for example, that pointed out that galaxy counts in the periphery of the LMC are unusually low, even though is no known dust or HI out there, and he postulated that in fact there was a lot of dust there and it was combined with cold H_2 on the periphery of the LMC. That's an example. This is still under intense debate. Most people would not believe it but it is marginally consistent with observations as they stand and is enormously important, so it should be discussed.

Hensler: When we try to discuss dark matter, we should try to fix some requirements for it. Why is dark matter really necessary? I see three points we should address. First, cosmology requires dark matter for $\Omega = 1$. The next point is the kinematics: in clusters of galaxies, and within galaxies themselves (the rotation curves). Another point is that mentioned in the beginning of the session —do we need dark matter for the cooling of a collapsing galaxy? The arguments come from virial temperatures, and I think those are now out of fashion because we know that the galaxies, if they start to collapse, have temperatures of several tens of thousands of degrees, so they will cool very rapidly and they will not achieve their virial temperature. That is one point that, in my opinion, helps us reject the dark matter requirement. Coming back to my second point, is there really a necessity from the kinematics? I would say that on the large scale we have no other ideas except that clusters are not virialised so that they show velocity dispersions which are much higher and are better attributed to their peculiar velocities without any gravitational binding of the clusters. Within galaxies, we could, perhaps, achieve something like this—that we need more cool matter, cool material—but another idea could also be brought up. Are the rotation curves of disk galaxies really an equilibrium situation? Do we really see an equilibrium? I only want to mention some work published by Christian Theis and myself in Astronomy & Astrophysics, where we found that if we have continued infall of material into the disk with angular momentum not matching that at the radius of the infall so that the angular momentum and also the mass in the disk has to be reorganised, we could achieve a flat rotation curve with variations v_{rad} of only a few $km\,s^{-1}$, so vanishingly small for observations. That could hint that dark matter on that scale is not necessary. There was also discussion in the Hague about the dark matter requirement in the solar vicinity. There was no conclusion—it could be zero, or it could be 30-or-so solar masses per square parsec. So a large amount could be required, but there was no firm conclusion. So I think we should put a large question mark about the necessity for dark matter.

Shore: First, it's not a free parameter, it's a requirement. Bruce [Elmegreen], if there really is a substantial amount of this cold stuff out there that happens to surround our Galaxy, has anybody bothered to look at the polarisation of external galaxies to see if there is an indication that somehow it's being passed through a screen? What are the optical depths supposed to be?

Elmegreen: This ties in nicely with quasar absorption lines, which I think drove the whole idea. That includes the metal-rich absorption lines.

Shore: But if there's dust there, you should be able to constrain it observationally.

Elmegreen: Not by polarisation–well, maybe.

Jones: My comment is just on dark matter, for which certainly on the scale of clusters and elliptical galaxies there is ample evidence from both the kinematics and from the X-ray observations. There's no way to explain the X-ray observations of 10^7 and 10^8 degree gas in these systems without dark matter. You can't do that as well in our Galaxy, but you can map it and you can can certainly not explain it. As far as having outlying gas in the Galaxy, I think the X-ray observations may tell you something. The soft component to the X-ray background is a local component. The hard component is an extragalactic component, and unless you want to bury this hot component right in the disk, you ought to see absorption, yet the X-ray background is very isotropic.

Lesch: Even on cluster scales, it is not necessary to have non-baryonic dark matter. What you showed us, the Ω values, mean that dark matter can still be baryonic. And it makes dark matter much less mystical.

Franco: But it's dark!

Jones: There are people like Monique Arnaud who in fact will argue that the dark matter is probably baryonic. There's no question that there's dark matter.

Elmegreen: But it can't be cold molecules, can it? They would just heat up in a cluster. Jupiters, maybe, but not cold molecules.

Jones: Yes, it could be Jupiters but not cold gas.

Palouš: I just wanted to add to this idea of little molecular clouds very far out from the galactic centre, as Combes and Pfenniger argue. They argue, for instance, that in merging galaxies, the increased star formation connected to the excess of molecules is because these molecules are pushed to the central star-forming regions due to the tidal forces. So this is the reservoir for star formation further out which is inactive in isolation but triggered when the two galaxies approach.

Chaboyer: I'd just like to comment on dark matter in clusters. A couple of groups have now made mass measurements using weak gravitational lensing. They get mass-to-light ratios two to three times higher than you get from the X-ray gas, or about 800 in the clusters.

Vietri: A comment to Gerhard [Hensler]: if you want to make the disk not in equilibrium, he ought to see if he can fit NGC 3198, which has a flat rotation curve out to 60 kpc, corresponding to about eight disk scale lengths, and on both sides of the galaxy. Very symmetric. I'd also like to ask a question. Since we have members of the Warsaw collaboration, could we have some comments on dark matter and microlensing?

Kaluzny: What we see at the moment can be explained just by disk stars and all events have relatively long timescales in the sense that it's about one to two weeks. There is really no need to introduce any special kinds of objects. It seems that simply the number of stars between us and the bulge is too large by a factor of about two compared to the Bahcall and Soneira model. There is really no need to introduce anything besides stars in the disk.

Vietri: Is this true also for Charles Alcock's group's results?

Kaluzny: Well there are claims that most of the events seen in the LMC are due to LMC stars. Actually, the same may apply to the lenses in the bulge. A significant fraction of the events we see could be due just to stars inside the bulge.

Palouš: I wanted to ask in the Hague that there have been no less than 10 detections in this microlensing experiment.

Kaluzny: Now it's 12.

Palouš: Someone said it could be explained by about 10% of the needed dark matter. Is that correct?

Kaluzny: I have no idea. But our group claims 12 events, and Alcock's group claims 25 events.

Shore: I think that, unfortunately, we now start to see that things begin to flow. There's this wonderful painting at the Met in New York that's called *The Expulsion from Paradise* that shows the local funding agency pushing the world from outside and Adam and Eve going out, like we are going: intellectually naked into the world. This is where we start to see that—the angular momentum is being pushed from the outside. I have a question: What is the relation between dark matter and angular momentum? If in these clusters of galaxies you have all these galaxies floating around—and they each have these halos, the halos are extended and maybe triaxial, and they're all interacting by tidal potentials—where the hell does the angular momentum come from in that case? It's not modelled; we don't have N-body distribution functions on that scale. Is this something anyone here would be interested in taking up as an interesting PhD thesis?

Carney: Do I see a volunteer? Mario [Vietri]?

Vietri: The only thing that I want to say is to remind everyone that in the early 1980s Lynden-Bell tried to apply the tidal spin-up picture to the local system. In that scenario, the orbital angular momentum of the Galaxy is equal, except for a minus sign, to the spin orbital angular momentum of Andromeda, and vice versa. He tried to reconstruct these four vectors and sum them up and see if they total to zero, and they do not. It comes out as an enormous number, and the obvious conclusion is that not only do we have spin-up because we interact with our nearest neighbours but also with our own dark halo. The dark halo ought to then contain a fair fraction of the total angular momentum of the local system. However, they rotate very little because the moment of inertia goes as MR^2, and M is very high for dark matter and R is even higher.

Shore: It's just that every time people have used the word "virialised", there's this image we carry in our heads that it's isotropic and virialised, and there's nothing that says that the tensor version of the virial theorem has to be isotropic when it's steady-state. There can be off-diagonal components. You can have triaxial ellipsoids. As a result, you can have extremely complicated tidal shapes to these outer potentials in these groups of galaxies. I just throw that out as something to start worrying about.

Franco: Igor Kolesnik has a poster on the evolution of triaxial figures. Would he like to comment?

Kolesnik: I had an idea that I should begin from the problem of polar-ring galaxy formation. I don't like the idea of merging—it seems very artificial and requires very specific conditions—so I became interested in how these galaxies form if the initial protogalaxy has a triaxial body. We know that if the gaseous figure is in near-equilibrium, as from the work by Martin Schwarzschild, then inside this figure two flows will be preferred: one around the short axis and the other around the longest axis. It's important, maybe, as initial motions that can form polar-ring structures. So it is an idea worthy of study. We began to simulate these processes by SPH codes that we developed. We then learned that if the protogalaxy is a triaxial body, it is easy to discern two effects. One, perhaps the warps are easy to understand because we have two planes, one connected with angular momentum and the other with the matter distribution. So then in the galaxies that form inherit an oscillation in these two planes so there will be differential precession. It's easy to understand. We can easily obtain grand-design spiral arms. So this is our first step, and now we are including star formation in the gas models and maybe we will be able to propose some new pictures of galaxy formation.

Carney: I'd like to change the focus of our discussions now from dark matter to something we know a little bit more about, but not necessarily very much more. This is the very nice set of calculations that Gerhard [Hensler] presented the other day, showing his model of the various constituents of the Galaxy: the halo, the disk, the bulge, and when they began to form stars, and maybe when the bulk of star formation occurred. What do we know observationally about these ages? Can we put good relative or absolute ages on the oldest or the bulk of the stellar populations? Can we rank these? I have my opinions, but I'd really like to hear the opinions of others. My own opinion is that some parts of the halo, perhaps those parts that eventually evolved into the disk—and we can argue about that if you wish—is the oldest, and that the bulge is younger, which is consistent with Gerhard [Hensler]'s models, but I welcome contradictions and arguments. As for the thick disk, I'm not sure where to put it, nor even the thin disk, although it's certainly younger than parts of the halo. Ken [Janes]'s and Randy [Phelps]'s results presented the other day for Berkeley 17 suggest that there are disk clusters that are as old as the halo, as best as we can tell, and they have not even completed their survey as yet.

Hensler: I have an additional question to this problem. Did I ever hear the term "Population III star"? Did somebody speak about them? Are they completely out of the race? Are they no longer something special? They're taken to be just the normal stars like we know them everyday astronomical life?

Carney: We haven't really defined what we mean by a Population III star, have we?

Hensler: A low-metallicity star, basically zero-metallicity.

Carney: Nobody has seen a zero-metallicity, or near-zero-metallicity star yet, but that may be because we have not yet sampled enough stars. If you consider a typical metallicity-distribution function, as you go to the lowest metallicities, the stars become rarer and rarer, and harder and harder to find. There is a good example, however, and I can't remember its telephone number, but its metallicity is quite low, and its abundance pattern is characteristic of nearly pure r-process, with very little sign of s-process contributions. So it's not a Population III star in the sense of being "aboriginal", but it is certainly a very early generation star—the son/daughter of Adam and Eve, as it were.

Shore: I'm not sure I remember the CPD number, but there are quite a number of ordinary main-sequence A stars that have r-process patterns that have nothing to do with nucleosynthesis. You've got to be really careful in just that range especially for the rare earths, which is what these are usually based on. All of the elements are mixed: you have r and s, and if you can't do isotopes, which you can't, you can't be sure.

Carney: You can be sure in some elements, like Europium, for example, which is a pure r-process element.

Shore: No that's not true. Europium has an s-process component. There's a way of getting to it.

Andersen: There is a late-type star with [Fe/H] just below −3 in which you can see both α-process and r-process elements, with a huge r-element enhancement that increases with Z (Sneden *et al.* 1994: *Ap.J.* **431**, L27).

Serrano: Yes, I think there is a lot debate about the ages of the halo and the bulge, and I think it comes from the mental picture one has. When we think about the halo, we think of the region far out from the centre of the Galaxy. Probably what is meant by halo in this talk is also the inner regions of the halo, because the central part of the halo should have a much shorter free-fall time. So I guess when we talk about the halo, we are also talking about the central parts, so first the central part of the halo, then the bulge formed. Is that true?

Carney: It's pretty hard to distinguish the inner halo from the bulge, so I don't really know how to answer your question.

Janes: I've always had a problem with this business of defining what the bulge is. If you have a halo, it's not going to have a hole in the middle. And if you have a disk, it's not going to have a hole in the middle. They all end up in the bulge, so the bulge represents the middle(s) of all of these things plus maybe stuff that isn't identifiable with the other components. I would think that the bulge almost has to be on your picture there with whatever else is the oldest one.

Carney: Are you asking me to scratch out the bulge?

Janes: No, because there may be a part that is not attributable to the other components.

Minniti: My favourite counter-argument to that is that if you look at the *COBE* pictures, you see a bulge and it doesn't go smoothly into the halo.

Hensler: From the collapse picture one might expect that the bulge is the end phase of the halo evolution, but, on the other hand, if one considers the longer timescale for the evolution, the bulge must be somehow constrained by the range of the interaction between outflow, star formation in a self-organised region, so I think there is a natural limit for the bulge where the information of the star formation there cannot reach out into the halo. Perhaps we can consider this as some natural range for the bulge where star formation decreases steeply, where outflow velocities decrease so that the information for further star formation that should be shown in the figure with a star-formation rate for a longer epoch in the bulge, where the star formation cannot communicate with the outer parts.

Elmegreen: In Fux's talk we just saw that what we've been calling a bulge looks identical to a bar and maybe only a bar with no bulge left over. In that case, it didn't come from the halo. It got puffed up from a formerly thinner disk. And you can't make a bar in a thicker disk. Where the bar ends, the disk stays thin. So we should look more at models where the bulge not only has non-isotropic velocities, but very severe streaming motions along it, and the bar has puffed up from the disk. It could still be old or young, but it's telling you when the inner part of the disk formed.

Shore: I would just echo that, Bruce. Stellar dynamics may be the best way to separate things here because that's the tracer of the potential energy. So instead of asking questions that are related to the star formation, here we have a way of doing the energetics which is what is coming out of these dissipationless models that at least give you a scale for how far you can pump stars up off the plane by these very eccentric orbits.

Minniti: How easy is it, then, to pump up the clusters in the bulge?

Majewski: Isn't this the usual argument against the "bottom-up" scenario? You can't pump the thick disk globulars up that high, for example.

Elmegreen: Is Fux here? These are resonant orbits, and they resonate with the bulge with a Z component. It's not individual scattering off of giant molecular clouds which could destroy a cluster. So it's possible that if a cluster is dense enough that it's tidally bound, it can begin to resonate also and puff up. Isn't that right?

Fux: Yes.

Shore: Are there any globular clusters that have orbits that are about the thickness of the bulge and look like they might be in resonance with it? That would be the way to answer your question.

Minniti: The proper motions are not measured well enough to do that.

Carney: There are some disk clusters with good enough proper motions, but they are not close enough in to the bulge to address your comment, Steve.

Shore: It sounds like a great key project.

Andersen: I think that there lots of uncertainties in determining even relative ages for different types of stars. For example, in globular clusters we use methods to determine ages that depend on HB luminosity, and these methods are sensitive to certain features of the models, like oxygen enhancement. Determining ages for field stars, metal-rich and metal-poor, is basically done from turn-off colours. I believe that as long as we have all the uncertainties in the models—of mixing length, detailed chemical composition and, worst of all, conversion of model parameters from the theoretical plane into colours and magnitudes—the results will not be very consistent. We saw some of the cluster fits, and they look very nice, but then maybe reddening is a real problem; maybe there are little deviations at the turn-off which tell you there is something significantly wrong with the models; and often the main sequence itself is a big mess due to unrecognised field stars and binaries. All in all, I don't think we're going to get real answers to these questions for a long time. You yourself, Bruce [Elmegreen], showed that if the RR Lyraes in metal-poor globular clusters are just a little bit evolved, their ages drop to become just about equal with the more metal-rich ones. So perhaps we should simply move on to another question!

Carney: That's why I left the table blank! But it's still a fundamental question, and we need to remember that and see what we can do to improve the situation. It's embarrassing that although we have very good evolutionary models for some metallicity regimes as well as very good observations, putting them all together has been extraordinarily difficult.

Alfaro: We have seen here today two very interesting results that are completely contradictory. The two results cover the age and metallicity of globular clusters, and they are practically perpendicular to each other. We also have another model to more or less estimate the age of the bulge. That is the one proposed by Young-Wook Lee to use the RR Lyraes' metallicity distributions in comparison with the K giants in the bulge. We need to discuss the theoretical aspect of this question, and we would also like to check what the differences and similarities are in our methodologies in both sets of age determinations.

Chaboyer: I think people have been unduly pessimistic. I agree that there are some problems and that you're never going to have a perfect stellar model. The error bars on the absolute ages are of the order of 3 or 4 Gyr. The relative ages can be determined better than that, I think, by using the $\Delta(B - V)$ techniques. So I think it is possible to get some good age estimates. People have looked long and hard at trying to find ways to change the ages, but when we add other things to the models we keep ending up with the same answers.

Hensler: I have only a short comment but a strong requirement for observers. Looking at the central region of the Galaxy, if the observers could determine the ages of the stars and find a trend such as that I showed, N/O *vs.* O/H first steepening and then declining. It is different as one goes further out in the radial direction, where this tendency decreases so that at the solar vicinity about 5 Gyr, a little bit later, it's totally different. If observers find this, then this model is correct.

Elmegreen: These models where the thick disk comes from a coalescence—Hernquist and somebody—if the coalescing companion galaxy is fairly massive, as it was in their case, it puffs up the whole disk. Then there should be no stars in the thin disk that are older than the stars in the thick disk. Isn't that right? Whereas if it's a low-mass companion and merely merges with our Galaxy and mixes its own stars, then it can make a thick disk that has a completely different population and leave that old thin disk behind.

Carney: Yes, that is a major part of the ongoing debate.

Janes: One of the things that we have been trying to do with our open clusters is to try to find the oldest open cluster, which might bear on this sort of thing. We do have this case of Berkeley 17, which might be as old as 12.5 billion years, although the data are still somewhat uncertain. There aren't a lot of star clusters that old but of course some people would say they were destroyed anyway. But the other evidence, as much as it exists, including things like white dwarfs that we have heard about here, also come up with about the same age: that there are no thin disk stars, or not very many anyway, older than about 10 billion years. So there still is a pretty long period of time, relative to the 15 billion year ages of the globulars, in which something could have gone "splash" and puffed up the thick disk. On the other hand, maybe it just took that long for the disk to get organised in the first place.

Carney: You raised a point during the break about the possible six- to ten-gigayear-old merger event identified by Preston, Shectman and Beers, and the possible effects that it might have had. Would you like to stick your neck out on that?

Janes: Thank you for reminding me of the Preston *et al.* study. We've had the persistent feeling in looking at our data that there seem to be a lot of star clusters in our sample that are in this six- to seven-billion-year age range, as though there was some kind of burst of star formation. There does not have to be a general burst in the overall star-formation rate, but simply an increase in the number of star clusters that have "nice orbits"—that is to say, ones that have significant Z-velocities.

Carney: While Ken [Janes] is searching for his viewgraph, Steve [Majewski], would you like to comment on this, since you authored the major stellar populations review article including the options for explaining the thick disk?

Majewski: First of all, I think that the work Schuster and his colleagues have been doing using Strömgren photometry has shown that the thick disk is at least as old as, or perhaps just a gigayear younger than, the halo. It seems then that if the thick disk has such an age, it is at least a couple of gigayears older than the thin disk. This will leave us with this problem of overlap in the mid-plane of the Galaxy. If you go searching for old stars in the mid-plane, you'll find them because even if you puff up into a thick disk you're still going to have some of its stars in the mid-plane. It's going to be a tough question to answer and we're probably going to have to look at age distributions.

Carney: As Isern noted earlier in this meeting, ultimately we hope to be able to use the white dwarfs to deduce the star-formation history, but there are some other indicators, such as Don Barry's K-line measure of chromospheric activity, that suggest at least an increase in the star-formation rate about 6–7 Gyr ago.

Majewski: I have that plot and will dig it out.

Carney: While you're doing that, Ken [Janes] has found his figure.

Janes: The question relates to the clusters in this range [he is showing a plot of the numbers of clusters *vs.* a colour-magnitude-diagram-based age indicator]. The distribution shows either a flattening or a hump or some other kind of change. There's quite a large number of our old clusters in this general age group. Since we produced this diagram, we have analysed more data and added even maybe five more clusters to this group, so that makes this a significant change. We keep wanting to obtain good enough statistics to be able to really say that something happened there. The current data are, at best, only marginally significant. The new clusters will improve the statistics, certainly. This is about the same time as the youngest stars in the Preston, Shectman and Beers sample. If that galaxy collided with us at that time, it obviously would shut off star formation in that galaxy.

Phelps: I'd like to point out that it's a lot more obvious if you normalise that to fit the clusters at the largest age range, because it's a somewhat arbitrary normalisation that we have here for the two exponentials, in the sense that you're trying to fit that central part with the five- to seven-billion-year-old clusters. So if you normalise it down at the other end, that will "pop those clusters up".

Janes: In our group, Randy [Phelps] is the optimist and I'm the pessimist. I see Janusz Kaluzny would like to comment.

Kaluzny: It happened that I observed more or less the same sample as you and it seems to me that we are just dealing with big selection effects. We found a lot of old clusters because we were choosing clusters which looked old, so the sample is not complete in the sense of volume. So I think we see a lot of old clusters here because that is what we were looking for.

Janes: That's true. However, the definition of what makes an "old cluster" in terms of looking for it on the Palomar Sky Survey images, which is what I've done, means that any of the clusters on my diagram will look about the same until you do the photometry. So I don't know of any mechanism in the selection that would get us to choose more clusters in the 6–7 Gyr region than in the 1–10 Gyr regions.

Kaluzny: I think that many of the clusters were selected already by Ivan King in the 1960s, and if you look at the Palomar charts and see a lot of red stars on the top of a starfield, that means you have a lot red giants, and you can already distinguish between intermediate age clusters and old clusters by just looking at the blue and red charts.

Janes: That's true, but all of these clusters in this figure are going to look the same on the Palomar charts. That's the point.

Kaluzny: The old ones are really far away, so the volume is much, much larger.

Carney: I see that Steve has found his figure (see Majewski 1993: *ARA&A*, **31**, 575). Let's turn back to the field stars for a moment.

Majewski: This is a plot by Noh & Scalo where they analyse the Ca II K-line emission stars studied by Don Barry, correcting for all kinds of selection effects. This is a summary of various other contributions to understanding the star-formation rate in the disk, and you can see that there are what I would call bursts of star formation in the disk. In fact, this gap here at 1–4 Gyr is called the Vaughan–Preston gap because it has been known for quite some time from K-line studies. The point is that it looks like the disk was built in a series of bursts. I'm not sure if this is a selection effect, that they haven't found stars out here (at the largest ages), but this is why I am interested in Gerhardt [Hensler]'s models. When he put out the paper by Truran and Hensler, they had these kind of "bursty" pauses and gaps and fits and starts in the formation of the disk. Gerhardt mentioned to me at the break that these have smoothed out a little bit, but is the thick disk another burst over here that isn't in this sample of stars? Is it just one out of a series of bursts that formed the disk? We need to find this out.

Shore: For the observers, this is a question that Ken [Janes]'s discussion brings up again: the calibration of the zero point for the age sequences. On the one hand, Ken, you said that this was a qualitative measure. On the other hand, you just tried to connect it to another sample that has a totally different age discriminator. Then we're hearing about variations in isochrones that for a colour variation that, to put it in very simple-minded terms, have colour-excess differences that are equivalent to the reddening to Orion, which most of us consider to be an "unreddened" OB association. You wind up changing metallicities by 50%. This winds up being a really serious problem when you're trying to compare what are fairly complicated and ill-constrained theoretical models for the evolution of the Galaxy, and I don't just means Gerhardt's models, I mean the stuff that Ferrini and his group have done. There are many free parameters. I had a referee say of one of my papers once that we had 17 free parameters, and with that we could fit anything, which is true. But somebody did fit this galaxy, and we need a timescale in order to be able to put all of these different measures on the same scale.

Carney: One real advantage of Ken [Janes]'s and Randy [Phelps]'s work is that their morphological age indicator is that it is reddening-independent, and can be calibrated in a sense by comparison with clusters which have had age estimates using some particular method and set of isochrones. Don Barry's K-line data are also reddening-independent, and again calibrated to cluster ages, though I doubt with the same set of isochrones. We do certainly need to try to put everything on the same scale.

Janes: There is certainly a part of your comment, Steve [Shore], that is very important because in fact the age scale is uncertain. But what has happened is that there have been a lot of people working on ages, and while there is not exactly uniformity in the derived ages, within surprisingly small limits, all of these diverse methods are tied to the same basic sets of models. They may all be wrong but they're then all wrong in the same way. So at some level it is possible to compare these very different data sets. I very specifically didn't do that because there are no free parameters in our material, just raw data, rather uninterpreted when you come right down to it. There is certainly no real dependence on models except in the crudest sense of the word for our indices.

Shore: The reason I'm asking is because this is one of the few times when we have so many people with so many diverse ways of doing the treatment of the observations together in one meeting. And something that would be very valuable for "those who will come after us" (if we want to call this a sort of John the Baptist meeting) is to leave behind in the proceedings of this meeting some kind of sets of either observations that need to be done to answer these questions, which is what Bruce [Elmegreen] is alluding to, or what kinds of measures we can say at the moment are the reference measures. So if people are going to do models, they have some set of measures, some set of parameters that they have to fit, some observations that everyone, or most everyone, can agree constrain all future models.

Carney: Well, we have moved on to the topic of the thin disk, and we don't have a lot of time before they throw us out of the building, and I would like to get on to some of the other issues in the thin disk. I am baffled by several of the issues that come up today and over the course of the week. There was a comment, by Pepe Franco, I think, about why isn't it more metal-rich and we got that addressed at some level. Did the disk form inside out? That's the model, but the data to support such a contention are meager, at best. The oldest open clusters, as Ken [Janes] pointed out, are in fact outside the solar circle, not inside. Now perhaps that suggests a destruction mechanism, but it would be nice to understand it better. Something on this list of mine that stumps me particularly is that we seem to have a shallow metallicity gradient as found from the HII regions in oxygen, nearly. It certainly does not seem to be as steep as had been claimed by Peter Shaver a decade ago. But we seem to have a relatively steep gradient in [Fe/H] in clusters that are somewhat older. In field stars, Johannes's work (Edvardsson *et al.* 1993) has revealed a fairly steep slope in [Fe/H] and even [O/H], and yet we also see this very large scatter and negligible slope in the field-star age–metallicity relation. I don't know how to explain it, and I hope somebody here can help us understand this. Help?

Franco: I got really confused because in the solar neighbourhood the metallicity that is found from the gas shows a well-defined trend—-a gradient. Ken [Janes] showed that in the solar neighbourhood the open clusters show basically no gradient.

Janes: Are you talking about gradient with radius or with age?

Franco: With radius. Near the solar neighbourhood it seems to be absent.

Janes: Yes, but there is quite a bit of scatter. If you stand back and look at our diagrams from a couple of feet away, you see a nice gradient. But if you get up close, you don't see a gradient near the solar position.

Franco: Okay! But if we take Emilio [Alfaro]'s data and his values for the globular clusters, the solar neighbourhood seems to be in a pocket of low metallicity. So the gradient is inverted, locally, if we look at globular clusters.

Carney: The field halo stars don't show much of a gradient when you select the halo on purely kinematical criteria. I don't know what the globular clusters are telling us, unless we're mixing samples somehow. But I was more concerned with the HII regions. The "current" gradient being near zero and yet the iron showing a steeper gradient in the past, as measured from the open clusters or field disk stars. And this very large scatter in metallicity at most ages—how do we explain this?

Tosi: I had not understood that the current HII region gradient is flat. What I have understood is from Vílchez is that the *outer* gradient is flat but the inner gradient—from, say, 3 to 12 kpc from the centre—is still steep, and it is at least as steep as that derived from open clusters of any age, and perhaps the HII region gradient is steeper. It is only in the outer regions, at the outskirts of the disk, that it is flat.

Janes: I think we can put all of the observations described at this meeting onto a single picture. From the solar circle and somewhat beyond, there is a considerable decline in the metallicity, but beyond about 12 kpc galactocentric or so, it levels out. This can be seen in the HII region work. In contrast, the Peimberts had found earlier that the oxygen gradient is steeper in the inner regions. As for our star clusters, we have a big batch of them locally and another group at large galactocentric distances, and the metallicity distribution almost looks discontinuous, although with a lot of scatter. We don't have very much information on the inner disk, but the bulge doesn't seem to be much higher in metallicity than the solar neighborhood. So the big change in the metallicities seems to be taking place just outside the solar position.

Andersen: I just want to make two comments. One is about the oxygen- abundance determination in stars, which is important since O is one of the first elements made, and it connects to the HII region work. There is a very considerable difference between the oxygen abundance one gets from the OI $\lambda 7770$ triplet lines and the [OI] lines at $\lambda\lambda 6300, 6363$. It's a factor of two or three. I think I fairly represent the opinion of Edvardsson *et al.*, which was that the forbidden-line results, which give the lower abundances, are about right, whereas the triplet gives too high abundances (but see King 1994: *PASP*, **106**, 423). So one should be very careful to check how the oxygen abundances were determined for the stars one is looking at. The other comment is that we said that hydrogen seems to be the variable element. This is right in the sense that the relative proportions of everything else seems to be constant, and it's the ratio of hydrogen to the rest that seems to vary. That does not necessarily mean that large amounts of hydrogen actually came crashing in—the explanation for that observation is still open to debate.

Shore: I would just hope that the observers will start to push the *Hubble Space Telescope* time-allocation committee and eventually some of the other telescopes—space telescopes—to try to work in the ultraviolet to work on these extragalactic problems, because that's where you're going to have the resonance lines, which are also necessary for many of these elements. What worries me here is that there could be a temeprature effect of the exciting source, and which may still be lurking because we're in a regime where metallicity is changing and we're not accounting properly for that in the illuminating stars. Vílchez has discussed this and people who do photo-ionisation models have discussed this, but it is something that is still lurking there and we do have to worry about as the models improve.

Carney: Interesting perspective. Working on stellar spectra, I would stay away from resonance lines like poison. How about the scatter in the age–metallicity relation, Johannes [Andersen]? Have you given that any more thought?

Andersen: No, all I can say is that it is real. Given the homogeneity of the data and the care that went into them—which you can read about in the paper—I am willing to say that the scatter is real, and I also think that we can say that the lack of much variation in the solar-circle sample between two and ten billion years is also real. Why that is so is an interesting and unsolved question.

Hensler: I think that's in agreement with observations of OB associations that also show a very large scatter in their abundances.

Serrano: I think that we can be more precise since the data are not only real but have a value. As Steve said, it is not chaotic. Any model must account for the dispersion as it is, at least a factor of four.

Carney: But the dispersion is large; it's a lot of atoms—a factor of four.

Serrano: But it's not forty!

Laird: I would point out the additional constraint that as you try to explain the variation in iron abundance at a given age, at the same time it would seem that you have to reproduce the same element-to-element ratio, the oxygen-to-iron for instance or magnesium-to-iron, seems to be the same in these stars of different overall iron abundance. That provides an additional and I think difficult constraint to explain what's going on.

Carney: Is it indeed difficult, Monica [Tosi]?

Tosi: Yes.

Andersen: But that's what makes it real fun.

Carney: And on that, I think we will bring this discussion—and the conference—to a close. It has been a wonderful week.